Analog Audio Amplifier Design

Analog Audio Amplifier Design introduces all the fundamental principles of analog audio amplifiers, alongside practical circuit design techniques and advanced topics. Covering all the basics of amplifier operation and configuration, as well as high-end audio amplifiers, this is a comprehensive guide with design examples and exercises throughout.

With chapters on single-device, operational, multi-stage, voltage buffer, power, line-stage and phono-stage amplifiers, *Analog Audio Amplifier Design* is a comprehensive and practical introduction that empowers readers to master a range of design techniques. This book also provides a variety of graphs and tables of key amplifying devices and properties of amplifier configurations for easy reference.

This is an essential resource for audio professionals and hobbyists interested in audio electronics and audio engineering, as well as students on electrical and audio engineering courses.

John C.M. Lam received a PhD in Electrical Engineering from the University of Washington. He has been employed in the semiconductor industries for over 13 years in various technical, sales and marketing management positions. He is the founder of JE Audio and was its Chief Design Engineer, designing and manufacturing high-end audio products. The firm holds four US patents for innovative amplifier designs. Even though he has now retired from JE Audio, John is still personally carrying out research on analog audio amplifier designs that are interesting to him.

Analog Audio Amplifier Design

John C.M. Lam

Routledge
Taylor & Francis Group

LONDON AND NEW YORK

Designed cover image: John C.M. Lam

First published 2024
by Routledge
4 Park Square, Milton Park, Abingdon, Oxon OX14 4RN

and by Routledge
605 Third Avenue, New York, NY 10158

Routledge is an imprint of the Taylor & Francis Group, an informa business

British Library Cataloguing-in-Publication Data
A catalogue record for this book is available from the British Library

ISBN: 978-1-032-43934-1 (hbk)
ISBN: 978-1-032-43933-4 (pbk)
ISBN: 978-1-003-36946-2 (ebk)

DOI: 10.4324/9781003369462

Typeset in Times New Roman
by Apex CoVantage, LLC

This book is dedicated to my dear wife Eunice.

Contents

Preface

The objective of the *Analog Audio Amplifier Design* is to develop in the reader the ability to analyze and design analog audio amplifiers using operational amplifiers, discrete semiconductor transistors, and vacuum tubes. The book is written to address many advanced topics and practical amplifier configurations. However, it also provides the necessary basics, examples and exercises that allow the readers to progressively absorb the materials of the book and gain confidence in using various design techniques.

There are several chapters (Chapters 2–5) intended as text for analog amplifier design courses for an undergraduate degree in electrical and electronics engineering. It should also prove to be useful for audio professionals and designers to refresh their knowledge. The rest of the book covers the advanced analog amplifier designs. However, it is not intended as a cookbook. Emphasis is put on understanding the concepts, amplifier configurations and variations of the circuit design techniques.

There are 14 chapters in this book, which cover amplifier designs ranging from basics of amplifiers, operational amplifiers, phono-stage amplifiers, line-stage amplifiers, and power amplifiers. The use of semiconductor transistors and vacuum tubes are fully discussed.

Chapter 1 covers the basics of the audio components that make up a typical Hi-Fi system. The audio components include the phono-stage amplifier, line-stage amplifier, and power amplifier, as well as the dc regulated power supply.

Chapter 2 covers the dc analysis for five commonly used amplifying devices. They are: solid-state bipolar junction transistor (BJT), junction field-effect transistor (JFET), enhancement type metal-oxide field-effect transistor (MOSFET), depletion type MOSFET, and vacuum tube triode. DC bias techniques, including fixed-bias, voltage-divider bias, and self-bias are discussed.

Chapter 3 covers the ac analysis of amplifying devices in response to a sinusoidal signal. Several small signal ac models are used to represent the devices. Hybrid-pi and T-model are two popular ac models used for BJTs and FETs. Thévenin and Norton models are used for triodes. By carrying out the ac analysis with an appropriate ac model, the voltage gain and input and output impedance can be determined for an amplifying device in three basic configurations.

The analysis of the operational amplifier and its applications is presented in Chapter 4. A number of audio applications for op-amps is discussed. They include phono-stage amplifier, tone-control amplifier, graphic equalizer, dc servo, dc regulated power supply, and composite op-amp.

Chapter 5 covers the analysis of feedback and frequency response for audio and linear amplifiers. Four negative feedback configurations are discussed in this chapter. They are (a) series-shunt, (b) shunt-shunt, (c) series-series and (d) shunt-series feedback. Negative feedback is a way to trade off gain for improvement to other properties of the amplifier.

Voltage buffer amplifier design is covered in Chapter 6, which starts with the discussion of the vacuum tube buffer amplifier. Then it moves on to buffer amplifiers implemented by using operational amplifiers and discrete transistors. A high current diamond buffer amplifier, which is capable of delivering over 15A current, is discussed in detail.

Chapter 7 covers the solid-state audio lines-stage amplifiers design. The topology of the line-stage amplifiers evolves from unbalanced input–unbalanced output into fully balanced. The solid-state devices used in the design include BJT, JFET, MOSFET, and op-amp. Various types of potentiometers for volume control are discussed. They include rotary, motorized, stepped attenuator, and digitally control potentiometers.

The vacuum tube line-stage amplifiers design is presented in Chapter 8. The chapter starts with the classic Marantz 7 amplifier design. Then unbalanced input–unbalanced output line-stage amplifiers employing single-stage compound amplifiers and two-stage amplifiers, without using negative feedback, are discussed. Finally, several fully balanced vacuum tube line-stage amplifiers are covered. The vacuum tubes discussed in this chapter include 6H30, 6922, 6N1P, 12AU7, and ECC99.

Chapter 9 covers the common noise that appears in electronic devices. The electronic devices include both passive and active, i.e., resistor, zener diode, transistor, op-amp and vacuum tube. Methods of determining the noise generated from an amplifier are discussed and examples are given. Basics of low noise design are given so as to minimize noise right from the start.

In Chapter 10, the phono-stage amplifiers design is covered. Active, semi-active, and passive RIAA equalization is discussed. The amplifying devices include vacuum tube, JFET, BJT, and op-amp. Examples for low noise unbalanced input–unbalanced output MM (30–40dB) and MC (60–70dB) phono amplifiers are discussed in detail.

Vacuum tube power amplifiers design is presented in Chapter 11. Emphasis is given to push–pull type vacuum tube power amplifiers. Various input stages, phase splitters, and output stages are discussed. How the ultra-linear output transformer connects to the power tubes is explained. Design examples of fully balanced vacuum tube power amplifiers are given.

Chapter 12 covers solid-state power amplifiers design. Various types of input stage, voltage amplifier stage, and output stage are fully covered. The operation principles for slew rate and V_{BE} multiplier over current protection are discussed. The pros and cons of choosing power BJTs and MOSFETs are given. Unbalanced and fully balanced power amplifiers are discussed in detail.

In Chapter 13, hybrid power amplifiers design is presented. The hybrid power amplifiers employ vacuum tubes for the input stage and voltage amplifier stage, while solid-state power devices are utilised for the output stage. Several examples of hybrid power amplifier design are given.

The final chapter, Chapter 14, covers dc regulated power supplies design. The chapter starts by describing three-terminal dc regulators, including fixed voltage and adjustable voltage type. Then shunt type and series type regulated power supplies employing discrete solid state devices, vacuum tubes, and a hybrid approach are discussed. Extensive practical design examples are given. The dc regulated voltages in the examples are ranging from ±3.3V, ±15V, ±45V up to −150V and +350V.

1 Introduction

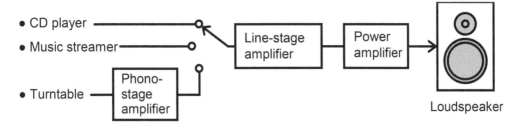

Figure 1.1 A typical Hi-Fi audio system containing several audio amplifiers: phono-stage amplifier, line-stage amplifier, and power amplifier

Figure 1.1 shows a typical Hi-Fi system that contains several audio amplifiers including a phono-stage amplifier, line-stage amplifier, and power amplifier. The line-stage amplifier also works as a control center that allows the user to switch between different audio sources. The audio sources may include CD player, music streamer, phono, DAC and tuner, etc.

A Hi-Fi system that contains separate audio amplifiers generally offers better performance than that of an integrated amplifier. An integrated amplifier is one that integrates a line-stage amplifier, power amplifier, and sometimes a phono-stage amplifier also, combining to form one single amplifier. Even though an integrated amplifier has limited space to contain everything, it can still offer excellent performance. However, if a user wants to get better audio performance, separate audio amplifiers are often the answer.

No matter whether the Hi-Fi system contains separate audio amplifiers or just one integrated amplifier, the system needs different types of amplifiers to perform the necessary tasks. Therefore, it is essential to understand the working principles of each audio amplifier type, such as the phono-stage amplifier, line-stage amplifier and power amplifier. In this chapter we briefly discuss each of the audio amplifier types. Then at least a full chapter is devoted to each of them, with Chapter 10 being given to phono-stage amplifiers, Chapters 7 and 8 to line-stage amplifiers, and Chapters 11, 12 and 13 to power amplifiers.

1.1 Phono-stage amplifiers

A moving magnet (MM) phono cartridge is connected to a phono-stage amplifier, shown in Figure 1.2(a). The electrical equivalent circuit of an MM cartridge is often modelled by means

DOI: 10.4324/9781003369462-1

of resistor Rc and inductor Lc. The value for Rc and Lc varies between different phono cartridge manufacturers. Rc often ranges from 500Ω to 800Ω, while the range is 300mH–500mH for Lc. In the phono-stage amplifier's side, capacitor C1 and resistor R1 shunt the input. R1 is often a 47kΩ resistor that provides damping for the MM cartridge. Capacitor C1 resonates with inductance Lc and a correct C1 will extend the frequency response of the cartridge. C1 includes the capacitance of the interconnecting cable and input stage of the phono amplifier. In general, C1 is around 100pF.

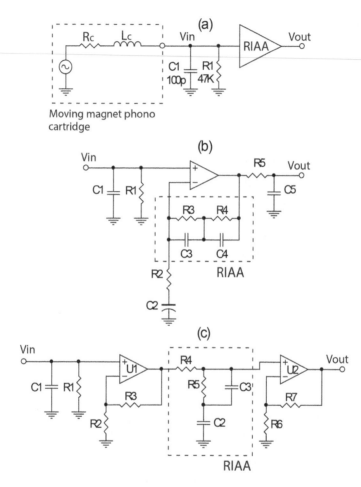

Figure 1.2 (a) An MM phono cartridge connecting to an MM phono-stage amplifier, (b) a phono-stage amplifier with active RIAA, and circuit (c) with passive RIAA

Since the output from an MM phono cartridge is typically around a few mV, in order to raise the signal amplitude, a phono-stage amplifier with sufficient voltage gain is needed. The phono-stage amplifier performs three major tasks: high voltage gain, low noise, and accurate RIAA equalization. A typical phono-stage amplifier produces a voltage gain of 30dB–40dB at 1kHz. On the other hand, since a moving coil (MC) cartridge produces output of 20–30dB lower than an MM cartridge, the voltage gain for an MC phono amplifier has to be around 60dB–70dB.

RIAA equalization has been the recording and playback standard for phonograph records since 1954. It was established by the Recording Industry Association of America (RIAA). The purposes of the equalization are to allow greater recording times (by decreasing the mean width of each groove), to improve sound quality, and to reduce the groove damage that might otherwise arise during playback. The RIAA equalization curve for recording is to reduce the low frequencies but boost the high frequencies. However, the RIAA equalization curve for playback performs in the opposite way.

Figure 1.2(b) shows a phono-stage amplifier realized by an op-amp and active RIAA equalization. The reason why it is called active equalization is because the RIAA network is placed in the feedback loop, which is connected between the output and the inverted input of the op-amp. It is a simple and popular design for an MM phono-stage amplifier that produces 30–40dB gain. However, since this is a single-stage amplifier, it cannot produce the 60–70dB gain required for an MC cartridge without compromising the noise and frequency response. In general, an MM phono-stage amplifier often produces a signal-to-noise ratio (SNR) better than 75dB and 60dB or better for an MC phono-stage amplifier.

Figure 1.2(c) shows a phono-stage amplifier realized by two op-amps employing passive RIAA equalization. Instead of placing the RIAA network in the feedback loop, the RIAA is now placed in between two op-amps. By using two op-amps, this phono amplifier may produce sufficient voltage gain for an MC cartridge.

In addition to active and passive RIAA equalization, there is also a semi-active arrangement, which contains both active and passive RIAA equalization. All different types of RIAA equalization are fully discussed in Chapter 10. Even though a phono-stage amplifier realized by op-amp works very well, in order to achieve the lowest possible noise level a semi-discrete transistor approach is often the preferred choice. Both op-amp and semi-discrete transistor approaches are discussed in Chapter 10. Basic working principles for op-amp are discussed in Chapter 4.

1.2 Line-stage amplifiers

The voltage gain for a line-stage amplifier often ranges from 5–10, i.e., 14dB–20dB. For this low voltage gain level, good SNR and low distortion can be easily achieved. Figure 1.3 shows a typical line-stage amplifier using discrete solid-state transistors. It has a three-stage configuration. The input stage is a *cascode differential amplifier* formed by transistors Q1 to Q4. A cascode amplifier produces broad bandwidth and low distortion, and so does the cascode differential amplifier. Transistors Q5 and Q6 form a current source that sets up the tail current for the cascode differential amplifier.

The input stage's output is directly coupled to the second stage, which is a cascode amplifier formed by Q7 and Q8. Transistors Q9 and Q10 provide a current source that works as an active load for transistor Q8. Since collector of Q9 has a high output impedance, transistor Q8 sees a high load impedance and, therefore, the cascode amplifier produces a very high voltage gain.

The output stage is formed by two pairs of complementary emitter followers by transistors Q11 to Q14. By cascading the emitter follower of Q11 to emitter follower Q13, this double emitter follower is often called a *Darlington pair*. It has a unity voltage gain but a very high current gain, which is a product of the two transistors. For instance, if Q11's current gain is 150 and 100 for Q13, the combined current gain for the Darlington pair becomes 150×100 = 15,000. This high current gain produces a high input impedance for the output stage so that it does not load down the second stage. This allows the second stage cascode amplifier to drive the output stage easily.

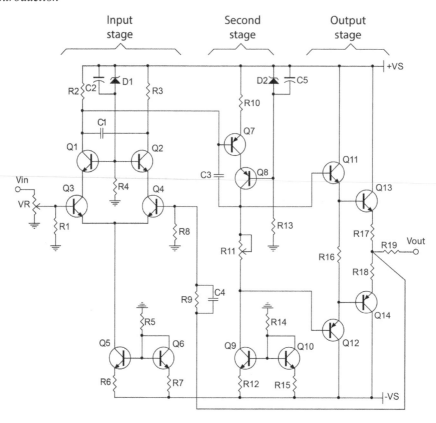

Figure 1.3 A simplified solid-state line-stage amplifier containing a three-stage configuration: input stage, second stage, and output stage

This three-stage amplifier produces a very high combined voltage gain, which is called the *open loop gain* of the entire amplifier. By applying feedback, which connects the output to the input stage via resistors R8 and R9, the overall voltage gain is reduced to around 5–10. The linearity of the amplifier is greatly improved with feedback. As a result, distortion is reduced and the bandwidth is extended. For a solid-state line-stage amplifier, the voltage gain is often set to around 10 (i.e., 20dB). In contrast, a vacuum tube line-stage amplifier may vary from 5 to 10.

Note that Figure 1.3 is an unbalanced line-stage amplifier – input is unbalanced and so is the output. Chapter 7 discusses solid-state line-stage amplifiers with unbalanced and fully balanced configurations.

Figure 1.4 shows two different vacuum tube line-stage amplifiers. The line-stage amplifier of Figure 1.4(a) has a three-stage configuration. The input stage is arranged in a *common cathode amplifier* configuration, which is similar to a *common emitter amplifier* configuration for a bipolar transistor, that produces high voltage gain. The output from the input stage is coupled to the second stage via capacitor C1. The second stage is also a common cathode amplifier configuration. The voltage gain from the first two stage produces the open loop gain for the entire amplifier. The output stage is a *cathode follower*, which is similar to an emitter follower, with unity voltage gain. Feedback is applied via resistors R12 and R4. When high amplifying factor triode 12AX7 is used for the first two stages, the amplifier produces a sufficiently high open loop gain so that overall voltage gain can be set to 10 or even higher if necessary. This type of three-stage design is popularized by the Marantz 7 line-stage amplifier.

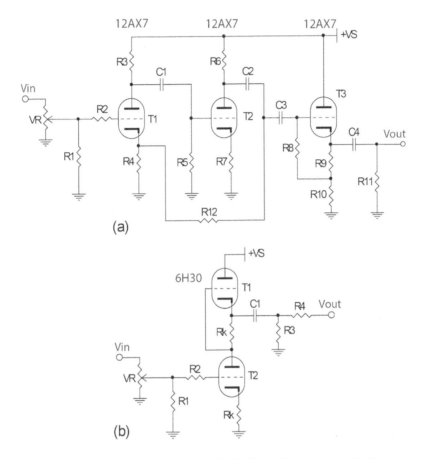

Figure 1.4 (a) A vacuum tube line-stage amplifier with feedback, (b) a vacuum tube line-stage amplifier without feedback

Figure 1.4(b) shows a single-stage vacuum tube line-stage amplifier without feedback. It is a *shunt regulated push-pull* (SRPP) amplifier. The voltage gain is determined by the cathode resistor RK and the *amplifying factor* of the vacuum tube. When low amplifying factor triode 6H30 is used, the voltage gain is commonly set to around 5–8. A cathode follower output stage, similar to the one in Figure 1.4(a), is often added so that the amplifier produces a low output impedance.

Again, it can be noted that Figure 1.4(a) and (b) show unbalanced amplifiers. Chapter 8 discusses vacuum tube line-stage amplifiers with unbalanced and fully balanced configurations. The basic amplifier configurations, dc and ac analysis of amplifying devices, and the use of feedback are discussed in Chapters 2, 3 and 5.

1.3   Power amplifiers

Power amplifiers – vacuum tube

There are several types of power amplifiers that are discussed in this book. These are vacuum tube power amplifiers, solid-state power amplifiers, and hybrid power amplifiers. Since a vacuum tube power amplifier has the longest history, let us first take a look of it. Figure 1.5 shows a typical vacuum tube power amplifier with a three-stage configuration.

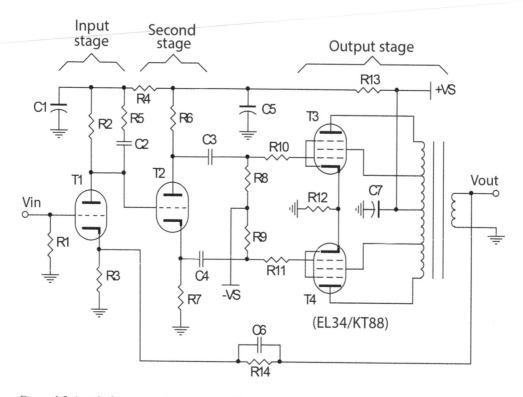

Figure 1.5 A typical vacuum tube power amplifier with a three-stage configuration

The input stage is formed by triode T1 in a common cathode configuration. The input stage amplifies the input signal as well as the feedback signal passing from resistor R14. The difference between the two signals is amplified and directly coupled to the second stage. The second stage is a *split-load phase splitter*. A phase splitter is an amplifying circuit that takes on an input signal and produces two out-of-phase output signals. There are many types of phase splitters. Most phase splitters employ two triodes and produce voltage gain. The split-load phase splitter is the simplest phase splitter that uses only one triode. However, the voltage gain is unity. In order to increase the open loop gain, T1 must be a triode with a high amplification factor or a high gain pentode. Another approach is, of course, to use a phase splitter that also produces voltage gain.

Outputs from the second stage are coupled to the output stage via capacitors C3 and C4. The output stage is a push-pull configuration employing an *ultra linear* output transformer. A pair of power tubes, such as EL34, 6550, or KT88, is often used. The vacuum tube power amplifier depicted in Figure 1.5 is operated in class AB, which achieves near to 50% efficiency. Vacuum tube power amplifiers are discussed in detail in Chapter 11.

Power amplifiers – solid-state

A typical solid-state power amplifier is shown in Figure 1.6. It represents a simple three-stage amplifier configuration. However, it provides the fundamentals that allow the configuration to be evolved into more advanced power amplifier designs. The power amplifier contains an input stage, voltage amplifier stage (VAS), and output stage. There are several sub-circuits also high-lighted in the amplifier, including input filter, output filter, and V_{BE} multiplier. These are briefly discussed in the following.

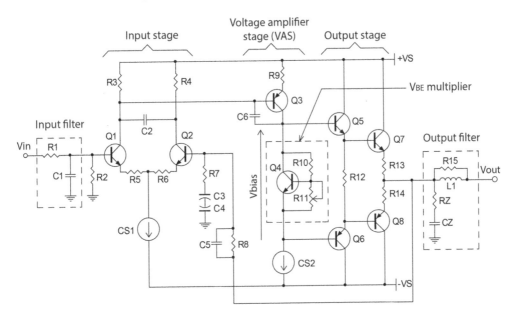

Figure 1.6 A typical solid-state power amplifier in a three-stage configuration

Input filter: R1 and C1 form a first order low-pass filter to keep out unwanted radio fre-quency noise. The corner frequency of the low-pass filter is determined by $f_c = 1/(2\pi R1C1)$. It is often set at $f_c = 200\text{kHz}$ or higher so that it has little effect on the audio band. Resistor R2 provides a current returning path for the transistor Q1. Since a *bipolar transistor* (BJT) has a base current, there is always a small dc input offset voltage present at the base of the Q1. Note that the small dc input–offset voltage may be carried over, creating a very small dc offset volt-age at the output unless a dc servo is used to eliminate it. Since a *junction-field-effect transistor* (JFET) does not have gate current, when a JFET is used for the input stage the dc input–offset voltage is equal to zero.

Input stage: The input stage is a differential amplifier formed by transistors Q1 and Q2. The current source CS1 sets the tail current for the differential amplifier. The function of the input stage is to amplify the signal Vin and the feedback signal from the output. The difference be-tween the two signals is amplified and then directly coupled to the second stage. R5 and R6 are degeneration resistors that serve two purposes. The first is to provide local feedback to the differential amplifier to improve the linearity of the input stage. As a result, bandwidth is increased and distortion is reduced. The second purpose is to improve *slew rate*. Slew rate is a measure of how fast the power amplifier can respond to a large signal. It is a parameter that indicates how well the power amplifier performs in a large signal condition. Therefore, the higher the slew rate, the better performance for a power amplifier when handling large signals.

Voltage amplifier stage (VAS): The second stage is an amplifier that contributes the most to the open loop gain of the entire power amplifier. Transistor Q3 works in a common emitter amplifier configuration with a small un-bypassed emitter resistor R9. In order to achieve high voltage gain, a current source (CS2) having high impedance is used as the active load for Q3. Capacitor C6 is the *Miller capacitor* that forms a dominant pole to stabilize the power amplifier at high frequency.

V_{BE} **multiplier**: Together with resistors R10 and R11, transistor Q4 forms the V_{BE} *multiplier*. It creates a dc voltage spread between the collector and the emitter of Q4. The dc voltage spread is to compensate the drop of V_{BE} across the driver and power transistors in the output stage. Thus, the output stage can be biased in class AB with a small collector dc quiescent current. If the output power transistors are not biased properly, the power amplifier may produce crossover distortion.

Output stage: The output stage of the power amplifier comprises complementary Darlington pairs, Q5–Q7 and Q6–Q8. If β_1 is the current gain for the driver transistors Q5 and Q6, and β_2 is that for power transistors Q7 and Q8, the current gain of the Darlington pair becomes $\beta_1 \times \beta_2$. Therefore, this provides a huge current gain for the output stage so that a small base current at Q5 and Q6 can control a high output current. Power transistors Q7 and Q8 must be mounted in a heatsink, together with the V_{BE} multiplier transistor Q4. Thus, any variation of V_{BE} from the power transistors due to temperature change can be properly compensated by the V_{BE} multiplier.

Output filter: Some power amplifiers may be unstable under no load conditions or when driving an inductive load. To prevent such unstable conditions happening, a *Zobel* network is placed at the output of a power amplifier. The Zobel network contains a shunt resistor RZ and a capacitor CZ, as shown in Figure 1.6. RZ is often a small resistor of several ohms. The CZ is a low capacitor around 0.01μF to 0.1μF. L1 is a small inductor, usually a few μH, while R15 is a small resistor of several ohms. The Zobel network stabilizes the power amplifier at high frequency.

Improvement can be brought to each stage of the power amplifier. For instance, a cascode differential amplifier and a complementary cascode differential amplifier can be used for the input stage. Cascode and complementary cascode amplifiers can be used for the VAS. A triple emitter follower and diamond buffer amplifier can be considered for the output stage. Details of solid-state power amplifiers are discussed in Chapter 12.

Power amplifiers – hybrid

Owing to the different uses of materials and construction processes that produce vacuum tube and solid-state devices, these two devices produce sound with different sonic characteristics. It is commonly believed that vacuum tube amplifiers produce good mid-range with a "warm sound" while solid-state amplifiers produce good bass. Thus, a hybrid power amplifier, which combines both vacuum tube and solid-state devices, may benefit from both worlds. Figure 1.7 shows a potential hybrid power amplifier that employs vacuum tubes for voltage amplification and a solid-state buffer amplifier for current amplification.

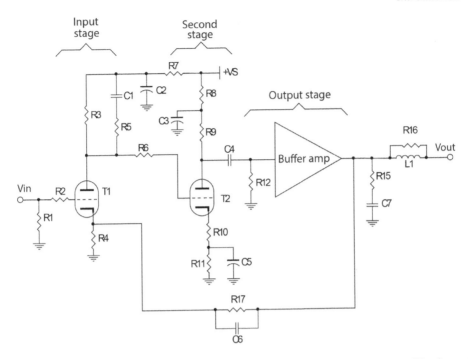

Figure 1.7 A potential hybrid power amplifier employing vacuum tubes for voltage amplification and a solid-state buffer amplifier for current amplification

The hybrid power amplifier shown in Figure 1.7 is again a three-stage configuration. The input stage is a common cathode amplifier formed by triode T1. The output from the input stage is directly coupled to the second stage, which is also a common cathode amplifier, formed by triode T2. The first two stages, therefore, produce the necessary open loop gain for the entire power amplifier. The output stage is a solid-state buffer amplifier, which has unity voltage gain but very high current gain.

Figure 1.8 shows an example of a solid-state buffer amplifier, which is referred to as a *floating-bias diamond buffer amplifier* in this book. It is noted that even though it is an all bipolar transistors (BJT) buffer amplifier, metal-oxide semiconductor field-effect transistors (MOSFETs) can also be used. A combination of BJT and MOSFET is called Bi-FET. Buffer amplifiers employing BJTs, Bi-FETs, and MOSFETs are discussed in Chapter 6. The buffer amplifier in Figure 1.8 has several properties, including (i) high input impedance, (ii) low output impedance, (iii) unity voltage gain, (iv) very high current gain, and (v) no output dc offset voltage.

Transistors Q2 and Q3 are working as complementary emitter followers. They form the *pre-driver* stage. The dc quiescent current for Q2 is set by the current source formed by Q1, while the dc quiescent current for Q3 is set by current source Q4. The outputs from the complementary emitter followers Q1 and Q2 are directly coupled to the second pair of complementary emitter followers formed by Q5 and Q6. They form the *driver* stage. The dc quiescent current of Q5 and Q6 is set by variable resistors VR1 and VR2, respectively. Transistor Q7 forms the V_{BE} multiplier circuit that sets the quiescent current for the power transistors Q8–Q11 of the output stage. The power transistors are operating in an emitter

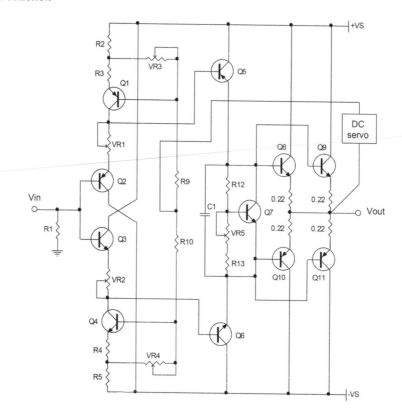

Figure 1.8 A solid-state buffer amplifier

follower configuration. Therefore, the buffer amplifier is a triple emitter arrangement, which is often called an output triple. Since a dc servo circuit is used, the output dc offset voltage is taken down to zero. Therefore, even without applying feedback from the power amplifier, the buffer amplifier can work independently and maintains its input and output at ground level. Details of buffer amplifiers are discussed in Chapter 6, and hybrid power amplifiers in Chapter 13.

1.4   DC regulated power supplies

Many people have found that it is difficult to design a good audio amplifier. Well, that is because it takes time to build up the skills and know-how. However, people have also found that it is even more difficult to make an audio amplifier that produces good sound. That is because, in order to design a good amplifier, it is necessary to rely on circuitry design techniques. However, in order to make an amplifier produce good sound, we must choose the right electronic components, including resistors, capacitors, transistors, op-amps, and vacuum tubes etc., for every part in the circuit. Additionally, there is one component that is often overlooked — a dc regulated power supply.

A dc regulated power supply is a must for low level amplifiers such as phono-stage and line-stage amplifiers. It improves the stability of the dc power supply against ac mains fluctuation. A good dc regulated power supply noticeably improves the low level amplifier's sound. Since the output from a low level amplifier is further amplified by a power amplifier, any imperfection in the low level amplifier will become prominent at the power amplifier's output. Thus, it is imperative to ensure the signal that appears early in the signal path is well taken care of by the low level amplifier. And it must be powered by a good dc regulated power supply. The imperfection of the sound is often due to a poorly designed amplifier, improper choice of electronic components, or a bad dc regulated power supply. Perhaps we may put it this way. A bad dc regulated power supply can easily ruin the sound from an otherwise good amplifier. On the other hand, an average amplifier having a good dc regulated power supply may produce sound far better than a well-designed amplifier having a poor dc regulated power supply. Details of dc regulated power supply designs are given in Chapter 14. Two brief examples are illustrated here.

Figure 1.9(a) is a simple 15V regulated power supply using a *3-terminal* regulator LM317. A 3-terminal regulator is a simple and easy to use device for the application of a general purpose power supply. Both fixed voltage and adjustable voltage regulators are available. LM317 is an adjustable regulator that requires only two external resistors, R1 and R2, for setting an output voltage up to 37V. There are 3-terminal adjustable regulators that offer higher output voltages. For example, LM317AHV delivers 57V while TL783 for 125V. In this example, the output is 15V for a current up to 1.5A.

Figure 1.9(b) is a *series type floating-mode* dc regulated power supply for a very high output at 250V. This type of high voltage dc regulated power supply is suitable for vacuum tube

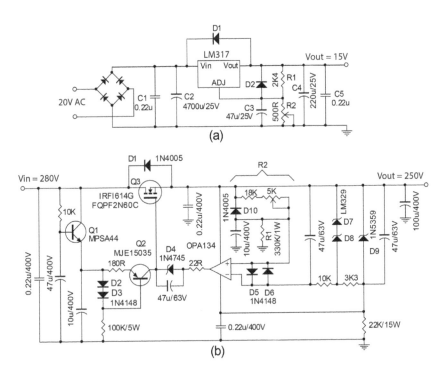

Figure 1.9 (a) A 3-terminal regulated power supply using LM317 for 15V output, (b) a series type floating-mode regulated power supply for 250V output

line-stage amplifier applications. The reason that it is called a series type is due to the fact that a power transistor (Q3) is placed in series between the unregulated input voltage Vin and regulated output voltage Vout. The transistor Q3 is called the *series pass transistor*. In this example, the series pass transistor is an N-channel MOSFET. It works as a source follower. The gate dc potential of the MOSFET is determined by the output of the op-amp OPA134, which is controlled by resistors R1 and R2, and reference zeners D7 and D8.

Zener D9 is used to ensure the op-amp operates within its maximum allowed supply voltage. For OPA134, the maximum supply voltage is 36V. In this example, the supply voltage for OPA134 is confined to the zener voltage of D9, i.e., 24V. Therefore, pin-4 of the op-amp is floating at 226V above the ground. As a consequence, Figure 1.9(b) is a series type *floating-mode* regulated power supply. By applying a similar floating technique, the 3-terminal regulator LM317 can also work for high output voltage. Discussion of various types of power supplies including shunt type, series type fixed-mode, and floating-mode employing solid-state devices, vacuum tubes, and hybrid devices is presented in Chapter 14.

2 The dc analysis for amplifying devices

2.1 Introduction

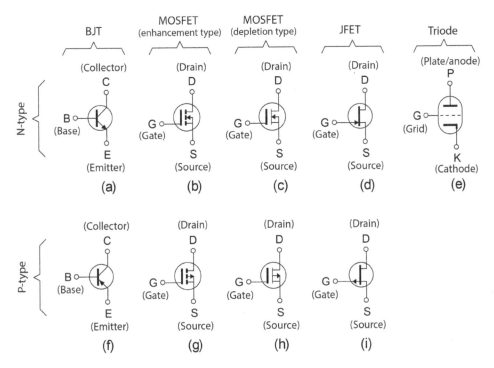

Figure 2.1 Symbols for BJT, enhancement and depletion type MOSFET, JFET, and triode

The dc analysis for five commonly used amplifying devices is discussed in this chapter. They include the solid-state bipolar junction transistor (BJT), the field-effect transistor (FET), and the vacuum tube triode. The FET is further broken down into the metal-oxide field-effect transistor (MOSFET) and the junction field-effect transistor (JFET). Furthermore, MOSFET transistors are divided into enhancement and depletion types. Thus, there are, in total, five different types of amplifying devices, as shown in Figure 2.1. It should be noted that some slightly different symbols for FETs are used in other literature that represent the same devices. However, in this book the symbols in Figure 2.1 are used consistently.

Note that solid-state amplifying devices contain complementary N-type and P-type devices that provide added flexibility to circuit design in such a way that a fully symmetrical circuit

DOI: 10.4324/9781003369462-2

becomes possible. In contrast, if a vacuum tube has electrical properties similar to an N-type (or N-channel) JFET, vacuum tubes with P-type properties do not exist. Owing to the absence of complementary tubes, the vacuum tube amplifier circuit design differs more than its solid-state counterpart. However, since the triode vacuum tube has much better linearity than solid-state devices, a few triodes will be sufficient to create a fine audio line-stage amplifier.

Before an amplifying device can perform as an amplifier, the device must be operated with the correct dc bias, i.e., the requisite amount of quiescent current and dc voltage so that the device operates in its active, or linear, region while the output signal swing is maximized. After the dc bias is correctly set up for the device, then it allows us to determine the small signal properties of the amplifier configuration in terms of voltage gain, input and output impedance.

This chapter is divided into two main sections. The first section is devoted to the discussion of dc analysis for each of the five amplifying devices [1–7]. The second section is focused on the dc analysis for cascade and compound amplifiers that have more than one amplifying device.

2.2   BJTs

The current conventions for the NPN and PNP transistors are shown in Figure 2.2. The collector current can be expressed as

$$I_C \approx I_S e^{(\frac{V_{BE}}{V_T})} \tag{2.1}$$

$$I_C = \alpha I_E + I_{CBO} \tag{2.2}$$

where I_S is a constant related to the intrinsic property of the transistor in the forward region; V_T is approximately equal to 26mV at 300°K; α is a constant close to unity; I_{CBO} is the collector to base current when the emitter is open. For small signal BJT transistors, I_{CBO} is insignificantly small at room temperature and it is often ignored. From Figure 2.2 we have,

$$I_E = I_B + I_C \tag{2.3}$$

Therefore, Eq. (2.2) can be rewritten as

$$I_C = \alpha(I_B + I_C) + I_{CBO}$$

Rearranging gives

$$I_C = \frac{\alpha I_B}{1-\alpha} + \frac{I_{CBO}}{1-\alpha} \tag{2.4}$$

Let us consider the case when $I_B = 0$, we have

$$I_C \big|_{I_{B=0}} = \frac{I_{CBO}}{1-\alpha} \equiv I_{CEO} \tag{2.5}$$

which is defined as the collector to emitter current when the base is open, i.e., $I_B = 0$.

For example, if $I_{CBO} = 1\mu A$, $\alpha = 0.995$, this gives $I_{CEO} = 1/(1-0.995)\mu A = 0.2mA$, which is not an insignificant figure that we can easily ignore. For a power transistor working at high temperature, the I_{CBO} can be higher than 50μA, giving I_{CEO} over 10mA, which becomes significant. In any case, when a BJT transistor operates at a collector current at or below I_{CEO}, the transistor is

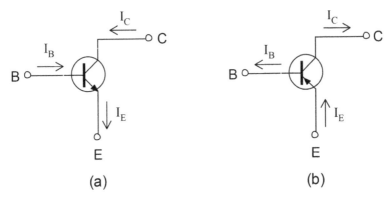

Figure 2.2 Current convention for the common-emitter configuration: (a) NPN transistor, (b) PNP transistor

operating in the cutoff region as shown in the shaded area of Figure 2.3(a) near to the horizontal axis. At the cut-off region, the base-emitter and base-collector junctions of a BJT transistor are both reverse biased.

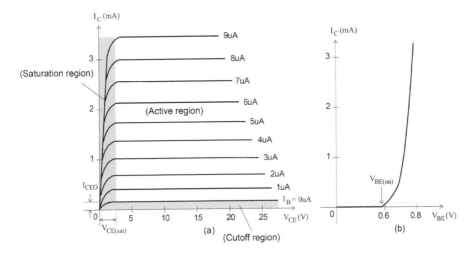

Figure 2.3 Characteristics of a BJT NPN transistor: (a) collector characteristics, (b) base characteristics

Another region in which a BJT transistor needs to avoid operating is the saturation region, as shown in the shaded area of Figure 2.3(a) near to the vertical axis. This occurs when the collector-emitter is operating below the $V_{CE(sat)}$. In the saturation region, the base-emitter and base-collector junctions are forward biased.

For a BJT transistor to work as an amplifier, it must be operated in the active region. In the active region, the base-emitter is forward biased while the base-collector is reverse biased. In order for the base-emitter to be forward biased, the transistor must be biased in such a way that V_{BE} is greater than the $V_{BE(on)}$, which is around 0.6 to 0.7V for small signal BJT transistors (see

Figure 2.3(b)). A dc bias network, which contains several resistors and is to be discussed later, will be needed to help the transistor to meet the base-emitter forward bias and base-collector reverse bias requirements.

Before moving to various dc bias techniques, let us take a look at BJT's current gain. The current gain is denoted by β. In fact, there is a dc current gain (β_{dc}) and ac current gain (β_{ac}). Even though they are not exactly equal, they are reasonably close and are often used interchangeably. In the manufacturer's data sheet, the dc current gain is often denoted by h_{FE}. The notation "h" is carried from the *h-parameter* model used in the early days for modelling the BJT transistor. Therefore, β_{dc} is equal to h_{FE} and h_{fe} is equal to β_{ac}. However, h_{fe} is not often published in data sheets these days. Therefore, we just take an approximation to treat h_{fe} as equal to h_{FE}. In this book, it is assumed that they are equal and simply denoted by β unless otherwise stated.

Current gain is expressed as $\beta = I_C/I_B$. When I_{CBO} is ignored from Eq. (2.2), we have $\alpha = I_C/I_E$. Therefore, Eq. (2.3) can be written as

$$\frac{I_C}{\alpha} = \frac{I_C}{\beta} + I_C$$

Eliminating I_C from the above expression yields

$$\alpha = \frac{\beta}{\beta+1} \text{ or } = \frac{\alpha}{1-\alpha} \tag{2.6}$$

This suggests that once either α or β is known, the other parameter can be determined. Since $\beta = I_C/I_B$, it is clear that Eq. (2.3) can also be written as

$$I_E = (\beta + 1) I_B \tag{2.7}$$

Fixed-bias configuration

Figure 2.4 shows two fixed-bias configurations. Figure 2.4(a) does not have an emitter resistor while an emitter resistor is used in Figure 2.4(b). Let us first examine Figure 2.4(a). C1 and C2 are the input and output dc blocking capacitors that block the dc from the amplifier to avoid affecting the input source, output load, and the dc bias of the circuit. In the following dc analysis, capacitors C1 and C2 are ignored.

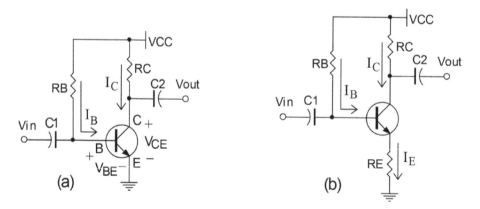

Figure 2.4 Fixed-bias configurations: (a) no emitter resistor, (b) with emitter resistor.

If we follow the input loop of Figure 2.4(a), starting from the power supply VCC to the resistor RB and base-emitter junction, we have

$$VCC - I_B RB - V_{BE} = 0$$

Therefore, the base current is given as

$$I_B = \frac{VCC - V_{BE}}{RB}$$

Since the collector current is $I_C = \beta I_B$, the voltage across the resistor RC becomes

$$V_{RC} = I_C RC = \beta I_B RC \tag{2.8}$$

If maximizing the output signal swing is of prime importance, the collector-emitter voltage V_{CE} should be equal to the voltage across resistor RC, so that

$$V_{CE} = V_{RC} = VCC/2 \tag{2.9}$$

For Figure 2.4(b), which has an emitter resistor, the dc bias is similar. To calculate the base current, we follow a similar route by looking into the input loop, starting from the power supply VCC to the resistor RB, base-emitter junction, and resistor RE. Thus, we have

$$VCC - I_B RB - V_{BE} - I_E RE = 0$$

and, since $I_E = (\beta + 1)I_B$, the base current is found to be

$$I_B = \frac{VCC - V_{BE}}{RB + (\beta + 1)RE} \tag{2.10}$$

The collector-emitter voltage V_{CE} can be determined by examining the output loop, starting from the power supply (VCC) to the voltage across collector resistor RC, across collector-emitter, across emitter resistor RE. We then have

$$VCC - I_C RC - V_{CE} - I_E RE = 0$$

and after rearranging yields

$$\begin{aligned} V_{CE} &= VCC - I_B \{\beta RC + (\beta + 1)RE\} \\ &\approx VCC - \beta I_B (RC + RE), \text{ for } \beta \gg 1 \end{aligned} \tag{2.11}$$

The addition of an emitter resistor affects several things. It improves the stability of the dc bias in a way that the dc bias currents and voltages remain very close to where they are set up by the circuit, despite any change of temperature or variation in transistor current gain β. It should be noted that the dc voltage across the resistor RE (i.e., V_{RE}) is often set to around 1–2V. To maximize the output signal swing, the voltage across the resistor RC (i.e., V_{RC}) is set equal to (VCC – V_{RE})/2. On the other hand, in the ac small signal sense, the emitter resistor RE improves the input impedance, frequency response, and distortion of the amplifier. But there is a trade-off, as voltage gain is reduced. Details of small signal ac analysis are given in Chapter 3.

Example 2.1

Given the fixed-bias configuration of Figure 2.5, determine:

a) I_B
b) I_C
c) V_B
d) V_C
e) V_E
f) V_{CE}
g) V_{BC}
h) V_{RC}

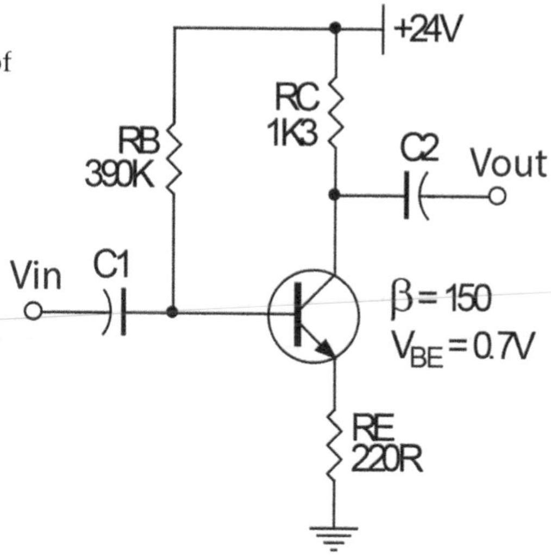

Solution:

a) Eq. (2.10):

Figure 2.5 A fixed-bias configuration for Example 2.1

$$I_B = \frac{VCC - V_{BE}}{RB + (\beta + 1)RE} = \frac{24V - 0.7V}{390k\Omega + (150 + 1)0.22k\Omega}$$

$$= 0.055mA$$

b) $I_C = \beta I_B = (150)(0.055)mA = 8.3mA$
c) $V_B = VCC - I_B RB = 24V - (0.055mA)(390k\Omega) = 2.55V$
d) $V_C = VCC - I_C RC = 24V - (8.3mA)(1.3\ k\Omega) = 13.2V$
e) $V_E = I_E RE = (\beta + 1)\ I_B RE = (150 + 1)(0.055mA)(0.22\ k\Omega) = 1.83V$
f) $V_{CE} = V_C - V_E = 13.2V - 1.83V = 11.4V$
g) $V_{BC} = V_B - V_C = 2.55V - 13.2V = -10.65V$ (reverse biased)
h) $V_{RC} = I_C RC = (8.3mA)(1.3\ k\Omega) = 10.8V$

It is noted that the base-emitter is forward biased while the base-collector is reverse biased. Therefore, the transistor is operating in the active region.

Voltage-divider bias configuration

In Figure 2.6, it can be seen that, for the fixed-bias configuration, the collector biasing currents I_C and V_{CE} are functions of the current gain β of the transistor. Because of the semiconductor's intrinsic property, β is temperature dependent. In addition, even transistors with the same part number have variations in β. Therefore, these make the fixed-bias sensitive to changes in temperature and β. Thus, it would be desirable to have a bias configuration that has less dependence on β and temperature changes. The voltage-divider bias configuration is the answer to this problem.

Figure 2.6(a) shows the voltage-divider bias configuration, which is different from the fixed-bias configuration of Figure 2.4(b) with an additional base resistor R2 connecting from base to ground. Because resistors R1 and R2 form a voltage divider network, this is where the name

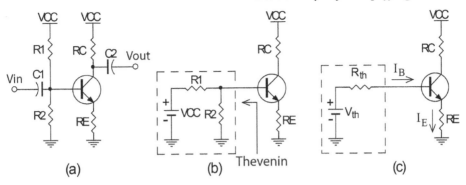

Figure 2.6 The voltage-divider bias configuration in circuit (a), simplified circuit, in (b); a Thévenin equivalent circuit for the base resistor network in (c)

"voltage-divider bias" comes from. The dc analysis for the voltage-divider bias configuration is similar to the fixed-bias configuration. Let us first redirect the circuit depicted in Figure 2.6(a) to circuit (b). To simplify the analysis, the voltage divider and the power supply can be replaced by a Thévenin equivalent circuit with a Thévenin power supply (V_{th}) and a Thévenin resistor (R_{th}), as shown in Figure 2.6(c).

To calculate the Thévenin equivalent resistor, R_{th}, we must assume that the voltage supply (VCC) is dead, i.e., shorted to the ground. (See Appendix A for an overview of Thévenin theorem.) Now the base of the transistor sees resistors R1 and R2 in parallel. This gives

$$R_{th} = (R1 \parallel R2) \tag{2.12}$$

To calculate the Thévenin equivalent power supply, V_{th}, it is what can be seen in the dc voltage supply at the base of the transistor in Figure 2.6(b). This is a dc voltage supply divided by resistors R1 and R2. Thus, this leads to

$$V_{th} = [R2/(R1 + R2)]VCC \tag{2.13}$$

When R_{th} and V_{th} are found, the base current can be determined by examining Figure 2.6(c), where we have

$$V_{th} - I_B R_{th} - V_{BE} - I_E RE = 0$$

Since $I_E = (\beta + 1)I_B$, rearranging the above equation yields

$$I_B = \frac{V_{th} - V_{BE}}{R_{th} + (\beta + 1)RE} \tag{2.14}$$

This expression has a form similar to Eq. (2.10), with VCC replaced by V_{th} and RB replaced by R_{th}. I_B can be determined by substituting Eqs. (2.12) and (2.13) for (2.14). Thus, the collector-emitter voltage is given as

$$\begin{aligned} V_{CE} &= VCC - I_C RC - I_E RE = VCC - I_B[\beta RC + (\beta + 1)RE] \\ &\approx VCC - \beta I_B(RC + RE), \text{ for } \beta \gg 1 \end{aligned} \tag{2.15}$$

Example 2.2

Given the voltage-divider bias configuration in Figure 2.7, determine:

a) R_{th}
b) V_{th}
c) I_B
d) I_C
e) V_{CE}

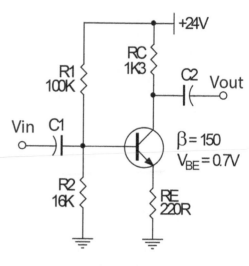

Solution:

a) Eq. (2.12):

$$R_{th} = (R1 \parallel R2) = (100 \times 16)/(100 + 16)k\Omega$$
$$= 13.8k\Omega$$

Figure 2.7 A voltage-divider bias configuration for Example 2.2

b) Eq. (2.13):

$$V_{th} = [R2/(R1 + R2)]VCC$$
$$= [16k\Omega/(100k\Omega + 16k\Omega)]\times24V = 3.31V$$

c) Eq. (2.14):

$$I_B = \frac{V_{th} - V_{BE}}{R_{th} + (\beta+1)RE} = \frac{(3.31 - 0.7)V}{13.8k\Omega + (150+1)0.22k\Omega} = 0.055mA$$

d) $I_C = \beta I_B = 8.3mA$
e) Eq. (2.15):
$$V_{CE} \approx VCC - \beta I_B(RC + RE) = 24V - 8.3mA(1.3k\Omega + 0.22k\Omega) = 11.4V$$

Example 2.3

Following Examples 2.1 and 2.2, we now reduce β from 150 to 75 and recalculate I_C and V_{CE}. The results are tabulated in the following.

	Fixed bias (Figure 2.5)		Voltage-divider bias (Figure 2.7)	
	$\beta = 75$	$\beta = 150$	$\beta = 75$	$\beta = 150$
I_C (mA)	4.3	8.3	6.4	8.3
V_{CE} (V)	17.5	11.4	14.3	11.4

When β is equal to 150, both fixed-bias and voltage-divider bias configurations give an identical operation point with the same I_C and V_{CE}. When β is reduced to 75, the operation points for both configurations are affected. However, the dc bias is less affected by the change of β in the voltage-divider bias configuration. On the other hand, the presence of an emitter resistor RE helps to reduce the effect caused by the change of V_{BE} due to temperature variation, which is not shown here. However, it should be noted that RE reduces the voltage gain of the amplifier, which is the subject matter of Chapter 3. Fortunately, voltage gain can be retained by a fully or partially bypassing the resistor RE with a large capacitor.

Load line analysis

The collector current I_C and V_{CE} that the transistor is biased to is often called the dc *quiescent point*, or Q-point. As discussed above, the Q-point can be determined analytically. Alternatively, it can also be found by the *method of load line*. Illustrated below is the latter method to determine the Q-point for the circuit depicted in Figure 2.7. The load line is configured by a line connecting the following two points, as shown in Figure 2.8:

$$\text{At } V_{CE} = 0V, I_C = \frac{VCC}{RC + RE} = \frac{24V}{1.3k\Omega + 0.22k\Omega} = 15.8mA$$

$$\text{At } I_C = 0, V_{CE} = VCC = 24V$$

In order for the transistor to deliver maximal output swing, the quiescent point has to be located near to the middle of the load line rather than in the cutoff region (near the horizontal axis) or the saturation region (near the vertical axis). For the load line shown in Figure 2.8, the middle point is located at $V_{CE} = VCC/2 = 12V$. The collector current I_C can be interpolated from Figure 2.8, giving 8.5mA. Therefore, the quiescent point $(V_{CE(Q)}, I_{C(Q)})$ is located at $V_{CE(Q)} = 12V$ and $I_{C(Q)} = 8.5mA$. If we now take into consideration the dc voltage across emitter

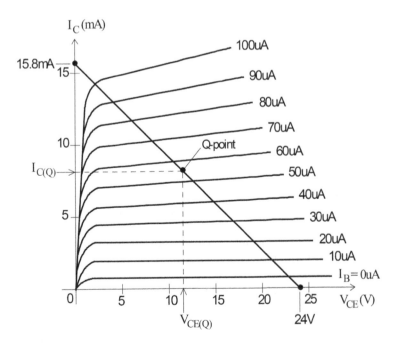

Figure 2.8 Defining the quiescent point for the voltage-divider bias configuration of Figure 2.7

resistor, $I_E RE \approx I_C RE = 8.5mA \times 0.22k\Omega = 1.87V$, the collector voltage of the transistor becomes $V_C = V_{CE} + 1.87V = 13.87V$. In contrast, the dc voltage across the collector resistor R_C is $24V - V_C = 24V - 13.87V = 10.13V$. Since the output signal is coming from the collector, the signal will swing up 10.13V before reaching the power supply rail, which is the limit at the top, and swing down $V_{CE} = 12V$ before reaching V_E, which is the limit at the bottom. These two amplitudes have a difference of $12V - 10.13V = 1.87V$. In other words, the output signal clips at the top 1.87V before the bottom cycle reaching its maximum level. This is illustrated in Figure 2.9. In low level applications where the output swing requirement is just a few volts, it is not a problem. Nevertheless, if optimizing the output swing is needed for this example, the V_{CE} quiescent point should be changed from $V_{CE} = VCC/2 = 12V$ to $V_{CE} = (VCC - V_E)/2 = (24V - 1.87V)/2 = 11.1V$.

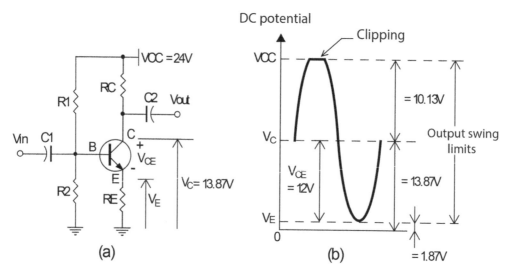

Figure 2.9 A voltage-divider bias configuration representing the Q-point of IC = 8.5mA and VCE = 12V in (a) while the maximum output swing is in (b)

2.3   Field-effect transistors (FETs)

Besides BJT, there is a broad class of solid-state FETs, as shown in Figure 2.10. The major difference between a BJT and an FET is that BJT is a current-controlled device while FET is a voltage-controlled device. By controlling the base current (I_B) of a BJT, the transistor generates an amplified collector current (I_C) and, therefore, a subsequent output signal is produced. However, there is no gate current required for FET. Therefore, the only means of controlling an FET is to vary the gate voltage with respect to the source (V_{GS}). By controlling the V_{GS} of an FET, the transistor generates an amplified drain current and, thus, the output signal is produced. Since there is no gate current in FET, it has a very high input impedance.

The FET family is divided into metal-oxide FET (MOSFET) and junction FET (JFET). MOSFET is also sub-divided into depletion and enhancement types. However, the depletion type MOSFET is not as popular as the enhancement type. Yet, there are some unique features of the depletion type MOSFET that make it difficult to be replaced.

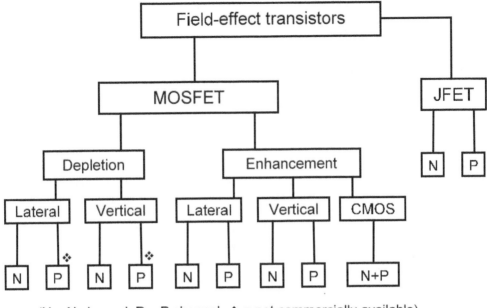

(N = N channel, P = P channel, ❖ = not commercially available)

Figure 2.10 Family tree of the field-effect transistors

JFET: A popular alternative to BJT in low power signal amplification. At room temperature, the reverse gate leakage current for JFET is compatible to MOSFET, but much lower than BJT. JFET has much lower flicker noise than MOSFET. Even though JFET has a higher input noise voltage density than BJT, JFET has very low input noise current density. JFET is suitable for low noise applications such as phono stage and line-stage amplifiers as well as the input stage for audio power amplifiers. On the other hand, JFET has electrical characteristics similar to the triode vacuum tube, making JFET a very popular choice for low level signal amplification. In recent years, some manufacturers (Infineon, SemiSouth, and United Silcon Carbide) have launched power JFET devices. They are intended for industrial applications. However, some power JFETs can also be used in audio power amplifiers. The power JFETs produced by SemiSouth have low $V_{GS(th)}$, making them very suitable for audio power amplifier applications. Unfortunately, SemiSouth closed its business in 2012. A list of representative power JFETs can be found in Chapter 13.

MOSFET (enhancement type): This is the most widely used device for all kinds of electronic applications. It is often described as the "work horse" of the electronic industry. The notable characteristics of the enhancement type MOSFET include no gate current, high input impedance, fast switching, low $R_{DS(ON)}$, high current and power handling capability. They are also capable of high scaling and miniaturization, and can be easily scaled down to smaller dimensions. They consume significantly less power and allow much higher density than BJTs. Enhancement type MOSFETs are responsible for increasing transistor density, enhancing performance and decreasing power consumption of integrated circuit chips and electronic devices. Even though most enhancement type MOSFETs are designed for switching applications, some manufacturers also offer products for linear applications from low to high power. Enhancement type power MOSFETs are popular for use in the output stage of audio power amplifiers.

However, since it has much higher flicker noise than JFET and BJT, MOSFET is not desirable for the input stage of an audio amplifier. Enhancement type MOSFETs can be built using either a vertical or a lateral structure.

MOSFET (depletion type): The depletion type MOSFET requires a special ion-implantation process that the enhancement device does not need. This makes the depletion type devices a little more difficult to manufacture and their characteristics harder to control than enhancement devices. It also results in depletion and enhancement type MOSFETs having some different electrical characteristics. The notable difference is that enhancement type devices are "normally off", while depletion type devices are "normally on". For instance, $V_{GS} < 0$, an N-channel depletion type MOSFET, has characteristics similar to a JFET, while $V_{GS} > 0$ has characteristics similar to an enhancement type device. In general, depletion type MOSFETs share many properties with the enhancement devices, such as no gate current, high input impedance, fast switching, low $R_{DS(ON)}$, high current and power handling capability. Depletion type MOSFETs can be also built using either a vertical structure or a lateral structure. Even though it is possible to manufacture both N- and P-channel depletion MOSFETs, discrete P-channel depletion type MOSFETs are not commercially available.

Vertical MOSFET: Figure 2.11(a) shows the structure of a vertical *double-diffused MOS* (DMOS). The name "double-diffused" comes from the fact that the P-doped substrate is first diffused and then followed by a highly doped N+ source diffusion. The term "vertical" is due primarily to the fact that the drain current is flowing vertically rather than horizontally, as for the lateral device. It should also be noted that the source and drain terminals are placed at the opposite sides of the device. Compared to the lateral devices, a vertical MOSFET supports higher breakdown voltage, lower $R_{DS(ON)}$ and higher current capability. The V_{GS} breakdown voltage can be over 1000V. There are two other vertical MOSFET structures used in the industry prior to the DMOS. The first is VMOS, which has the gate in a V-shaped structure, and the second is UMOS, with a gate in a U-shaped structure.

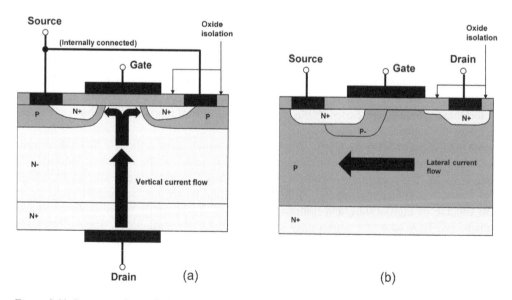

Figure 2.11 Structure of a vertical MOSFET (a) and lateral MOSFET (b)

Lateral MOSFET: Figure 2.11(b) shows the structure of a lateral MOSFET. Note that the lateral structure has its drain, source, and gate terminals located on the top surface of the device. The current flows laterally from the drain to the source. Compared with the vertical structure, the lateral MOSFETs have lower breakdown voltage, higher $R_{DS(ON)}$, and lower current capability. As a result, the lateral MOSFETs are not as popular as the vertical MOSFETs in industrial applications. However, lateral MOSFETs have been widely used in audio power amplifiers. Examples of lateral power MOSFETs for audio amplifiers include 2SK216/2SJ79 for the driver stage, and 2SK134/2SJ49, 2SK1058/2SJ162, and 2SK2221/2SJ352 for output power stage. They have been popularized by Hitachi since the 1970s, although some of these devices are no longer in production. In general, production of lateral MOSFETs is much lower than the vertical devices.

CMOS: *Complementary MOSFET* (CMOS) was introduced primarily for digital circuit design. CMOS uses complementary symmetrical pair of P- and N-channel MOSFET. In order to improve speed, power consumption, and area reduction of digital integrated circuits, the feature size of CMOS has been continuously shrunk in the last few decades. CMOS has become the dominant MOSFET fabrication process in *VLSI (Very Large Scale Integration)*. It is believed that over 99% of integrated circuits (ICs), including digital, analog and mixed signals ICs, are fabricated using the CMOS technology. Besides digital applications, CMOS technology is also used in analog applications, i.e., CMOS op-amps, RF circuits and mixed-signal applications. While digital ICs are benefited from the feature size shrinking, analog CMOS op-amps do not have such advantages due to the intrinsic properties of analog design. Because there is an intrinsic gain reduction of short channel transistors that affects the overall voltage gain of the amplifier.

2.4   MOSFETs – enhancement type

The current conventions for the N-channel and P-channel MOSFETs are shown in Figure 2.12. Since there is no gate current, we must have

$$I_G = 0, \text{ and}$$
$$I_D = I_S \tag{2.16}$$

The relationship between drain current I_D and gate-source voltage V_{GS} is given as

$$I_D = k[V_{GS} - V_{GS(th)}]^2 \tag{2.17}$$

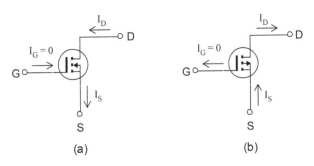

Figure 2.12 Current convention for an enhancement type MOSFET in a common-source configuration: (a) N-channel MOSFET, (b) P-channel MOSFET

The constant k is a function of the intrinsic properties of the device. It can be determined by Eq. (2.17) with a given point from the characteristic curve of the device, say, $I_{D(on)}$ and $V_{GS(on)}$. $V_{GS(th)}$ is the threshold voltage for V_{GS} when $I_D = 0$. For instance, Figure 2.13 shows the characteristic curves of an N-channel MOSFET. Figure 2.13(a) is the transfer characteristic curve of the device for common source configuration. Figure 2.13(b) shows the curves of I_D versus V_{DS} for various V_{GS}. To determine the constant k, first notice that $V_{GS(th)} = 3V$. Then pick a point in the curve, say, $V_{GS} = 8V$ and $I_D = 1.87mA$. Therefore, rearranging Eq. (2.17) yields

$$k = \frac{I_{DS(on)}}{[V_{GS(on)} - V_{GS(th)}]^2} \tag{2.18}$$

$$= \frac{1.87mA}{(8V - 3V)^2} = 0.075mA/V^2$$

When the constant k is found, Eq. (2.17) permits I_D to be calculated for a given V_{GS}. For instance, at the point $V_{GS} = 5V$, the drain current is determined by Eq. (2.17) such that

$$I_D = k[V_{GS} - V_{GS(th)}]^2 = 0.075[5V - 3V]^2 \, mA = 0.3mA$$

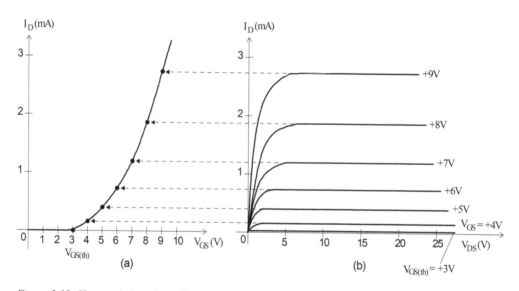

Figure 2.13 Characteristics of an enhancement type N-channel MOSFET for a common source configuration: (a) gate transfer characteristics, (b) drain characteristics

Voltage-divider bias configuration

As discussed earlier for BJT, a voltage-divider is a very stable dc bias configuration. This also applies to MOSFET devices, shown in Figure 2.14. In fact, this configuration works equally well for common source, common gate, and source follower (often called common drain) amplifying configurations. Details about the ac small signal analysis of these amplifying configurations are discussed in the next chapter.

Since there is no gate current, the dc potential at the gate is equal to

$$V_G = \frac{R2}{R1+R2} VDD \qquad (2.19)$$

Therefore, in the input loop we have

$$V_G - V_{GS} - I_D RS = 0,$$

after rearranging yields

$$I_D = \frac{V_G - V_{GS}}{RS} = \frac{1}{RS}\left(\frac{R2}{R1+R2} VDD - V_{GS}\right) \qquad (2.20)$$

In the output loop, we have

$$VDD - I_D RD - V_{DS} - I_D RS = 0, \text{ or}$$
$$V_{DS} = VDD - I_D(RD + RS) \qquad (2.21)$$

Eqs. (2.17) and (2.20) form equations with two unknowns, I_D and V_{GS}. Therefore, they can be solved either analytically or numerically. After I_D and V_{GS} are found, V_{DS} can be determined by Eq. (2.21). We illustrate it in the following example.

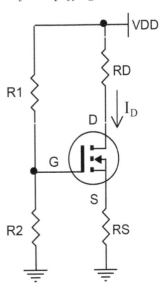

Figure 2.14 Voltage-divider bias configuration

Example 2.4

Determine the dc quiescent point, I_D and V_{DS}, for the circuit in Figure 2.15.

From the data sheet for the MOSFET 2N7000, we have $V_{GS(th)}$ max = 3V, $I_{D(on)}$ = 75mA at $V_{GS(on)}$ = 4.5V.

Eq. (2.19):

$$V_G = \frac{R2}{R1+R2} VDD = \frac{1}{3.3+1} 24V = 5.58V$$

Eq. (2.18):

$$k = \frac{I_{DS(on)}}{[V_{GS(on)} - V_{GS(th)}]^2}$$

$$= \frac{75mA}{[4.5V - 3V]^2} = 33.3mA/V^2$$

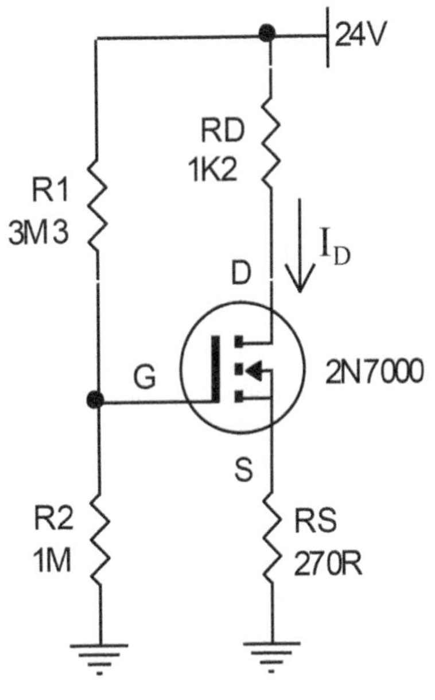

In the input loop, we have

$$V_G - V_{GS} - I_D RS = 0, \text{ or}$$

$$I_D = \frac{V_G - V_{GS}}{RS} = \frac{5.58 - V_{GS}}{0.27} mA = (20.67 - 3.7V_{GS})mA \qquad (2.22)$$

Figure 2.15 A voltage divider bias configuration for Example 2.4

Eq. (2.17):

$$I_D = k[V_{GS} - V_T]^2$$

$$= 33.3[V_{GS} - 3]^2 mA \qquad (2.23)$$

Thus, Eqs. (2.22) and (2.23) form equations with two unknowns, I_D and V_{GS}. They can be solved by analytical and numerical methods.

Analytical method

After Eq. (2.22) is substituted by Eq. (2.23), we have

$$20.67 - 3.7V_{GS} = 33.3[V_{GS} - 3]^2$$

and rearranging yields

$$33.3V_{GS}^2 - 196.1V_{GS} + 279 = 0$$

The above quadratic equation can be written in the standard form

$$aV_{GS}^2 + bV_{GS} + c = 0$$

where

$$a = 33.3, b = -196.1, c = 279.$$

The above quadratic equation has two solutions

$$V_{GS1} = \frac{-b + \sqrt{b^2 - 4ac}}{2a} = 3.48V \text{ and } V_{GS2} = \frac{-b - \sqrt{b^2 - 4ac}}{2a} = 2.4V$$

However, since 2.4V is lower than $V_{GS(th)} = 3V$, the transistor resides in the cut-off region. Therefore, V_{GS2} is not a solution. The valid solution is $V_{GS1} = 3.48V$. From Eq. (2.22), we find $I_D = 7.8mA$. V_{DS} can be determined by Eq. (2.21):

$$V_{DS} = VDD - I_D(RD + RS) = 24V - 7.8mA(1.2k\Omega + 0.27k\Omega) = 12.5V.$$

Numerical method

Alternatively, Eqs. (2.22) and (2.23) can be solved numerically. With the help of a spreadsheet, we can plot Eq. (2.22) for various values of I_D and V_{GS}. The plot is the straight line shown in Figure 2.16. However, the plot of Eq. (2.23) is a curve. The intersection between the line and the curve is the quiescent point for the voltage-divider bias configuration of Figure 2.15. It is the same quiescent point that is previously determined by the analytical method, i.e., $I_D = 7.8mA$ and $V_{GS} = 3.48V$. Thus, V_{DS} is also 12.5V.

The analytical and numerical methods lead to the same results. However, both methods take considerable time to arrive at the solution. If the solution is accurate, it is absolutely worth the

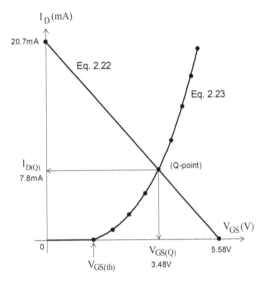

Figure 2.16 A graphical presentation of Eqs. (2.22) and (2.23)

effort. However, the accuracy often depends on the data given for the MOSFET, $V_{GS(th)}$. The calculated quiescent point can differ from the values measured from the actual circuit by more than 20%. For example, the quiescent point measured for a circuit built according to Figure 2.15 is: $I_D = 10.9mA$ and $V_{GS} = 2.7V$. Comparing that to the calculated values $I_D = 7.8mA$ and $V_{GS} = 3.48V$, the difference is over 20%.

The major factor that contributes to the differences is $V_{GS(th)}$. MOSFETs have a wider range of $V_{GS(th)}$ than the $V_{BE(on)}$ of BJTs. It is noted that $V_{BE(on)}$ is very well defined and it mostly varies from 0.6V to 0.7V. The variation in $V_{BE(on)}$ is around 20% or so. However, the variation of $V_{GS(th)}$ for a MOSFET is a lot higher. In this example, the $V_{GS(th)}$ specified in the manufacturer's data sheet for the 2N7000 MOSFET varies from 0.8V (min) to 3V (max), which has a difference of over 300%. We have chosen $V_{GS(th)} = 3V$ (max) for the above example. Since both constant k and I_D is a function of $V_{GS(th)}$, if $V_{GS(th)}$ is not accurately defined in the beginning, this leads to inaccurate constant k and I_D. As a result, the accuracy of the quiescent point is enormously affected.

Therefore, it's worth exploring an approximated method, which is simple and easy to use, to determine the quiescent point of the voltage-divider bias configuration with acceptable accuracy. Let us revisit the circuit shown in Figure 2.15 and now test the circuit for six different MOSFETs from different manufacturers. The resistors R1, R2, RD, and RS remain unchanged. Each time we plug in a transistor and then measure the V_{GS} and drain current I_D. The results of the measured V_{GS} are shown in Figure 2.17.

In Figure 2.17, the straight line represents that the measured V_{GS} is equal to $V_{GS(th)\,max}$, i.e., in a 1:1 ratio. As shown in the Figure, the measured V_{GS} is always lower than $V_{GS(th)\,max}$. Some are closer to, and some are farther away from, the straight line. But they are not too far away. On average, the V_{GS} is roughly within 20% of the $V_{GS(th)\,max}$. Therefore, this result suggests that we can use $V_{GS(th)\,max}$ as the first order approximation to V_{GS}. Here, we call this an *approximated method*. In this method, the V_{GS} is no longer a variable that requires a quadratic equation in V_{GS} to be solved by either of the methods, analytical or numerical, discussed earlier. V_{GS} is now assumed to be $V_{GS(th)\,max}$, which is a constant. It is treated like the $V_{BE(on)}$, a constant. Thus, the quiescent point of the voltage-divider bias configuration for a MOSFET can be worked out in a similar

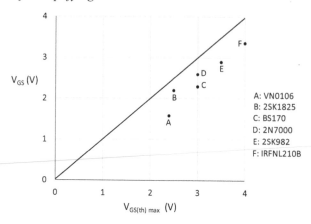

Figure 2.17 Measured VGS in the circuit shown in Figure 2.15 for six MOSFETs

fashion to that for a BJT. Only simple circuit analysis is needed. The approximated method is illustrated in the following.

Approximated method

Figure 2.18 shows a MOSFET in a voltage-divider bias configuration. The resistors R2, RD, and RS are the same as in Figure 2.15 for Example 2.4. We keep the same drain resistor RD and source RS because these two resistors determine the voltage gain of the amplifier. 1MΩ is kept for R2 because it helps to preserve a high input impedance to this amplifier. It is now necessary to find a suitable R1 resistor so that the gate has the correct dc voltage to set up the desired dc quiescent point. Here, the required quiescent drain current is 10mA.

The approximated method assumes that V_{GS} is now $V_{GS(th)\,max}$, which is a constant. Therefore, the drain current can be determined directly. The drain current is given by Eq. (2.20) and it is rewritten as follows.

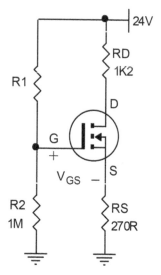

Figure 2.18 Change R1 to suit 6 different MOSFETs

$$I_D = \frac{V_G - V_{GS}}{RS} \approx \frac{1}{RS}\left(\frac{R2}{R1+R2}VDD - V_{GS(th)max}\right) \quad (2.24)$$

where R2 = 1MΩ, RD = 1.2kΩ, RS = 0.27kΩ, VDD = 24V and $V_{GS} = V_{GS(th)\,max}$.

When the target drain current I_D is 10mA, the required R1 can be easily determined by Eq. (2.24) for a given $V_{GS(th)\,max}$. Figure 2.19(a) shows the calculated and actual values for R1. For instance, if the calculated resistor is 3.7MΩ, since this resistance is not a practical resistor value, the actual resistor 3.9MΩ is used. The circuit is tested for the actual resistor R1 and drain current is measured for each MOSFET. As shown in Figure 2.19(a) and (b), the measured drain currents are not far from the target. On average, the measured drain current is within 20% of the target.

Let us recall Example 2.4. The transistor is 2N7000. The same resistor R1 (3.3MΩ) and $V_{GS(th)\,max}$ (3V) are used for the analytical method. The analytical solution gives I_D = 7.8mA. However,

Part number	$V_{GS(th)\,max}$ (V)	R1 (Calculated)	R1 (Actual)	Measured for actual R1				
				V_G (V)	V_D (V)	V_S (V)	V_{GS} (V)	I_D (mA)
VN0106	2.4	3.7M	3.9M	4.93	9.11	3.33	1.6	12.4
2SK1825	2.5	3.6M	3.6M	5.16	10.7	2.95	2.2	11.1
BS170	3	3.2M	3.3M	5.58	9.68	3.2	2.4	11.9
2N7000	3	3.2M	3.3M	5.58	10.9	2.92	2.7	10.9
2SK982	3.5	2.87M	3M	5.96	10.8	2.95	3.0	11.0
IRFNL210B	4	2.58M	2.7M	6.43	10.8	2.94	3.5	11.0

(a)

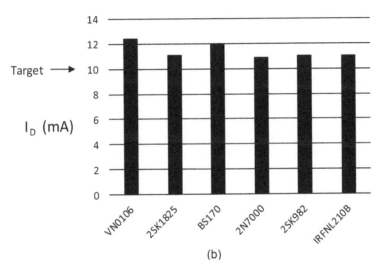

(b)

Figure 2.19 Values for R1, measured voltages and currents are given in (a). The measured drain currents are stacked against the target value

the actual measured drain current is $I_D = 10.9$mA. The analytical solution differs from the actual measurement by 28%. Therefore, the accuracy of the solution by the approximated method is on par with the analytical method. However, the approximated method is easy to use and time saving. The drain current I_D can be easily determined by Eq. (2.24), by assuming $V_{GS} = V_{GS(th)\,max}$. The accuracy is not the best in class, but it is acceptable. And the accuracy will be improved for a large power supply when $VDD \gg V_{GS(th)\,max}$.

2.5   MOSFETs – depletion type

The current conventions for depletion type MOSFET are shown in Figure 2.20. Since there is no gate current, we must have:

$$I_G = 0$$

and

$$I_D = I_S$$

The relationship between I_D and V_{GS} is given as

$$I_D = I_{DSS}\left[1 - \frac{V_{GS}}{V_{GS(th)}}\right]^2 \qquad (2.25)$$

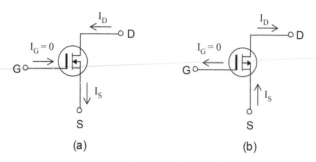

(a)   (b)

Figure 2.20 Current convention for depletion type MOSFET in common-source configuration: (a) N-channel MOSFET, (b) P-channel MOSFET

The constant I_{DSS} is the drain current I_D at $V_{GS} = 0$. As suggested by Eq. (2.25), a depletion type MOSFET is characterized by I_{DSS} and $V_{GS(th)}$. They are given in the data sheet provided by the manufacturer. When these two parameters are known, we can easily figure out the drain current I_D and V_{GS} characteristics. Figure 2.21(a) is the transfer characteristic curve of a depletion type MOSFET in a common source configuration. Figure 2.21(b) shows the curves of I_D versus V_{DS} for various V_{GS}.

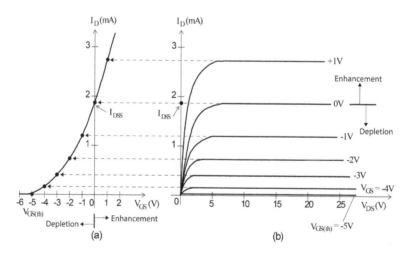

Figure 2.21 Characteristics of a depletion type N-channel MOSFET for a common source configuration: (a) gate transfer characteristics, (b) drain characteristics

As shown in Figure 2.21, in the region of $V_{GS} < 0$, the MOSFET is operating in the depletion mode, while in the region of $V_{GS} > 0$ it is in the enhancement mode. It can be seen later that the depletion mode's transfer characteristic is similar to that of a JFET and a vacuum tube triode. Given this "triode-like" transfer characteristic, they can be operated in a unique self-bias configuration. However, the self-bias configuration cannot apply to a BJT and an enhancement type MOSFET. In addition to the self-bias configuration, the depletion type MOSFET can also be operated in the fixed-bias and voltage-divider bias configuration. Likewise, these two biasing configurations can also apply to the JFET and triode.

Fixed-bias configuration

Figure 2.22 shows the fixed-bias configuration for a depletion type MOSFET. The fixed-bias configuration requires two power supplies, VDD and VGG. RG is a large resistor that feeds the dc voltage VGG to the gate but prevents the input signal from being shorted to VGG. In the input loop, since $I_G = 0$ we have,

$$V_{GS} = -VGG \qquad (2.26)$$

In the output loop, we have,

$$V_{DS} = VDD - I_D RD \qquad (2.27)$$

Substituting Eq. (2.26) for Eq. (2.25) yields

$$I_D = I_{DSS} \left[1 + \frac{VGG}{V_{GS(th)}} \right]^2 \qquad (2.28)$$

Eq. (2.28) can be solved for I_D. Then V_{DS} is determined by Eq. (2.27).

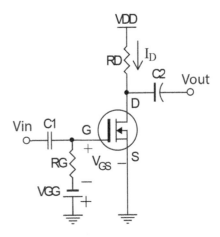

Figure 2.22 Fixed-bias for an N-channel depletion mode MOSFET in a common source configuration

Self-bias configuration

Figure 2.23 shows the self-bias configuration for an N-channel depletion type MOSFET. Note that the self-bias configuration is unique to the depletion type MOSFET but it does not apply to the enhancement type MOSFET. It is because, in order for an enhancement type MOSFET to work in the active region, V_{GS} must be greater than $V_{GS(th)}$, which is greater than 0. An N-channel enhancement type MOSFET becomes cut off in the depletion region where $V_{GS} < 0$.

As shown in Figure 2.23, the self-bias configuration requires a gate resistor RG connecting to the ground. Unlike the fixed bias with a negative dc supply connecting to the gate, the self-bias renders the source biased at a higher dc potential than the gate, as the drain current flows down to RS. As a result, V_{GS} becomes negative as required. Thus, in the input loop, we have

$$V_{GS} = -I_D RS, \text{ or}$$

$$I_D = -\frac{V_{GS}}{RS} \qquad (2.29)$$

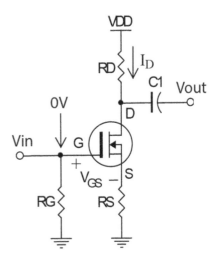

Figure 2.23 Self-bias for an N-channel depletion type MOSFET in a common source configuration.

Since the drain current is governed by Eq. (2.25), together with Eq. (2.29), they form equations with two unknowns, I_D and V_{GS}. They can be solved either analytically or numerically. If solving by the analytical method, Eq. (2.29) is first substituted for Eq. (2.25):

$$-\frac{V_{GS}}{RS} = I_{DSS}\left[1 - \frac{V_{GS}}{V_{GS(th)}}\right]^2 \tag{2.30}$$

Eq. (2.30) can be expressed in the form of

$$aV_{GS}^2 + bV_{GS} + c = 0$$

The above quadratic equation has two solutions for V_{GS}:

$$V_{GS1} = \frac{-b + \sqrt{b^2 - 4ac}}{2a} \text{ and } V_{GS2} = \frac{-b - \sqrt{b^2 - 4ac}}{2a}$$

One of the solutions is invalid because it is below $V_{GS(th)}$, i.e., in the cut-off region. The valid solution must satisfy the condition $V_{GS(th)} < V_{GS} < 0$, as the MOSFET has to be operated in the depletion mode region. Since the gate is at dc ground level, the self-bias configuration eliminates the need for an input dc blocking capacitor that may affect the low frequency response of the amplifier. This is an attractive feature for the self-bias configuration that is shared among depletion-type MOSFET, JFET and triodes.

Example 2.5

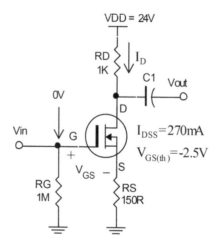

Figure 2.24 A self-bias configuration for a depletion type MOSFET

The quiescent point for the depletion type MOSFET in Figure 2.24 is determined as follows. Eq. (2.29):

$$I_D = -\frac{V_{GS}}{RS} = -\frac{V_{GS}}{0.15} \text{ mA} = -6.67 V_{GS} \text{ mA} \tag{2.31}$$

Eq. (2.25):

$$I_D = I_{DSS}\left[1 - \frac{V_{GS}}{V_{GS(th)}}\right]^2 = 270\left[1 - \frac{V_{GS}}{-2.5V}\right]^2 \text{ mA}$$

$$= 43.2[2.5 + VGS]^2 \text{ mA} \tag{2.32}$$

Substituting Eq. (2.31) with Eq. (2.32) and rearranging yields,

$$aV_{GS}^2 + bV_{GS} + c = 0$$

where a = 43.2, b = 222.7, c = 270. The solutions to the quadratic equation are given as:

$$V_{GS1} = -1.95V \text{ and } V_{GS2} = -3.2V$$

V_{GS2} is rejected because it is below $V_{GS(th)}$, i.e., –2.5V. The valid solution is V_{GS1}. Substituting $V_{GS} = -1.95V$ for Eq. (2.31) yields

$$I_D = -6.67V_{GS} \text{ mA} = (-6.67)(-1.95) \text{ mA} = 13mA$$

From the output loop, we have

$$V_{DS} = VDD - I_D(RD + RS) = 24V - 13mA(1k\Omega + 0.15k\Omega) = 9V$$

Thus, the quiescent point is given by $I_D = 13mA$ and $V_{DS} = 9V$.

Voltage-divider bias configuration

In addition to the fixed- and self-bias configurations, the voltage-divider bias configuration also works for depletion type MOSFET, as shown in Figure 2.25. As there is no gate current, the dc potential at the gate is simply determined by the voltage divider formed by resistors R1 and R2. Since the input impedance of a MOSFET is very high, in order to avoid loading down the input impedance, R1 and R2 are often large resistors in the MΩ range. Given such a high resistor, the input coupling capacitor C1 can be a small capacitor without affecting the low frequency response. When the capacitor is small, good quality film capacitors such as polypropylene and polyester are often used instead of electrolytic capacitor.

The dc potential at the gate is equal to

$$V_G = \frac{R2}{R1+R2}VDD$$

In the input loop we have

$$V_G - V_{GS} - I_D RS = 0,$$

after rearranging yields

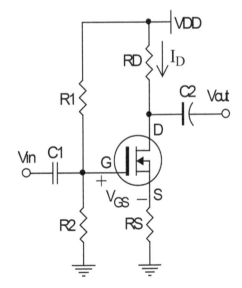

Figure 2.25 A voltage-divider bias configuration for N-channel depletion type MOSFET

$$I_D = \frac{V_G - V_{GS}}{RS} = \frac{1}{RS}\left(\frac{R2}{R1+R2}VDD - V_{GS}\right) \tag{2.33}$$

Eq. (2.25) and (2.33) form equations with two unknowns, I_D and V_{GS}. Therefore, they can be solved either analytically or numerically. The analytical solution can be found by substituting Eq. (2.33) for Eq. (2.25) so that

$$\frac{1}{RS}\left(\frac{R2}{R1+R2}VDD-V_{GS}\right)=I_{DSS}\left[1-\frac{V_{GS}}{V_{GS(th)}}\right]^2 \tag{2.34}$$

Eq. (2.34) can be expressed in the form of

$$aV_{GS}^2+bV_{GS}+c=0$$

Again, the quadratic equation has two solutions for V_{GS}. One of the solutions is invalid because it is below $V_{GS(th)}$, i.e., cut-off region. The valid solution must satisfy the condition $V_{GS}>V_{GS(th)}$. After I_D and V_{GS} are found, V_{DS} can be determined by considering the output loop,

$$VDD-I_DRD-V_{DS}-I_DRS=0, \text{ or}$$
$$V_{DS}=VDD-ID(RD+RS) \tag{2.35}$$

On the other hand, the approximated method that was introduced for enhancement type MOSFET does not work for depletion type MOSFET.

Example 2.6

The quiescent point for the depletion type MOSFET in Figure 2.26 is determined as follows. Eq. (2.33):

$$I_D=\frac{1}{RS}\left(\frac{R2}{R1+R2}VDD-V_{GS}\right)$$

$$=\frac{1}{0.47k\Omega}\left(\frac{1M\Omega}{33M\Omega+1M\Omega}24V-V_{GS}\right)mA$$

$$=(1.5-2.13V_{GS})\,mA \tag{2.36}$$

Eq. (2.25):

$$I_D=I_{DSS}\left(1-\frac{V_{GS}}{V_{GS(th)}}\right)^2$$

$$=270\left(1-\frac{V_{GS}}{(-2.5)}\right)^2 mA$$

$$=43.2[2.5+V_{GS}]^2\,mA \tag{2.37}$$

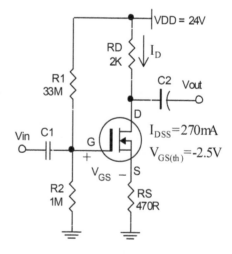

Figure 2.26 A voltage-divider bias configuration example

Substituting Eq. (2.36) with Eq. (2.37) and rearranging yields

$$aV_{GS}^2 + bV_{GS} + c = 0$$

where a = 43.2, b = 218.1, c = 268.5
The quadratic equation has two solutions for V_{GS}:

$$V_{GS1} = \frac{-b + \sqrt{b^2 - 4ac}}{2a} = -2.13V$$

$$V_{GS2} = \frac{-b - \sqrt{b^2 - 4ac}}{2a} = -2.9V$$

V_{GS2} is rejected because it is lower than $V_{GS(th)}$, −2.5V. Substituting V_{GS} = −2.13V for Eq. (2.36) yields

$$I_D = [1.5–2.13V_{GS}] \text{ mA} = [1.5–2.13(–2.13)] \text{ mA} = 6mA$$

From the output loop, we have

$$V_{DS} = 24V - I_D(RD + RS) = 24V - 6mA[2k\Omega + 0.47k\Omega] = 9.2V$$

2.6   JFETs

The current conventions for JFET are shown in Figure 2.27. Since there is no gate current, we must have:

$$I_D = I_S$$

The relationship between I_D and V_{GS} is given as

$$I_D = I_{DSS}\left[1 - \frac{V_{GS}}{V_{GS(th)}}\right]^2 \tag{2.38}$$

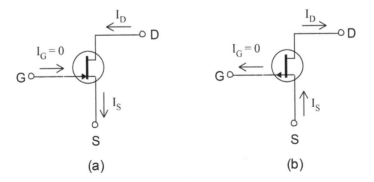

Figure 2.27 Current convention for JFET in a common-source configuration: (a) N-channel JFET, (b) P-channel JFET

The constant I_{DSS} is the drain current I_D at $V_{GS} = 0$. Compared with Eq. (2.25) for a depletion type MOSFET, Eq. (2.38) has the identical form, which is also characterized by I_{DSS} and $V_{GS(th)}$. However, Eq. (2.25) works for both depletion mode (i.e., $V_{GS} < 0$ for a N-channel MOSFET), and enhancement mode (i.e., $V_{GS} > 0$ for an N-channel MOSFET). But JFET only works in the region equivalent to the depletion mode of a MOSFET (i.e., $V_{GS} < 0$ for a N-channel JFET).

Both I_{DSS} and $V_{GS(th)}$ are given in a data sheet provided by the manufacturer. When these two parameters are known, we can easily figure out the drain current I_D and V_{GS} characteristics. Figure 2.28(a) is the transfer characteristic curve of a JFET device in common source configuration. Figure 2.28(b) shows the curves of I_D versus V_{DS} for various V_{GS}. It is noted from Figure 2.28 that the active region for an N-channel is when $V_{GS(th)} < V_{GS} < 0$. For the region $V_{GS} < V_{GS(th)}$, JFET is in the cut-off region. The region $V_{GS} > 0$ is prohibited, as it may force a current to run into the gate that will damage the device.

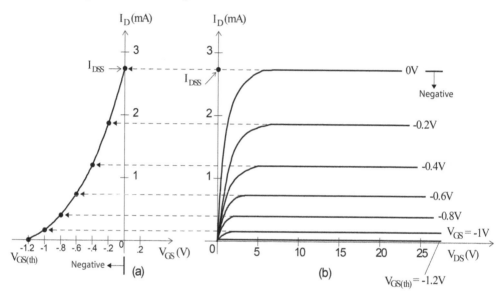

Figure 2.28 Characteristics of an N-channel JFET in common source configuration: (a) gate transfer characteristics, (b) drain characteristics

Since JFET and depletion type MOSFET have identical characteristics given by Eqs (2.38) and (2.25), they share the same dc bias configurations: fixed-bias, self-bias, and voltage-divider bias. The techniques for determining the quiescent point are also the same.

Fixed-bias configuration

Figure 2.29 shows the fixed-bias configuration for a JFET. The fixed-bias configuration requires two power supplies: VDD and VGG. In the input loop, since $I_G = 0$, we have,

$$V_{GS} = -VGG \tag{2.39}$$

In the output loop, we have,

$$V_{DS} = VDD - I_D RD \tag{2.40}$$

Substituting Eq. (2.39) for Eq. (2.38) yields

$$I_D = I_{DSS} \left[1 + \frac{V_{GG}}{V_{GS(th)}} \right]^2 \qquad (2.41)$$

Eq. (2.41) can be solved for I_D. Then V_{DS} is determined by Eq. (2.40). This approach is identical to the depletion MOSFET. Since it requires two power supplies, VDD and VGG, this makes the fixed-bias configuration an unpopular choice for dc bias. Self-bias and voltage-divider bias configurations are more popular, as they only require one single power supply, VDD. Again, RG is a large resistor that feeds the dc voltage VGG to the gate but prevents the input signal from being shorted to VGG.

Self-bias configuration

As shown in Figure 2.30, the self-bias configuration requires a gate resistor RG connecting to the ground. Again, RG is a large resistor. Since there is no gate current, the dc potential at the gate is 0V. Thus, looking at the input loop, we have

$$V_{GS} = -I_D RS, \text{ or}$$

$$I_D = -\frac{V_{GS}}{RS} \qquad (2.42)$$

Since the drain current is governed by Eq. (2.38), together with Eq. (2.42), they form equations with two unknowns, I_D and V_{GS}. Thus, they can be solved either analytically or numerically. If solving by the analytical method, Eq. (2.42) is first substituted for Eq. (2.38) then rearranged to give a quadratic equation in the form:

$$a V_{GS}^2 + b V_{GS} + c = 0$$

The quadratic equation has two solutions for V_{GS}:

$$V_{GS1} = \frac{-b + \sqrt{b^2 - 4ac}}{2a} \text{ and } V_{GS2} = \frac{-b - \sqrt{b^2 - 4ac}}{2a}$$

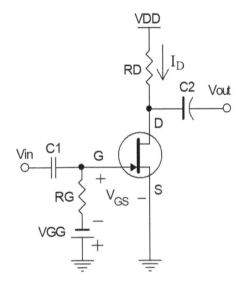

Figure 2.29 Fixed-bias for an N-channel JFET in common source configuration

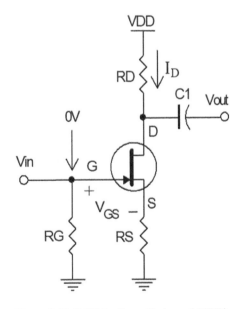

Figure 2.30 Self-bias for an N-channel JFET in common source configuration

One of the solutions is invalid because it is below $V_{GS(th)}$, i.e., in the cut-off region. To be a valid solution it must satisfy the condition $V_{GS(th)} < V_{GS} < 0$. Since the gate is at dc ground level, the self-bias configuration eliminates the need for an input dc blocking capacitor that may affect the low frequency response of the amplifier. This is an attractive feature for the self-bias configuration that is shared among depletion-mode MOSFET, JFET, and triode.

In the special case when $V_{GS} = 0$, from Eq. (2.38) we have $I_D = I_{DSS}$. As the drain current is always equal to I_{DSS}, this arrangement makes the JFET operate as a constant current source, as shown in Figure 2.31(a). Since this JFET is not intended to work as an amplifier for amplifying the ac signal, the gate resistor RG is not needed. The current source can be simplified by eliminating RG, as shown in Figure 2.31(b). For practical applications, it is often best to insert a small variable resistor VR into the source, as shown in Figure 2.31(c), so that fine current adjustment is made possible. Since depletion type MOSFET has similar property to JFET, depletion type MOSFET can also work as constant current source, as shown in Figure 2.31(d). Depletion type MOSFET is a better choice than JFET for high current and high voltage current source application.

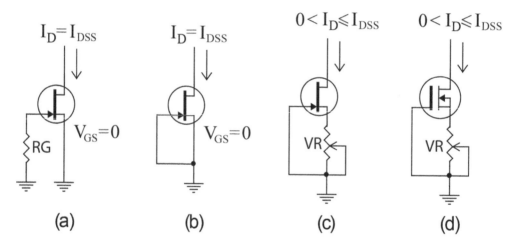

Figure 2.31 Constant current source: (a) JFET with a gate resistor RG, (b) RG eliminated, (c) JFET with a variable resistor VR for fine adjusting the current, (d) depletion type MOSFET with a variable resistor VR for finely adjusting the current

A typical application for the current source is to set the tail current for a differential amplifier, as shown in Figure 2.32(a). A differential amplifier is one that amplifies differential inputs (Vin1 and Vin2) and generates differential outputs (Vout1 and Vout2). If BJT's Q1 and Q2 are closely matched, the constant current I_o is split equally, supplying the collector currents. Another application for a JFET current source is to work as an active load Q2, shown in Figure 2.32(b). If JFETs Q1 and Q2 are closely matched, and same source resistors RS are used, output is biased at dc ground level. Thus, in this circuit, no dc blocking capacitors are needed for input and output. This circuit operates as a buffer amplifier with unity voltage gain.

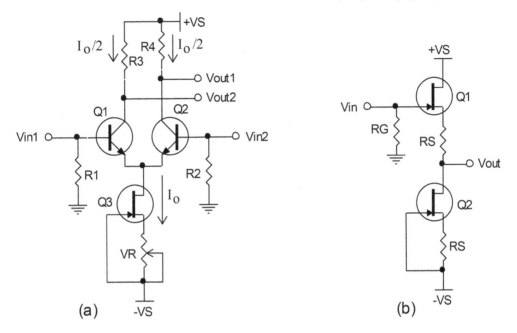

Figure 2.32 Q3 works as a current source to set the tail current for the differential amplifier in circuit (a). Q2 works as active load for the JFET buffer amplifier in circuit (b)

Voltage-divider bias configuration

Figure 2.33 shows the voltage-divider bias configuration. As the input impedance of MOSFET is very high, in order to avoid loading down the input impedance, R1 and R2 are often large resistors in the MΩ range. Since there is no gate current, the dc potential at the gate is equal to

$$V_G = \frac{R2}{R1+R2} \text{VDD}$$

Therefore, in the input loop we have

$$V_G - V_{GS} - I_D RS = 0$$

after rearranging yields

$$I_D = \frac{V_G - V_{GS}}{RS} = \frac{1}{RS}\left(\frac{R2}{R1+R2} \text{VDD} - V_{GS} \right) \tag{2.43}$$

Substituting Eq. (2.43) for Eq. (2.38) and rearranging gives a quadratic equation in the form:

$$aV_{GS}^2 + bV_{GS} + c = 0$$

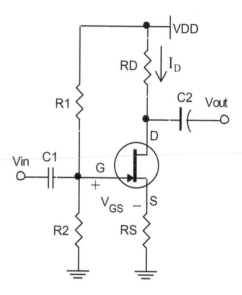

Figure 2.33 Voltage-divider bias for an N-channel JFET in common source configuration

Again, this quadratic equation can be solved analytically, giving two solutions for V_{GS}. One of the solutions is invalid because it is below $V_{GS(th)}$, i.e., in the cut-off region. To be a valid solution, it must satisfy the condition $V_{GS(th)} < V_{GS} < 0$ for an N-channel JFET.

You can refer to Examples 2.5 and 2.6 for calculating the quiescent point of self-bias and voltage-divider bias configurations. There is no difference between JFET and depletion type MOSFET as far as the method of calculating the dc quiescent point is concerned.

Note that there is a dc bias voltage V_G present at the gate of the JFET. In order to prevent the dc voltage from affecting the input signal source, a dc blocking capacitor C1 must be used. If R1 and R2 are large resistors, C1 can be a small capacitor. As noted earlier, the fixed-bias configuration also needs a dc blocking capacitor. Additionally, a fixed-bias configuration needs two dc power supplies: one positive and one negative. These make fixed-bias, perhaps, the least popular biasing configuration for JFET. However, self-bias configuration does not require an input dc blocking capacitor. This makes the self-bias the most popular configuration among the three.

2.7   Triodes

The current conventions for triode are shown in Figure 2.34. Since there is no grid current, we must have:

$$I_P = I_K$$

The plate current I_P can be expressed in terms of plate and grid voltage as follows,

$$I_P = k[V_P + \mu V_G]^{\frac{3}{2}} \tag{2.44}$$

where
    k = a constant, also called *perveance*,
    V_P = plate voltage,
    V_G = grid voltage,
    μ = *amplification factor*.

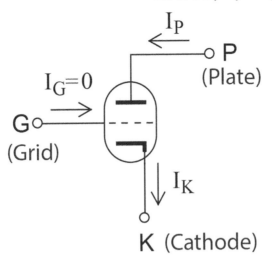

Figure 2.34 Current conventions for a triode

The amplification factor is related to *transconductance*, g_m, and intrinsic plate impedance, r_p, of the triode, such that

$$\mu = g_m r_p \tag{2.45}$$

The amplification factor is always given in the vacuum tube's data sheet. However, the constant k is not given. On the other hand, even if the constant k is given, the fractional power that appears in Eq. (2.44) will immediately discourage people from solving for I_p or V_G analytically. Fortunately, from the vacuum tube's published characteristic curves, there is sufficient information given to determine the dc bias for design work. In determining the dc bias, Eq. (2.45) is not required. But it will be needed for ac analysis when determining voltage gain, input and output impedances. The ac analysis for solid-state devices and triodes is discussed in Chapter 3.

Figure 2.35(a) shows the typical transfer characteristics of a triode. In this example, it only features a plate voltage $V_p = 100V$. However, it should be noted that a wider range of V_p is often published in the data sheet. For low power small signal triodes such as 12AT7, 12AU7, and 12AX7, the transfer characteristic is published for V_p ranging from 50V to 300V.

Fixed-bias configuration

Figure 2.36 shows the fixed-bias configuration for a common cathode amplifier. It is best to illustrate via an example. Suppose that the triode's characteristics are given in Figure 2.35. If the desired plate voltage is 100V and plate current is 2mA, we need to determine VGG, VPP and RP.

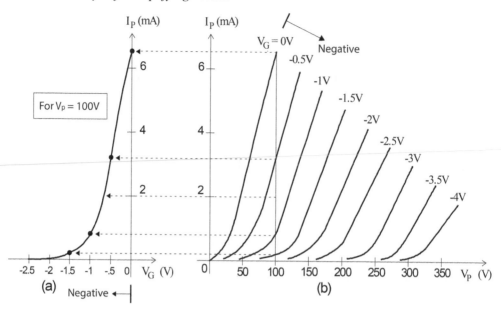

Figure 2.35 Typical characteristics of a triode in a common source configuration: (a) transfer characteristics, (b) I_p and V_p characteristics for various V_G

From Figure 2.35, it can be seen that at plate voltage 100V and plate current 2mA, the grid voltage is about –0.7V. Therefore, we must have VGG = –0.7V. RG is a large resistor around 1MΩ that feeds the dc voltage VGG to the grid but prevents the input signal from being shorted to VGG. Suppose that the power supply VPP is 150V. If the desired plate voltage is 100V, then the plate resistor is RP = (150V – 100V)/2mA, i.e., 25kΩ. Similarly, if VPP = 200V, RP is 50kΩ.

Again, since it requires two dc power supplies, the fixed-bias configuration is not as popular as the self-bias configuration that only requires one dc power supply and is discussed next. However, for power triodes and power tetrodes and pentodes, fixed-bias is often used for the push–pull output stage in a vacuum tube power amplifier, as shown in Figure 2.37. A pair, or multiple pairs, of power tubes are required. The push–pull configuration is discussed further in Chapter 11.

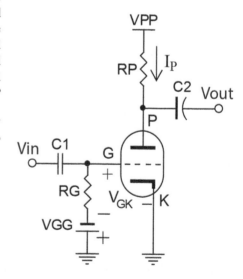

Figure 2.36 Fixed-bias configuration for a common cathode amplifier

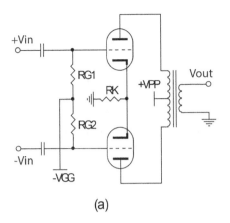

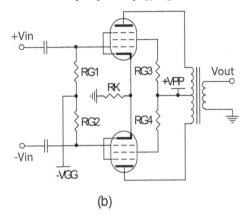

(a)   (b)

Figure 2.37 Fixed-bias is used for the push–pull output stage, (a) using power triodes; (b) using power tetrodes or pentodes

Self-bias configuration

Figure 2.38 shows the self-bias configuration for a common cathode amplifier. The steps for determining the biasing condition of a self-bias configuration are very much a reverse process of the fixed-bias configuration. Here a second dc power supply is not needed. Since there is no grid current, the dc potential at the grid is at dc ground level. When current flows to the cathode resistor RK, a dc potential is built across resistor RK. Since the grid is at dc ground level, the voltage between the grid and cathode, V_{GK}, is therefore the inverse of cathode resistor voltage,

$$V_{GK} = -I_p RK \tag{2.46}$$

For example, if the triode is intended to be biased at 2mA plate current and 100V plate-cathode voltage, we know from Figure 2.35 that it requires a grid voltage of –0.7V. Therefore, we have

$$-0.7V = -I_p RK = (-2mA)RK$$

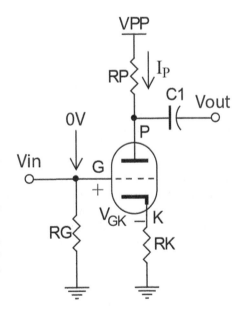

Figure 2.38 Self-bias for a triode in a common cathode configuration

The cathode resistor is equal to 0.7V/2mA = 350Ω. If the dc power supply VPP is 150V and the plate-cathode voltage is 100V, then the plate resistor is determined by

$$RP = (VPP - V_{PK} - I_p RK)/I_p$$

Since the cathode voltage I_pRK is around a few volts, in this example $2mA \times 350\Omega = 0.7V$, which is small compared to VPP and V_{PK}, it is often ignored in the calculation. Thus, we have $RP = (VPP - V_{PK})/I_p = (150V - 100V)/2mA = 25k\Omega$. Similarly, if VPP = 200V, we have $RP = 50k\Omega$.

However, since the grid is at dc ground level, the input does not require a dc blocking capacitor, which is needed by the fixed-bias configuration. In other words, the self-bias configuration saves two components: a dc block capacitor and a negative power supply VGG. It is easy to understand why self-bias is a more popular choice than the fixed-bias and the voltage-divider bias configuration, which requires a dc blocking capacitor at the input.

Voltage-divider bias configuration

Figure 2.39 shows the voltage-divider bias for a common cathode amplifier. Since there is no grid current, the voltage at the grid is given as

$$V_G = \frac{R2}{R1+R2} VPP \tag{2.47}$$

If we look at the input loop, we have

$$V_G = V_{GK} + I_pRK \tag{2.48}$$

For a desired plate current I_p and plate voltage V_p, we can determine the grid voltage V_{GK} from the triode's transfer characteristic curves in a way similar to that for the fixed-bias and self-bias configurations discussed earlier. After V_{GK} is known, then we must choose a suitable value for the cathode resistor RK. Then the dc voltage at the grid, V_G, can be determined by Eq. (2.48). Thus, the resistors R1 and R2 can be determined by Eq. (2.47). Since there are two unknowns, R1 and R2, in Eq. (2.48), a common practice is to assign one of the resistors to a certain value, say R2 = 1MΩ, or a suitable value that suits the application. Then R1 can be easily determined.

Since the grid is now biased at a dc voltage V_G, in order to avoid the possibility of affecting the input signal source, a dc blocking capacitor C1 must be used at the input.

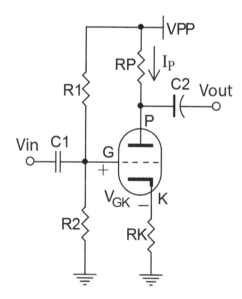

Figure 2.39 Voltage-divider bias for a triode in a common cathode configuration

Example 2.7

Figure 2.40 shows a popular common cathode amplifier using 12AX7. If the desired plate voltage is $V_p = 150V$ and plate current $I_p = 1mA$, let us determine the required plate resistor RP and cathode resistor RK. The transfer characteristics for 12AX7 are given in Figure 2.41.

From Figure 2.41, it is found that at $V_p = 150V$, $I_p = 1mA$, the dc quiescent point is shown in the diagram. It corresponds to $V_{GK} = -1.1V$. Considering the input loop, we have

$$0 = V_{GK} + I_p RK$$

Therefore, the cathode resistor is given as

$$RK = \frac{-V_{GK}}{I_p} = \frac{-(-1.1V)}{1mA} = 1.1k\Omega$$

Then the plate resistor RP is determined as

$$RP = \frac{VPP - V_P}{I_p} = \frac{250V - 150V}{1mA} = 100k\Omega$$

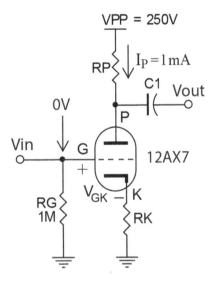

Figure 2.40 A common cathode amplifier in a self-bias configuration

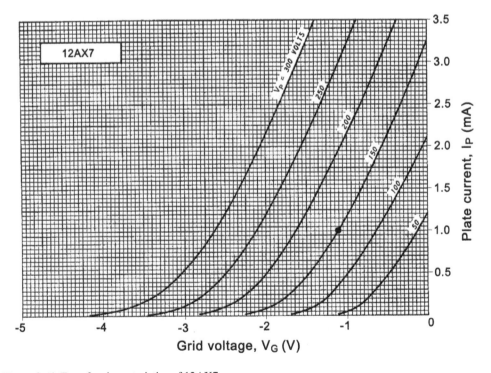

Figure 2.41 Transfer characteristics of 12AX7

2.8   Characteristics summary

The transfer characteristics are important for each device from BJT, FET to triode. Table 2.1 summarizes the differences among them. A clear understanding of the transfer characteristics in

Table 2.1 Summary of transfer characteristics for BJT, enhancement and depletion type MOSFET, JFET and triode

Type	Basic relationships	Transfer curve
BJT (NPN)	$I_B > 0$ $I_E = I_B + I_C$ $I_C \approx I_s e^{(\frac{V_{BE}}{V_T})}$ (V_{BE} must be forward biased)	
MOSFET enhancement type (N-channel)	$I_G = 0$ $I_D = I_S$ $I_D = k\,[V_{GS} - V_{GS(th)}]^2$ (V_{GS} must be forward biased)	
MOSFET depletion type (N-channel)	$I_G = 0$ $I_D = I_S$ $I_D = I_{DSS}\left(1 - \frac{V_{GS}}{V_{GS(th)}}\right)^2$ (V_{GS} can be reverse or forward biased)	
JFET (N-channel)	$I_G = 0$ $I_D = I_S$ $I_D = I_{DSS}\left(1 - \frac{V_{GS}}{V_{GS(th)}}\right)^2$ (V_{GS} must be reverse biased)	
Triode	$I_G = 0$ $I_K = I_P$ $I_P = k\,[V_P + \mu V_G]^{\frac{3}{2}}$ (V_{GK} must be reverse biased)	

the table forms the basis for the dc analysis of amplifying devices. It also helps to perform the ac analysis that is discussed in the next chapter.

2.9 The dc analysis of composite and compound amplifiers

In the preceding sections, we have discussed three commonly used dc bias configurations and their application to a single active device. They are fixed-bias, self-bias, and voltage-divider bias. The active device can be a BJT, MOSFET, JFET, or triode. The most common configuration is the voltage-divider bias and it works for all active devices. However, a self-bias configuration works more effectively and requires fewer components than the voltage-divider bias configuration for depletion type MOSFET, JFET, and triode.

In this section, various dc bias arrangements in composite and compound amplifiers are discussed. Here, a composite amplifier is one created by cascading two or more individual amplifiers together, while a compound amplifier is a single amplifier formed from two active devices. Examples of compound amplifiers include *cascode* amplifiers, differential amplifiers, *SRPP* amplifiers, and *White cathode follower*. Also discussed are various current sources and their applications in the composite and compound amplifiers.

Cascading two amplifiers

Figure 2.42 shows some examples of composite amplifiers created by cascading two amplifiers together. Figure 2.42(a) is a composite amplifier formed by cascading two common emitter amplifiers via dc blocking capacitor C2. Capacitor C2 blocks the dc between the two common emitter amplifiers and, therefore, allows the dc bias in the two amplifiers to work independently. In other words, the dc bias for the two amplifiers can be first determined separately. Then, by cascading them together to form a composite amplifier, that meets the desired voltage gain and input–output impedance requirements.

The merits of this design include: (i) the output swing can be maximized and (ii) The dc bias quiescent points for the two transistors can be determined independently. The downside of this composite amplifier is the use of a dc blocking capacitor C2, which may affect the low frequency response. Even though capacitor C2 can be a large one that permits low frequency extending far below 20Hz, the capacitor is in the signal path that may cause coloration to the audio signal. However, we cannot completely avoid using capacitor coupling in vacuum tube amplifiers because there is no complementary N-type and P-type vacuum tubes that allow the use of circuit symmetry and dual dc power supplies to bring the output dc down to ground level.

Note that a triode has characteristics similar to a N-channel JFET. Owing to the physical construction of a vacuum tube, a triode with characteristics similar to a P-channel JFET does not exist. Nevertheless, it can be seen that when good quality film capacitors are used, vacuum tube amplifiers perform as well as many solid-state amplifiers. However, good quality film capacitors do come with a high cost.

Figure 2.42(b) is a composite amplifier created by cascading a common emitter amplifier to an emitter follower. For illustrative purposes, the common emitter amplifier is taken from Example 2.2. The quiescent collector current for Q1 was found to be $I_C = 8.3$mA. To determine the emitter resistor RE2, we assume that the base current for Q2 is low so that it does not affect the dc potential at the collector of Q1, i.e., voltage at point A. The dc potential at point A is given as

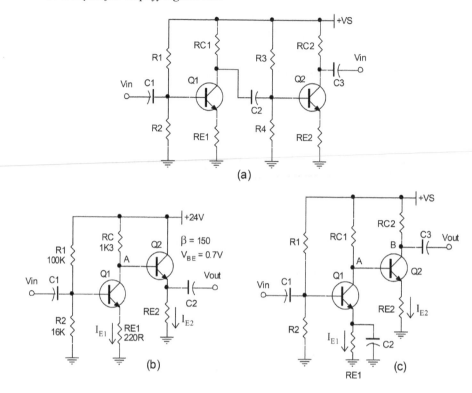

Figure 2.42 Cascading two amplifiers: (a) capacitor coupling between two common emitter amplifiers; (b) direct coupling from a common emitter amplifier to an emitter follower, and (c) direct coupling from a common emitter amplifier to a second common emitter amplifier

$$V_A = 24V - I_{C(Q1)}RC = 24V - (8.3mA)(1.3k\Omega) = 13.2V$$

In the input loop for Q2, we have

$$V_A - V_{BE(Q2)} - I_{E2}RE2 = 0$$

After rearranging yields

$$RE2 = (V_A - V_{BE(Q2)})/I_{E2} = (13.2V - 0.7V)/I_{E2} = 12.5V/I_{E2}$$

If the desired Q2 emitter current is 10mA, the emitter resistor is given as

$$RE2 = 12.5V/10mA = 1.25k\Omega$$

In the above calculation, Q2's base current has been ignored. If we now take into account the base current, it will reduce the potential V_A by the following amount,

$$\Delta V_A = I_{B2}RC = [I_{E2}/(\beta + 1)]RC = [10mA/(151)]1.3k\Omega = 0.086V$$

Since V_A is 13.2V, which is much greater than ΔV_A, ignoring Q2's base current from the above calculation is justified.

Figure 2.42(c) shows another example. Output from the first common emitter amplifier is directly coupled to the second common emitter amplifier. When the emitter resistor RE1 is bypassed by a large capacitor, the voltage gain of the first common emitter amplifier is increased. When the two common emitter amplifiers cascade together, the composite amplifier produces a very high voltage gain, which is the combined product of the voltage gain from two amplifiers. If the voltage gain of the first amplifier is 100 while that of the second one is 40, the combined voltage gain becomes 100×40 = 4000. In applications where high voltage gain is needed, a composite amplifier built by cascading two or more amplifiers easily meets the requirement.

The method of determining the emitter current I_{E2} of Figure 2.42(c) is identical to Figure 2.42(b). Again, first the dc potential at A is calculated, then, from the input loop of Q2, the emitter current I_{E2} is determined.

From these two examples, it is clear that it is important to set the dc quiescent point for the first amplifier correctly. This is because it not only affects the first amplifier, but, due to direct coupling, it will also affect the dc quiescent point of the second amplifier. When the dc quiescent point for the first amplifier is set, the second amplifier will follow from the point where the first amplifier is directly coupled to the second amplifier.

Cascode amplifier

Figure 2.43 shows a *cascode* amplifier that employs two transistors. In this example, BJT is used. Transistor Q2 is a common emitter amplifier and its output (collector) is directly coupled to the Q1, which is a common base amplifier. Since the input is connected to the base of Q2 while the output is taken from the collector of Q1, the *Miller effect capacitor* of the cascode amplifier is very small. As a result, a cascode amplifier has a very wide bandwidth. The Miller effect capacitor is discussed in Chapter 5.

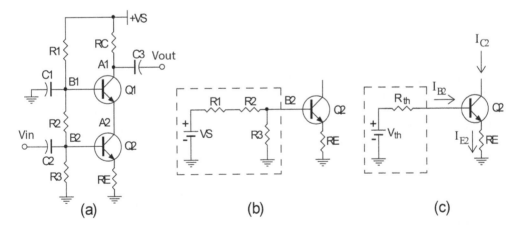

Figure 2.43 Circuit (a): a cascode amplifier contains two BJT transistors in a totem-pole arrangement. Circuit (b): the dc voltage power supply to the base of Q2. Circuit (c): a Thévenin equivalent circuit for the base resistor network

Figure 2.43(a) shows that voltage-divider bias is used. The dc analysis is similar to that for a common emitter amplifier that was discussed in Section 2.2. First, the voltage-divider network is simplified in Figure 2.43(b). The dc power supply at the base of Q2 can be further simplified by the Thévenin equivalent circuit, as shown in Figure 2.43(c). Thus, the Thévenin equivalent V_{th} and R_{th} are determined as

$$V_{th} = \frac{R3}{R1 + R2 + R3} VS \tag{2.49}$$

$$Rth = R3 \| (R1 + R2) \tag{2.50}$$

By considering the input loop of Figure 2.43(c), we have

$$V_{th} = I_{B2} R_{th} + V_{BE} + I_{E2} RE = V_{BE} + I_{E2} [R_{th}/(\beta + 1) + RE]$$

and after rearranging yields

$$I_{E2} = \frac{V_{th} - V_{BE}}{\left(\dfrac{R_{th}}{\beta + 1} + RE \right)} \tag{2.51}$$

Thus, the emitter current I_{E2} is determined by Eq. (2.51). Then the collector current I_{C1} of Q1 can be approximated by $I_{C1} \approx I_{E2}$, so that collector resistor RC can be determined. In common practice, the dc potential is often set across the collector-emitter of Q1 and the collector-emitter of Q2, each equal to one-third of the power supply VS. Thus, the dc potential across the collector resistor RC is also one-third of VS. The dc potentials across the output loop become

$$VS - V_{A1} = V_{A1} - V_{A2} = V_{A2} = VS/3.$$

The dc voltage across emitter resistor RE (i.e., V_{RE}) is typically around 1–2V.

Special JFET configurations

Figure 2.44 shows several special JFET circuits that have a unique dc bias arrangement. Figure 2.44(a) is the current source formed by connecting a JFET in a self-bias configuration. When a second JFET is added in Figure 2.44(b), it performs as an improved current source with higher output impedance [8]. In order for this current source to work properly, both JFETs must be operated with sufficient drain-gate voltage, V_{DG}, such that $V_{DG} > V_{GS(th)}$. Q1 and Q2 are two different JFETs. Since the drain-source of Q1 is biased by the gate-source of Q2, it is best to choose $V_{GS(th)\,Q2} > V_{GS(th)\,Q1}$. For example, 2N4340 ($V_{GS(th)} = 3V$ max) is used for Q1 and 2N4341 ($V_{GS(th)} = 6V$ max) for Q2. Alternatively, depletion mode MOSFET can be used to replace JFET Q2, as shown in Figure 2.44(c).

On the other hand, if we add a gate resistor R1 and a drain resistor R2, the improved current source can be transformed into a cascode amplifier, as shown in Figure 2.44(d). Again, self-bias is employed for Q1 and it sets up the required dc bias current for this amplifier. This is perhaps the simplest cascode amplifier that can be constructed from any active devices. No voltage-divider bias is needed for Q2. In comparison, the BJT cascode amplifier depicted in Figure 2.43 requires a few more resistors to form the voltage-divider network.

The cascode amplifier shown in Figure 2.44(d) can be modified to form a cascode differential amplifier, as can be seen in Figure 2.44(e). This circuit was popularized by Borbely [9–11] and

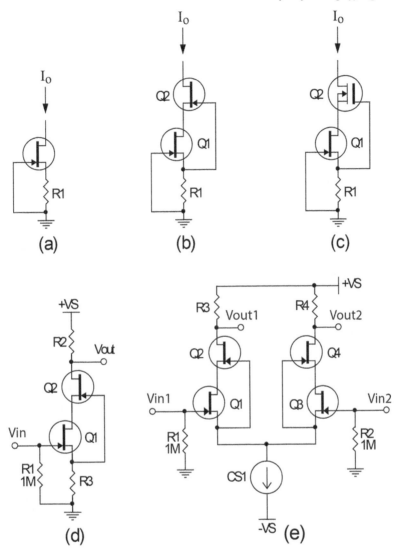

Figure 2.44 Special dc bias arrangements for N-channel JFET: (a) simple current source, (b) an improved current source using two JFETs, (c) an improved current source using one JFET and one depletion type MOSFET, (d) a cascode amplifier, (e) a cascode differential amplifier

was used in many of his audio amplifier designs. When a current source CS1 is used to set the tail current, it makes the dc bias for the differential amplifier simple.

Several JFET differential amplifiers are shown in Figure 2.45. A typical JFET differential amplifier using a dual dc power supply is shown in Figure 2.45(a). It employs a simple tail resistor R5 to set the tail current I_o. When the JFET's Q1 and Q2 are closely matched, and R3 = R4, the JFETs split the tail current equally so that the drain current for each transistor is $I_o/2$.

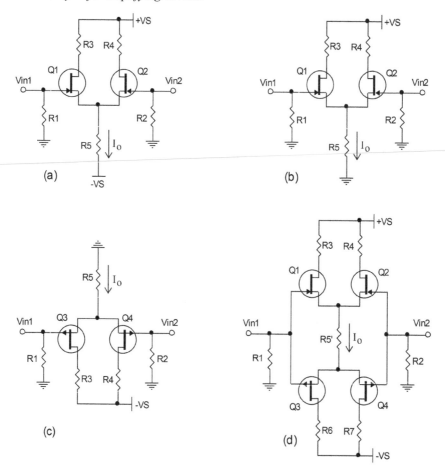

Figure 2.45 JFET differential amplifier: (a) a dual dc power supply ±VS for an N-channel JFET, (b) a single power supply (+VS) for an N-channel JFET in self-bias, (c) a single power supply (−VS) for a P-channel JFET in self-bias, (d) a dual dc power supply ±VS for the combined differential amplifiers (b) and (c) in self-bias

However, JFET can also be operated in a self-bias arrangement, as shown in Figure 2.45(b), where the tail resistor is connected to the ground. As is discussed in Chapter 3, a differential amplifier with a large tail resistor has a better *common-mode rejection ration* (CMRR) than a small tail resistor. An ideal differential amplifier has a CMRR approaching infinite. Therefore, when both inputs (Vin1 and Vin2) are connected to a common signal, the differential output (Vout1 − Vout2) vanishes. In other words, an ideal differential amplifier completely rejects common-mode signal. In reality, the CMRR is finite. It should be noted that the differential amplifier of Figure 2.45(a) has a larger tail resistor R5 than Figure 2.45(b) because a dual ±VS power supply is used. Therefore, Figure 2.45(a) produces a higher CMRR than 2.45(b).

When a differential amplifier is used to amplify a fully balanced signal, it is best that the differential amplifier has a high CMRR. However, when the differential amplifier handles a single-ended (i.e., unbalanced) signal, whether the amplifier has a high CMRR or not is not as important. Therefore, the differential amplifier in Figure 2.45(b), using an N-channel JFET and that in Figure 2.45(c) using a P-channel JFET are suitable for handling a single-ended signal. Given the fact that the tail current flows in the opposite direction to the ground in Figure 2.45(b) and (c), the ground can be removed by combining two tail resistors R5 together, as in Figure 2.45(d). The I_{DSS} of Q1 to Q4 must be closely matched for best performance. The result is a complementary differential amplifier, and R5' is the combined tail resistor. This unique JFET differential amplifier is a very elegant design popularized by John Curl.

The differential amplifier and current sources

Figure 2.46 shows two typical differential amplifiers: a tail resistor is used in (a) while a current source is used in (b). For a desired tail current I_o, the resistance of the tail resistor R5 is limited by the power supply. Although we know a large R5 improves CMRR, there is not much by which we can increase the tail resistor. For example, if the desired tail current is $I_o = 2mA$ and the power supply is dual $\pm24V$, the tail resistor R5 can be calculated as follows.

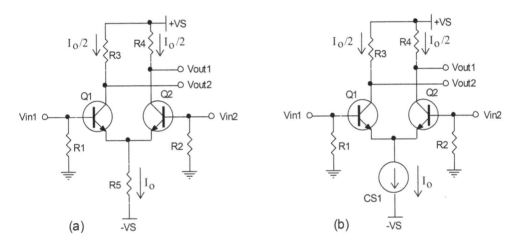

Figure 2.46 A typical BJT differential amplifier using a tail resistor R5 in (a) while the tail resistor R5 is replaced by a current source in (b)

Assume Q1 and Q2 are closely matched transistors, R3 = R4 and R1 = R2, then the collector current on each transistor is $I_o/2 = 2mA/2 = 1mA$. From the input loop of Q1, we have

$$-I_B R1 - V_{BE} - I_o R5 = -24V$$

Since the base current I_B is low, we can ignore it in the calculation. Assuming V_{BE} is 0.7V, we have

$$R5 = (24V - V_{BE})/I_o = (24V - 0.7V)/2mA = 11.6k\Omega$$

Even though R5 is not too small in this example, a larger tail resistor is always welcome. Since the current source has high output impedance, it is common practice to use a current source to replace the tail resistor, as shown in Figure 2.46(b). In addition to a high output impedance, some current sources are less affected by power supply fluctuation, giving better stability. In the following, a number of commonly used current sources are discussed. They are shown in Figure 2.47.

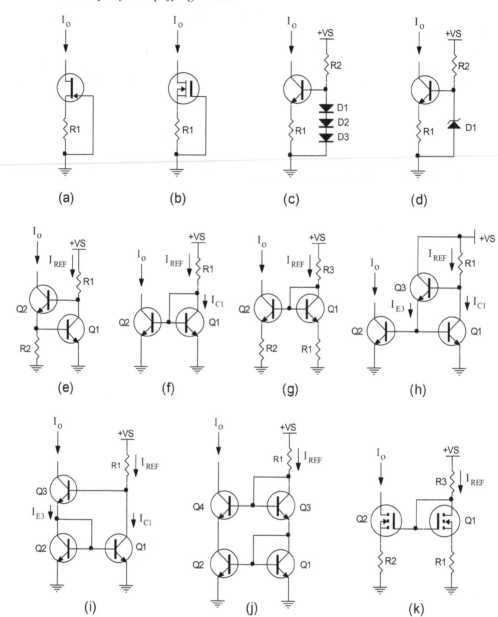

Figure 2.47 Current sources: (a) JFET, (b) depletion type MOSFET, (c) BJT and diodes, (d) BJT and zener, (e) independent supply, (f) simple current source, (g) current source with emitter resistors, (h) improved current source, (i) Wilson current source, (j) cascode current source, (k) simple current source using MOSFET

Figure 2.47(a) and (b): These are current sources formed by a JFET and depletion type MOSFET, respectively. They have been discussed earlier.

Figure 2.47(c): A current source is formed by a BJT and diodes. The diodes are used to turn on the transistor and set up the emitter current. Resistor R2 is chosen so that there are several mA dc bias currents flowing into the diodes. If V_F is the forward voltage of a diode, which is usually around 0.7V to 1V, the emitter current can be determined as follows. We have $3V_F = V_{BE} + I_E R1$. This gives $I_E = (3V_F - V_{BE})/R1 \approx I_o$.

Figure 2.47(d): A current source is formed by a BJT and a zener diode. A zener diode can be used to replace the diodes in Figure 2.47(c). Therefore, if V_R is the reverse breakdown voltage of the zener diode, we have $I_E = (V_R - V_{BE})/R1 \approx I_o$. For the same level of desired current I_o, if a zener diode with a high V_R is employed, R1 becomes a large resistor. When a large emitter resistor is used, it helps to create a high output impedance for the current source. Hence, it improves the CMRR. However, it should be noted that a zener diode with $V_R > 7V$ generates predominately avalanche noise that may cause a noise issue in the amplifier. Fortunately, avalanche noise can be suppressed by a suitable capacitor (usually around 10μF to 100μF) connecting across the zener diode.

Figure 2.47(e): It is a so-called supply-independent current source formed by two BJTs. In order for this current source to work, there must be sufficient current going to bias transistor Q1. The I_{REF} can be determined by the following

$$VS = I_{REF}R1 + V_{BE(Q2)} + V_{BE(Q1)}$$

Rearranging yields

$$I_{REF} = [VS - V_{BE(Q2)} - V_{BE(Q1)})]/R1 \tag{2.52}$$

The V_{BE} for BJT is around 0.6V to 0.7V. I_{REF} is often set to around 1mA. However, I_{REF} is not crucial so long as it is sufficient to turn on the transistor Q2. When Q2 is turned on, the emitter current of Q2 is equal to

$$I_{E2} = V_{BE(Q1)}/R2 \approx I_o \tag{2.53}$$

Eq. (2.53) suggests that the current I_o is independent of the power supply. However, this is not fully supply independent because the $V_{BE(Q1)}$ will still change very slightly with power supply voltage.

Figure 2.47(f): This is called a *current mirror* and it is commonly found in analog integrated circuits. When transistor Q1 and Q2 are closely matched, they have equal base-emitter voltage. Therefore, the collector currents are equal, $I_{C1} = I_{C2}$. If we examine the currents at the collector of Q1, we have

$$I_{REF} = I_{C1} + I_{B1} + I_{B2}$$

Since $I_{C1} = I_{C2}$, we must have $I_{B1} = I_{B2} = I_{C1}/\beta$. The above expression can be rearranged as

$$I_{REF} = I_{C1} + 2\frac{I_{C1}}{\beta} = I_{C1}\left(1 + \frac{2}{\beta}\right) \tag{2.54}$$

If the current gain β is high, we have

$$I_{REF} \approx I_{C1} = I_{C2} = I_o$$

Note that I_{REF} can be determined as follows:

$$I_{REF} = (VS - V_{BE})/R1 \tag{2.55}$$

In other words, after I_{REF} is set according to Eq. (2.55), a current source with a "mirror" current $I_o \approx I_{REF}$ is produced.

Figure 2.47(g): This is the same current source as is shown in Figure 2.47(f), with additional emitter resistors. If r_o is the output impedance for a common emitter amplifier, it can be shown that the output impedance is increased to $r_o(1 + g_m RE)$ when emitter resistor RE is added. In this case, R2 is the emitter resistor RE. To make the design simple, the same emitter resistors are used, R1 = R2. Transconductance g_m is given by $I_{C(Q2)}/V_T$, where $V_T = 26$mV at room temperature.

The emitter resistor also improves the thermal stability of the current source. The base-emitter V_{BE} has a negative temperature coefficient of -2.2mV/°C. In other words, V_{BE} decreases when the temperature increases. If power supply VS and resistor R1 remain unchanged while V_{BE} decreases, currents I_o and I_{REF} increase according to Eq. (2.55). However, if emitter resistors are now in place, the increased current will create a dc voltage across the emitter resistor that opposes further increase of the currents I_o and I_{REF}. Thus, the thermal stability of the current source is improved by the addition of an emitter resistor.

For example, we want to develop a current source depicted in Figure 2.47(g) for VS = 24V and I_o = 2mA. The dc voltage across the emitter resistor R1 is $V_{R1} = I_{E1}R1 \approx I_{REF}R1$. A few volts are usually sufficient for V_{R1} to establish a large emitter resistor. Let us choose $V_{R1} = 3$V. Therefore, we have R1 = 3V/2mA = 1.5kΩ. To determine resistor R3, we consider the loop from the power supply VS, to R3, $V_{BE(Q1)}$ and R1. We have

$$VS = I_{REF}R3 + V_{BE(Q1)} + V_{R1}$$

Therefore, we have

$$R3 = \frac{VS - V_{BE(Q1)} - VR1}{I_{REF}} \tag{2.56}$$

$$= \frac{24V - 0.65V - 3V}{2mA} = 10.1k\Omega$$

Let us pick R3 = 10kΩ so that I_{REF} is slightly greater than 2mA. Then we choose R2 = R1 = 1.5kΩ giving $I_o \approx I_{REF} = 2$mA.

Figure 2.47(h): This is a 3-transistor current source. In this current source, the error in the approximation $I_o \approx I_{REF}$ is greatly reduced by the addition of the third transistor Q3. Here, we assume Q1 and Q2 are closely matched transistors so that $I_{C1} = I_{C2}$. And three transistors have the same current gain β. The emitter current of Q3 is equal to

$$I_{E3} = I_{B1} + I_{B2} = 2\frac{I_{C2}}{\beta}$$

The base current of Q3 is equal to

$$I_{B3} = \frac{I_{E3}}{\beta + 1} = \frac{2}{\beta(\beta + 1)}I_{C2} \tag{2.57}$$

If we examine the currents at the collector of Q1, we have

$$I_{REF} = I_{C1} + I_{B3} = I_{C1} + \frac{2}{\beta(\beta + 1)}I_{C2} \tag{2.58}$$

Since I_{C1} and I_{C2} are equal, Eq. (2.58) rearranges to

$$I_{C2} = \frac{I_{REF}}{[1 + \dfrac{2}{\beta(\beta+1)}]} = I_o \tag{2.59}$$

Thus, I_o and I_{REF} are now different by a factor of $1/\beta^2$. If the current gain β is large, this makes the error of approximating $I_o \approx I_{REF}$ much smaller than the current source of Figure 2.47(f) and (g). Note that I_{REF} is given as $(VS - 2V_{BE})/R1$. When I_{REF} is set, then so is the output current I_o. Again, if an emitter resistor is added to Q1 and Q2, it improves the output impedance as well as thermal stability of this current source.

Figure 2.47(i): This is a *Wilson current source* [3] that has a very high output impedance and a very small error in the approximation $I_o \approx I_{REF}$. If we follow a similar approach to Figure 2.47(h), the output current is given as

$$I_o = I_{REF}\left(1 - \frac{2}{\beta^2 + 2\beta + 2}\right) \tag{2.60}$$

If the current gain β is large, it is clear that $I_o \approx I_{REF}$ is a very good approximation with only a very small error of $1/\beta^2$. Note that I_{REF} is given as $(VS - 2V_{BE})/R1$. On the other hand, if r_o is the intrinsic output impedance of transistor Q3, the effective output impedance of the Wilson current source [3] is equal to

$$Z_{out} \approx (\beta r_o)/2 \tag{2.61}$$

This is equal to the increase of the output impedance of Q3 by a factor of $\beta/2$. The high output impedance makes the Wilson current source a very desirable one. In practice, an emitter resistor may also be added to Q1 and Q2.

Figure 2.47(j): This is a *cascode current source*. It is also a desirable current source with a very high output impedance, which is identical to the Wilson current source, i.e., $Z_{out} \approx (\beta r_o)/2$. However, it takes four transistors to form the cascode current source while only three for the Wilson current source. In practice, an emitter resistor may also be added to Q1 and Q2.

Figure 2.47(k): This is a MOSFET equivalent to the BJT current source shown in Figure 2.47(g). In fact, the Wilson and cascode current sources can also employ MOSFET in place of BJT. The method of determining the source resistors R1, R2 and drain resistor R3 is similar that for BJT. For example, assuming we have VS = 24V and the desired current $I_o = 2$mA, assume the dc voltage across the source resistor is 3V. Then R1 = R2 = 1.5kΩ. We use Eq. (2.56) to calculate the reference current I_{REF}. To approximate the solution, V_{BE} is replaced by $V_{GS(th)\,max}$ so that we have

$$R3 \approx \frac{VS - V_{GS(th)\,max} - VR1}{I_{REF}} \tag{2.62}$$

If a MOSFET is specified with $2V < V_{GS(th)} < 4V$, we substitute $V_{GS(th)\,max} = 4V$ for Eq. (2.62). Thus, we have

$$R3 \approx \frac{24V - 4V - 3V}{2mA} = 8.5k\Omega$$

Since 8.5kΩ is not a standard value, we have to choose a practical value for R3. It can be a resistor slightly lower or higher than 8.5kΩ. When the power supply VS $\gg$ V$_{GS(th)\,max}$, the error in the approximation in Eq. (2.62) is small.

A practical composite amplifier

The dc analysis discussed in sections 2.2 to 2.7 applies to a single active device. Most of them are common emitter (BJT), common source (JFET and MOSFET), and common cathode (triode) configurations. These amplifier configurations produce high voltage gain. However, a practical audio amplifier, such as a line stage amplifier and power amplifier, requires more than just voltage gain. It also needs high input impedance, low output impedance, wide bandwidth, low noise, and low distortion. Therefore, it takes various types of amplifier configurations to make a practical audio amplifier.

For instance, in addition to the common emitter configuration, BJT devices also have common collector (often called emitter follower) and common base configurations. Each of these configurations is an amplifier in its own right and has its own merits and shortcomings. The next chapter shows that a common emitter amplifier produces high voltage gain, medium input impedance and high output impedance. An emitter follower produces high input impedance, low output impedance and unity gain. Therefore, an emitter follower is suitable as a buffer amplifier. A common base amplifier produces high voltage gain, wide bandwidth, high output impedance, but low input impedance.

In addition to these three basic amplifiers, the differential amplifier is also a very popular and important amplifier configuration. A differential amplifier has two inputs and they are biased at the dc ground level, at least very close to ground level. A differential amplifier is often used in the input stage of most modern solid-state audio amplifiers. One of the two inputs handles the input signal while the second input takes care of the feedback signal. However, in a vacuum tube amplifier design, a differential amplifier (or long-tail amplifier) is not a popular configuration for the input stage. The input stage is usually a simple common cathode amplifier. More details feature in later chapters. In the following, we are going to see how to make use of these amplifier configurations to form a practical amplifier. Attention is paid to how the dc bias is determined for each amplifier configuration in the composite amplifier.

Figure 2.48 shows the input stage and second stage of a typical amplifier. In Figure 2.48(a), the input stage is a differential amplifier. A current source CS1 is used to set the tail current I$_o$. Although BJT is shown, JFET is a popular alternative for use in the input stage. Since MOSFET has higher flicker noise than BJT and JFET, MOSFET is not commonly used for the input stage. When closely matched transistors are chosen for Q1 and Q2, the transistors have the same collector current and it is equal to I$_o$/2. After the collector current has been established, the dc potential at the collector of Q1 is given as

$$V_{C1} = VS - \frac{I_o R3}{2} \tag{2.63}$$

The collector current of transistor Q3 at the second stage (I$_1$) is approximately equal to the emitter current, which is determined by considering the loop from power supply VS, to R5, V$_{BE(Q3)}$ and V$_{C1}$. Thus we have

$$VS - I_1 R5 - V_{BE(Q3)} = V_{C1} \tag{2.64}$$

Substituting Eq. (2.63) by Eq. (2.64) and rearranging yields provides

$$I_1 = \frac{\dfrac{I_o R3}{2} - V_{BE(Q3)}}{R5} \tag{2.65}$$

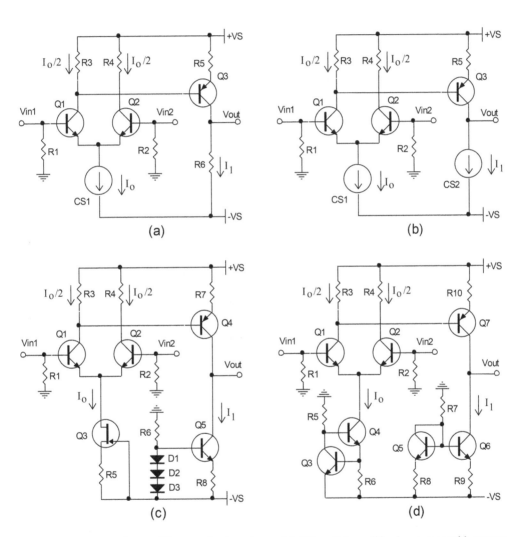

Figure 2.48 A composite amplifier contains two stages: (a) differential amplifier input stage with current source and common emitter second stage with collector load resistor, (b) differential amplifier input stage with current source and common emitter second stage with current source as an active load, (c) differential amplifier input stage with JFET current source and common emitter second stage with a BJT-diode current source as active load, (d) differential amplifier input stage with a supply-independent current source and common emitter second stage with a BJT current source as active load

If it is desired to bias the collector of Q3 near to dc ground level, the collector load resistor R6 should be chosen such that $I_1R6 = VS$. Generally speaking, the voltage gain of the common emitter second stage is proportional to the collector load resistor R6. The larger resistor R6, the higher the voltage gain will be. Owing to the constraint imposed by the power supply, bias current I_1 and the desired dc potential at the collector of Q3, there is a limitation to how large the resistor R6 can be. Alternatively, the resistor load (R6) can be replaced by a current source (CS2), as shown in Figure 2.48(b). As the current source has very high output impedance, the amplifier in Figure 2.48(b) produces a much higher voltage gain than Figure 2.48(a).

Figure 2.48(c) and (d) shows how the current sources of Figure 2.48(c) are realized. No particular preference is given to these current sources. Each has its own merits. Here, they are chosen merely to show the flexibility of introducing the current source to an amplifier. In Figure 2.48(c), a simple JFET current source is used in the input stage. In practice, R5 is a variable resistor, so that the precise tail current I_0 can be set. A BJT-diode current source is used in the second stage. Two diodes are generally enough to bias Q5. However, the use of three diodes will produce a larger emitter resistor R8, creating a larger output impedance for the current source. I_1 is the current generated by this source. In order to avoid the input stage and second stage fighting for dominance to bias the transistor Q4, the current I_1 should more or less satisfy the condition expressed in Eq. (2.65).

In Figure 2.48(d), a supply-independent current source (Q3 and Q4) is used for the input stage. It offers a stable current and it is almost unaffected by the variation in the power supply. In the second stage, a simple BJT current source (Q5 and Q6) is used to operate as the active load for the common emitter amplifier formed by Q7. In order to avoid the input stage and second stage fighting for dominance to bias the transistor Q7, the current I_1 should satisfy the condition expressed in Eq. (2.65). Note that the dc potential at the output (Vout) is not well defined in either Figure 2.48(c) or (d).

Figure 2.49(a) is a two-stage amplifier identical to that shown in Figure 2.48(c). However, feedback is now taken from the output of the second stage and fed to the input stage. Details of feedback principles can be found in Chapter 5. It is well known that when negative feedback is applied, the overall voltage gain is reduced. However, the voltage gain is traded for feedback so that the amplifier has lower distortion, lower noise, and broader bandwidth, etc. The feedback is created by resistors R2 and R11. In this arrangement, the voltage gain of the amplifier is (1 + R11/R2). Since both inputs Vin1 and Vin2 are biased at dc ground level (or nearly ground level), the output Vout is also forced by the feedback to stay at ground level. Thus, dc blocking capacitors are not required for the input and output. However, this amplifier has a problem. It has a rather high output impedance because the output impedance of the current source Q5 is high. When this amplifier is used to drive a low impedance load, the voltage gain will drop and distortion increases. This is not desirable. Therefore, a buffer amplifier must be added to the output.

A buffer is added to the output, as shown in Figure 2.49(b). The buffer output stage is provided by a complementary emitter follower Q6 and Q7. For an emitter follower, the output impedance is low. At the same time, the emitter follower has a high input impedance and, therefore, it does not excessively load down the output impedance of the current source Q5. Diodes D4 to D6 are used to create a dc voltage spread to bias the transistors Q6 and Q7. When negative feedback is applied via resistor R11, the output Vout stays at dc ground level.

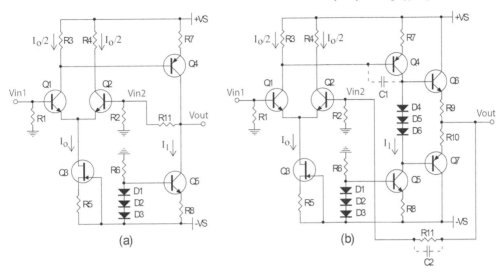

Figure 2.49 Applying feedback: (a) feedback is taken from the output of the second stage and fed to the input stage, (b) feedback is taken from the third stage (output stage) and fed to the input stage

Figure 2.49(b) is a typical audio amplifier consisting of three stages. The input stage is a differential amplifier (Q1 and Q2) with a current source, Q3, setting the tail current. The output of the input stage is directly coupled to the common emitter amplifier second stage (Q4) with an active load (Q5). The output from the second stage is again directly coupled to the buffer output stage formed by the complementary emitter follower (Q6 and Q7). All the dc quiescent points for each stage are well defined. The amplifier is stable in the dc sense. However, it may present a high frequency instability problem.

Frequency compensation capacitors must be added to the circuit to ensure stability at high frequency. To do that, a small capacitor, C1, is added to the base-collector of Q4 and a small capacitor, C2, is connected across the feedback resistor R11. These frequency compensation capacitors ensure the voltage gain of the amplifier is reduced to below unity when the phase of the output signal shifts by 180 degrees. This is because, when the phase is shifted by 180 degrees, the original non-inverted output signal becomes inverted. Therefore, the supposed negative feedback now becomes positive feedback. If the amplifier gain is still greater than unity with a 180 degrees phase shift, the amplifier becomes unstable.

Composite and compound triode amplifiers

Figure 2.50 shows a composite amplifier and three compound amplifiers employing two triodes. The first example is an amplifier created by cascading a common cathode amplifier (T1) to a cathode follower (T2), as shown in Figure 2.50(a). Since the dc bias for the triode T2 depends on the plane voltage of the first triode, i.e., point A, it is important to first set up a proper dc bias to triode T1. If we follow the Example 2.7 and carry it over here, we have T1 = 12AX7,

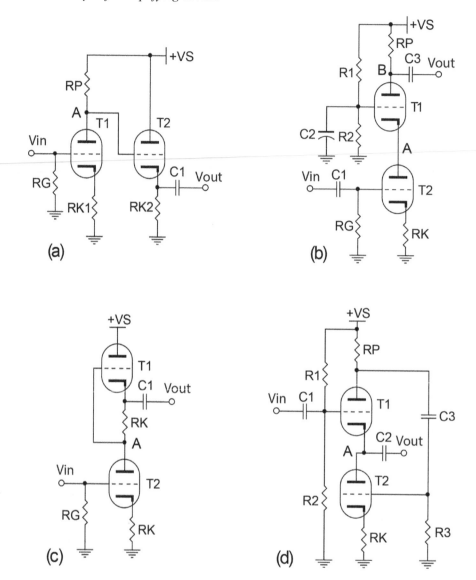

Figure 2.50 A composite amplifier using triodes: (a) common cathode amplifier directly coupled to a cathode follower, (b) a cascode amplifier, (c) an SRPP amplifier, (d) a White cathode follower

$VS = 250V$, $RP = 100k\Omega$, $RK1 = 1.1k\Omega$. The plate current for triode T1 is 1mA. If T2 is a 6922 triode and the desired plate current is $I_{P2} = 5mA$, the cathode resistor RK2 can be determined as follows.

Since there is no grid current, if we examine the loop from point A, V_{GK} of T2 and voltage across RK2, we have

$$V_A = V_{GK(T2)} + I_{P2}RK2$$

After rearranging yields,

$$RK2 = \frac{V_A - V_{GK(T2)}}{I_{P2}} \qquad (2.66)$$

Since $V_{GK(T2)}$, the grid-cathode voltage of T2, is just a few volts for triode 6922, it can be ignored from Eq. (2.66) when $V_A \gg V_{GK(T2)}$. In this case, $V_A = 150V$, Eq. (2.66) can be simplified as

$$RK2 \approx \frac{V_A}{I_{P2}} = \frac{150V}{5mA} = 30k\Omega$$

A more accurate RK2 can be found as follows. From the transfer characteristic of the triode 6922 as shown in Figure 2.51, at 5mA plate current and plate voltage of 90V, the grid voltage is –2.2V, which is the grid-cathode voltage V_{GK} of this example. If we extrapolate the curve for a plate voltage of 100V, V_{GK} is approximately equal to –2.5V. From Eq. (2.66), it gives RK2 = 30.5kΩ, which is, in fact, close to the approximated value 30kΩ found earlier.

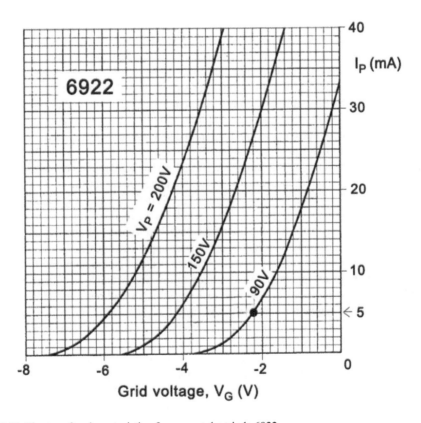

Figure 2.51 The transfer characteristic of vacuum tube triode 6922

Figure 2.50(b) is a cascode amplifier, which is a compound amplifier formed by two triodes in a single-stage arrangement. A self-bias configuration is used for T2 and a voltage-divider bias configuration for T1. As there is no grid current, T1's plate current is equal to T2's plate current, which is determined by the self-bias configuration. T1's grid voltage is determined by the voltage-divider bias configuration. Regarding how to use the self-bias configuration for setting the bias current, it can be seen in Example 2.7. On the other hand, the dc voltage distributed among the triodes is often set to one-third of the power supply VS. Therefore, we have

$$V_A = \frac{1}{3}VS$$

$$V_B = \frac{2}{3}VS$$

Thus, the remaining VS/3 is distributed across the plate resistor RP. Given the dc potential V_A is equal to VS/3, the voltage-divider network is arranged so that

$$\frac{R2}{R1+R2}VS \approx \frac{1}{3}VS$$

After simplifying yields,

$$R1 \approx 2R2 \tag{2.67}$$

Figure 2.50(c) is a *Shunt Regulated Push–Pull* (SRPP) amplifier, which is a compound amplifier formed by two triodes in a single-stage arrangement. The triode T2 on the bottom operates as a common cathode amplifier while the triode on the top, T1, produces an approximated push–pull output. For the SRPP amplifier to work properly, the same cathode resistor RK is used for both triodes. The SRPP amplifier produces a decent voltage gain while the output impedance is relatively low. The SRPP was a popular amplifier in color TV systems and for driving 75Ω cables where high output current and low load impedance are of prime concern. There is more discussion of SRPP amplifier in Chapters 3 and 8.

Since there is no grid current, the same plate current is running through both triodes. However, it is the triode T2 at the bottom that actually sets up the plate current for the compound amplifier. Self-bias is applied to T2. The method of determining the dc bias current of a self-bias configuration is similar to that shown in Figure 2.50(a) and (b). Details can be found by referring to Example 2.7. When the same cathode resistor RK and the closely matched triodes (T1 and T2) are used, the dc potential at T2's plate (i.e., point A) is half of the power supply voltage, i.e, $V_A = VS/2$.

Figure 2.50(d) is the *White cathode follower*, which is a compound amplifier formed by two triodes in a single-stage arrangement. As a cathode follower, it produces a voltage gain close to unity and a low output impedance. The plate resistor RP is a small resistor that creates a small output signal and it is capacitor coupled to the triode T2 at the bottom, which operates as a common cathode amplifier. The output of T2 is phase inverted and it is in phase with the output coming from the cathode follower T1.

Note that as far as the dc bias is concerned, the White cathode follower closely resembles the cascode amplifier of Figure 2.50(b). The triode T2 at the bottom is in a self-bias arrangement while the triode T1 on the top is in a voltage-divider bias arrangement. However, T2's plate voltage (at point A) is set to equal to, or slightly below, half the supply voltage, i.e., $V_A \leq VS/2$. Both triodes T1 and T2 have the same plate current, which is determined by the self-bias arrangement of T2.

Distinguished engineer/inventor

Lee de Forest [17] (August 26, 1873–June 30, 1961) was an American inventor. He was born in Council Bluffs, Iowa. He is often called the "father of radio" and a pioneer in the development of sound-on-film recording used for motion pictures. He had over 180 patents.

His most famous invention was the three-element *"Audion"* vacuum tube developed in 1906. It was a thermionic triode vacuum tube — a three-element electronic "valve". It was the first practical amplification device in history. The triode was the foundation of the field of electronics, making possible radio broadcasting, long distance telephone lines, and sound-on-film recording, among many other applications.

In 1912, de Forest conceived the idea of cascading a series of Audion tubes so as to amplify high-frequency radio signals far beyond what could be accomplished by merely increasing the plate voltage on a single tube. It allowed for an enormous amplification of a signal that was originally very weak. This was a foundation development in the field of radio and telephonic long-distance communication.

2.10 Exercises

Ex 2.1 Determine the resistor R3 of the JFET cascode amplifier in Figure 2.44(d). Assume that the transfer characteristic for Q1 is given by Figure 2.28. It is also given that VS = 20V, R2 = 4.3kΩ, and the required drain current is I_D = 1mA. We can assume that $V_{GS(th)}$ of Q2 is least twice of $V_{GS(th)}$ of Q1, so that Q1 has sufficient V_{DS} to operate in the active region.

Ex 2.2 This exercise requires having a pair of closely matched 2SK170 or any small signal JFET on hand. The pair of JFETs is used to build the circuit in Figure 2.32(b), which is a buffer amplifier with unity gain. Use a 24V power supply, i.e., VS = 24V. Find the source resistor RS, such that the drain current is equal to 1mA. Then measure the Dc voltage at the output. Note that if JFETs with closely matched I_{DSS} are used, the output dc level should be very close to ground. Repeat above for drain current equal to 2mA.

Ex 2.3 Figure 2.52 shows a current source that supports two output currents, I_{o1} and I_{o2}. Assume closely matched BJTs are used so that they all have $V_{BE(on)}$ = 0.65V. If it is required that I_{o1} =1 mA and I_{o2} = 3mA, determine resistors R1 to R4.

Ex 2.4 For Figure 2.48(d), assume that the following are given: ±VS = ±24V, R1 = R2 = 33kΩ, R3 = R4 = 1.5kΩ and all $V_{BE(on)}$ = 0.65V. It is required that I_o = 2mA and I_1 = 4mA. Determine resistors R5 to R10.

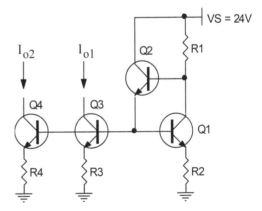

Figure 2.52 A current source for Ex 2.3

Ex 2.5 For Figure 2.50(c), assume that we are given the following: VS = 200V, T1 = T2 = 6H30. It is required that the plate current is 10mA. The transfer characteristic for 6H30 is given in Figure 2.53. Determine the cathode resistor RK.

Ex 2.6 Figure 2.54 shows a three stage composite amplifier employing BJTs. The first two stages are formed by two differential amplifiers. The third stage is a simple emitter follower that operates

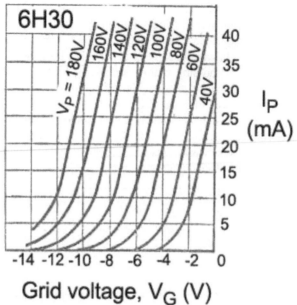

Figure 2.53 6H30's transfer characteristic

as an output buffer. Assume all BJTs have $\beta = 150$ and $V_{BE(on)} = 0.7V$. Diodes D1 to D3 have forward voltage $V_F = 0.6V$. It is required that the dc bias current for the diodes is 1mA while $I_1 = 1mA$, $I_2 = 2mA$ and $I_3 = 5mA$. It is also required that the outputs are biased at dc ground level, i.e., the dc potential at emitter of Q6 and Q7 is ground level. Determine resistors R5 to R11.

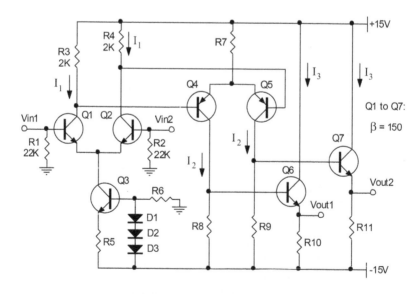

Figure 2.54 A three stage directly coupled amplifier for Ex 2.6

Ex 2.7 Figure 2.55 shows a two stage composite amplifier employing triodes. Both stages are formed in a common cathode configuration. Triode 12AX7 is used for T1 and 6922 for T2. It is required to set the dc bias current I_{p1} = 1mA, I_{p2} = 5mA; the dc potentials V_{p1} = 120V and V_{pk2} = 90V. Using the transfer characteristics in Figure 2.41 and Figure 2.51 for T1 and T2, respectively, determine resistors RP1, RK1, RP2, and RK2.

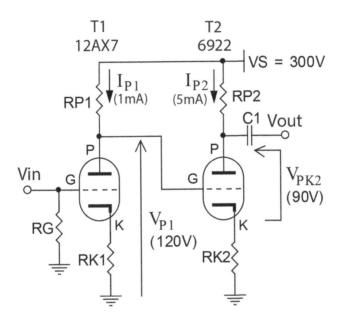

Figure 2.55 A two stage composite amplifier for Ex 2.7

References

[1] Sedra, A. and Smith, K., *Microelectronic circuits*, Oxford, 7th edition, 2017.

[2] Boylestad, R. and Nashelsky, L., *Electronic devices and circuit theory*, Pearson, 10th edition, 2009.

[3] Gray, P. and Meyer, R., *Analysis and design of analog integrated circuits*, Wiley, 3rd edition, 1992.

[4] Millman, J., *Microelectronics, digital and analog circuits and systems*, McGraw-Hill, 1979.

[5] Terman, F. E., *Radio engineers' handbook*, McGraw-Hill, 1943.

[6] Jones, M., *Valve amplifiers*, Newnes, 3rd edition, 2003.

[7] Blencowe, M., *Designing high-fidelity tube preamps*, Merlin Blencowe, 2016.

[8] "The FET constant-current source/limiter," Application note AN103, Siliconix.

[9] Borbely, E., "JFET: the new frontier, part 1," Audio Electronics, May 1999.

[10] Borbely, E., "JFET: the new frontier, part 2," Audio Electronics, June 1999.

[11] Borbely, E., "The all-FET line amp," AudioXpress, May 2002.

[12] "RCA receiving tube Manual," reprinted by Antique Electronic supply.

[13] Rodenhuis, E., "Valves for audio frequency amplifiers," reprinted by Audio Amateur Press.

[14] Valley & Wallman, "Vacuum tube amplifiers," MIT Radiation Laboratory Series, reprinted by Audio Amateur Press.

[15] "Essential characteristics," General Electric Company, 1973.

[16] Smith, W. and Buchanan, B., *Tube substitution handbook*, Prompt publications, 1992.
      de Forest, L., Wikipedia.

[17] Wikipedia, "Lee de Forest," https://en.wikipedia.org/wiki/Lee_de_Forest, n.d.

3 The ac analysis for amplifying devices

3.1 Introduction

The basic characteristics and dc bias techniques of amplifying devices are discussed in Chapter 2. They include BJTs, FETs and triodes. In this chapter, we discuss the ac response of the amplifying devices to a sinusoidal signal.

Several small signal ac models were developed to represent the devices. Hybrid-pi and T-model are two popular ac models used for BJTs and FETs. On the other hand, Thévenin and Norton models are used for triodes. By carrying out the ac analysis with an appropriate ac model, the voltage gain, input and output impedance can be determined for an amplifying device in three basic configurations, which are, for BJTs, a common emitter, a common base and an emitter follower. FETs and triodes also have similar amplifying configurations.

Figure 3.1 shows a three-terminal amplifying device in response to no input signal and a sinusoidal signal. The amplifying device can be a solid-state transistor (e.g., BJT and FET) or a triode vacuum tube. The device is properly dc biased so that terminal-2 has a dc potential of V2. When a sinusoidal signal (Vin) is applied to the input, an amplified sinusoidal signal is generated at the output (Vout). The voltage gain of the amplifier configuration is defined as

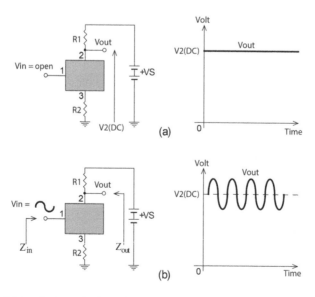

Figure 3.1 An amplifying device: (a) with no input signal and (b) with a sinusoidal input signal

DOI: 10.4324/9781003369462-3

A_v = Vout/Vin. If the amplifier produces a non-inverted output, of which input and output are in phase, A_v is positive. If the amplifier produces an inverted output, of which input and output are out-of-phase by 180 degrees, A_v is negative. Z_{in} is the impedance looking into the input of the amplifier. Z_{out} is the impedance looking into the output of the amplifier. For an ideal voltage amplifier, Z_{in} is infinite and Zout is zero. In this chapter, we discuss how to determine the voltage gain, input and output impedances for single-device amplifiers and some compound amplifiers.

3.2 BJT amplifiers

Hybrid-pi and T-model

Figure 3.2 shows two small signal ac models for a NPN transistor [1–4]. It is noted that lower case letters are used in all ac models. Therefore, i_e denotes the emitter ac current while I_E is the emitter dc current, which is related to the dc analysis discussed in Chapter 2. The first ac model is the hybrid-pi (or hybrid-π) as shown in Figure 3.2 (b). It has a dependent current source βi_e. r_{bb} is the base spread resistor, which is in the range from few ohms to around 100Ω. Since resistor generates thermal noise, which is proportional to $\sqrt{R}$, a transistor with high r_{bb} inevitably produces high thermal noise. For low noise application, it is important to choose BJTs with low r_{bb}. Details of low noise amplifier design will be discussed in Chapter 9.

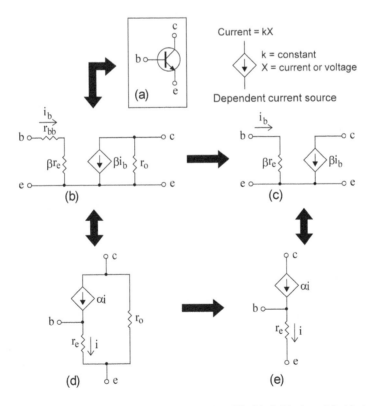

Figure 3.2 (a) NPN transistor; (b) hybrid-pi model; (c) simplified hybrid-pi model; (d) the T-model (or re-model); (e) simplified T-model

The input impedance looking into the base of the transistor is $r_{bb} + \beta r_e$, where β = ac current gain and $r_e = 26\text{mV}/I_E$ (emitter diode resistance) at room temperature. $g_m = 1/r_e$ is transconductance. For small signal BJTs, β is ranging from 100 to 500. For instance, if $I_E = 1\text{mA}$, r_e is 26Ω. Thus, βr_e is ranging from 2.6kΩ to 13kΩ because $\beta r_e \gg r_{bb}$, r_{bb} is often ignored in ac analysis. On the other hand, r_o is the collector-emitter resistor, which is ranging from 40kΩ to 100kΩ. If the collector load resistor is much lower than r_o, it can be ignored in ac analysis. When both r_{bb} and r_o are ignored, the hybrid-pi model is simplified to Figure 3.2(c). This simplified hybrid-pi model is extensively used in ac analysis.

Figure 3.2(d) is the T-model (or re-model), which is also a popular ac model for BJT. It has a dependent current source αi, $\alpha \approx 1$. The base spread resistor r_{bb} is also ignored in the T-model. When the collector-emitter resistor r_o is ignored, the T-model is simplified to Figure 3.2(e). It should be noted that both hybrid-pi and T-model lead to identical results in ac analysis. Since their ac model arrangements are different, in certain configurations hybrid-pi is easier to use than the T-model, and vice versa.

There are three amplifier configurations for BJT, as shown in Figure 3.3. They are the common emitter, common collector (often called the emitter follower), and common base. They all have different ac characteristics in terms of voltage gain, input and output impedance. If we consider the two practical dc bias arrangements, fixed-bias and voltage-divider bias, the corresponding amplifier configurations are shown in Figure 3.4. It should be noted that even though both fixed-bias and voltage-divider bias are shown, in practice the voltage-divider bias is a more popular bias arrangement because it is more stable and less affected by the transistor's temperature and current gain variation. Details are discussed in Chapter 2.

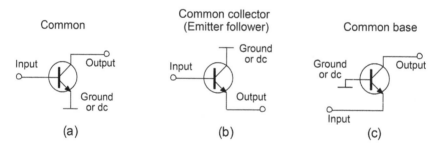

Figure 3.3 Three amplifier configurations for a NPN BJT: (a) common emitter, (b) emitter follower, (c) common base

Common emitter configuration

Let us begin with the ac analysis for the fixed-bias common emitter amplifier of Figure 3.4(a). It is now redrawn in Figure 3.5(a). Assume that the ideal dc power supply (+VCC) is used so that it does not have internal impedance. Therefore, no signal can exist in the power supply. Thus, in the ac sense, the dc power supply is shorted to ground as shown in Figure 3.5(b). The NPN transistor is now replaced by the simplified hybrid-pi model of Figure 3.2(c). The ac equivalent circuit of the entire fixed-bias common emitter amplifier is shown Figure 3.5(c).

Looking at the input loop of Figure 3.5(c), we have

$$\text{Vin} = i_b \beta r_e + i_e \text{RE, or}$$
$$\text{Vin} = i_b \beta r_e + (\beta+1)i_b \text{RE} \tag{3.1}$$

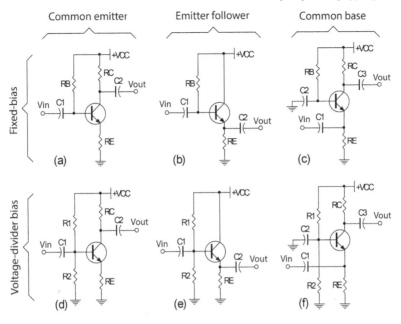

Figure 3.4 The fixed-bias and voltage-divider bias arrangements used for 3 BJT amplifier configurations

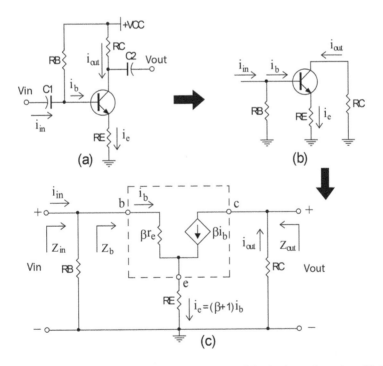

Figure 3.5 A fixed-bias common emitter amplifier: (a) the amplifier in dc configuration, (b) the amplifier in ac configuration, (c) ac equivalent circuit for the amplifier with the transistor replaced by the hybrid-pi model

The input impedance looking into the base of the transistor is

$$Z_b = Vin/i_b = \beta r_e + (\beta+1)RE \approx \beta(r_e + RE) \tag{3.2}$$

Therefore, the input impedance of the common emitter amplifier is given as

$$Z_{in} = RB \| Z_b = RB \| \beta(r_e + RE) \approx RB \| \beta RE \tag{3.3}$$

In contrast, the output impedance of an amplifier is determined when Vin = 0. For Vin = 0, the input is shorted to ground. Therefore, $i_b = 0$ and the dependent current source βi_b becomes an open circuit. Thus, the impedance looking into output of the amplifier in Figure 3.5(c) is given as

$$Z_{out} = RC \tag{3.4}$$

If we had included the collector-emitter resistor r_o in the hybrid-pi model, the output impedance would be given as

$$Z_{out} = r_o \| RC \tag{3.5}$$

It is clear that if $r_o \gg RC$, Eq. (3.5) is approximated to Eq. (3.4).
    To determine the voltage gain, first we have

$$V_{out} = -i_{out}RC = -\beta i_b RC \tag{3.6}$$

If Eq. (3.2) is rearranged, we have

$$i_b = \frac{Vin}{\beta(r_e + RE)} \tag{3.7}$$

Substituting Eq. (3.7) for Eq. (3.6) yields

$$V_{out} = \frac{-RC}{r_e + RE} V_{in}$$

Thus, the voltage gain is defined and given as

$$A_v \equiv \frac{Vout}{Vin} = \frac{-RC}{r_e + RE} \tag{3.8}$$

The negative sign indicates that the output is inverted, i.e., 180 degrees output of phase with respect to the input. If $RE \gg r_e$, the voltage gain can be approximated to

$$A_v \approx \frac{-RC}{RE} \tag{3.9}$$

Figure 3.6 shows a voltage-divider bias common emitter amplifier with and without an emitter by-passed capacitor. In both cases, the simplified ac circuits are similar to the fixed-bias common emitter amplifier of Figure 3.5(b). Therefore, it is expected to see that the voltage gain, input and output impedance that we have derived above can be applied to them in a similar fashion.

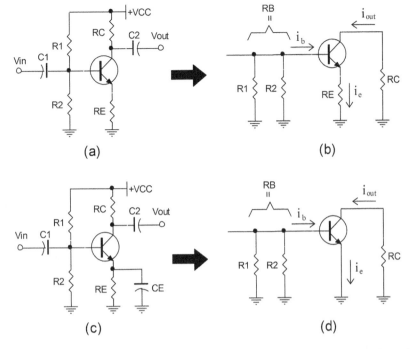

Figure 3.6 A voltage-divider bias common emitter amplifier: (a) un-bypassed emitter resistor, (b) simplified ac circuit for (a), (c) bypassed emitter resistor, (d) simplified ac circuit for (c).

Voltage-divider bias common emitter amplifier with un-bypassed emitter resistor:

$$Z_{in} \approx RB \parallel \beta RE, \text{ where } RB = R1 \parallel R2 \tag{3.10}$$
$$Z_{out} = r_o \parallel RC \approx RC \tag{3.11}$$

$$A_v = \frac{-RC}{r_e + RE} \tag{3.12}$$

Voltage-divider bias common emitter amplifier with bypassed emitter resistor:

$$Z_{in} = RB \parallel \beta r_e, \text{ where } RB = R1 \parallel R2 \tag{3.13}$$
$$Z_{out} = r_o \parallel RC \approx RC \tag{3.14}$$

$$A_v = \frac{-RC}{r_e} \tag{3.15}$$

Since $r_e \ll RE$, it is clear that the common emitter amplifier with a bypassed emitter resistor
produces much greater voltage gain than is the case with an un-bypassed emitter resistor. However, the input impedance is just the opposite. What the bypass capacitor CE does is to short the
emitter resistor RE to ground from mid to high frequency. Since the impedance of a capacitor
is inversely proportional to frequency, the lower the frequency the higher the impedance of the
capacitor becomes. Therefore, in order to keep the capacitor impedance low at low frequency,
capacitor CE must be chosen such that

$$CE > \frac{1}{2\pi f_o RE} \tag{3.16}$$

where $f_o = 20Hz$. For instance, if $RE = 100\Omega$, CE is found to be 79μF. In practice, choose a much larger capacitor, for example 330uF.

Example 3.1

Figure 3.7 is a voltage-divider bias common emitter amplifier taken from Example 2.2. Let us determine the voltage gain, input and output impedance of the amplifier for a bypassed and an un-bypassed emitter resistor.

From Example 2.2, the base current is found to be $I_B = 0.055mA$. Therefore, we have

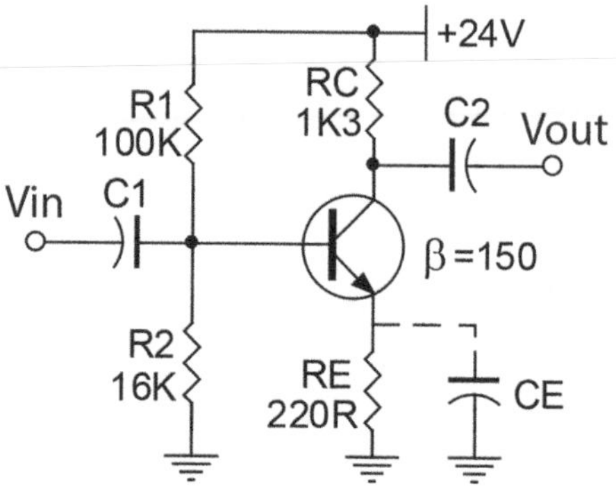

$$I_E = (\beta+1)I_B = 151 \times 0.055mA = 8.3mA$$
$$r_e = 26mV/I_E = 3.1\Omega$$

Figure 3.7 A common emitter amplifier for Example 3.1

(a)   For un-bypassed RE,

Eq. (3.10) to Eq. (3.12):

$$Z_{in} \approx RB \parallel \beta RE = (R1 \parallel R2) \parallel \beta RE = \left(\frac{100k\Omega \times 16k\Omega}{100k\Omega + 16k\Omega}\right) \parallel (150 \times 220\Omega) = 9.7k\Omega$$

$$Z_{out} \approx RC = 1.3k\Omega$$

$$A_v = \frac{-RC}{r_e + RE} = \frac{-1.3k\Omega}{3.1\Omega + 220\Omega} = -5.8$$

(b)   For bypassed RE,

Eq. (3.13) to Eq. (3.16):

$$Z_{in} = RB \parallel \beta r_e = (R1 \parallel R2) \parallel \beta r_e = \left(\frac{100k\Omega \times 16k\Omega}{100k\Omega + 16k\Omega}\right) \parallel (150 \times 3.1\Omega) = 0.45k\Omega$$

$$Z_{out} \approx RC = 1.3k\Omega$$

$$A_v = \frac{-RC}{r_e} = \frac{-1.3k\Omega}{3.1\Omega} = -419$$

$$CE > \frac{1}{2\pi f_o RE} = \frac{1}{2\pi 20 \times 220} = 36\mu F, \text{ choose } CE = 100\mu F \text{ or larger}$$

An emitter bypassed amplifier produces high voltage gain. However, it should be noted that the collector-emitter resistance, r_o, has been ignored when the voltage gain is formulated.

Furthermore, the voltage gain calculated above is under no-load conditions. When r_o and the load are taken into consideration, the voltage gain will be lower than the above figures.

Emitter follower configuration

Figure 3.8 shows a fixed-bias emitter follower and its ac equivalent circuit. Note that the output is now produced from the emitter. In general, the collector resistor RC is not required in an emitter follower. Since the input is the same as the common emitter amplifier, the input impedance is determined in the same manner as described earlier, so we have

$$Z_{in} = RB \| Z_b \tag{3.17}$$

where

$$Z_b = \beta r_e + (\beta+1)RE \approx \beta(r_e + RE) \tag{3.18}$$

To determine the output impedance, let us first consider the base current i_b,

$$Vin = i_b Z_b \text{ or } i_b = Vin/Z_b \tag{3.19}$$

and

$$i_e = (\beta+1)i_b \tag{3.20}$$

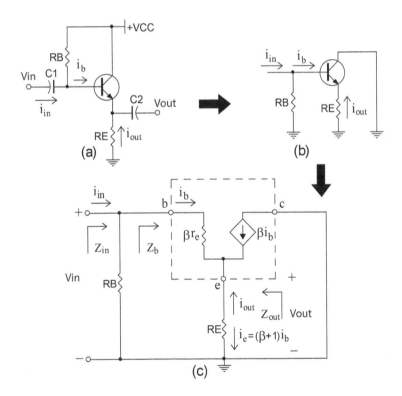

Figure 3.8 A fixed-bias emitter follower: (a) the amplifier in dc configuration, (b) the amplifier in ac configuration, (c) an ac equivalent circuit for the amplifier with the transistor replaced by the hybrid-pi model

Substituting Eq. (3.18) and (3.19) for (3.20) yields

$$i_e \approx (\beta+1)\frac{Vin}{\beta(r_e+RE)} \approx \frac{Vin}{r_e+RE} \tag{3.21}$$

An equivalent circuit can be constructed to represent Eq. (3.21). It is shown in Figure 3.9(a). To determine the output impedance, Vin is set to zero, as shown in Figure 3.9(b).

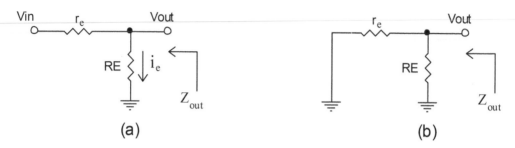

(a)　　　　　　　　　　　　　　(b)

Figure 3.9 (a) An equivalent circuit representing Eq. (3.21); (b) Vin setting to zero

Therefore, the output impedance is given as

$$Z_{out} = r_e \| RE$$

Since RE is typically much greater than r_e, the output impedance is approximated as:

$$Z_{out} \approx r_e \tag{3.22}$$

To determine the voltage gain, from Figure 3.8(c) we have

$$V_{out} = -i_{out}\,RE = i_e\,RE \tag{3.23}$$

Substituting Eq. (3.21) for Eq. (3.23) yields

$$Vout = \frac{RE}{r_e+RE}\,Vin$$

Thus, the voltage gain is given as

$$A_v = \frac{RE}{r_e+RE} \approx \frac{RE}{RE} = 1, \ \text{ as } RE \gg r_e \tag{3.24}$$

The above ac analysis also applies to an emitter follower with voltage-divider bias, giving the same results for output impedance and voltage gain. For input impedance, the base resistor RB is replaced by R1‖R2, where resistors R1 and R2 form the voltage-divider network.

Common base configuration

Figure 3.10 shows a common base amplifier and its ac equivalent circuits. In the ac analysis for the common base amplifier, the T-model is used for the BJT as shown in Figure 3.10(c). After rearranging the circuit of Figure 3.10(c), a simpler circuit is shown in Figure 3.10(d). If we examine Figure 3.10(d), the input and output impedances are given as:

$$Z_{in} = r_e \| RE \approx r_e \tag{3.25}$$
$$Z_{out} = RC \tag{3.26}$$

On the other hand, the output is given by:

$$Vout = -i_{out} RC = -\alpha i_e RC \tag{3.27}$$

By examining the input loop, we have

$$Vin = -i_e r_e \text{ or rearranging to } i_e = -Vin/r_e \tag{3.28}$$

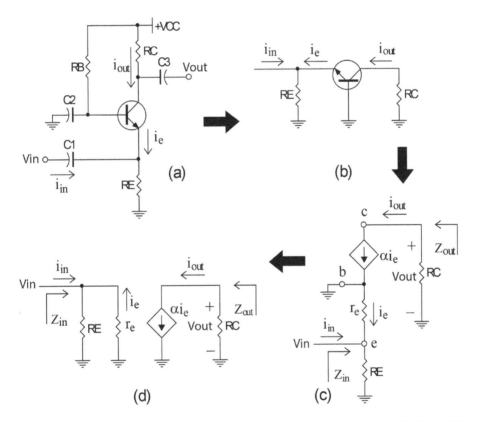

Figure 3.10 A fixed-bias common base amplifier: (a) the amplifier in dc configuration, (b) the amplifier in ac configuration, (c) an ac equivalent circuit for the amplifier with the transistor replaced by the T-model, (d) rearranging from (c)

Substituting Eq. (3.28) for Eq. (3.27) yields

$$Vout = -\alpha i_e RC = -\alpha \left(\frac{-Vin}{r_e} \right) RC$$

Thus, the voltage gain is given by

$$A_v = \alpha\left(\frac{RC}{r_e}\right) \approx \frac{RC}{r_e} \tag{3.29}$$

where $\alpha \approx 1$. Note that the voltage gain is equal to the bypassed common emitter amplifier. But the common base amplifier's output is phase non-inverted, while the common emitter amplifier is phase inverted. The common base amplifier has a low input impedance, while the common emitter amplifier has a high input impedance. As the input is now connected to the emitter while the base is grounded, it has very little impact on the Miller effect capacitor, which is between the base and the output collector. Thus, the common base amplifier has an excellent high frequency response, much better than the common emitter amplifier.

Example 3.2

Figure 3.11 is a voltage-divider bias common base amplifier modified from Example 3.1. We have the following:

$I_E = 8.3\text{mA}$
$r_e = 26\text{mV}/I_E = 3.1\Omega$

Eq. (3.25):

$$Z_{in} = r_e \| RE \approx r_e = 3.1\Omega$$

Eq. (3.26):

$$Z_{out} = RC = 1.3\text{k}\Omega$$

Eq. (3.29):

$$A_v \approx \frac{RC}{r_e} = \frac{1.3\text{k}\Omega}{3.1\Omega} = 419$$

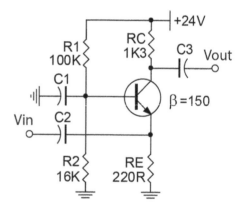

Figure 3.11 A voltage-divider bias common base amplifier

Note that the common base amplifier has a very high voltage gain. In this example, it is 419. However, in reality it is a lot lower. This is due to the fact that the input impedance of the common base amplifier is very small. In this example it is 3.1Ω. Let us say the source has a 1Ω output impedance. There is already a 25% attenuation to the input signal, making the output much lower than it is expected to be.

The h-parameter model

A hybrid parameter model, which is often called an h-parameter model was a popular low frequency model for BJT ac analysis before the hybrid-pi model was introduced. There are four essential parameters used in the model, as shown in Figure 3.12(b). They are given as

h_{ie}:   input impedance (ohms)
h_{re}:   reverse-open-circuit voltage amplification (dimensionless)
h_{fe}:   small signal current gain (dimensionless)
h_{oe}:   output conductance (mhos or S)

Since the value of h_{re} is very small, it is often ignored, so that it leads to a simplified h-parameter model, as shown in Figure 3.12(c). When the simplified h-parameter model is compared to the simplified hybrid-pi model in Figure 3.12(d), we can easily find that the parameters are related as follows:

$$h_{ie} = \beta r_e, \; h_{fe} = \beta, \; h_{oe} = 1/r_o \; (r_o \text{ is inverted as } h_{oe} \text{ is conductance}) \tag{3.30}$$

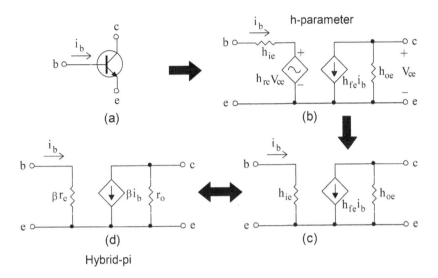

Figure 3.12 (a) An NPN transistor, (b) and h-parameter model, (c) a simplified h-parameter model, (d) a hybrid-pi model

Most BJTs data sheets printed before the mid-1980s published the h-parameters. They were published either in tabular or graphical form, or both. For example, Tables 3.1 and 3.2 show the h-parameters for the BC550 series transistor in tabular form. These tables are extracted from the *Siemens Transistor Data Book 1980/1981*. Table 3.2 shows the transistors grouped or classified according to the dc current gain, h_{FE}. Group A offers a dc current gain ranging from 110 to 220, Group B from 200 to 450, and Group C from 420 to 820. Today, this current gain classification is still being used for the BC550 series. Table 3.3 shows the h-parameters for the transistor at the stated quiescent conditions: $I_C = 2\text{mA}$; $V_{CE} = 5\text{V}$, $f = 1\text{kHz}$ and 25°C. Note that h_{re} is indeed a small value, in the order of 10^{-4}, which is often ignored in the ac model so as to simplify the ac analysis.

Table 3.1 Grouping hFE for BC546 to BC550 (taken from Siemens transistor data book 1980/1981).

Parts	BC546 BC547 BC548			BC546 BC547, BC549 BC548, BC550			BC548 BC549 BC550		
h_{FE} Group	A			B			C		
	Min	Typ	Max	Min	Typ	Max	Min	Typ	Max
h_{FE} @2mA I_C	110	180	220	200	290	450	420	500	800

h_{FE} at 25°C and $V_{CE} = 5\text{V}$

Table 3.2 h-parameter for BC546 to BC550 (taken from Siemens transistor data book 1980/1981).

Parts	BC546 BC547 BC548			BC546 BC547, BC549 BC548, BC550			BC548 BC549 BC550			
h_{FE} Group	A			B			C			
	Min	Typ	Max	Min	Typ	Max	Min	Typ	Max	
h_{ie}	1.6	2.7	4.5	3.2	4.5	8.5	6	8.7	15	kΩ
h_{re}		1.5			2			3		x10^{-4}
h_{fe}		220			330			600		-
h_{oe}		18	30		30	60		60	110	µS

Dynamic characteristics measured at I_c=2mA, 25°C, V_cE = 5V, f = 1kHz

Figure 3.13(a) shows the BC550's h-parameters in graphical form. The h-parameters are normalized to unity at the same quiescent condition, i.e., I_C = 2mA and V_{CE} = 5V. The curve is plotted with the collector current ranging from 0.1mA to 10mA. Since the normalized h-parameters at I_C=2mA are already given in Table 3.3, they can be easily determined for other collector currents. Figure 3.13(b) shows the 2SC2240's h-parameters in graphical form. The quiescent condition is I_C = 1mA, f = 270Hz. The curve is plotted with a collector-emitter voltage ranging from 1V to 50V. It can be seen from both Figure 3.13(a) and (b), the ac current gain h_{fe} is quite flat for various collector current and collector-emitter voltages.

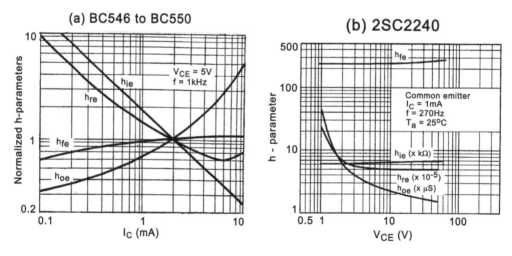

Figure 3.13 h-parameters in graphical form: (a) normalized h-parameters for BC546 to BC550, (b) 2SC2240

We could find the h-parameters printed in BJTs' data sheets published in the early 1980s. At least the data sheets continue to publish the h-parameters for the existing devices. Unfortunately, h-parameters are no longer available in data sheets today, even for legacy device such as the BC550 series. We can occasionally find one of the h-parameters still printed in data sheets. It is the ac current gain, h_{fe}. However, still not all manufacturers publish it. The only thing that is certain to be found is the dc current gain, h_{FE}. In situations where the ac current gain is not published, we just have to use the approximation $h_{fe} \approx h_{FE}$. In fact, they are very close. Some commonly used complementary NPN-PNP transistors can be found in Table 3.3.

Table 3.3 Some commonly used complementary NPN-PNP transistors.

Part number				V_{CEO}	$I_{C(max)}$	h_{FE}	h_{FE}	$@ I_C$	C_{ob}	f_T
NPN		*PNP*								
TO-92	*SOT-23*	*TO-92*	*SOT-23*	*(V)*	*(mA)*	*(min)*	*(max)*	*(mA)*	*(pF)*	*(MHz)*
2N5088	MMBT5088	2N5087	MMBT5087	30	50	300	900	0.1	4	50
2N4401	MMBT4401	2N4403	MMBT4403	40	600	100	300	150	6.5	250
2N3904	MMBT4904	2N3906	MMBT3906	40	200	100	300	10	4	300
BC337	BC817	BC327	BC807	45	500	100	600	100	3	100
BC550	BC850	BC560	BC860	50	100	110	800	2	3.5	300
2SC2240	2SC3324	2SA970	2SA1312	120	100	200	700	2	3	100
2N5551	MMBT5551	2N5401	MMBT5401	160	100	80	250	10	6	100
MPSA42	MMBTA42	MPSA92	MMBTA92	300	500	40	-	10	3	50

3.3   FET amplifiers

Figure 3.14 shows two ac models for FETs including JFET, depletion type and enhancement type MOSFET. The first is the hybrid-pi model as shown in Figure 3.14(d). Since FETs have very high gate impedance, in order to simplify the ac analysis the gate current is assumed to be zero. As a result, the input impedance is infinite — an open circuit between gate and source. The r_d represents the drain-source impedance, which is equivalent to the collector-emitter impedance r_o of a BJT. Since r_d is very high, it is often removed from the model to simplify the ac analysis. The result is a simplified hybrid-pi model as shown in Figure 3.14(e).

Figure 3.14(f) shows the second ac model for FETs — the T-model. Again, r_d is often removed. The result is a simplified T-model as shown in Figure 3.14(g).

Since the FETs share the same ac models, the results obtained from the ac analysis are also shared among them. JFETs, depletion type and enhancement type MOSFETs share the same ac properties in terms of voltage gain, input and output impedance.

There are three basic amplifier configurations for FETs. They are common source, common drain (often called a source follower) and common gate. They have ac properties similar to the BJT's common emitter, emitter follower and common base amplifier configurations. Figure 3.15 shows the amplifier configurations for the N-channel FETs. Each of the configurations is discussed in the following.

Common source configuration

Figure 3.16 shows N-channel FET common source amplifiers in different dc bias configurations. Since the FETs share the same ac small signal model (hybrid-pi or T-model), they have the same ac small signal properties in terms of voltage gain, input and output impedance. Therefore, it will suffice to analyze the ac model of Figure 3.16(e) for determining the ac properties of the FET common source amplifier.

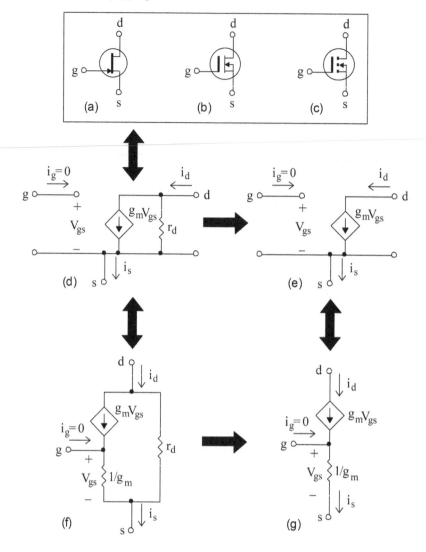

Figure 3.14 (a) JFET, (b) a depletion type MOSFET, (c) an enhancement type MOSFET, (d) a hybrid-pi model, (e) a simplified hybrid-pi model, (f) a T-model, (g) a simplified T-model

The ac equivalent circuit for the common source amplifier is shown in Figure 3.17(c). Since the gate and the output network is open-circuit, the input impedance is given as

$$Z_{in} = R_G \tag{3.31}$$

For the voltage-divider bias configuration, we have $R_G = R1\|R2$. But for fixed-bias and self-bias, R_G is just a single gate resistor. To determine the output impedance, again we set the input source $V_{in} = 0$. Therefore, we have

$$V_{gs} = -i_s R_S = -(g_m V_{gs})R_S \tag{3.32}$$

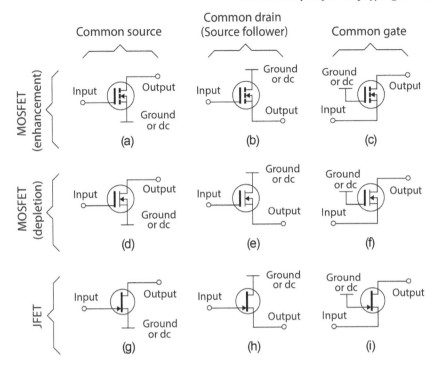

Figure 3.15 Three amplifier configurations for FETs: common source, source follower, and common gate. The FETs include JFET, depletion type and enhancement type MOSFET

Rearranging Eq. (3.32) yields

$$V_{gs}(1 + g_mRS) = 0$$

Since g_m and RS are finite values, the only way to satisfy the above condition is when $V_{gs} = 0$. Therefore, the dependent current source is open circuit when Vin = 0. Thus, the output impedance is simply equal to the drain resistor:

$$Z_{out} = RD \qquad (3.33)$$

To determine the voltage gain, let us examine the input loop in Figure 3.17(c). We have

$$Vin = V_{gs} + i_sRS$$

where the source current is equal to i_{out}. Therefore, the above expression can be written as

$$Vin = V_{gs} + i_{out}RS$$

After rearranging, this yields

$$V_{gs} = Vin - i_{out}RS \qquad (3.34)$$

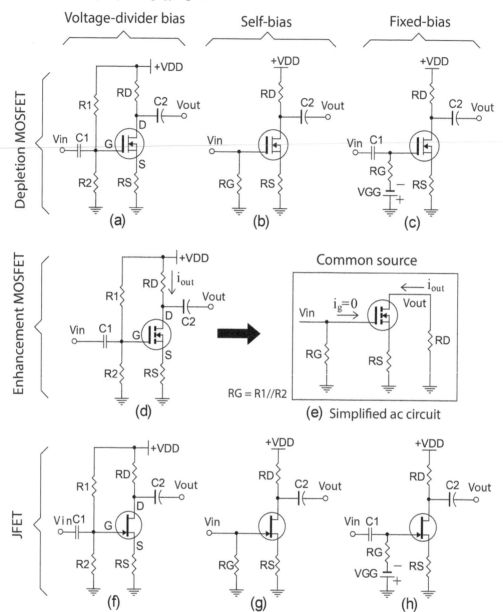

Figure 3.16 A common source amplifier in different dc bias configurations. They share the same simplified ac circuit as shown in (e)

From the output loop, we notice that

$$i_{out} = g_m V_{gs}$$

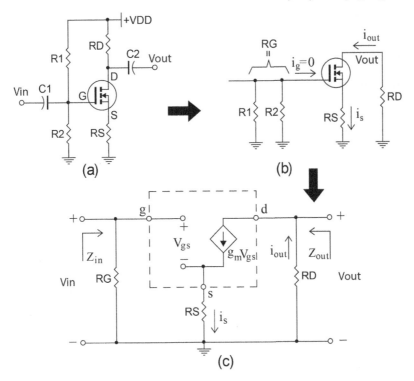

Figure 3.17 Common source amplifier: (a) the amplifier in dc configuration, (b) the amplifier in ac configuration, (c) the ac equivalent circuit for the amplifier with transistor replaced by the hybrid-pi model

Substituting Eq. (3.34) for the above expression yields

$$i_{out} = g_m(Vin - i_{out}RS)$$

which, after rearranging, yields

$$i_{out} = \frac{g_m Vin}{1 + g_m RS} \tag{3.35}$$

The output is given as

$$V_{out} = -i_{out}RD = \frac{-g_m Vin}{1 + g_m RS}RD$$

Thus, the voltage gain for an un-bypassed common source amplifier becomes

$$A_v = \frac{-g_m RD}{1 + g_m RS} \tag{3.36}$$

In contrast, when the source resistor RS is bypassed by a large capacitor, the voltage gain becomes

$$A_v = -g_m RD \qquad (3.37)$$

The negative sign indicates that the output is phase inverted. The phase inverted output is something to be expected, as the common source amplifier is equivalent to a common emitter amplifier of BJT. However, transconductance (g_m) of an FET is smaller than a BJT. As a result, the voltage gain of a common source amplifier is lower than that of a common emitter amplifier.

Example 3.3

Figure 3.18 is a voltage-divider bias common source amplifier taken from Example 2.4. Here, we calculate the voltage gain, input and output impedance of the amplifier for a bypassed and un-bypassed source resistor.

$Zin = R1 \| R2 = 767k\Omega$
$Zout = RD = 1.2k\Omega$

Eq. (3.36):

$$A_{V(un-bypassed)} = \frac{-80\mathrm{x}10^{-3} \times 1200}{1+80\mathrm{x}10^{-3} \times 270} = -4.2$$

Eq. (3.37):

$$A_{v(bypassed)} = -80 \times 10^{-3} \times 1200 = -96$$

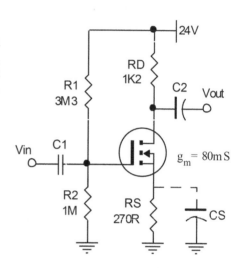

Figure 3.18 A common source amplifier with $g_m = 80mS$

Source follower configuration

Figure 3.19 shows N-channel FET source follower amplifiers in different dc bias configurations. Since the FETs share the same ac small signal model (hybrid-pi or T-model), they have the same ac properties in terms of voltage gain, input and output impedance. Therefore, it will suffice to analyze the ac model of Figure 3.19(e) for determining the ac properties of the source follower amplifier.

The ac equivalent circuit for the common source amplifier is given in Figure 3.20(c). Since the gate and the output network is open-circuit, the input impedance is given as

$$Z_{in} = RG \qquad (3.38)$$

For the voltage-divider bias configuration, we have RG = R1||R2. But for fixed-bias and self-bias, RG is just a single gate resistor. To determine the output impedance, again we set the input source Vin = 0. The ac equivalent circuit is now simplified to Figure 3.20(d). It is noted that

$$V_{gs} = -Vout \qquad (3.39)$$

and applying KCL at the resistor RS yields

$$\frac{Vout}{RS} = g_m V_{gs} + i_{out} \qquad (3.40)$$

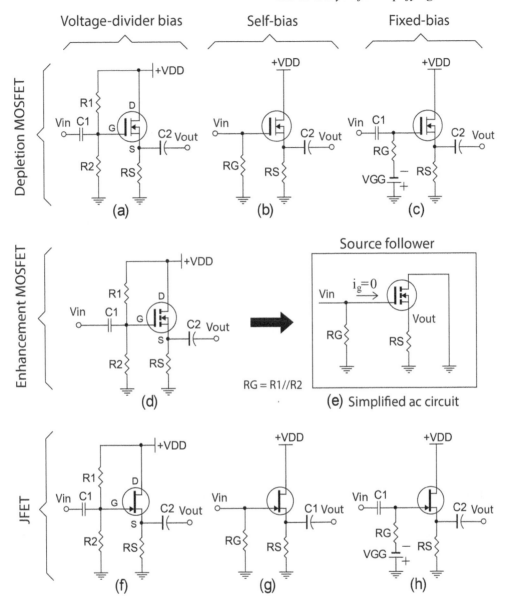

Figure 3.19 A source follower amplifier in different dc bias configurations. They share the same simplified ac circuit as shown in (e).

Substituting the expression for V_{gs} from Eq. (3.39) for (3.40) yields

$$V_{out}\left(\frac{1}{RS} + g_m\right) = i_{out}$$

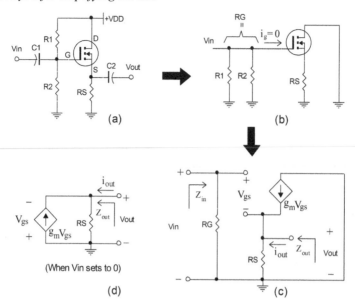

Figure 3.20 Source follower amplifier: (a) the amplifier in dc configuration, (b) simplified ac circuit, (c) ac equivalent circuit for the amplifier with the transistor replaced by the hybrid-pi model, (d) setting Vin = 0 for determining output impedance

Therefore, the output impedance is given as

$$Z_{out} \equiv \left[\frac{Vout}{i_{out}} \right]_{(at\ Vin=0)} = \frac{1}{\frac{1}{RS} + g_m} = RS \| \left(\frac{1}{g_m} \right) \tag{3.41}$$

To determine the voltage gain, let us examine the ac equivalent circuit in Figure 3.20(c). We have

$$Vout = g_m V_{gs} RS \tag{3.42}$$

and

$$Vin = V_{gs} + Vout \tag{3.43}$$

Substituting the expression for V_{gs} from Eq. (3.43) for (3.42) yields

$$Vout\ (1 + g_m RS) = g_m RS\ Vin$$

Thus, the voltage gain is given as

$$A_v = \frac{g_m RS}{1 + g_m RS} \approx 1 \tag{3.44}$$

Since the denominator of Eq. (3.44) is greater than the numerator by a factor of one, the voltage gain for the source follower is always less than unity. It is similar to the emitter follower of BJT.

Common gate configuration

Figure 3.21 shows the N-channel FET common gate amplifier in different dc bias configurations. Since the FETs share the same ac small signal model (hybrid-pi or T-model), they share the same ac properties in terms of voltage gain, input and output impedance. Therefore, it will suffice to analyze the ac model of Figure 3.21(e) to determine the ac properties of the common gate amplifier.

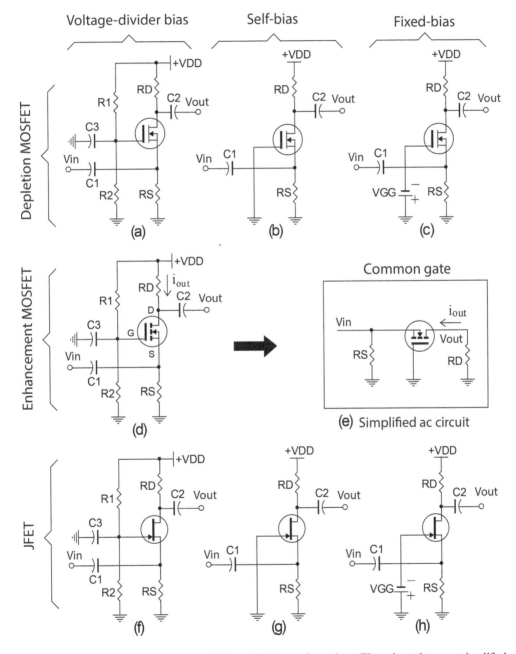

Figure 3.21 Common gate amplifier in different dc bias configurations. They share the same simplified ac circuit as shown in (e)

The ac equivalent circuit for the common gate amplifier is shown in Figure 3.22(d). Since the gate and the output network are disconnected, the input impedance is given as

$$Z_{in} = RS \| (1/g_m)$$ (3.45)

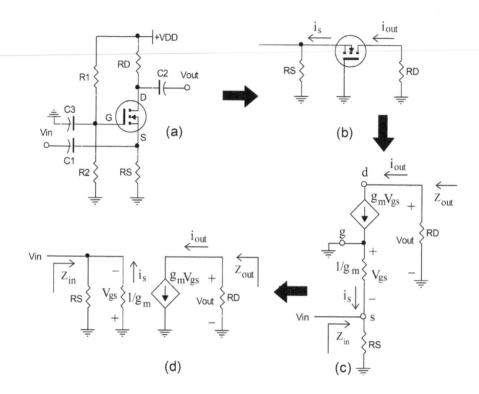

Figure 3.22 An N-channel FET common gate amplifier: (a) the amplifier in dc configuration, (b) the amplifier in ac configuration, (c) ac equivalent circuit for the amplifier with the transistor replaced by the T-model, (d) a rearrangement from circuit (c)

To determine the output impedance, again we set the input source at Vin = 0. This makes $V_{gs} = 0$ and, as a result the dependent current source in Figure 3.22(d) becomes an open circuit. Thus, the output impedance is given as

$$Z_{out} = RD$$ (3.46)

To determine the voltage gain, we notice from Figure 3.22(d) that

$$V_{gs} = -Vin$$

and

$$Vout = -i_{out}RD = -g_m V_{gs}RD = -g_m(-Vin)RD = g_m RDVin$$

Thus, the voltage gain is given as

$$A_v = g_m RD \qquad\qquad (3.47)$$

It should be noted that the voltage gain of a common gate amplifier Eq. (3.47) is similar to that of a bypassed common source amplifier Eq. (3.37). Since the voltage gain Eq. (3.47) is a positive value, the output of a common gate amplifier is non-inverted. However, the common source is an inverted amplifier. On the other hand, as the input is now connected to the source while the gate is grounded, it has very little impact on the Miller effect capacitor, which is between the gate and the output drain. Thus, a common gate amplifier has an excellent high frequency response, much better than the common source amplifier.

Table 3.4 A list of representative low power N-channel enhancement mode MOSFET.

Parts	$V_{DS(BR)}$ (V)	P_D (W)	I_D (A)	$V_{GS(th)}$ (V)	Y_{fs} (S)	C_{iss} (pF)	C_{oss} (pF)	C_{rss} (pF)	Package
BS170	60	0.35	0.5	0.8 to 3	0.2	60	-	-	TO-92
2N7002	60	0.36	0.5	1.0 to 2.5	> .08	50	25	5	SOT-23
2SK982	60	0.4	0.2	2 to 3.5	0.1	55	40	13	TO-92
FDN5630	60	0.5	1.7	1 to 3	6	400	65	27	SOT-23
ZVN2106A	60	0.7	0.45	0.8 to 2.4	0.3	75	45	20	TO-92
VN2106	60	1	0.3	0.8 to 2.4	0.4	35	13	4	TO-92
VN0106	60	1	0.35	0.8 to 2.4	0.45	55	20	5	TO-92
DMN6140LQ	60	1.3	2.3	1 to 3	2.2	315	18	16	SOT-23
Si2308BDS	60	1.6	1.9	1 to 3	5	190	26	15	SOT-23
CPH6445	60	1.6	3.5	1.2 to 2.6	2	310	40	25	SOT-26
SQ2308CES	60	2	2.3	1.5 to 2.5	5.5	164	22	14	SOT-23
FQT13N06	60	2.1	2.8	2 to 4	3	240	90	15	SOT-223
IRLL014	60	3.1	2.7	1 to 2	3.2	400	170	42	SOT-223
PCP1403	60	3.5	4.5	1.2 o 2.6	2.5	310	40	25	SOT-89
SQ3426EV	60	5	7	1.5 to 2.5	21	560	85	55	TSOP-6
VN2110	100	0.36	0.2	0.8 to 2.4	0.4	35	13	4	SOT-23
BSS123N	100	0.5	0.19	0.8 to 1.8	0.41	15.7	3.4	2.1	SOT-23
ZVNL110A	100	0.7	0.32	0.75 to 1.5	0.23	75	25	8	TO-92
ZXMN10A07F	100	0.625	0.8	2 to 4	1.6	138	12	6	SOT-23
CPH3462	100	1	1	1.2 o 2.6	2.5	155	12	6	SOT-23
FQT7N10L	100	2	1.7	1 to 2	2.75	220	55	12	SOT-223
ZXMN10A11G	100	2	1.7	2 to 4	4	274	21	11	SOT-223
IRLL110	100	3.1	1.5	1 to 2	0.57	250	80	15	SOT-223
FDN86246	150	0.6	1.6	2 to 4	4	168	21	1.6	SOT-23
ZXMN15A27K	150	4.2	1.7	2 to 4	2.8	169	65	23	DPAK
Si4848ADY	150	5	5.5	2 to 4	5	335	70	3	SO-8
TN0620	200	1	0.25	0.6 to 1.6	0.4	110	40	10	TO-92
FQT4N20L	200	2.2	0.85	1 to 2	1.42	240	36	6	SOT-223
ZVN4424A	240	0.75	0.26	0.8 to 1.8	0.75	110	15	3.5	TO-92
BSS87	240	1	0.26	0.8 to 1.8	0.33	77.5	11.2	5.8	SOT-89
BSP89	240	1.8	0.35	0.8 to 1.8	0.36	80	11.2	5.2	SOT-223
ZVN4424G	240	2.5	0.5	0.8 to 1.8	0.75	110	15	3.5	SOT-223
TN5325	250	0.36	0.15	0.6 to 2	0.15	110	60	23	SOT-23
TN5325	250	1.6	0.316	0.6 to 2	0.15	110	60	23	SOT-89
FQT4N25F	250	2.5	0.83	3 to 5	1.28	155	35	5	SOT-223
PCP1405	250	3.5	0.6	0.4 to 1.3	1.4	140	8	3	SOT-89

Table 3.5 A list of representative low power P-channel enhancement mode MOSFET.

Parts	$V_{DS(BR)}$ (V)	P_D (W)	I_D (A)	$V_{GS(th)}$ (V)	Y_{fs} (S)	C_{iss} (pF)	C_{oss} (pF)	C_{rss} (pF)	Package
BSS84P	-60	0.36	-0.17	-1 to -2	0.13	15	6	2	SOT-23
2SJ148	-60	0.4	-0.2	-2 to -3.5	0.1	73	48	15	TO-92
FDN5618P	-60	0.5	-1.25	-1 to -3	4.3	430	52	19	SOT-23
ZVP2106A	-60	0.7	-0.28	-1.5 to -3.5	0.15	100	60	20	TO-92
ZVP2106A	-60	0.7	-0.28	-1.5 to -3.5	0.15	100	60	20	TO-92
DMP63505	-60	0.72	-1.5	-1 to -3	-	206	15	11	SOT-23
VP2106	-60	1	-0.25	-1.5 to -3.5	0.2	45	22	3	TO-92
VP0106	-60	1	-0.25	-1.5 to -3.5	0.19	45	22	3	TO-92
CPH6354	-60	1.6	-4	-1.2 to -2.6	4.8	600	60	50	SOT-26
Si2309CDS	-60	1.7	-1.6	-1 to -3	2.8	210	28	20	SOT-23
SQ2309ES	-60	2	-1.7	-1.5 to -2.5	1.8	211	30	21	SOT-23
NVF2955	-60	2.3	-2.6	-2 to -4	1.77	492	165	50	SOT-223
IRFL9014	-60	3.1	-1.8	-2 to -4	1.3	270	170	31	SOT-223
PCP1302	-60	3.5	-3	-1.2 to 2.6	3.2	262	29	19	SOT-89
SQ3427EV	-60	5	-5.3	-1.5 to -2.5	9	700	90	50	TSOP-6
VP2110	-100	0.36	-0.12	-1.5 to -3.5	0.2	45	22	3	SOT-23
ZXMP10A13F	-100	0.63	-0.5	-2 to -4	1.2	141	13	11	SOT-23
ZVP2110A	-100	0.7	-0.23	-1.5 to -3.5	0.13	100	35	10	TO-92
CPH3362	-100	1	-0.7	-1.2 to -2.6	1	142	12	7.3	SOT-23
FQT5P10	-100	2	-1	-2 to -4	1.4	190	70	18	SOT-223
ZXMP10A17G	-100	2	-1.7	-2 to -4	2.8	424	37	30	SOT-223
IRFL9110	-100	3.1	-1.1	-2 o -4	0.82	200	94	18	SOT-223
ZXMP10A16K	-100	4.24	-3	-2 to -4	4.7	717	55	46	DPAK
FDN86265P	-150	0.6	-0.8	-2 to -4	1.5	158	17	1.6	SOT-23
Si235DS	-150	0.75	-0.53	-2.5 to -4.5	2.2	340	30	16	SOT-23
SQ325ES	-150	3	-0.84	-2.5 to -3.5	2.2	200	20	17	SOT-23
Si4455DY	-150	5.9	-2.8	-2 o -4	12	1190	61	42	SO-8
TP0620	-200	1	0.175	-1 to -2.4	0.15	85	30	10	TO-92
FQT3P20	-200	2.5	-0.67	-3 to -5	0.7	190	45	7.5	SOT-223
TP5322	-220	0.36	-0.12	-1 to -2.4	0.25	110	45	20	SOT-23
TP5322	-220	1.6	-0.26	-1 to -2.4	0.25	110	45	20	SOT-89
ZVP4424A	-240	0.75	-0.2	-0.7 to -2	0.13	100	18	5	TO-92
ZVP4424G	-240	2.5	-0.48	-0.7 to -2	0.13	100	18	5	SOT-223
BSS192P	-250	1	-0.19	-1 to -2	0.38	83	13	6	SOT-89
BSP92P	-250	1.8	-0.26	-1 to -2	0.57	83	13	6	SOT-223
FQT2P25	-250	2.5	-0.55	-3 to -5	0.6	190	40	6.5	SOT-223

Table 3.6 A list of representative low power N-channel depletion mode MOSFET.

Parts	$V_{DS(BR)}$ (V)	P_D (W)	I_D (A)	$V_{GS(th)}$ (V)	Y_{fs} (S)	C_{iss} (pF)	C_{oss} (pF)	C_{rss} (pF)	Package
DN1509	90	0.49	0.2	-1.8 to -3.5	0.2	70	20	6	SOT-23
DN1509	90	1.6	0.36	-1.8 to -3.5	0.2	70	20	6	SOT-89
CPC3710	250	1.4	0.22	-1.4 to -3.9	0.23	100	30	15	SOT-89
CPC3703	250	1.1	0.36	-1.6 to -3.9	0.23	327	51	27	SOT-89
DN2530	300	0.74	0.175	-1 to -3.5	0.3	300	30	5	TO-92
DN2530	300	1.6	0.2	-1 to -3.5	0.3	300	30	5	SOT-89
DN2540N3	400	1	0.15	-1.5 to -3.5	0.33	200	12	1	TO-92
DN2540N5	400	15	0.5	-1.5 to -3.5	0.33	200	12	1	TO-220

Parts	$V_{DS(BR)}$ (V)	P_D (W)	I_D (A)	$V_{GS(th)}$ (V)	Y_{fs} (S)	C_{iss} (pF)	C_{oss} (pF)	C_{rss} (pF)	Package
IXTY02N50D	500	25	0.25	-2.5 to -5	0.15	120	25	5	TO-220
CPC3960	600	1.8	0.1	-1.4 to -3.1	0.1	100	6.8	4.2	SOT-89
CPC3982	800	0.4	0.02	-1.4 to -3.1	0.02	20	2.2	1.3	SOT-23
IXTP01N100D	1000	25	0.4	-2 to -4.5	0.2	100	12	2	TO-220

Table 3.7 A list of representative low power N-channel JFET.

Parts	Single	Dual	$V_{GS(BR)}$ (V)	P_D (mW)	I_{DSS} (mA)	$V_{GS(th)}$ (V)	Y_{fs} (mS)	C_{iss} (pF)	C_{rss} (pF)	Package
MMBFJ211	●		25	225	20	-2.5 to -4.5	12	-	-	SOT-23
MMBFJ309L	●		25	225	30	-1 to -4	18	5	2.5	SOT-23
J211	●		25	350	20	-2.5 to -4.5	12	-	-	TO-92
J107	●		25	625	100	-0.5 to -4.5	0.1	38	22	TO-92
LSK589		●	25	500	40	-1.5 to -5	7	5	1.2	SO-8, TO-71
MMBF4393L	●		30	225	30	-0.5 to -3	10	14	3.5	SOT-23
MMBFJ112	●		35	350	5	-1 to -5	10	-	-	SOT-23
MMBFJ111	●		35	350	20	-3 to -10	10	-	-	SOT-23
J112	●		35	625	5	-1 to -5	10	-	-	TO-92
J111	●		35	625	20	-3 to -10	10	-	-	TO-92
2SK170	●		40	400	20	-0.2 to 1.5	22	30	6	TO-92
LSK170X-1	●		40	400	50	-0.2 to -2	10	20	5	SOT-23, TO-92
MMBF4119	●		40	250	0.6	-2 to -6	0.1	3	1.5	SOT-23
2N5462	●		40	310	16	-1.8 to 9	6	7	2	SO-92
2N4119	●		40	350	0.6	-2 to -6	0.1	3	1.5	TO-92
SST202	●		40	350	4.5	-0.8 to -4	> 1	-	-	SOT-23
J202	●		40	350	4.5	-0.8 to -4	1	4.5	1.3	TO-92
LSK389B		●	40	400	12	-0.15 to 2	20	25	5.5	SO-8, TO-71
2SK208	●		50	100	6.5	-0.4 to 5	1.2	8.2	2.6	SOT-23
2SK209	●		50	150	14	-0.2 to -1.5	15	13	3	SOT-23
2SK246	●		50	300	14	-0.7 to -6	1.5	9	2.5	TO-92
LS846	●		60	300	15	-0.5 to -3.5	1.5	8	3	SOT-23, TO-92
LSK189	●		60	300	15	-0.5 to -3.5	1.5	8	3	SOT-23, TO-92
LSK489		●	60	500	15	-1.5 to -3.5	1.5	8	3	SO-8, TO-71
2SK373	●		100	400	6.5	-0.4 to -3.5	4.6	13	3	TO-92

Table 3.8 A list of representative low power P-channel JFET.

Parts	Single	Dual	$V_{GS(BR)}$ (V)	P_D (mW)	I_{DSS} (mA)	$V_{GS(th)}$ (V)	Y_{fs} (mS)	C_{iss} (pF)	C_{rss} (pF)	
2SJ74	●		-25	400	-20	0.15 to 2	22	105	32	TO-92
LSJ74	●		-25	400	-30	0.15 to 2	22	105	32	TO-92
LS94	●		-25	400	-30	0.15 to 2	22	103	32	TO-92
SST74	●		-25	400	-30	0.15 to 2	22	105	32	SOT-89
SST94	●		-25	400	-30	0.15 to 2	22	103	32	SOT-89
MMBFJ175	●		-30	225	-60	3 to 6	-	-	-	SOT-23
MMBFJ176	●		-30	225	-25	1 to 4	-	-	-	SOT-23

(Continued)

Table 3.8 (Continued)

Parts	Single	Dual	$V_{GS(BR)}$ (V)	P_D (mW)	I_{DSS} (mA)	$V_{GS(th)}$ (V)	Y_{fs} (mS)	C_{iss} (pF)	C_{rss} (pF)	
MMBFJ177	●		-30	225	-20	0.8 to 2.5	-	-	-	SOT-23
J175	●		-30	350	-60	3 to 6	-	-	-	TO-92
J176	●		-30	350	-25	1 to 4	-	-	-	TO-92
MMBFJ270	●		-30	225	-15	0.5 to 2	15	-	-	SOT-23
MMBFJ271	●		-30	225	-15	0.5 to 2	18	-	-	SOT-23
MMBF5460	●		-40	225	-5	0.75 to 6	2	5	1	SOT-23
MMBF5461	●		-40	225	-9	1 to 7.5	1.5	5	1	SOT-23
MMBF5462	●		-40	225	-16	1.8 to 9	1	5	1	SOT-23
2SJ103	●		-50	300	-14	0.3 to 6	4	18	3.6	TO-92
LSJ289	●		-50	300	-30	1.5 to 5	1.5	8	3	SOT-23, TO-92
LSJ689		●	-50	500	-30	1.5 to 5	1.5	8	3	SO-8, TO-71

3.4   Triode amplifiers

Figure 3.23 shows two ac models for a triode. The first is the Thévenin ac model in Figure 3.23 (b) and the second is the Norton ac model in Figure 3.23(c). In practice, a triode is biased with no dc grid current. It is similar to an FET with no dc gate current. As a result, the input impedance is assumed to be infinite – an open circuit between the grid and the cathode. If we compare the hybrid-pi model for a FET, Figure 3.14(d), to the Norton ac model for a triode, Figure 3.23(c), they have identical forms. As the drain-source impedance (r_d) is high, it is often ignored to simplify the ac analysis. However, the plate impedance for a triode can be very

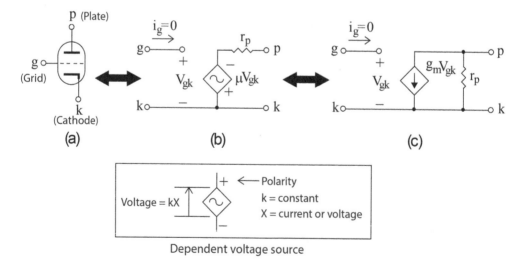

Figure 3.23 (a) A triode, (b) a Thévenin ac model for triode, (c) a Norton ac model for a triode

low. Therefore, the plate impedance is always kept in the ac analysis. It should be noted that the plate impedance r_p is related to the triode's amplification factor (μ) and transconductance (g_m) as follows:

$$\mu = g_m r_p, \ g_m = \frac{\mu}{r_p}, \ r_p = \frac{\mu}{g_m} \tag{3.48}$$

These are simple and very useful expressions that relate the three important parameters of a triode. Figure 3.24 shows how the triode amplifiers are classified in three different configurations: common cathode, cathode follower and common grid. Each of these configurations is discussed below.

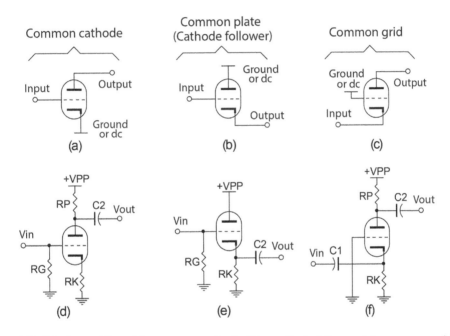

Figure 3.24 Triode amplifiers: (a) a common cathode, (b) a cathode follower, (c) a common grid, (d) a self-bias common cathode, (e) a self-bias cathode follower, (f) a self-bias grounded grid configuration

Common cathode configuration

Figure 3.25 shows a self-biased common cathode amplifier with un-bypassed cathode resistor RK. It is a very popular triode amplifier configuration which gives a low to medium voltage gain. Since there is no grid current, it can be easily seen from Figure 3.25(c) that the input impedance is given as

$$Zin = RG \tag{3.49}$$

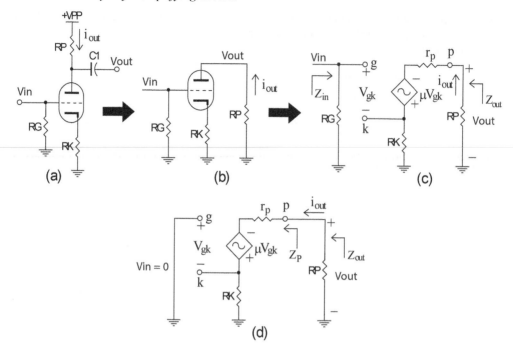

Figure 3.25 A common cathode amplifier: (a) self-bias configuration, (b) simplified ac circuit, (c) ac equivalent circuit for the amplifier with the triode replaced by the Thévenin model, (d) setting Vin = 0 for determining output impedance

To determine the output impedance, we first set the input to zero as shown in Figure 3.25(d). Note that the grid is connected to ground. Therefore, we have

$$V_{gk} = -i_{out}RK \tag{3.50}$$

From the output loop, we have

$$V_{out} = i_{out}r_p - \mu V_{gk} + i_{out}RK \tag{3.51}$$

Substituting Eq. (3.50) for (3.51) and rearranging yields

$$Vout = i_{out}[r_p + (\mu + 1)RK]$$

The impedance, looking into the plate of the triode, is given as

$$Z_p = Vout/i_{out} = r_p + (\mu + 1)RK$$

The output impedance for the un-bypassed cathode resistor is, therefore, given as

$$Z_{out} = RP\|Z_p = RP\|[r_p + (\mu + 1)RK]$$

$$= \frac{RP[\ r_p + (\mu + 1)RK]}{RP + r_p + (\mu + 1)RK} \ , \ \text{(un-bypassed cathode resistor)} \tag{3.52}$$

If the cathode resistor RK is bypassed by a large capacitor, the output impedance becomes,

$$Z_{out} = RP \parallel r_p = \frac{r_p RP}{r_p + RP} \ , \ \text{(bypassed cathode resistor)} \tag{3.53}$$

To determine the voltage gain, we apply KVL to the output loop in Figure 3.25(c).

$$-i_{out} RP - i_{out} r_p + \mu V_{gk} - i_{out} RK = 0$$

and rearranging yields

$$i_{out}[RP + r_p + RK] = \mu V_{gk} \tag{3.54}$$

From the input loop, we have

$$Vin = V_{gk} + i_{out} RK$$

and rearranging yields

$$V_{gk} = Vin - i_{out} RK \tag{3.55}$$

Substituting Eq. (3.55) for (3.54) and rearranging yields

$$i_{out} = \frac{\mu}{r_p + RP + (\mu + 1)RK} Vin \tag{3.56}$$

Therefore, the output is given as

$$V_{out} = -i_{out} RP = \frac{-\mu RP}{r_p + RP + (\mu + 1)RK} Vin$$

Thus, the voltage gain for an un-bypassed common cathode amplifier is given as

$$A_v = \frac{-\mu RP}{r_p + RP + (\mu + 1)RK} \ , \ \text{(un-bypassed cathode resistor)} \tag{3.57}$$

If the cathode resistor is bypassed by a large capacitor, the voltage gain becomes

$$A_v = \frac{-\mu RP}{r_p + RP} \ , \ \text{(bypassed cathode resistor)} \tag{3.58}$$

The negative sign indicates that the output is phase inverted with respect to the input.

Example 3.4

Example 3.4 is illustrated by a common cathode amplifier using a triode (Figure 3.26).

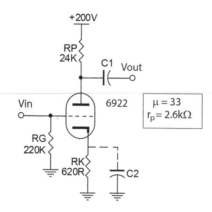

Figure 3.26 A common cathode amplifier using triode 6922

Eq. (3.49):

$$Z_{in} = RG = 220k\Omega$$

Eq. (3.52):

$$Z_{out\ (un\text{-}bypassed)}$$

$$= \frac{RP[\ r_p + (\mu + 1)RK]}{RP + r_p + (\mu + 1)RK}$$

$$= \frac{24k\Omega[2.6k\Omega + (33+1)0.62k\Omega]}{24k\Omega + 2.6k\Omega + (33+1)0.62k\Omega} = 11.9k\Omega$$

Eq. (3.53):

$$Z_{out\ (bypassed)}$$

$$= \frac{r_p RP}{r_p + RP} = \frac{2.6k\Omega \times 24k\Omega}{2.6k\Omega + 24k\Omega} = 2.3k\Omega$$

Eq. (3.57):

$$A_{v(un\text{-}bypassed)} = \frac{-\mu RP}{r_p + RP + (\mu+1)RK} = \frac{-33 \times 24k\Omega}{2.6k\Omega + 24k\Omega + (33+1)0.62k\Omega} = -16.6$$

Eq. (3.58):

$$A_{v(bypassed)} = \frac{-\mu RP}{r_p + RP} = \frac{-33 \times 24k\Omega}{2.6k\Omega + 24k\Omega} = -29.8$$

Cathode follower configuration

Figure 3.27 shows a self-bias cathode follower amplifier. Similar to the emitter follower BJT amplifier and the source follower FET amplifier, the cathode follower triode amplifier also has a voltage gain close to unity and a very low output impedance. Therefore, a cathode follower is often used in the output stage on a line-stage amplifier. To determine the ac properties of the cathode follower amplifier, let us examine the ac equivalent circuit shown in Figure 3.27(c).

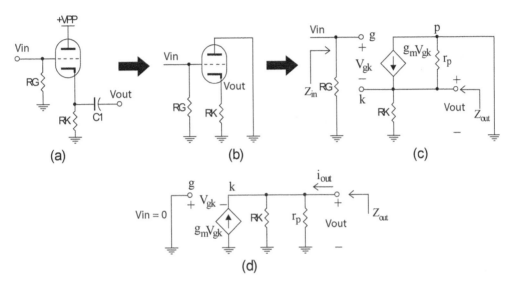

Figure 3.27 A cathode follower amplifier: (a) in self-bias configuration, (b) in simplified ac circuit, (c) ac equivalent circuit for the amplifier with the triode replaced by the Thévenin model, (d) setting Vin = 0 for determining output impedance

Since there is no grid current, it can be easily seen that the input impedance is equal to,

$$Z_{in} = RG \tag{3.59}$$

To determine the output impedance, we first set the input to zero as shown in Figure 3.27(d). Note that when Vin = 0, we have

$$V_{gk} = -Vout \tag{3.60}$$

Examining the currents in the output loop, we have

$$I_{out} = \frac{Vout}{r_p} + \frac{Vout}{RK} - g_m V_{gk} \tag{3.61}$$

Substituting Eq. (3.60) with (3.61) and rearranging yields

$$I_{out} = \left(\frac{1}{r_p} + \frac{1}{RK} + g_m \right) Vout$$

Therefore, the output impedance is given as

$$Z_{out} \equiv \left[\frac{Vout}{Iout} \right]_{(Vin=0)} = \frac{1}{\dfrac{1}{r_p} + \dfrac{1}{RK} + g_m} = \frac{r_p RK}{r_p + (\mu + 1)RK} \tag{3.62}$$

If $(\mu + 1)RK \gg r_p$, the output impedance can be approximated to

$$Z_{out} \approx \frac{r_p}{\mu + 1} \approx \frac{1}{g_m} \tag{3.63}$$

To determine the voltage gain, we examine the input loop from Figure 3.27(c) to get

$$Vin - V_{gk} = Vout \tag{3.64}$$

By considering the currents at the output, we have

$$g_m V_{gk} = \frac{Vout}{r_p} + \frac{Vout}{RK} \tag{3.65}$$

Substituting the expression for V_{gk} from Eq. (3.64) with (3.65) and rearranging yields

$$g_m V_{in} = \left[\frac{1}{r_p} + \frac{1}{RK} + g_m \right] Vout$$

Thus, the voltage gain is given as

$$A_v = \frac{g_m}{\dfrac{1}{r_p} + \dfrac{1}{RK} + g_m} = \frac{\mu RK}{r_p + (\mu + 1)RK} \tag{3.66}$$

Since the denominator of Eq. (3.65) is greater than the numerator, the voltage gain for the cathode follower is always less than unity. It is similar to the emitter follower of BJT and the source follower of FET.

Effect with a plate resistor RP

Figure 3.28(a) shows a cathode follower with a plate resistor RP. The ac equivalent circuit is shown in Figure 3.28(b). It is clear that the input impedance remains the same, i.e., $Z_{in} = RG$. To determine the voltage gain, from the input loop we have

$$Vin = V_{gk} + i_{out} RK \tag{3.67}$$

Applying KVL to the output loop, we have

$$i_{out}(r_p + RK + RP) - \mu V_{gk} = 0 \tag{3.68}$$

Substituting the expression for V_{gk} from Eq. (3.67) with Eq. (3.68) and rearranging yields

$$i_{out} = \frac{\mu}{r_p + RP + (\mu + 1)RK} Vin$$

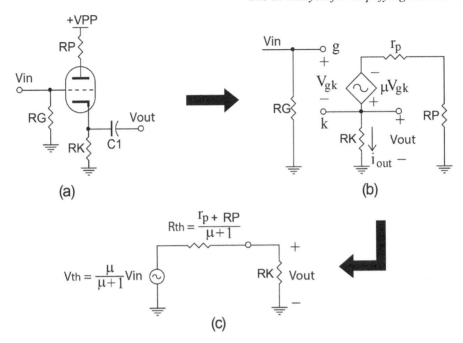

Figure 3.28 A cathode follower amplifier with a plate resistor RP: (a) the amplifier in dc configuration, (b) the ac equivalent circuit for the amplifier with the triode replaced by the Thévenin model, (c) the Thévenin equivalent circuit for (a).

The output becomes

$$V_{out} = i_{out} RK = \frac{\mu RK}{r_p + RP + (\mu + 1) RK} Vin \tag{3.69}$$

It is obvious that Eq. (3.69) leads to the expression for voltage gain,

$$A_v = \frac{\mu RK}{r_p + RP + (\mu + 1) RK} \tag{3.70}$$

In comparison with Eq. (3.66), the denominator of (3.70) has an additional factor of RP. Therefore, the addition of a plate resistor RP will slightly reduce the voltage gain. On the other hand, Eq. (3.69) can be rewritten as

$$V_{out} = \left(\frac{\mu}{\mu + 1}\right)\left(\frac{(\mu + 1) RK}{r_p + RP + (\mu + 1) RK}\right) Vin = \left(\frac{RK}{RK + Rth}\right) Vth \tag{3.71}$$

where

$$Vth = \left(\frac{\mu}{\mu + 1}\right) Vin \tag{3.72}$$

$$Rth = \frac{r_p + RP}{\mu + 1} \tag{3.73}$$

Equations (3.71) to (3.73) suggest that they can be used to represent a Thévenin equivalent circuit for the cathode follower amplifier with plate resistor RP. The Thévenin equivalent circuit is shown in Figure 2.28(c). Therefore, the output impedance becomes

$$Z_{out} = RK \parallel Rth = RK \parallel \left(\frac{r_p + RP}{\mu + 1} \right)$$

and rearranging yields

$$Z_{out} = \frac{(r_p + RP)RK}{r_p + RP + (\mu + 1)RK} = \frac{r_p' RK}{r_p' + (\mu + 1)RK} \tag{3.74}$$

It is perhaps the simplest approach to determine the output impedance. Alternatively, we can use the conventional method by setting Vin = 0 in Figure 3.28(b). It will lead to the same result as given in Eq. (3.74). In comparing to Eq. (3.62), Eq. (3.74) has the identical form if $(r_p + RP)$ is replaced by r_p'. This suggests that the presence of a plate resistor increases the effective plate impedance from r_p to r_p'. As a consequence, the output impedance is also slightly increased.

Example 3.5

Figure 3.29(a) shows a common cathode amplifier (T1) directly coupled to *a split-load phase splitter* (T2). For a split-load phase splitter application, a plate resistor is chosen equal to the cathode resistor, RP = RK. This is a popular phase splitter circuit often found in vacuum tube power amplifiers. The two out-of-phase outputs drive a pair of power tubes, which are arranged in a push–pull configuration.

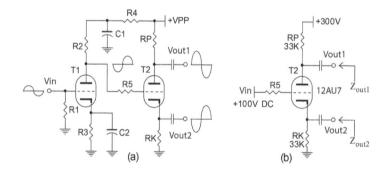

Figure 3.29 (a) A common cathode amplifier is directly coupled to a split-load phase splitter, (b) a split-load phase splitter is biased with a grid dc potential of 100V

Suppose that the split-load phase splitter is dc biased, as shown in Figure 3.29(b). The grid dc potential is 100V, which is the dc voltage carried from triode T1's plate. If we ignore the grid-cathode voltage, which is several volts, triode T2's cathode dc voltage is also around 100V, as there is no grid current. Therefore, T2's plate current is equal to 100V/33kΩ = 3mA. Let us determine the voltage gains and output impedances for the split-load phase splitter.

Triode 12AU7: $\mu = 20$, $r_p = 6.5k\Omega$

Eq. (3.52):

$$Z_{out1} = \frac{RP[\, r_p + (\mu + 1)RK\,]}{RP + r_p + (\mu + 1)RK} = \frac{33k\Omega[\, 6.5k\Omega + (20 + 1)33k\Omega\,]}{33k\Omega + 6.5k\Omega + (20 + 1)33k\Omega} = 31.5k\Omega$$

Eq. (3.57):

$$A_{v1} = \frac{Vout1}{Vin} = \frac{\mu RP}{r_p + RP + (\mu+1)RK} = \frac{20 \times 33k\Omega}{6.5k\Omega + 33k\Omega + (20+1)33k\Omega} = -0.9$$

Eq. (3.74):

$$Z_{out2} = \frac{(r_p + RP)RK}{r_p + RP + (\mu+1)RK} = \frac{(6.5k\Omega + 33k\Omega)33k\Omega}{6.5k\Omega + 33k\Omega + (20+1)33k\Omega} = 1.78k\Omega$$

Eq. (3.70):

$$A_{v2} = \frac{Vout2}{Vin} = \frac{\mu RK}{r_p + RP + (\mu+1)RK} = \frac{20 \times 33k\Omega}{6.5k\Omega + 33k\Omega + (20+1)33k\Omega} = 0.9$$

Note that when we choose RP = RK, the voltage gains for the two outputs are equal to the opposite phase. However, the output impedances are quite different. Since the split-load phase splitter only employs one triode, the output impedances are compromised. If we want to achieve more closely matched output impedance and greater than unity voltage gain for a phase splitter, we need to use two or more triodes. For example, a cathode coupled phase splitter (see Figure 3.48) and floating paraphase phase splitter (see Figure 10.12) employs two triodes. Details of phase splitters are discussed in Chapter 11.

Grounded grid configuration

Figure 3.30 shows a self-biased grounded grid amplifier. Similar to the common base BJT amplifier and the common gate FET amplifier, the grounded grid amplifier has a low input impedance and a non-inverted voltage gain. Because of low input impedance, the ground grid seldom works as a stand-alone amplifier. It often works with a common cathode amplifier or cathode follower to form a compound amplifier. For example, a cascode amplifier is formed by a common grid amplifier and a common cathode amplifier. And a cathode coupled amplifier is formed by a grounded grid amplifier and a cathode follower. It should be noted that, in the ac analysis, the common grid amplifier and the grounded grid amplifier are identical. The cascode amplifier and cathode coupled amplifier are shown in Figure 3.31.

To determine the input impedance, let us consider Figure 3.30(d), where the cathode resistor RK is temporarily removed to simplify the analysis. From the input loop, we have

$$V_{gk} = -Vin \tag{3.75}$$

From the output loop, we have

$$Vin = \mu V_{gk} + i_{in}(r_p + RP) \tag{3.76}$$

Substituting the expression for V_{gk} from Eq. (3.75) with (3.76) yields

$$Vin(\mu + 1) = i_{in}(r_p + RP)$$

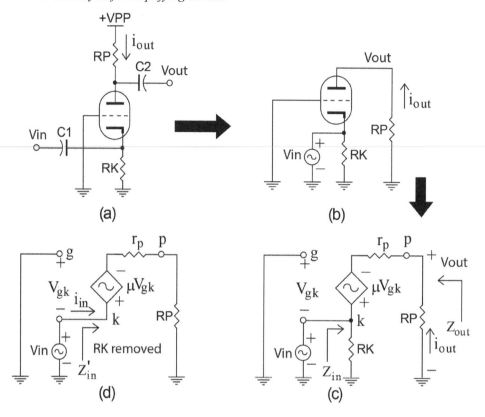

Figure 3.30 A grounded grid amplifier: (a) a self-bias configuration, (b) a simplified ac circuit, (c) an ac equivalent circuit for the amplifier with the triode replaced by the Thévenin model, (d) when RK is removed from (c)

Therefore, we have

$$Z_{in}' \equiv \frac{Vin}{i_{in}} = \frac{r_p + RP}{\mu + 1}$$

Thus, when we put back RK, the input impedance is given as

$$Z_{in} = Z_{in}' \parallel RK = \frac{(r_p + RP)RK}{r_p + RP + (\mu + 1)RK} \tag{3.77}$$

It is interesting to note that Eq. (3.77) has the same expression for the output impedance of a cathode follower with a plate resistor, Eq. (3.74).

To determine the output impedance, we set Vin = 0 in Figure 3.30(c), then consider the output loop. First we have V_{gk} = 0. Therefore, the dependent voltage source μV_{gk} is dead. Thus, the dependent voltage source is replaced by a short circuit. As a result, the output impedance is simply equal to:

$$Z_{out} = r_p \parallel RP = \frac{r_p RP}{r_p + RP} \qquad (3.78)$$

To determine the voltage gain, from the input loop of Figure 3.30(c) we have

$$V_{gk} = -Vin \qquad (3.79)$$

In the output loop, we have

$$Vin - \mu V_{gk} = -i_{out}(r_p + RP) \qquad (3.80)$$

Substituting Eq. (3.79) for (3.80) and rearranging yields

$$i_{out} = -\frac{\mu+1}{r_p + RP} Vin$$

Thus, the output is given as

$$Vout = -i_{out} RP = \frac{(\mu+1)RP}{r_p + RP} Vin$$

Hence the voltage gain is given as

$$A_v = \frac{(\mu+1)RP}{r_p + RP} \qquad (3.81)$$

Since the voltage gain Eq. (3.81) is a positive value, the output of a grounded grid amplifier is non-inverted. On the other hand, as the input is now connected to the cathode while the grid is grounded, it has very little impact on the Miller effect capacitor, which is between the grid and the output plate. Thus, a grounded grid amplifier has an excellent high frequency response, much better than the common cathode amplifier.

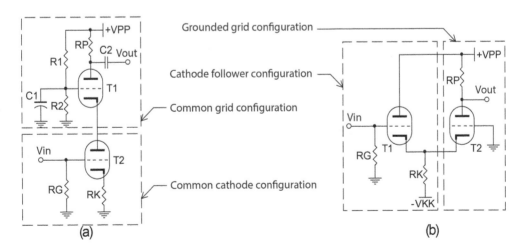

Figure 3.31 (a) a cascode amplifier, (b) a cathode coupled amplifier

Table 3.9 A list of representative small signal triodes.

Tube 83	Typical value			Maximum value		Heater voltage/current			
	Amplifying factor, μ	Transconductance, g_m (mA/V)	Plate resistance, r_p (kΩ)	Plate voltage (V)	Plate dissipation (W)	6.3V	Current (A)	12.6V	Current (A)
ECC81/ 12AT7	60	4 to 5.5	15 to 11	300	2.5	●	0.3	●	0.15
ECC82/ 12AU7	20	2.2 to 3.1	7.7 to 6.5	300	2.75	●	0.3	●	0.15
ECC83/ 12AX7	100	1.2 to 1.6	80 to 62	330	1.2	●	0.3	●	0.15
ECC88/ 6922	33	11.5 to 12.5	2.6 to 2.9	250	1.5	●	0.3		
ECC99	22	9.5	2.3	400	5	●	0.8	●	0.4
6H30	15±3	18±5	0.52 to 1.3	250	4	●	0.83		
6N1P	35±8	3.5 to 5.5	4.9 to 12	300	2.2	●	0.6		
6SL7	70	1.6	44	300	1	●	0.3		
6SN7	20	2.6 to 3	7.7 to 6.7	450	5	●	0.6		
12AY7	40	1.75	22.8	300	1.5	●	0.3	●	0.15
12BH7	16.5	3.1	5.3	300	3.5	●	0.6	●	0.3

3.5   The ac analysis for compound amplifiers

Cascode amplifiers

BJT cascode amplifier

Figure 3.32 shows a BJT cascode amplifier with a voltage-divider bias configuration. When the amplifier is represented by the simplified ac circuit, as shown in Figure 3.32(b), it is clear that transistor Q1 is working in a common base configuration while transistor Q2 is in a common emitter configuration. Therefore, the voltage gain of the cascode amplifier can be determined as a product of the two amplifiers.

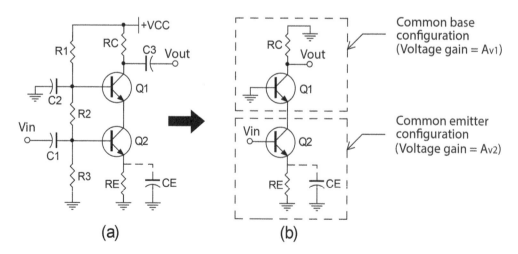

Figure 3.32 (a) A BJT cascode amplifier, (b) a simplified ac circuit of the cascode amplifier

From Eq. (3.8), the voltage gain for a common emitter amplifier is given as

$$A_{v2} = \frac{Vout2}{Vin} = \frac{-RC2}{r_{e2} + RE}$$

where RC2 is the collector load resistor seen by transistor Q2. In this case, RC2 is the input impedance of the common base amplifier, which is given by Eq. (3.25)

$$RC2 = r_{e1}$$

Therefore, the A_{v2} becomes

$$A_{v2} = \frac{-r_{e1}}{r_{e2} + RE} \qquad (3.82)$$

From Eq. (3.29), the voltage gain for the common base amplifier is given as

$$A_{v1} = \frac{RC}{r_{e1}} \qquad (3.83)$$

At the same dc quiescent current, we have $r_{e1} = r_{e2}$. Thus, the voltage gain for a cascode amplifier becomes

$$A_v = A_{v1}A_{v2} = \frac{RC}{r_e}\frac{-r_e}{r_e + RE} = \frac{-RC}{r_e + RE} \text{ (un-bypassed emitter resistor)} \qquad (3.84)$$

$$A_v = \frac{-RC}{r_e} \text{ (bypassed emitter resistor)} \qquad (3.85)$$

The negative sign indicates that a cascode amplifier produces an inverted output. The voltage gain, input and output impedance are very much like that of a common emitter amplifier.

FET cascode amplifier

Figure 3.33 shows two FET cascode amplifier configurations. The first is a self-bias JFET cascode amplifier in Figure 3.33(a). The second is a voltage-divider bias cascode amplifier using an enhancement type MOSFET in Figure 3.33(b). A depletion type MOSFET can be used in both configurations. As far as the voltage gain and output impedance are concerned, both configurations have the same ac properties.

To determine the voltage gain, we separate the cascode amplifier into two parts. The first part is a common gate amplifier on the top with a common source amplifier in the bottom. From Eq. (3.47), the voltage gain for a common gate amplifier is given as

$$A_{v1} = g_{m1}RD$$

From Eq. (3.36), the voltage gain for a common source amplifier is given as

$$A_{v2} = \frac{-g_{m2}RD2}{1 + g_{m2}RS} \qquad (3.86)$$

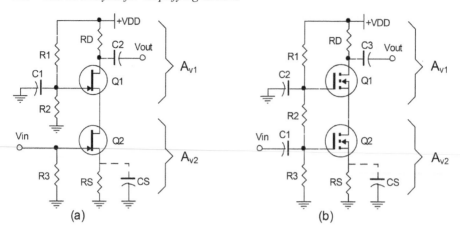

Figure 3.33 (a) A JFET cascode amplifier in a self-bias configuration, (b) an enhancement type MOSET cascode amplifier in a voltage-divider bias configuration

where RD2 is the load impedance looking from the drain of Q2, which is now the input impedance looking into the source of Q1. The input impedance of a common gate amplifier is given by Eq. (3.45). Therefore, we have

$$RD2 \approx 1/g_{m1} \tag{3.87}$$

If closely matched transistors are used for Q1 and Q2, we have $g_{m1} = g_{m2}$. Thus the voltage gain for the cascode amplifier with un-bypassed source resistor is given as

$$A_v = A_{v1} A_{v2}$$
$$= g_m RD \frac{-g_m}{g_m(1+g_m RS)} = \frac{-g_m RD}{1+g_m RS} \quad \text{(un-bypassed source resistor)} \tag{3.88}$$
$$A_v = -g_m RD \quad \text{(bypassed source resistor)} \tag{3.89}$$

The output impedance is the same as that for the two cascode amplifier configurations in Figure 3.33, which is equal to RD. However, the input impedance for Figure 3.33(a) is equal to R3 while R2‖R3 is for Figure 3.33(b).

Triode cascode amplifier

Figure 3.34(a) shows a triode cascode amplifier formed by T1 and T2. Triode T2 is in a self-bias configuration while T1 is in a voltage-divider bias configuration. Figure 3.34(b) is the simplified ac circuit for the triode cascode amplifier. It is clear that triode T1 is working in a common grid configuration while T2 is in a common cathode configuration. The voltage gain of the cascode amplifier can be determined as a product of the two amplifiers.

From Eq. (3.81), the voltage gain of a common grid amplifier is given as

$$A_{v1} = \frac{(\mu_1 + 1)RP}{r_{p1} + RP}$$

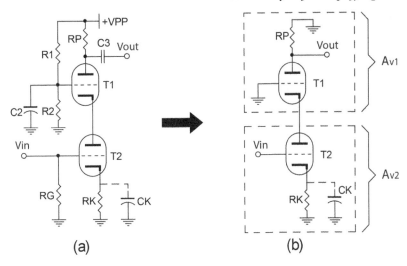

Figure 3.34 (a) A triode cascode amplifier, (b) a simplified ac circuit of the cascode amplifier

where r_{p1} and μ_1 are the plate impedance and amplification factor of triode T1. From Eq. (3.57), the voltage for the common cathode amplifier is given as

$$A_{v2} = \frac{-\mu_2 RP2}{r_{p2} + RP2 + (\mu_2 + 1)RK}$$

where r_{p2} and μ_2 are the plate impedance and amplification factor of triode T2. RP2 is the load resistor looking from the plate of T2. In this situation, it is the input impedance of the common grid amplifier formed by T1. From Eq. (3.77), RP2 is given as

$$RP2 = \frac{r_{p1} + RP}{\mu_1 + 1}$$

Since the input impedance of a common grid amplifier is small, A_{v2} can be approximated as

$$A_{v2} \approx \frac{-\mu_2 RP2}{r_{p2} + (\mu_2 + 1)RK} \approx \frac{-\mu_2}{r_{p2} + (\mu_2 + 1)RK}\left(\frac{r_{p1} + RP}{\mu_1 + 1}\right)$$

If closely matched triodes are used for T1 and T2, we have $\mu_1 = \mu_2$ and $r_{p1} = r_{p2}$. Thus, the voltage gain for the cascode amplifier with an un-bypassed cathode resistor becomes

$$A_v = A_{v1}A_{v2} \approx \left(\frac{(\mu+1)RP}{r_p + RP}\right)\frac{-\mu}{r_p + (\mu+1)RK}\left(\frac{r_p + RP}{\mu+1}\right)$$

$$= \frac{-\mu RP}{r_p + (\mu+1)RK} \quad \text{(un-bypassed cathode resistor)} \tag{3.90}$$

$$A_v = \frac{-\mu RP}{r_p} = -g_m RP \quad \text{(bypassed cathode resistor)} \tag{3.91}$$

Example 3.6

A BJT cascode amplifier is shown in Figure 3.35(a). Before evaluating the voltage gain of the amplifier, we need to determine the dc quiescent current of the transistors. Following a similar procedure to that discussed in Chapter 2, we first formulate the dc bias Thevenin equivalent circuit as shown in Figure 3.35(b).

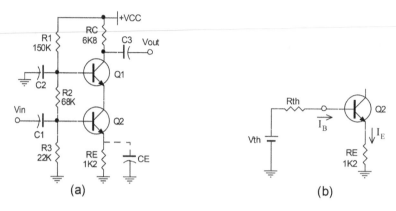

Figure 3.35 (a) A BJT cascode amplifier in a voltage-divider bias configuration, (b) the Thevénin equivalent circuit is used for dc analysis

The Thevénin equivalent resistor and voltage source are given as

$$Rth = (R1 + R2) \| R3 = 20k\Omega$$

$$Vth = \frac{R3}{R1+R2+R3} 30V = 2.75V$$

In the input loop of Figure 3.35(b), we have

$$2.75V = I_B Rth + V_{BE} + I_E RE = I_B[Rth + (\beta+1)RE] + V_{BE}$$

Substituting the values for Rth, RE, V_{BE} and β with the above expression and solving for I_B yields $I_B = 0.01mA$ and $I_E = (\beta+1)I_B = 1.53mA$. Thus, we have

$$r_e = \frac{26mV}{1.53mA} = 17\Omega$$

From Eqs. (3.84) and (3.85), the voltage gain for a cascode amplifier is given as

$$A_v = \frac{-RC}{r_e + RE} = \frac{-6.8k\Omega}{0.017k\Omega+1.2k\Omega} = -5.6 \text{ (un-bypassed emitter resistor)}$$

$$A_v = \frac{-RC}{r_e} = \frac{-6.8k\Omega}{0.017k\Omega} = -400 \text{ (bypassed emitter resistor)}$$

Input impedance is equal to $R2 \| R3 \| \beta(r_e + RE) = 15.2k\Omega$ and $R2 \| R3 \| \beta r_e = 2.2k\Omega$ for an un-bypassed and a bypassed emitter resistor, respectively.

Follower compound amplifiers

Figure 3.36 shows compound amplifiers formed by two BJTs, FETs and triodes. They are an *emitter coupled amplifier*, a *source coupled amplifier* and a *cathode coupled amplifier*, as shown in Figure 3.36(a,b,c), respectively. In each compound amplifier there is a resistor connected right at the junction between the two devices, i.e., REE, RSS, and RKK. Therefore, in Figure 3.36 (a), Q1 and REE can be considered as forming an emitter follower amplifier and its output is directly coupled to Q2, which works as a common base amplifier. Since an emitter follower and a common base are both non-inverted amplifiers, the emitter coupled amplifier is also non-inverted. All these compound amplifiers in Figure 3.36 are non-inverted. They also have a good frequency response, similar to cascode amplifiers.

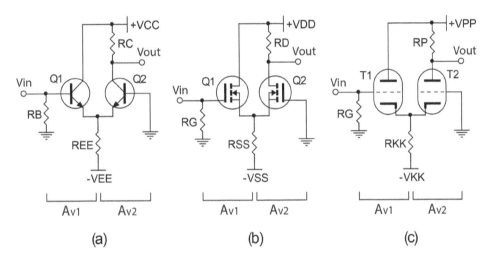

Figure 3.36 Follower compound amplifiers: (a) BJT emitter coupled, (b) FET source coupled, (c) triode cathode coupled

In Figure 3.36(a), in order for the emitter follower of Q1 to work well, it is desirable to make the emitter resistor (REE) as large as possible. For improvement, it is best to replace resistor REE, RSS and RKK with the current source.

Emitter coupled amplifier

To determine the voltage gain of the emitter coupled amplifier of Figure 3.36(a), we first determine the voltage gain for the emitter follower (A_{v1}) and that for the common base amplifier (A_{v2}) separately. Then the overall voltage gain is the sum of those two gains. From Eq. (3.24), the voltage gain for the emitter follower is given as

$$A_{v1} = \frac{RE}{r_{e1} + RE}$$

where RE is the emitter resistor. In this circuit, RE is equal to REE in parallel to the input impedance of the common base amplifier T2. Therefore, we have

$$RE = REE \parallel r_{e2} \approx r_{e2} \text{ (for REE} \gg r_{e2})$$

leading to

$$A_{v1} = \frac{r_{e2}}{r_{e1} + r_{e2}}$$

From Eq. (3.29), the voltage gain for a common base amplifier is given as

$$A_{v2} = \frac{RC}{r_{e2}}$$

If closely matched transistors are used for Q1 and Q2, we have $r_{e1} = r_{e2}$. Therefore, the overall voltage gain for the emitter coupled amplifier becomes

$$A_v = A_{v1}A_{v2} \approx \frac{RC}{2r_e} \tag{3.92}$$

Source coupled amplifier

We follow a similar approach to the emitter coupled amplifier for determining the voltage gain of the source coupled amplifier. From Eq. (3.44), the voltage gain for a source follower is given as

$$A_{v1} = \frac{g_{m1}RS}{1 + g_{m1}RS}$$

and as given by Eq. (3.45) $RS = RSS \| (1/g_{m2}) \approx 1/g_{m2}$ (for $RSS \gg 1/g_{m2}$).
    From Eq. (3.47), the voltage gain for the common gate amplifier is given as

$$A_{v2} = g_{m2}RD$$

If closely matched transistors are used for Q1 and Q2, we have $g_{m1} = g_{m2}$. Therefore, the overall voltage gain for the source coupled amplifier becomes

$$A_v = A_{v1}A_{v2} \approx \frac{g_m RD}{2} \tag{3.93}$$

Cathode coupled amplifier

Again, we follow a similar approach to that for the emitter coupled amplifier to determine the voltage gain for the cathode coupled amplifier. From Eq. (3.66), the voltage gain for a cathode follower is given as

$$A_{v1} = \frac{\mu_1 RK}{r_{p1} + (\mu_1 + 1)RK}$$

where RK = input impedance of the grounded grid amplifier. It is given by Eq. (3.77) and rewritten as

$$RK \approx RKK \| \frac{r_{p2} + RP}{\mu_2 + 1} \approx \frac{r_{p2} + RP}{\mu_2 + 1} \quad (\text{for } RKK \gg \frac{r_{p2} + RP}{\mu_2 + 1})$$

From Eq. (3.81), the voltage gain for the grounded grid amplifier is given as

$$A_{v2} = \frac{(\mu_2 + 1)RP}{r_{p2} + RP}$$

When closely matched triodes are used for T1 and T2, we have $\mu_1 = \mu_2$ and $r_{p1} = r_{p2}$. Therefore, the overall voltage gain for the cathode coupled amplifier becomes

$$A_v = A_{v1}A_{v2} \approx \frac{\mu RP}{2r_p + RP} \tag{3.94}$$

Example 3.7

Figure 3.37 shows a cathode coupled amplifier with a Wilson current source for setting the dc bias current. As discussed in Chapter 2, a Wilson current source has a very high impedance. Therefore, it is a good choice to replace the resistor RKK of Figure 3.36(c).

The dc biasing current is determined as follows. From the loop starting from R4, to Q1, Q2 and R2, we have

$$I_o(R2 + R4) + 2V_{BE} = 20V$$

For $V_{BE} = 0.65V$, $R2 = 680\Omega$, $R4 = 1.2k\Omega$, I_o is equal to 10mA. Therefore, each triode carries a 5mA plate current. Thus, the dc plate voltage for T2 is $200V - (5mA)$ $(20k\Omega) = 100V$.

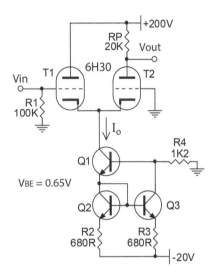

Figure 3.37 A cathode coupled amplifier with a Wilson current source

Triode 6H30: $\mu = 15$, $r_p = 1.3k\Omega$
From Eq. (3.94), the voltage gain is given by

$$A_v \approx \frac{\mu RP}{2r_p + RP} = \frac{15 \times 20k\Omega}{2 \times 1.3k\Omega + 20k\Omega} = 13.3$$

From Eq. (3.78), the output impedance is given as

$$Z_{out} = r_p \parallel RP = \frac{r_p RP}{r_p + RP} = \frac{1.3k\Omega \times 20k\Omega}{1.3k\Omega + 20k\Omega} = 1.2k\Omega$$

Transistors MJE15030 and MJE15032 are suitable for implementing the Wilson current source, Q1–Q3, for Figure 3.37. A cathode coupled amplifier is a popular choice for a line-stage amplifier. It has a non-inverted output and wide bandwidth.

Differential amplifiers

Here we discuss *differential amplifiers* that employ BJTs, FETs and triodes. A triode differential amplifier is often called a long-tail pair. A differential amplifier is one that amplifies two inputs and produces two outputs. It is an important building block in modern solid-state audio amplifier design. When a dual dc power supply is used, the inputs are biased at or near to dc ground level. Therefore, when feedback is taken from the output of the amplifier and connected to one of the two inputs, direct coupling is made possible. Thus, no dc blocking capacitor is required. Together with complementary N-type and P-type BJTs and FETs, solid-state amplifiers can be developed to amplify the signal down to dc level. Such an amplifier is often called a *dc amplifier*. For example, op-amps are dc amplifiers.

There are several properties associated with differential amplifiers. They are *differential voltage gain* (A_{dif}), *common-mode voltage gain* (A_{com}) and *common-mode rejection ratio* (CMRR). In the following discussion, the two inputs are denoted by Vin1 and Vin2 while the outputs are Vout1 and Vout2. Then the definition for the voltage gains and CMRR are given as follows:

$$A_{dif} \equiv \frac{Vout1 - Vout2}{Vin1 - Vin2} \text{ or } \frac{Vout2 - Vout1}{Vin2 - Vin1}$$

(3.95)

$$A_{com} \equiv \frac{Vout1 + Vout2}{2Vin}$$

(3.96)

$$CMRR \equiv \frac{A_{dif}}{A_{com}}$$

(3.97)

For a good differential amplifier, it produces a high differential gain (A_{dif}) and a very low common-mode gain (A_{com}). In other words, the differential amplifier rejects the common-mode signal. An ideal differential amplifier has zero common-mode gain and, therefore, it produces no common-mode output signal. The common-mode rejection ratio (CMRR) can be viewed as a factor that measures how good the differential signal amplification is in comparison to the common-mode signal rejection.

BJT differential amplifier

Figure 3.38 shows a BJT differential amplifier with no emitter degeneration resistor. A BJT differential amplifier with an emitter degeneration resistor will be discussed later. Figure 3.38 is perhaps the simplest differential amplifier using a tail resistor REE for setting the dc biasing current. In real applications, the tail resistor is usually replaced by a current source. It is because the current source has a much higher impedance than a normal resistor at a given negative supply, –VEE. It can be seen later that the higher the tail resistor is, the better the CMRR will be.

The differential amplifier shown in Figure 3.38 is very similar to the emitter coupled amplifier of Figure 3.36(a). The difference is that only one collector resistor (RC) is used in Figure 3.36(a) while two RCs are used in Figure 3.38. Additionally, there is only one input and one output in Figure 3.36(a). However, there are two inputs (Vin1 and Vin2) as well as two outputs (Vout1 and Vout2) in the differential amplifier of Figure 3.38.

With no emitter degeneration resistor

To determine the output from the differential amplifier with two input sources, we apply the *superposition*. The superposition states that, for all linear systems, the net response caused by two or more sources is an algebraic sum of the responses that would have been enacted by each source individually. (See Appendix B for an overview of superposition principle.) Thus, if there are two inputs, Vin1 and Vin2, we first enable Vin1 and disable Vin2 to determine the output,

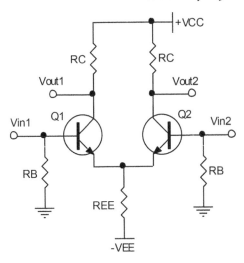

Figure 3.38 A BJT differential amplifier with no emitter degeneration resistor

V_{o1}. Then we disable Vin1 and enable Vin2 to determine output, V_{o2}. Then the total output is a sum of V_{o1} and V_{o2}. Therefore, by applying the superposition to the differential amplifier of Figure 3.38, which has two inputs and two outputs, the differential amplifier is now broken down into four cases, as shown in Figure 3.39. In each case, there is only one input and one output. The outputs are given as:

$$\text{Vout1} = \text{Vin1 } A_{11}\Big|_{(\text{Vin2}=0)} + \text{Vin2 } A_{21}\Big|_{(\text{Vin1}=0)} \tag{3.98}$$
$$\text{Vout2} = \text{Vin1 } A_{12}\Big|_{(\text{Vin2}=0)} + \text{Vin2 } A_{22}\Big|_{(\text{Vin1}=0)} \tag{3.99}$$

where A_{11}, A_{12}, A_{21} and A_{22} are the voltage gains for each of the cases.

Figure 3.39(a) illustrates a common cathode amplifier (Q1) with emitter resistor REE in parallel to the input impedance of a common base amplifier (Q2). From Eq. (3.8), we have

$$A_{11}\big|_{(\text{Vin2}=0)} = \frac{-RC}{r_{e1} + RE}$$

where RE = REE$\|$Zin, Zin = r_{e2} (input impedance of a common base amplifier). Therefore, we have RE $\approx r_{e2}$ (as REE $\gg r_{e2}$). If closely matched transistors are used for Q1 and Q2, we have $r_{e1} = r_{e2} = r_e$. Therefore, the above expression for A_{11} becomes

$$A_{11}\big|_{(\text{Vin2}=0)} = \frac{-RC}{2r_e} \tag{3.100}$$

Figure 3.39(b) shows an emitter coupled amplifier similar to that depicted in Figure 3.36(a). It is notable that Figure 3.39(b) has two collector resistors (RC) while there is only one RC in Figure 3.36(a). It can be shown that the additional resistor RC in transistor Q2, which operates

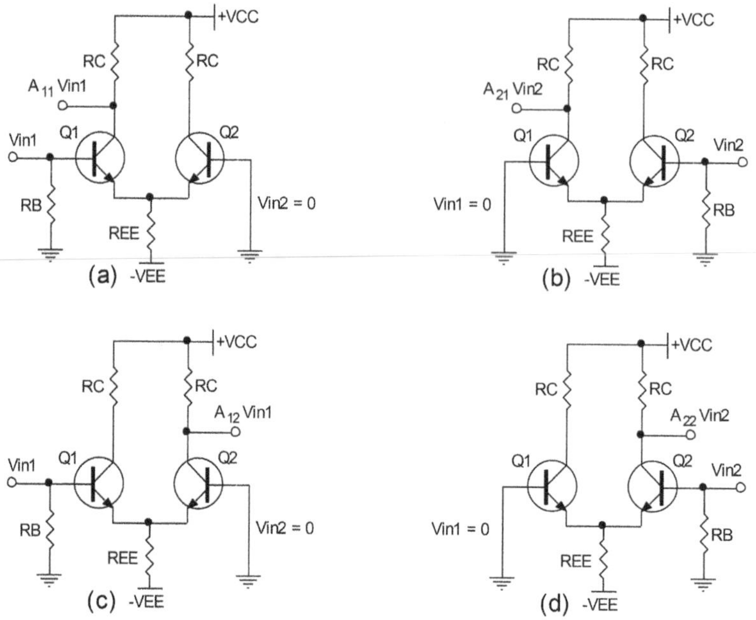

Figure 3.39 Four different cases where only one signal source is active in each case

as an emitter follower, does not affect the emitter follower's voltage gain and output impedance. Therefore, the voltage gain A_{21} is given by Eq. (3.92):

$$A_{21}\big|_{(Vin1=0)} = \frac{RC}{2r_e} \tag{3.101}$$

Figure 3.39(c) depicts an emitter coupled amplifier in the exact same configuration of Figure 3.39(b). Therefore, the voltage gain A_{12} is given as

$$A_{12}\big|_{(Vin2=0)} = \frac{RC}{2r_e} \tag{3.102}$$

Figure 3.39(d) shows a common emitter amplifier (Q2) with an emitter resistor REE in parallel to the input impedance of a common base amplifier (Q1). It is the exact same configuration as that of Figure 3.39(a). Therefore, the voltage gain A_{22} is given as

$$A_{22}\big|_{(Vin1=0)} = \frac{-RC}{2r_e} \tag{3.103}$$

Thus, Eq. (3.98) and (3.99) can be written as

$$Vout1 = Vin1\left(\frac{-RC}{2r_e}\right) + Vin2\left(\frac{RC}{2r_e}\right) = \left(\frac{-RC}{2r_e}\right)(Vin1 - Vin2) \tag{3.104}$$

$$Vout2 = Vin1\left(\frac{RC}{2r_e}\right) + Vin2\left(\frac{-RC}{2r_e}\right) = \left(\frac{RC}{2r_e}\right)(Vin1 - Vin2) \qquad (3.105)$$

Subtracting Eq. (3.104) from Eq. (3.105) yields

$$Vout1 - Vout2 = \left(\frac{-RC}{r_e}\right)(Vin1 - Vin2)$$

Hence, the differential voltage gain for the differential amplifier is given as

$$A_{dif} = \frac{-RC}{r_e} \qquad (3.106)$$

To determine the common-mode voltage gain, the two inputs are connected to the same signal source as shown in Figure 3.40(a). If each transistor has a small signal emitter current i_e, the total current flowing to the tail resistor REE is $2i_e$. Since the same collector resistor is used for each transistor, if Q1 and Q2 are closely matched, the outputs are equal, Vout1 = Vout2. The differential amplifier can be split into two halved circuits, as shown in Figure 3.40(b). The two amplifiers are common emitter amplifiers with an emitter resistor of 2×REE. Therefore, from Eq. (3.12), the output for the common emitter amplifier is given as

$$Vout1 = Vout2 = \frac{-RC}{r_e + RE}\,Vin = \frac{-RC}{r_e + 2REE}\,Vin \ \ (\text{where } RE = 2\times REE)$$

Thus, the common-mode gain is given as

$$A_{com} \equiv \frac{Vout1 + Vout2}{2Vin} = \frac{-RC}{r_e + 2REE} \qquad (3.107)$$

Hence, from Eq. (3.97) the common-mode rejection ratio is given as

$$CMRR = \frac{A_{dif}}{A_{com}} = \frac{\dfrac{-RC}{r_e}}{\dfrac{-RC}{r_e + 2REE}} = 1 + \frac{2REE}{r_e} = 1 + 2g_m REE \qquad (3.108)$$

where $g_m = 1/r_e$ is the transconductance of the transistor. Eq. (3.108) appears to be a simple expression for determining the CMRR of a differential amplifier that does not have an emitter degeneration resistor. It is clear that the CMRR is directly proportional to REE. Thus, the higher tail resistor REE is, the better the CMRR will be.

With an emitter degeneration resistor (RE)

Figure 3.41 shows a BJT differential amplifier with an emitter degeneration resistor RE. We can follow a similar approach to that employed for Figure 3.38 to determine differential voltage gain, except that Q1 and Q2 have an equivalent resistor $r_e' = r_e + RE$.

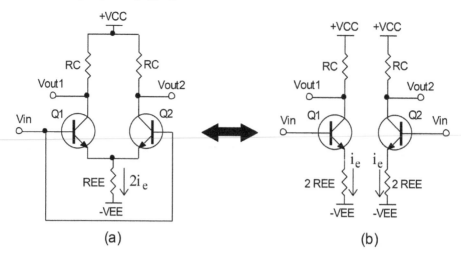

Figure 3.40 (a) The inputs of the differential amplifier are connected to a common signal source, (b) the differential amplifier is split into two halved circuits

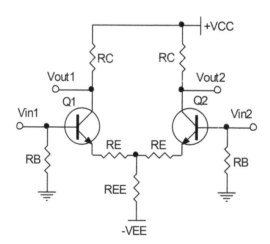

Figure 3.41 A BJT differential amplifier with an emitter degeneration resistor RE

$$A_{dif} = \frac{-RC}{r_e + RE} \tag{3.109}$$

$$A_{com} = \frac{-RC}{r_e + RE + 2REE} \tag{3.110}$$

$$CMRR = \frac{\dfrac{-RC}{r_{e+RE}}}{\dfrac{-RC}{r_e + RE + 2REE}} = \frac{r_e + RE + 2REE}{r_e + RE} \approx \frac{RE + 2REE}{RE} = 1 + \frac{2REE}{RE} \tag{3.111}$$

In the above expressions, there is an RE factor in the denominator. Therefore, with the addition of an emitter degeneration resistor RE, A_{dif}, A_{com} and CMRR are lower than a differential amplifier without an RE, since RE > r_e and REE > RE, A_{com} is less affected than A_{dif} and CMRR. A degeneration emitter resistor provides local feedback to the differential amplifier. As a result, the voltage gain is reduced. However, it improves the bandwidth, the slew rate, and reduces distortion of the amplifier. A differential amplifier with an emitter degeneration resistor is often used as the input stage of an audio power amplifier. Details of slew rate and solid-state audio power design are discussed in Chapter 12.

Example 3.8

Figure 3.42 shows two differential amplifiers with the same collector resistor RC and dc biasing current I_o. The one depicted in Figure 3.42(a) has a tail resistor REE while there is a current source in Figure 3.42(b). Let us determine the A_{dif}, A_{com} and CMRR for both amplifiers.

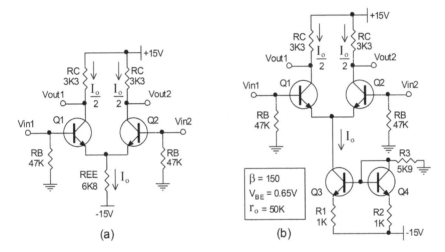

Figure 3.42 (a) A BJT differential amplifier with a tail resistor REE, (b) a BJT differential amplifier with a current source

For Figure 3.42(a), applying KVL to the input loop yields

$$I_B RB + V_{BE} + I_o REE - 15V = 0$$

where $I_B = I_E/(\beta + 1) = I_o/2(\beta + 1)$ therefore, we have

$$I_o\left(\frac{RB}{2(\beta + 1)} + REE\right) = 15V - V_{BE}$$

Thus, we find $I_o = 2.06mA$. As a result, the emitter current is given as $I_E = 1.03mA$ and $r_e = 26mV/I_E = 25.2\Omega$. From Eq. (3.106) and (3.107), we have

$$A_{dif} = \frac{-RC}{r_e} = \frac{-3.3 \times 10^3 \Omega}{25.2\Omega} = -131$$

$$A_{com} = \frac{-RC}{r_e + 2REE} = \frac{-3.3 \times 10^3 \Omega}{25.2\Omega + 2 \times 6.8 \times 10^3 \Omega} = 0.24$$

$$CMRR = A_{dif}/A_{com} = 545.8 \text{ (or 54.7dB)}$$

For Figure 3.42(b), the dc biasing current is determined by the loop starting from R3 to Q3, R1, and the negative dc power supply as follows:

$$I_o(R1 + R3) + V_{BE} - 15V = 0$$

By solving the above equation for I_o, we have $I_o = 2.08mA$ and emitter current $I_E = 1.04mA$, which is about the same emitter current as shown in Figure 3.42(a). Similarly, we have $r_e = 26mV/I_E = 25\Omega$, $g_m = 1/r_e = 0.04\Omega^{-1}$. As discussed, when looking at Figure 2.47(g) of Chapter 2, the output impedance of the current source is given as

$$REE = r_o(1 + g_m RE) = 50k\Omega(1 + 0.04\Omega^{-1} \times 1000\Omega) = 2{,}050k\Omega$$

where r_o = collector-emitter impedance, which is $50k\Omega$ in this example, and $RE = R1 = 1k\Omega$. Therefore, from Eq. (3.106) and (3.107), we have

$$A_{dif} = \frac{-RC}{r_e} = \frac{-3.3 \times 10^3 \Omega}{25\Omega} = -132$$

$$A_{com} = \frac{-RC}{r_e + 2REE} = \frac{-3.3 \times 10^3 \Omega}{25\Omega + 2 \times 2050 \times 10^3 \Omega} = -8.05 \times 10^{-4}$$

Thus, the CMRR for the differential amplifier shown in Figure 3.42(b) is given as

$$CMRR = \frac{A_{dif}}{A_{com}} = \frac{132}{8.05 \times 10^{-4}} = 163{,}975 \text{ (or 104.3dB)}$$

The use of a current source has greatly improved the CMRR from 54.7dB to 104.3dB.

With an active load

Figure 3.43 shows that a simple current source is used as an active load for a common emitter amplifier. As we know, the voltage gain for a common emitter with a collector resistor RC is given as $A_v = -RC/r_e$ (or $-g_m RC$). Therefore, the larger the collector resistor is, the higher the voltage gain we achieve. Owing to the limited power supply voltage, the resistance for collector resistor RC is limited. However, a current source has high output impedance. Thus, a current source can be used as an active load to replace the collector resistor RC.

Note that we need to determine the output impedance for the collector of Q2 because it is the load resistor for Q1. Similar to that which we did to determine the output impedance of a single transistor amplifier, we set the input for Q2 to zero. Knowing that Q2 operates in a common emitter configuration, we set $V_{B2} = 0$, so that $i_{b2} = 0$, as shown in the ac equivalent circuit in Figure 3.43(b). Thus, the dependent current source βi_{b2} is zero. As a result, the dependent current source becomes an open circuit. Now Figure 3.43(b) is simplified to Figure 3.43(c). The effective impedance at the collector of Q1 and Q2 becomes $Z_C = r_{o1} \| r_{o2}$. When complementary

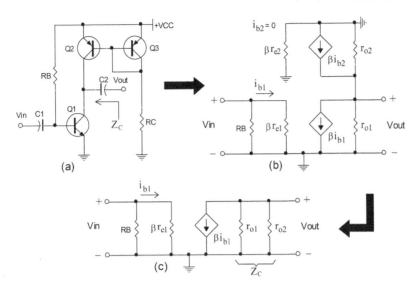

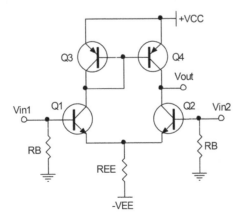

Figure 3.43 (a) An active load is used in a common emitter amplifier; (b) the ac equivalent circuit for (a); (c) the final ac equivalent circuit

transistors with similar properties are used for Q1 and Q2, we assume $r_{o1} = r_{o2} = r_o$. Thus, the voltage gain for the common emitter amplifier with an active load is given as $A_v = -g_m Z_C = -g_m(r_{o1} \| r_{o2}) = -g_m r_o/2 = -r_o/2r_e$.

A differential amplifier with an active load is shown in Figure 3.44. Following a similar ac analysis, the voltage gain can be found as follows:

$$A_{dif} \equiv \frac{Vout}{Vin1 - Vin2} = g_m(r_{o4} \| r_{o2}) = \frac{g_m r_o}{2} = \frac{r_o}{2r_e} \tag{3.112}$$

where r_{o2} and r_{o4} are the collector-emitter impedances for transistor Q2 and Q4, respectively. When Q2 and Q4 have similar ac properties, we assume $r_{o2} = r_{o4} = r_o$.

Figure 3.44 A differential amplifier with an active load

Example 3.9

Figure 3.45(a) shows a differential amplifier with an active load and its output is directly coupled to a common emitter amplifier formed by Q5. The differential amplifier also has emitter degeneration resistors R3 and R4. We first determine the tail current I_o by applying KVL to the input loop,

$$I_B RB + V_{BE} + (I_o/2)R3 + I_o REE - 15V = 0$$

where $I_B = I_E/(\beta + 1) = I_o/2(\beta + 1)$. Rearranging the above expression yields

$$I_o = \frac{15V - V_{BE}}{\dfrac{RB}{2(\beta+1)} + \dfrac{R3}{2} + REE} = \frac{15V - 0.65V}{\dfrac{47k\Omega}{2(150+1)} + \dfrac{0.1k\Omega}{2} + 6.8k\Omega} = 2.04mA$$

Therefore, we have $I_E = 1.02mA$ and $r_e = 26mV/I_E = 25.5\Omega$, $g_m = 1/r_e = 0.039\Omega^{-1}$.

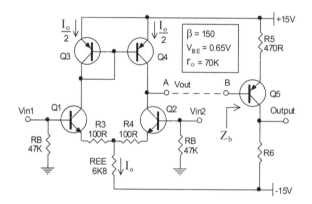

Figure 3.45(a) A differential amplifier with an active load and directly coupled to a common emitter amplifier

Now we determine the voltage gain Vout/(Vin1 – Vin2) when the two amplifiers are disconnected, i.e., point A and B are open. From Eq. (3.112), the voltage gain for a differential amplifier with an active load but no emitter degeneration resistor is given as

$$\frac{Vout}{Vin1 - Vin2} = \frac{r_o}{2r_e}$$

When an emitter degeneration resistor is added, the voltage gain becomes

$$\frac{Vout}{Vin1 - Vin2} = \frac{r_o}{2(r_e + RE)}$$

$$= \frac{70 \times 10^3 \Omega}{2(25.5\Omega + 100\Omega)} = 279 \text{ (or 49dB)}$$

This is the voltage gain for the differential amplifier when points A and B are open circuit. If points A and B are now connected, the output of the differential amplifier is directly coupled to the common emitter amplifier formed by Q5. The input impedance of the common emitter amplifier, Z_b, will create a loading effect to the differential amplifier. The input impedance for the common emitter amplifier is given as

$$Z_b \approx \beta R5 = (150)(0.47k\Omega) = 70.5 \text{ k}\Omega$$

If we take into account the impedance Z_b, the voltage gain for the differential amplifier becomes

$$\frac{Vout}{Vin1 - Vin2} = \frac{r_o \parallel Z_b}{2(r_e + RE)} = \frac{r_o \parallel \beta R5}{2(r_e + RE)} = 140 \text{ (or 43dB)}$$

It is clear that the voltage gain of the differential amplifier is reduced by the loading effect from the input impedance of the common emitter amplifier. To minimize the loading effect, a buffer amplifier is often inserted between the differential amplifier and the common emitter amplifier. The buffer amplifier is often an emitter follower, which has high input and low output impedance.

Figure 3.45(b) shows an emitter follower formed by a PNP transistor Q5. Alternatively, since FET has very high input impedance, FET can be used to replace Q5 of Figure 3.45(a). Therefore, the common emitter amplifier Q5 becomes a common source amplifier in Figure 3.45(a). It should be noted that the voltage gain of a common source amplifier is generally lower than a common emitter amplifier. It is because the transconductance (g_m) of FET is lower than that of BJT. As a consequence, the overall voltage gain of Figure 3.45(a), in which the common emitter amplifier Q5 is replaced by a common source amplifier, is slightly lower than the voltage gain of Figure 3.45(b). If it is needed to increase the overall voltage gain of the amplifier, an active load can be used to replace the collector resistor R6 of Figure 3.45(a) and R7 of Figure 3.45(b).

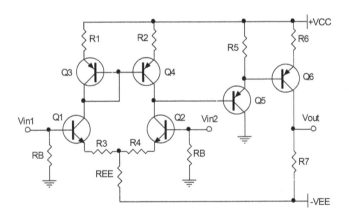

Figure 3.45(b) An emitter follower (Q5) is inserted between the differential amplifier (Q1–Q4) and the common emitter amplifier (Q6) to minimize the loading effect from Q6

FET differential amplifier

With no source degeneration resistor

Figure 3.46 shows the FET differential amplifiers using JFET, enhancement type and depletion type MOSFET. Since JFET, enhancement type and depletion type MOSFET have the same ac equivalent circuit, they share the same voltage gain. By following the same technique that is used to derive the voltage gain and CMRR for the BJT differential amplifier in the previous section, we have the following.

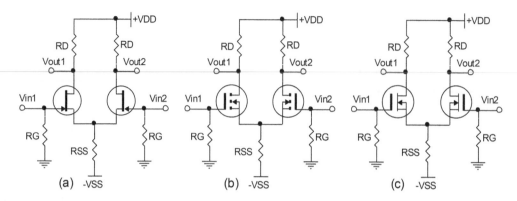

Figure 3.46 A differential amplifier with no source degeneration resistor: (a) JFET, (b) enhancement type MOSFET, (c) depletion type MOSFET

$$A_{dif} = -g_m RD \tag{3.113}$$

$$A_{com} = \frac{-g_m RD}{1 + 2g_m RSS} \tag{3.114}$$

$$CMRR = 1 + 2g_m RSS \tag{3.115}$$

With source degeneration resistor (RS)

When source degeneration resistor RS is added, the voltage gain and CMRR are given as follows.

$$A_{dif} = \frac{-g_m RD}{1 + g_m RS} \tag{3.116}$$

$$A_{com} = \frac{-g_m RD}{1 + g_m RS + 2g_m RSS} \tag{3.117}$$

$$CMRR = 1 + \frac{2g_m RSS}{1 + g_m RS} \tag{3.118}$$

Triode differential amplifier

Figure 3.47 shows the triode differential amplifiers with and without cathode degeneration resistor RK. The differential voltage gain for a differential amplifier with cathode resistor RK has a similar form to the differential amplifier with no cathode resistor but having an effective plate impedance $r_p' = r_p + (\mu + 1)RK$.

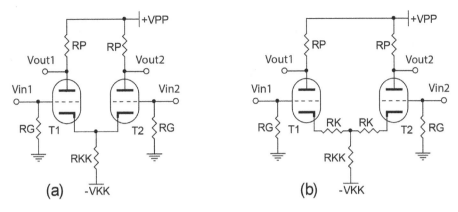

Figure 3.47 A triode differential amplifier: (a) with no cathode degeneration resistor, (b) with a cathode
degeneration resistor RK

With no cathode degeneration resistor

$$A_{dif} = \frac{-\mu RP}{RP + r_p} \tag{3.119}$$

$$A_{com} = \frac{-\mu RP}{RP + r_p + 2(\mu+1)RKK} \tag{3.120}$$

$$CMRR = \frac{RP + r_p + 2(\mu+1)RKK}{RP + r_p} \tag{3.121}$$

With cathode degeneration resistor (RK)

$$A_{dif} = \frac{-\mu RP}{RP + r_p + (\mu+1)RK} \tag{3.122}$$

$$A_{com} = \frac{-\mu RP}{RP + r_p + (\mu+1)(RK+2RKK)} \tag{3.123}$$

$$CMRR = \frac{RP + r_p + (\mu+1)(RK+2RKK)}{RP + r_p + (\mu+1)RK} \tag{3.124}$$

Cathode coupled phase splitter

Figure 3.48(a) shows a *cathode coupled phase splitter*. It can be also viewed as a differential
amplifier with a single input and differential outputs. It is a popular phase splitter, often used
in vacuum tube power amplifiers, that enhances the output from a common cathode amplifier,

and generates two out-of-phase outputs, as shown in Figure 3.48(b). Since the common cathode amplifier is directly coupled to the phase splitter, the grids of the phase splitter (T2 and T3) carry the dc potential from the plate of T1. In this example, it is 100V. Given this dc potential, the tail resistor RKK has a reasonably high value, 16kΩ. Consequently, both triodes T2 and T3 run at a plate current of about 3mA. The dc voltage across RP is 150V, as is the voltage across the plate-cathode for T2 and T3.

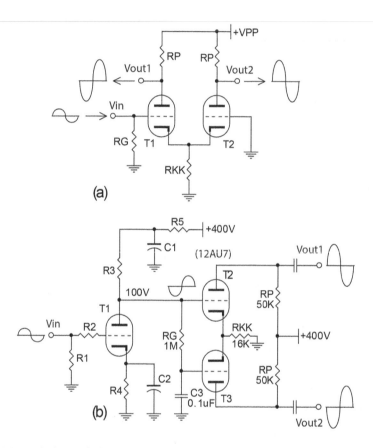

Figure 3.48 (a) A cathode coupled phase splitter, (b) a common cathode amplifier (T1) is directly coupled to a cathode coupled phase splitter (T2 and T3)

The voltage gain corresponding to output Vout1 and Vout2 of Figure 3.48(a) is

$$A_{v1} \equiv \frac{Vout1}{Vin} = \frac{-\mu RP}{2(RP + r_p)} \tag{3.125}$$

$$A_{v2} \equiv \frac{Vout2}{Vin} = \frac{\mu RP}{2(RP + r_p)} \tag{3.126}$$

For the cathode coupled phase splitter example of Figure 3.48(b), T2 and T3 are twin triodes 12AU7. At dc plate current 3mA and plate-cathode voltage of 150V, we find from the 12AU7 data sheet: μ = 15 and r_p = 12kΩ. Thus, the voltage gains are given as

$$A_{v1} = \frac{-\mu RP}{2(RP + r_p)} = \frac{-(15)(50 \times 10^3)}{2(50 \times 10^3 + 12 \times 10^3)} = -6, \text{ and } A_{v2} = 6$$

The voltage gain is not high for the given choice of the vacuum tube 12AU7. It should be noted that these gains are determined for the phase splitter alone (T2 and T3). The gain of the first stage (T1) has not been taken into account.

SRPP amplifier

Figure 3.49 shows a *shunt regulated push–pull* (SRPP) amplifier. It also has other names, such as "totem-pole" amplifier [5], "series-balanced" amplifier [6] and "bootstrap" amplifier [7]. It is a simple compound tube amplifier that comprises two triodes. Figure 3.49(a) has a bypassed cathode resistor while there is an un-bypassed cathode resistor in Figure 3.49(b). The triode on the bottom, T2, operates in a common cathode configuration. The triode on the top, T1, produces an approximated push–pull output. The equal cathode resistor RK is often used so that the dc potential at point A is equal to one-half of the power supply, i.e., $V_A = VPP/2$. Since the output is taken from the cathode of T1, the output impedance is low. In addition, an SRPP amplifier offers a medium voltage gain and high input impedance. SRPP is gaining popularity in audio line-stage amplifier and vacuum tube power amplifier applications. The voltage gain and output impedance [8–11] are given as follows.

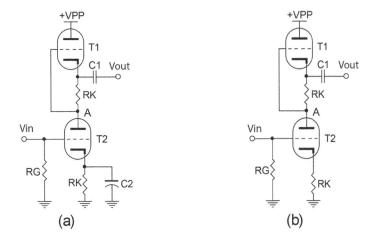

Figure 3.49 An SRPP amplifier: (a) bypassed cathode resistor, (b) un-bypassed cathode resistor

BYPASSED CATHODE RESISTOR

$$A_v \approx \frac{-\mu(r_p + \mu RK)}{2r_p + (\mu + 1)RK} \tag{3.127}$$

$$Z_{out} \approx \frac{r_p(r_p + RK)}{2r_p + (\mu + 1)RK} \tag{3.128}$$

UN-BYPASSED CATHODE RESISTOR (RK)

$$A_v \approx \frac{-\mu(r_p + \mu RK)}{2r_p + 2(\mu + 1)RK} \tag{3.129}$$

$$Z_{out} \approx \frac{(r_p + 2RK)[r_p + (\mu + 2)RK]}{2r_p + 2(\mu + 2)RK} \tag{3.130}$$

Several SRPP amplifier design examples are given in Chapter 8.

White cathode follower

Figure 3.50 shows a *White cathode follower*. It is a compound amplifier formed by a cathode follower (T1) and an active load (T2). The cathode follower T1 has a small plate resistor RP. The output from the plate of T1 is capacitor coupled to the grid of T2. Therefore, the output signal from the plate of T1 sees T2 working as a common cathode amplifier. In other words, triode T2 is working with dual roles: as an active load for T1 as well as a common cathode amplifier for the signal coming to T2's grid. As a common cathode amplifier produces phase inverted output, the output from the plate of T1 is phase inverted and so is the output from the plate of T2. Since the signal from the plate of T2 has been inverted twice, it becomes in phase with respect to the output coming from the cathode of T1. Thus, triodes T1 and T2 produce an approximated push–pull output. This push–pull operation is similar to that of the SRPP amplifier. In addition, the White cathode follower produces a very low output impedance. These features make White cathode followers an attractive choice for the output stage of a line-stage amplifier and even for a low power headphone amplifier. White cathode followers may also form a building block for an output transformer-less (OTL) vacuum tube power amplifier. If closely matched triodes T1

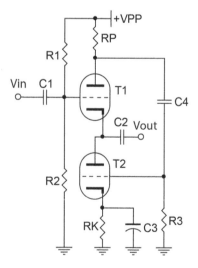

Figure 3.50 A white cathode follower

and T2 are used, the voltage gain and output impedance for the White cathode follower shown in Figure 3.50 are given as follows [11, 12].

$$A_v = \frac{\mu(r_p + \mu RP)}{(\mu + 1)\left(r_p + \mu RP\right) + r_p + RP} \tag{3.131}$$

$$\approx \frac{\mu(r_p + \mu RP)}{(\mu + 1)\left(r_p + \mu RP\right)} \quad \text{for } (\mu + 1)(r_p + \mu RP) \gg r_p + RP$$

$$= \frac{\mu}{\mu + 1} \approx 1$$

$$Z_{out} = \frac{r_p(r_p + RP)}{r_p + RP + (\mu + 1)\left(r_p + \mu RP\right)} \quad \text{(for bypassed cathode resistor)} \tag{3.132}$$

$$Z_{out} = \frac{(r_p + RK)(r_p + RP)}{r_p + RP + (\mu + 1)\left(r_p + RK\right) + \mu(\mu + 1)RP} \tag{3.133}$$

(for un-bypassed cathode resistor RK).

Example 3.10

The *Shigeru Wada* line-stage amplifier is a variant of the Marantz 7 preamplifier, which is discussed in Chapter 8. See Figure 8.2 for the complete circuit. The White cathode follower is employed for the output stage. It is now redrawn in Figure 3.51(a). From the given dc biasing condition, the plate current is 8mA. To simplify the calculation, we assume the plate-cathode voltages for the two triodes are equal (slightly below 140V) and they have equal amplification factors and plate impedances. Figure 3.51(b) shows a simple cathode follower with the same dc biasing condition. In the following, we compare the output impedances between them.

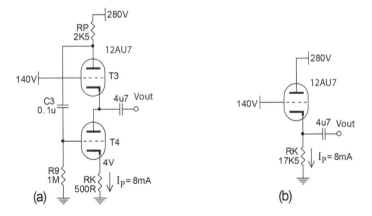

Figure 3.51 (a) The White cathode follower taken from the Shigeru Wada amplifier of Figure 8.2, (b) a simple cathode follower with the same dc biasing condition

Table 3.10 Summary of ac properties for BJT, FET and vacuum tube

Configuration	$A_v = V_{out}/V_{in}$	Z_{in}	Z_{out}
+VCC, RC, C2 Vout, RB, Vin C1, RE	Low $$= \frac{-RC}{r_e + RE}$$ $$\approx -\frac{RC}{RE}$$	High $$= RB \,\|\, [\beta(r_e + RE)]$$ $$\approx RB \,\|\, \beta RE$$ $$(RE \gg r_e)$$	Medium $$= RC$$
+VCC, R1, RC, C2 Vout, Vin C1, R2, RE, C3	Low: (un-bypassed RE) $$= \frac{-RC\|r_o}{r_e + RE}$$ $$\approx \frac{-RC}{RE}$$	High $$= RB \,\|\, [\beta(r_e + RE)]$$ $$RB = R1 \,\|\, R2$$	Medium $$= RC \,\|\, r_o$$ $$\approx RC$$
	High: (bypassed RE) $$= \frac{-RC\|r_o}{r_e}$$ $$\approx \frac{-RC}{r_e}$$	Medium $$= RB \,\|\, \beta r_e$$ $$RB = R1 \,\|\, R2$$	Medium $$= RC \,\|\, r_o$$ $$\approx RC$$
+VCC, R1, RC, C2 Vout, Vin C1, R2, RE1, RE2, C3	Low to medium $$= \frac{-RC\|r_o}{r_e + RE1}$$ $$\approx \frac{-RC}{r_e + RE1}$$	Medium to high $$= RB \,\|\, [\beta(r_e + RE1)]$$ $$RB = R1 \,\|\, R2$$	Medium $$= RC \,\|\, r_o$$ $$\approx RC$$
+VCC, RB, Vin C1, C2 Vout, RE	Low $$= \frac{RE}{r_e + RE}$$ $$\approx 1$$	High $$= RB \,\|\, [\beta(r_e + RE)]$$	Low $$= RE \,\|\, r_e$$ $$\approx r_e$$
+VCC, R1, Vin C1, C2 Vout, R2, RE	Low $$= \frac{RE}{r_e + RE}$$ $$\approx 1$$	High $$= RB \,\|\, [\beta(r_e + RE)]$$ $$RB = R1 \,\|\, R2$$	Low $$= RE \,\|\, r_e$$ $$\approx r_e$$

Configuration	$A_v = V_{out}/V_{in}$	Z_{in}	Z_{out}
	High $$\approx \frac{RC}{r_e}$$	Low $= RE \parallel r_e$ $\approx r_e$ $(RE \gg r_e)$	Medium $= RC$
JFET and depletion type MOSFET	Low (un-bypassed RS) $$\approx \frac{-g_m RD}{1 + g_m RS}$$	High $= RG$	Medium $\approx RD$
	Medium (bypassed RS) $\approx -g_m RD$	High $= RG$	Medium $\approx RD$
All FETs	Low (un-bypassed RS) $$\approx \frac{-g_m RD}{1 + g_m RS}$$	High $= R1 \parallel R2$	Medium $= RD \parallel r_d$ $\approx RD$
	Medium (bypassed RS) $\approx -g_m RD$	High $= R1 \parallel R2$	Medium $= RD \parallel r_d$ $\approx RD$
JFET and depletion type MOSFET	Low $$\approx \frac{g_m RS}{1 + g_m RS}$$ ≈ 1	High $= RG$	Low $= RS \parallel r_d \parallel (1/g_m)$ $\approx RS \parallel (1/g_m)$

(Continued)

Table 3.10 (Continued)

Configuration	$A_v = V_{out}/V_{in}$	Z_{in}	Z_{out}
All FETs 	Low $\approx \dfrac{g_m RS}{1+g_m RS}$ ≈ 1	High $= RG$ $(RG = R1 \parallel R2)$	Low $= RS \parallel r_d \parallel (1/g_m)$ $\approx RS \parallel (1/g_m)$
JFET and depletion type MOSFET 	Medium $\approx g_m RD$	Low $\approx RS \parallel (1/g_m)$	Medium $= RD$
All FETs 	Medium $\approx g_m RD$	Low $\approx RS \parallel (1/g_m)$	Medium $= RD$
	Low (un-bypassed RK) $= \dfrac{-\mu RP}{r_p + RP + RK(\mu + 1)}$	High $= RG$	Medium $= RP \parallel [r_p + RK(\mu+1)]$ $= \dfrac{RP[r_p + RK(\mu + 1)]}{RP + r_p + RK(\mu + 1)}$
	Medium (bypassed RK) $= \dfrac{-\mu RP}{r_p + RP}$	High $= RG$	Medium $= RP \parallel r_p$ $= \dfrac{r_p RP}{r_p + RP}$

Configuration	$A_v = V_{out}/V_{in}$	Z_{in}	Z_{out}
	Low $$= \frac{\mu RK}{r_p + (\mu + 1)RK}$$ $$\approx \frac{\mu}{\mu+1} \approx 1$$	High $$= RG$$	Low $$= \frac{r_p RK}{r_p + (\mu + 1)RK}$$
	Low $$= \frac{\mu RK}{r_p + RP + (\mu + 1)RK}$$ $$\approx \frac{\mu}{\mu+1} \approx 1$$	High $$= RG$$	Low $$= \frac{(r_p + RP)RK}{r_p + RP + (\mu + 1)RK}$$
	Medium $$= \frac{(\mu+1)RP}{r_p+RP}$$	Low $$= \left(\frac{r_p+RP}{\mu+1}\right) \| RK$$	Medium $$= RP \| r_p$$ $$= \frac{r_p RP}{r_p+RP}$$

Configuration	A_{dif} = differential gain	A_{com} = common-mode gain
	High $$= \frac{-RC}{r_e}$$	Low $$= \frac{-RC}{r_e+2REE}$$

(*Continued*)

Table 3.10 (Continued)

Configuration	A_{dif} = differential gain	A_{com} = common-mode gain
	Medium $$= \frac{-RC}{r_e + RE}$$	Low $$= \frac{-RC}{r_e + RE + 2REE}$$
	Medium $$= -g_m RD$$	Low $$= \frac{-g_m RD}{1 + 2g_m RSS}$$
	Low to medium $$= \frac{-g_m RD}{1 + g_m RS}$$	Low $$= \frac{-g_m RD}{1 + g_m RS + 2g_m RSS}$$
	Low to medium $$= \frac{-\mu RP}{RP + r_p}$$	Low $$= \frac{-\mu RP}{RP + r_p + 2(\mu+1)RKK}$$
	Low to medium $$= \frac{-\mu RP}{RP + r_p + (\mu+1)RK}$$	Low $$= \frac{-\mu RP}{RP + r_p + (\mu+1)(RK+2RKK)}$$

(Note: All FETs = JFET, depletion and enhancement mode MOSFET)

From Eq. (3.133), the output impedance for the White cathode follower shown in Figure 3.51(a) is given as

$$Z_{out} = \frac{(r_p + RK)(r_p + RP)}{r_p + RP + (\mu+1)(r_p + RK) + \mu(\mu+1)RP}$$

From the data sheet for triode 12AU7, we find $r_p \approx 8k\Omega$ and $\mu \approx 18$. Substituting these for the above expression, we have

$$Z_{out} = \frac{(8x10^3\,\Omega + 500\Omega)(8\times10^3\,\Omega + 2.5\times10^3\,\Omega)}{8\times10^3\,\Omega + 2.5\times10^3\,\Omega + (19)(8\times10^3\,\Omega + 500\Omega) + (18)(19)2.5\times10^3\,\Omega} = 87\Omega$$

From Eq. (3.62), the output impedance for the cathode follower shown n Figure 3.51(b) is given as

$$Z_{out} = \frac{r_p RK}{r_p + (\mu+1)RK} = \frac{(8\times10^3\,\Omega)(17.5\times10^3\,\Omega)}{8\times10^3\,\Omega + (19)(17.5\times10^3\,\Omega)} = 411\Omega$$

It is obvious that, under the same dc bias condition, the White cathode follower produces a much lower output impedance than a simple cathode follower.

3.6 Summary table

The ac analysis is important for determining the ac properties of the amplifying device from BJT and FET to triode. Table 3.10 shows the differences between them in terms of voltage gain and input and output impedances. A clear understanding of these ac properties is helpful for us to choose the right amplifier configurations for forming a composite amplifier with the desired voltage gain and input and output impedances. These ac properties apply to audio amplifiers as well as linear amplifiers.

Distinguished engineer/inventor

Shortly after the war ended in 1945, Bell Labs formed a solid-state physics group, led by William Shockley [13] and chemist Stanley Morgan, which included John Bardeen, Walter Brattain, physicist Gerald Pearson, chemist Robert Gibney, electronics expert Hilbert Moore, and several technicians. Their assignment was to seek a solid-state alternative to fragile and bulky glass vacuum tube amplifiers.

The first point-contact bipolar transistor was successfully demonstrated on December 23, 1947, at Bell Laboratories in Murray Hill, New Jersey. The three individuals credited with the invention of the transistor were William Shockley, John Bardeen, and Walter Brattain. Shockley introduced the improved bipolar junction transistor (BJT) in 1948, which entered production in the early 1950s and led to the first widespread use of transistors. The three scientists were jointly awarded the 1956 Nobel Prize in Physics.

The principle of a field-effect transistor (FET) was proposed by Julius Edgar Lilienfeld in his 1925 patent. He was an Austro-Hungarian (later became American) physicist and electrical engineer. Lilienfeld was credited with the first patent on the FET. However, because of his failure to publish articles in learned journals and the lack of high-purity semiconductor materials at the time, his FET patent never received wide attention.

The metal–oxide–semiconductor field-effect transistor (MOSFET) was invented by Mohamed Atalla and Dawon Kahng at Bell Labs in 1959. As MOSFETs consume less power and can operate at higher switching speed compared with BJTs, they are the favorable device for use in a wide range of industrial and consumer electronic applications. CMOS uses a complementary symmetrical pair of P- and N-channel MOSFET. In order to improve speed, power consumption, and area reduction of digital integrated circuits, the feature size of CMOS has been continuously shrunk in the past few decades. It is believed that over 99% of integrated circuits (ICs), including digital, analog and mixed signals ICs, are fabricated using the CMOS technology. MOSFET has now become the most widely manufactured solid-state semiconductor device.

3.7   Exercises

Ex 3.1   Each of the composite amplifiers depicted in Figure 3.52 contains a common source amplifier and a common emitter amplifier. The load impedance is 10kΩ. Determine and compare the voltage gain and input and output impedances between Figure 3.52(a) and (b).

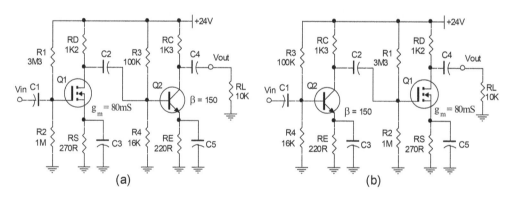

Figure 3.52 The same common cathode amplifier and common emitter amplifier are cascaded in (a) and (b) in two different orders

Ex 3.2   Figure 3.53 is a cascode amplifier formed by a BJT and JFET. Determine the voltage gain in terms of RC, RS, and transconductance of Q2: (i) bypassed cathode resistor and (b) unbypassed cathode resistor RS.

Ex 3.3   Figure 3.54 is a cascode differential amplifier formed by four BJTs. Are the differential gain and common-mode gains identical to those of the differential amplifier in Figure 3.38?

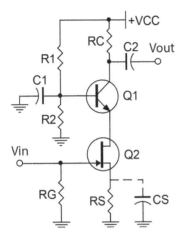

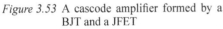

Figure 3.53 A cascode amplifier formed by a BJT and a JFET

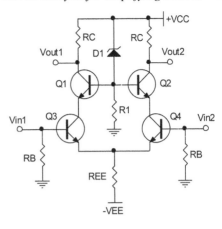

Figure 3.54 A cascode amplifier formed by a BJT and a JFET

Ex 3.4   Determine the differential voltage gain of the composite amplifier of Figure 3.55. Note that the input impedance of a BJT differential amplifier is the same as a common emitter amplifier, i.e., $Z_{in} = \beta r_e$. Assume $V_{BE} = 0.7V$.

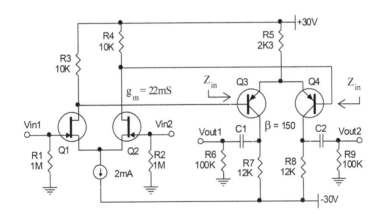

Figure 3.55 A composite amplifier is formed by cascading a JFET differential amplifier with a BJT differential amplifier

Ex 3.5   Figure 3.56 shows a three-stage composite amplifier. Assume $g_m = 33mS$ for Q1 and Q2, $\beta = 150$, $V_{BE(on)} = 0.65V$ for Q3 to Q7. Determine (a) the dc bias current for Q1 to Q7, (b) voltage gain Vout/Vin, (c) output impedance, (d) the dc potential at the output, (e) the maximum output signal voltage swing before clipping, and (f) whether the output signal is phase inverted or not.

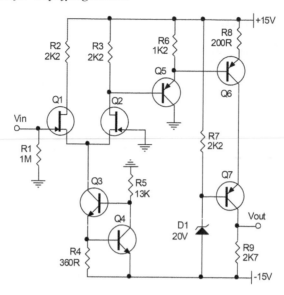

Figure 3.56 A three-stage composite amplifier for Ex 3.5

Ex 3.6   It is noted that a grounded grid amplifier has a very low input impedance. If the source has a high output impedance, it will affect the voltage gain and output impedance. Assuming that the source in Figure 3.57 has an output impedance RS, show that

$$A_v = \frac{(\mu+1)RP}{r_p + RP + (\mu+1)RS}$$

$$Z_{out} = [r_p + (\mu + 1)RS] \| RP$$

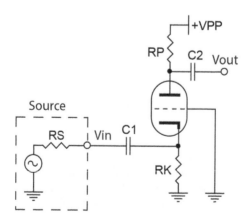

Figure 3.57 A grounded grid amplifier is connected to a source with impedance RS

Ex 3.7   Figure 3.58 shows a two-stage composite amplifier. Assume $\mu = 33$, $r_p = 2.7k\Omega$ for T1 and T2; $\mu = 15$, $r_p = 1k\Omega$ for T3; $V_{BE(on)} = 0.65V$ for Q1–Q3. Determine (a) the plate current for T1 to T3, (b) voltage gain Vout/Vin, (c) output impedance, (d) maximum output signal voltage swing before clipping.

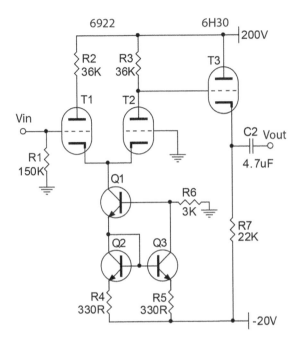

Figure 3.58 A two-stage composite amplifier for Ex 3.7

References

[1] Sedra, A. and Smith, K., *Microelectronic circuits*, Oxford, 7th edition, 2017.

[2] Boylestad, R. and Nashelsky, L., *Electronic devices and circuit theory*, Pearson, 10th edition, 2009.

[3] Gray, P. and Meyer, R., *Analysis and design of analog integrated circuits*, Wiley, 3rd edition, 1992.

[4] Millman, J., *Microelectronics, digital and analog circuits and systems*, McGraw-Hill, 1979.

[5] Millman, J. and Taub, H., *Pulse and digital circuits*, p. 100, McGraw-Hill, 1956.

[6] Artzt, M., "Survey of DC amplifiers," pp. 212–218, *Electronics*, August 1945.

[7] Keen, A.W., "Bootstrap circuit technique," *Electronic and Radio Engineer*, pp. 345–354, September 1958.

[8] Valley, G.E. and Wallman, H., "Vacuum tube amplifiers," pp. 456–64, reprinted by Audio Amateur Press, 2000.

[9] Blencowe, M., "The optimized SRPP amp (part 1)," pp. 13–19, Audio Xpress, 2010.

[10] Blencowe, M., "The optimized SRPP amp (part 2)," pp. 18–21, Audio Xpress, 2010.

[11] Blencowe, M., *Designing high-fidelity tube preamps*, Merlin Blencowe, 2016.

[12] Jones, M., *Valve amplifiers*, Newnes, 3rd edition, 2003.

[13] Wikipedia, "William Shockley," available at: https://en.wikipedia.org/wiki/William_Shockley, n.d.

[14] "RCA receiving tube Manual," reprinted by Antique Electronic supply.

[15] Rodenhuis, E., "Valves for audio frequency amplifiers," reprinted by Audio Amateur Press.

[16] "Essential characteristics," General Electric Company, 1973.

[17] Smith, W. and Buchanan, B., *Tube substitution handbook*, Prompt publications, 1992.

[18] Lurch, E.N., *Fundamentals of electronics*, Wiley, 2nd edition, 1971.

[19] Holt, C.A., *Electronic circuits digital and analog*, Wiley, 1978.

4   Operational amplifiers

4.1   Introduction

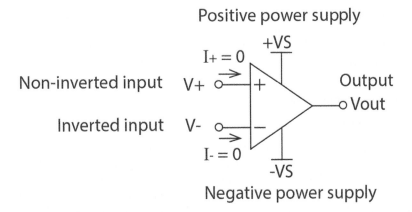

Figure 4.1 Diagram of an operational amplifier (op-amp)

Figure 4.1 shows a diagram of an operational amplifier, which is often called an op-amp. An op-amp has two inputs (inverted and non-inverted), one output, and dual power supply terminals. Some op-amps also offer two more terminals for connecting a trimmer to reduce dc offset voltage. Op-amp is a versatile electronic device that is found in almost every electronic application. A good understanding of op-amp and its applications is essential to perform the design work for audio and linear application.

An ideal op-amp has the following properties [1–3]:

* infinite differential voltage gain;
* infinite input impedance;
* zero output impedance;
* zero input offset voltage;
* zero input bias current.

DOI: 10.4324/9781003369462-4

Therefore, the differential inputs of an ideal op-amp must be zero, i.e., V+ – V- = 0, and the input current is also zero, i.e., I+ = 0, I- = 0. We can easily see why these must be true. If A_v denotes the differential voltage gain of the op-amp, the output is given as

Vout = (V+ – V-) A_v

From a mathematical point of view, if the differential voltage gain A_v approaches infinite, the output will also approach infinite. In order to confine the output to a finite value for practical use, the term (V+ – V-) must be zero. In other words, the non-inverted input and inverted input must be equal at all times, V+ = V-. On the other hand, since the input impedance is infinite, it can be easily understood that the input current must be zero, I+ = 0 and I- = 0. These properties lead to the so-called *golden rules* for op-amp application analysis:

- **non-inverted and inverted inputs are equal at all times, V+ = V-**
- **zero input current at all times, I+ = 0 and I- = 0**

Even though an ideal op-amp does not exist in real life, the golden rules help tremendously to simplify the op-amp application analysis. After all, the results that are determined by applying the golden rules are sufficiently accurate for almost all applications. In the following, some audio and linear applications using an op-amp are discussed.

4.2   Basic linear applications

Non-inverted amplifier

A *non-inverted amplifier* is shown in Figure 4.2(a). To simplify the circuit, the dual power supply terminals for +VS and –VS are omitted here. The voltage gain of the non-inverted amplifier is set by two resistors, R1 and R2. To determine the voltage gain, let us examine Figure 4.2(b). Since there is no input current flowing into the inverted input of the op-amp, the voltage at the inverted input is determined by the divider network formed by the two resistors. Therefore, we have

$$V_- = \frac{R1}{R1+R2}\text{Vout} \tag{4.1}$$

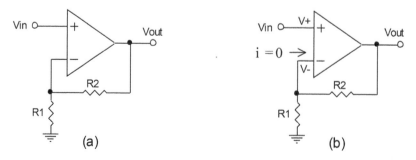

(a)   (b)

Figure 4.2 (a) A non-inverted amplifier containing two resistors and one op-amp; (b) showing the input voltages and current in a non-inverted amplifier

The non-inverted input is simply equal to the input,

$$V+ = Vin \qquad (4.2)$$

as stated by the golden rules, we must have $V+ = V-$. Substituting Eq. (4.2) for Eq. (4.1) yields

$$Vin = \frac{R1}{R1+R2} Vout$$

Therefore, the voltage gain is given by

$$A_v \equiv \frac{Vout}{Vin} = \frac{R1+R2}{R1} = 1 + \frac{R2}{R1} \qquad (4.3)$$

Since the above expression is a positive quantity, it is a non-inverted amplifier. For example, if $R1 = 1k\Omega$ and $R2 = 10k\Omega$, the voltage gain is equal to $1 + 10/1 = 11$. The input impedance is simply equal to the input impedance of the op-amp, which is very high, in the order of $1M\Omega$ or higher. Resistor R2 works as a feedback resistor that encloses the output and input. Negative feedback reduces output impedance. As a result, the output impedance of the non-inverted amplifier depicted in Figure 4.2 is in the order of $m\Omega$. Details about feedback and its effect on the input and output impedance of an amplifier is discussed in Chapter 5.

Inverted amplifier

An *inverted amplifier* is shown in Figure 4.3(a). Again, the voltage gain is simply set by two resistors, R1 and R2. To determine the voltage gain, let us examine Figure 4.3(b). The current i_1 and i_2 are first determined, and they are given as follows.

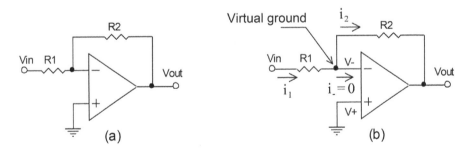

Figure 4.3 (a) An inverted amplifier containing two resistors and one op-amp; (b) showing the input voltages and current in an inverted amplifier

$$i_1 = \frac{Vin - V_-}{R1} \qquad (4.4)$$

$$i_2 = \frac{V_- - Vout}{R2} \qquad (4.5)$$

Since there is no current flowing into the op-amp, $i_- = 0$, we must have

$$i_1 = i_2$$

On the other hand, the inputs to the op-amp are equal, V- = V+ = 0. By equating Eq. (4.4) and Eq. (4.5) and putting V- = 0, we have

$$\frac{Vin}{R1} = \frac{-Vout}{R2}$$

Therefore, the voltage gain is given as

$$A_v \equiv \frac{Vout}{Vin} = -\frac{R2}{R1} \tag{4.6}$$

Since the above expression has a negative sign, it is an inverted amplifier. In other words, output is 180 degrees out-of-phase with respect to the input signal. For example, if R1 = 1kΩ and R2 = 10kΩ, the voltage gain is equal to −10/1 = −10. In contrast, as V+ is connected to ground, and V- = V+, V- is said to be a *virtual ground*. Thus, the input impedance is simply equal to the resistor R1. In comparison with the non-inverted amplifier, the input impedance of the inverted amplifier is lower. Again, resistor R2 works as a feedback resistor that encloses the output and input. The output impedance of the inverted amplifier is very small, and is similar to a non-inverted amplifier.

Voltage buffer

In applications where we want to improve the output driving capability while maintaining same voltage gain, a *voltage buffer* is often used. A voltage buffer is required to have high input impedance, low output impedance, wide bandwidth and unity voltage gain. Figure 4.4(a) shows a voltage buffer using a single op-amp. No resistor is needed in this configuration. Feedback is 100% taken from the output. Therefore, it is clear that V- = Vout and V+ = Vin. Since V+ = V-, we have Vout = Vin, as expected for a voltage buffer. As the input Vin is directly fed to the non-inverted input V+, the input impedance of the voltage buffer is very high. It is equal to the input impedance of the op-amp.

Figure 4.4(b) shows an inverted amplifier giving a voltage gain of −1. It may be considered as an inverted voltage buffer. However, the input impedance is equal to resistor R1, which is

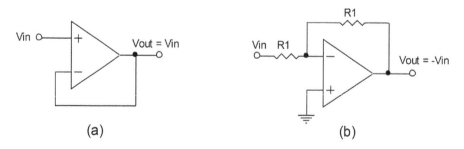

(a)   (b)

Figure 4.4 (a) A voltage buffer with high input impedance; (b) an inverted amplifier with unity gain

generally much lower than the voltage buffer depicted in Figure 4.4(a). If R1 is low, it will cause a loading effect to the source. If R1 is very high, it generates a considerable amount of thermal noise (known as Johnson noise). Since the output is inverted, and the input impedance is usually low to avoid thermal noise, the inverted amplifier of Figure 4.4(b) is not a popular choice for a voltage buffer application.

Integrator

A simple RC *integrator* is shown in Figure 4.5(a). The current flowing through the resistor R is given as

$$I_R = \frac{Vin - V_-}{R} \tag{4.7}$$

In the time domain, the current flow through the capacitor is given as

$$I_C = C\frac{d(V_- - Vout)}{dt} \tag{4.8}$$

Since there is no current flowing into the inverted input V- of the op-amp, we must have $I_R = I_C$. Note also that V- = V+ = 0. By equating Eqs. (4.7) and (4.8), we have

$$\frac{Vin}{R} = -C\frac{d(Vout)}{dt}$$

Rearranging the above expression yields

$$d(Vout) = -\frac{Vin}{RC}dt \tag{4.9}$$

Integrating both sides yields

$$Vout(t) = -\frac{1}{RC}\int_0^t Vin(t)dt + Vout(0) \tag{4.10}$$

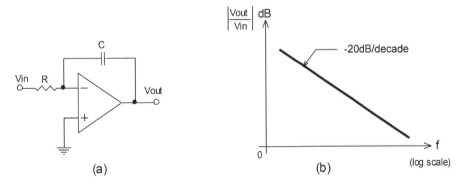

Figure 4.5 (a) A simple RC integrator, (b) the frequency response of the integrator

where Vout(0) is the value of the output at t = 0, which depends on the amount of charge initially stored in the capacitor. Eq. (4.10) indicates the output is proportional to the integration of Vin with a constant $-1/RC$.

In order to see the frequency response of the integrator, we express the integrator in the frequency domain in terms of complex variable s. The impedance of the capacitor is expressed as $Z_C = 1/sC$. As there is no current flowing into the inverted input V-, we have

$$\frac{Vin}{R} = \frac{-Vout}{Z_C}$$

(See Appendix B for an overview of complex variables and transfer functions.)
Substituting $Z_C = 1/sC$ and rearranging yields

$$\frac{Vout(s)}{Vin(s)} = \frac{-1}{sRC} \tag{4.11}$$

Putting $s = j\omega = j2\pi f$, where f = frequency, Eq. (4.11) can be rewritten as

$$\frac{Vout(j\omega)}{Vin(j\omega)} = \frac{-1}{j(2\pi f)RC} \tag{4.12}$$

Thus, the amplitude of the integrator transfer function is given as

$$\left|\frac{Vout(j\omega)}{Vin(j\omega)}\right| = \frac{1}{(2\pi f)RC} \tag{4.13}$$

It is clear from Eq. (4.13) that the amplitude of the integrator is inversely proportional to the frequency. The frequency response of the integrator is shown in Figure 4.5(b). The amplitude of the integrator rolls off at a rate of -20dB/decade. Because of the roll-off characteristic, which

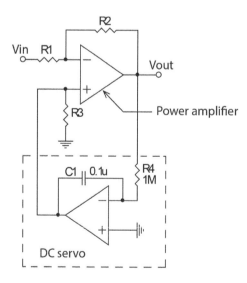

Figure 4.6 An integrator working as a dc servo to eliminate output dc offset voltage in a power amplifier

behaves like a low pass filter, the integrator is often used in a dc servo circuit to eliminate output dc offset voltage in a power amplifier, as shown in Figure 4.6. More details of dc servo circuits are discussed later in this chapter.

Summing amplifier

Figure 4.7 shows a *summing amplifier* with three inputs: Vin1, Vin2 and Vin3. Generally speaking, there is no restriction to the number of inputs. For example, the number of inputs for an audio octave equalizer, which is discussed later in this chapter, can be 10 or even more. Basically, the output of the summing amplifier is an algebraic sum of all inputs. Each input may carry a different weight that is determined by the ratio of the feedback resistor R4 to the input resistor R1 to R3. To see how this works, let us consider the input current from each input. We first notice that V- = V+ = 0, therefore, we have

$$i_1 = \frac{Vin1}{R1}, i_2 = \frac{Vin2}{R2}, i_3 = \frac{Vin3}{R3}$$

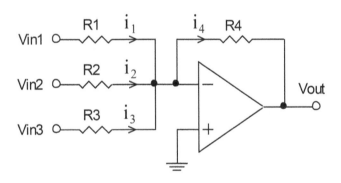

Figure 4.7 A summing amplifier with three inputs

As there is no current flowing into the op-amp, we must have

$$i_4 = i_1 + i_2 + i_3$$

The summing amplifier's output is given as

$$Vout = (-i_4)R4 = -R4\left[\frac{Vin1}{R1} + \frac{Vin2}{R2} + \frac{Vin3}{R3}\right] \tag{4.14}$$

The weights for the inputs Vin1, Vin2 and Vin3 are given by R4/R1, R4/R2 and R4/R3, respectively. If the resistors are chosen so that R1 = R2 = R3 = R4, the output is simplified to

$$Vout = -[Vin1 + Vin2 + Vin3] \tag{4.15}$$

The summing amplifier forms the basic building block for implementing a multi-band graphic equalizer and a multi-channel mixer. Note that a summing amplifier is an inverted amplifier.

Balanced input–unbalanced output amplifier

Figure 4.8(a) shows a balanced input–unbalanced output amplifier. It is a very popular amplifier that is used to deal with a balanced input signal. It is often called a *difference amplifier*. The voltage gain is determined by considering the voltage at the inverted and non-inverted input on the op-amp. They are given as:

$$V_+ = \frac{R2}{R1 + R2} Vin2$$

$$V_- = \frac{R2}{R1 + R2} Vin1 + \frac{R1}{R1 + R2} Vout \quad \text{(by the method of superposition)}$$

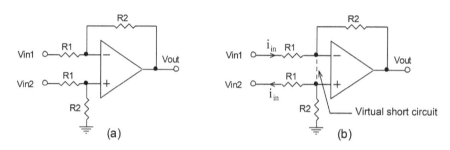

Figure 4.8 (a) A balanced input–unbalanced output amplifier, (b) a virtual short circuit between the inverted and non-inverted input for determining the balanced input impedance

Since V+ = V-, by equating the above two equations we have

$$\frac{R2}{R1 + R2} Vin2 = \frac{R2}{R1 + R2} Vin1 + \frac{R1}{R1 + R2} Vout$$

Thus, rearranging the above equation leads to the required voltage gain

$$A_v \equiv \frac{Vout}{Vin2 - Vin1} = \frac{R2}{R1} \tag{4.16}$$

The differential voltage gain is easily set by two resistors, R1 and R2. To determine the input impedance, we assume that there is an input current i_{in} flowing into and out of the differential inputs, as shown in Figure 4.8(b). In the differential input loop, assuming there is no current flowing into the op-amp, we have

$$Vin2 - Vin1 = i_{in} R1 + 0 + i_{in} R1$$

The V+ and V- inputs are considered as virtual ground. Thus, the (differential) input impedance is given as

$$Z_{in} \equiv \frac{Vin2 - Vin1}{i_{in}} = 2R1 \tag{4.17}$$

The input impedance is limited by the resistor used for R1. A high resistor R1 is not desirable, as it produces high thermal noise. A low resistor R1, on the other hand, may create a loading effect to the signal source. Therefore, if very high input impedance is of prime importance, the difference amplifier may not be the ideal solution. Fortunately, this problem can be solved by the so-called *instrumentation amplifier* described below.

Balanced input–balanced output amplifier

Figure 4.9(a) shows a balanced input–balanced output amplifier. This is also known as a *fully balanced* amplifier, which amplifies balanced input signals and produces balanced outputs. To determine the differential voltage gain, let us consider the output loop in Figure 4.9(b). We have

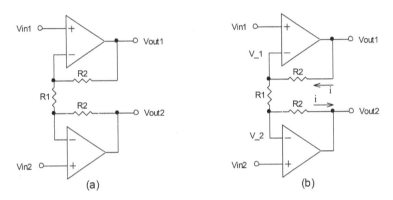

Figure 4.9 (a) A balanced input–balanced output amplifier, (b) showing current i and voltages at the inverted inputs

$$i = \frac{\text{Vout1} - \text{Vout2}}{\text{R1} + 2\text{R2}} \tag{4.18}$$

and also

$$i = \frac{V_{-1} - V_{-2}}{\text{R1}} \tag{4.19}$$

As we must have V-1 = Vin1 and V-2 = Vin2, by equating Eq. (4.18) and (4.19) we have

$$\frac{\text{Vout1} - \text{Vout2}}{\text{R1} + 2\text{R2}} = \frac{\text{Vin1} - \text{Vin2}}{\text{R1}}$$

Thus, the differential voltage gain is given as

$$A_v \equiv \frac{\text{Vout2} - \text{Vout1}}{\text{Vin2} - \text{Vin1}} = \frac{\text{Vout1} - \text{Vout2}}{\text{Vin1} - \text{Vin2}} = 1 + \frac{2\text{R2}}{\text{R1}} \tag{4.20}$$

The voltage gain is again easily set by two resistors, R1 and R2. The input impedance of the fully balanced amplifier is equal to the impedance of the op-amp. Therefore, the input impedance is very high.

Instrumentation amplifier

Figure 4.10 illustrates the well-known instrumentation amplifier. It is formed by cascading the fully balanced amplifier shown in Figure 4.9 to the difference amplifier depicted in Figure 4.8. Therefore, the voltage gain of the instrumentation amplifier is a product of the two amplifiers:

$$A_v \equiv \frac{Vout}{Vin2 - Vin1} = \left(1 + \frac{2R2}{R1}\right)\left(\frac{R4}{R3}\right) \tag{4.21}$$

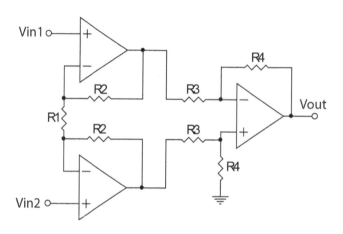

Figure 4.10 An instrumentation amplifier formed by a fully balanced amplifier and a difference amplifier

Even though both the difference amplifier in Figure 4.8 and the instrumentation amplifier in Figure 4.10 handle balanced input and produce unbalanced output, the instrumentation amplifier has several advantages over the difference amplifier. The advantages include: (i) high input impedance, (ii) high voltage gain and (iii) high common-mode rejection ratio. These make an instrumentation amplifier a better choice to deal with a balanced input signal.

Unbalanced input–balanced output amplifier

Figure 4.11 shows an unbalanced input–balanced output amplifier [4]. It comprises three op-amps (U1, U2 and U3). The first op-amp (U1) works as a buffer amplifier. Op-amps U2 and U3 are cross-coupled to generate the balanced output. It is demonstrated in the following that the voltage gain is determined by R2/R1. This amplifier is often used as a line driver when the amplifier is set to unity gain by choosing R1 = R2. To determine the voltage gain, let us first examine the inverted and non-inverted inputs of op-amp U2:

$$V_{-2} = \frac{R2}{R1 + R2} Vin + \frac{R1}{R1 + R2} Vo1 \text{ (the small } 50\Omega \text{ resistor is ignored)}$$

$$V_{+2} = \frac{R1}{R1 + R2} Vo2$$

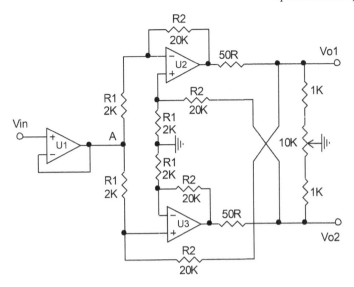

Figure 4.11 An unbalanced input–balanced output amplifier

Since we must have $V_{+2} = V_{-2}$, equating the above two expressions yields

$$\frac{R1}{R1 + R2} Vo2 = \frac{R2}{R1 + R2} Vin + \frac{R1}{R1 + R2} Vol$$

Rearranging the above expression, the voltage gain becomes

$$A_v \equiv \frac{Vo2 - Vol}{Vin} = \frac{R2}{R1} \tag{4.22}$$

Alternatively, voltage gain can be determined by looking at the inverted and non-inverted inputs of op-amp U3. It leads to the same voltage gain as stated in Eq. (4.22). From the given values for R1 and R2 in Figure 4.11, the voltage gain is 10. Note that there is a 50Ω resistor added to the output of the op-amp that improves the stability of the amplifier when driving capacitive load. In addition, there is a 10kΩ variable resistor connected between the two outputs. It helps to produce a symmetrical output swing between Vo1 and Vo2.

Current-to-voltage amplifier

Figure 4.12(a) shows a current-to-voltage (I-to-V) amplifier. The input is now in the form of current while the output is in voltage. It is obvious that the output is given as

$$Vout = (-R)I_{in} \tag{4.23}$$

The gain, –R, is negative because of the way the current direction is specified in the circuit. The most popular audio application for an I-to-V amplifier is to amplify the current signal from a DAC, as shown in Figure 4.12(b). The purpose of the op-amp is to amplify the signal while buffering the source, DAC, with the op-amp's high input impedance.

Figure 4.12 (a) A current-to-voltage (I-to-V) amplifier, (b) amplifying the output from a digital-to-analog converter

Voltage-to-current amplifier

A voltage-to-current (V-to-I) amplifier is shown in Figure 4.13. It is also called a *Howard current generator* after the name of the inventor. It looks similar to the difference amplifier that contains four resistors. However, in a difference amplifier, resistor R4 is connected to ground while the V-to-I amplifier R4 is connected to the output of the op-amp. Therefore, the V-to-I amplifier employs both negative feedback, via resistor R2, and positive feedback, via resistor R4. To determine the relationship between the input voltage and the output current, we look at the currents at the junction of the non-inverted input:

$$I_o = \frac{Vin - Vout}{R3} + \frac{Vo - Vout}{R4} \tag{4.24}$$

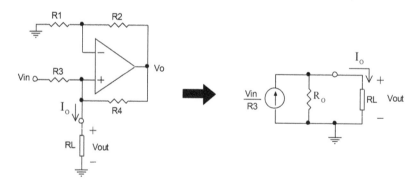

Figure 4.13 (a) A voltage-to-current amplifier, (b) equivalent circuit for the V-to-I amplifier for R4 = R2 and R3 = R1

On the other hand, V- = V+ = Vout, thus we have

$$Vo = \left(1 + \frac{R2}{R1}\right) Vout \tag{4.25}$$

Substituting Eq. (4.25) for (4.24) and rearranging yields

$$I_o = \frac{Vin}{R3} - \frac{Vout}{Ro} \tag{4.26}$$

where

$$Ro = \dfrac{\dfrac{R4}{R2}}{\dfrac{R2}{R1} - \dfrac{R4}{R3}} \qquad (4.27)$$

Eq. (4.26) can be viewed as the output of the current generator equivalent circuit in Figure 4.13(b). Thus, Ro is the equivalent output impedance of the current generator. If the resistors are chosen so that R4/R3 = R2/R1, the denominator of Eq. (4.27) becomes zero and, therefore, Ro becomes infinite. In practice, we know Ro is not infinite, but it is at least a very high output impedance for the V-to-I current amplifier. Thus, when the resistors satisfy the condition

$$\dfrac{R4}{R3} = \dfrac{R2}{R1} \qquad (4.28)$$

Then Eq. (4.26) is simplified to

$$I_o = \dfrac{Vin}{R3} \qquad (4.29)$$

which is independent of the load resistor RL. Thus, the circuit approaches a true V-to-I convertor (or amplifier), which is controlled by the input voltage Vin and resistor R3. The output current is generally limited by the maximum current handling capability of the op-amp, which is usually less than 30mA for a general purpose op-amp.

Inverted amplifier with Tee feedback network

An inverted amplifier with a Tee-shaped feedback network is shown in Figure 4.14. Note that it is an inverted amplifier, but the single feedback resistor is now replaced by three resistors in a Tee-shaped arrangement. Here we refer to it as a Tee feedback network inverted amplifier, or simply *TFN inverted amplifier*. It can be seen later that the TFN inverted amplifier permits

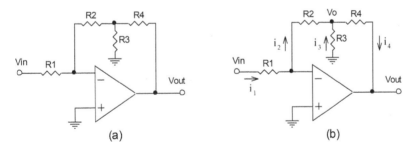

Figure 4.14 (a) An inverted amplifier with Tee feedback network, (b) showing the currents in the feedback network

smaller resistors to be used in the feedback network to achieve the same voltage gain of a conventional inverted amplifier having a large feedback resistor. In applications where high input impedance and high voltage gain is required, the TFN inverted amplifier is a better choice than the conventional inverted amplifier. We will see how this can be achieved.

Figure 4.14(b) shows the current flow in each of the resistors. Since the inverted and non-inverted inputs of the op-amp are equal, we must have V- = V+ = 0. Therefore, current i_1 is given as

$$i_1 = \dfrac{Vin}{R1} \qquad (4.30)$$

Since there is no current flowing into the op-amp, we must have

$$i_2 = i_1 \tag{4.31}$$

The voltage at the node connecting R2, R3 and R4 is given as

$$Vo = V_- - i_2 R2 = -\left(\frac{Vin}{R1}\right)R2 \tag{4.32}$$

This allows us to determine the current i_3 such that

$$i_3 = \frac{-Vo}{R3} = \frac{R2}{R1R3}Vin \tag{4.33}$$

At the node connecting R2, R3 and R4, the currents are related as

$$i_4 = i_2 + i_3 = \frac{Vin}{R1} + \frac{R2}{R1R3}Vin \tag{4.34}$$

Thus, the output is given as

$$Vout = Vo - i_4 R4 = -\left(\frac{Vin}{R1}\right)R2 - \left(\frac{Vin}{R1} + \frac{R2Vin}{R1R3}\right)R4$$

Finally, the voltage gain is found to be

$$A_v \equiv \frac{Vout}{Vin} = -\frac{R2}{R1}\left(1 + \frac{R4}{R2} + \frac{R4}{R3}\right) \tag{4.35}$$

Example 4.1

The voltage gain for the TFN inverted amplifier of Figure 4.15(a) is given by

$$A_v = -\frac{R2}{R1}\left(1 + \frac{R4}{R2} + \frac{R4}{R3}\right) = -\frac{10k\Omega}{10k\Omega}\left(1 + \frac{10k\Omega}{10k\Omega} + \frac{10k\Omega}{102\Omega}\right) = -100$$

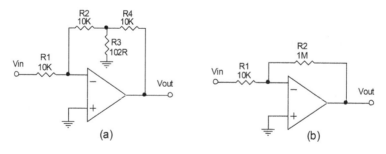

Figure 4.15 (a) A TFN inverted amplifier with a gain of -100, (b) a conventional inverted amplifier with a gain of -100

For the same voltage gain (-100) and input impedance (10kΩ), the conventional inverted amplifier of Figure 4.15(b) must have a 1MΩ feedback resistor that produces considerable amount of thermal noise.

4.3   Practical op-amp dc limitations

Input dc offset voltage

For an ideal op-amp, shorting the two inputs together should give zero voltage at the output. However, for a real-life op-amp, because of an imperfection in the integrated circuit fabrication, V+ and V- inputs will lead to a non-zero output even though they are shorted together. In order to force the output to zero, a suitable correcting dc voltage must be applied to the input. This voltage is called the input offset voltage, Vos, as shown in Figure 4.16(a). Figure 4.16(b) shows that the op-amp is now arranged in a non-inverted amplifier configuration with the input shorted to ground. The real-life op-amp now produces a dc output offset voltage given by

$$\text{Verr} = \left(1 + \frac{R2}{R1}\right)\text{Vos}$$

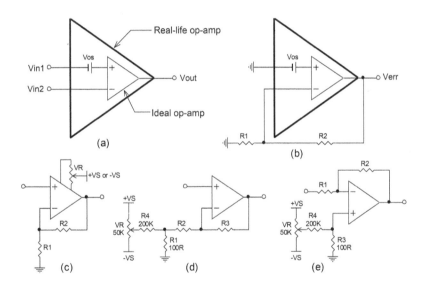

Figure 4.16 (a) A real-life op-amp has a dc offset voltage, Vos, (b) an op-amp in a non-inverted amplifier configuration, (c) an op-amp has two external terminals for nulling the dc offset voltage, (d) using an external circuit for nulling dc offset voltage for a non-inverted amplifier, (e) using an external circuit for nulling dc offset voltage for an inverted amplifier

In general, input offset Vos is ranging from 0.003mV, for a precision op-amp, to greater than 10mV, for a general purpose op-amp. If the gain of the amplifier is set to 10, the output dc offset voltage can be somewhere between 0.03mV and 100mV. A 0.03mV output dc offset is generally acceptable in most applications. However, it is not the case for 100mV. Fortunately, there are several methods to reduce the unwanted output dc offset voltage.

Some op-amps provide two external terminals for nulling the output dc offset as shown in Figure 4.16(c). By connecting a variable resistor to the two terminals and to a power supply, +VS or –VS, the output dc offset voltage can be significantly reduced to an acceptable level.

However, if the op-amp does not provide two terminals for trimming down the output dc offset, we need to add a few more external components to the amplifier circuit. Figure 4.16(d) shows an external circuit added to a non-inverted amplifier. As the trimmer VR is connected between the dual power supply ±VS, by turning the VR the output dc offset can be reduced to nearly zero. Since R2 is in series with R1 before connecting to ground, the voltage gain for the non-inverted amplifier becomes

$$A_v = 1 + \frac{R3}{R1 + R2}$$

Similarly, an external circuit can be added to an inverted amplifier for reducing the output dc offset as shown in Figure 4.16(e). Note that the external circuit does not alter the voltage gain of the amplifier, which is still given by –R2/R1.

Input bias current

The assumption that no current flows into the input of an op-amp simplifies the analysis work, because there are fewer unknown currents in the analysis that we have to deal with. However, the inputs of a practical op-amp do sink (or source, depending on the op-amp type) a tiny current. This tiny current is called an input bias current: IB1 in the inverted input, and IB2 in the non-inverted input, as shown in Figure 4.17(a). The input bias current, after being amplified by the op-amp, may produce an unacceptable output dc offset for some applications. Therefore, it is important to understand where the input bias current originates from and what we can do to reduce it.

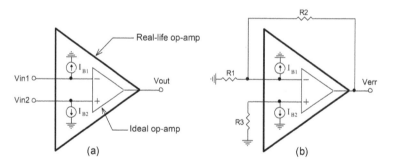

Figure 4.17 (a) Input bias currents I_{B1} and I_{B2} in a real-life op-amp, (b) op-amp is connected to resistive feedback network

Figure 4.17(b) shows an inverted amplifier with its input shorted to ground. Resistor R3 is introduced to the non-inverted input. The reason for adding a dummy resistor R3 will be clear soon. If the op-amp is ideal, there is no input bias current and, therefore, the output is zero. Let us see what the output is with the presence of input bias current I_{B1} and I_{B2}.

From Figure 4.17(b), the non-inverted input is given as

$$V+ = -I_{B2} R3 \qquad (4.36)$$

The currents at the inverted input node are given as

$$\frac{0 - V_-}{R1} + \frac{Verr - V_-}{R2} = I_{B1} \tag{4.37}$$

Since we have V+ = V-, eliminating them from the above two equations yields

$$Verr = \left(1 + \frac{R2}{R1}\right)[(R1 \| R2)I_{B1} - I_{B2}R3] \tag{4.38}$$

Verr is the output error voltage (or dc offset voltage) produced by the input bias current I_{B1} and I_{B2}. If it is an ideal op-amp, Verr is zero for input bias current $I_{B1} = I_{B2} = 0$. However, Eq. (4.38) suggests that we can choose the dummy resistor R3 in such a way that Verr approaches zero. Assume that $I_{B1} \approx I_{B2}$, the right-hand side of Eq. (4.38), is equal to zero when

$$R3 = R1 \| R2 \tag{4.39}$$

In other words, if resistor R3 has a value equal to R1 and R2 in parallel, the output offset vanishes. This is the reason that the dummy resistor R3 is introduced in the above analysis.

Figure 4.18 shows three examples. Figure 4.18(a and b) are an inverted and a non-inverted amplifier, respectively. To minimize the offset due to the input bias current, R3 is chosen to be R1‖R2 = 9.09kΩ. Figure 4.18(c) is a simple integrator with a capacitor as the feedback element. Since the impedance (or reactance) of a capacitor is equal to $Zc = 1/(j\omega C1)$, at dc ($\omega = 0$) the impedance of the capacitor becomes infinite: $|Zc| \to \infty$. As a result, we choose R2 = R1. When an integrator is used in a dc servo circuit, R1 is often a 1MΩ resistor. Therefore, R2 is also a 1MΩ resistor. More details about dc servo circuits are discussed later in this chapter.

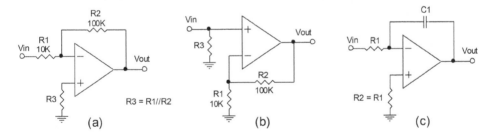

Figure 4.18 (a) An inverted amplifier with a dummy resistor, R3, (b) a non-inverted amplifier with a dummy resistor R3, (c) an integrator with a dummy resistor, R2

4.4 Audio applications

MM and MC phono stage amplifiers

The first audio application is a phono-stage amplifier. There are three areas that are of prime importance to a phono stage amplifier: high voltage gain, low noise, and accurate RIAA equalization. A gain of 30–40dB at 1kHz is generally required for a moving magnet (MM) phono stage amplifier. However, since a moving coil (MC) cartridge produces output 20–30dB lower than an MM cartridge, the voltage gain for an MC phono stage amplifier has to be around 60–70dB so as to produce same level of output.

Figure 4.19 shows the standard RIAA equalization curve from dc to 100kHz. The gain at dc is normalized to 0dB. This curve has three breakpoint frequencies at 50Hz, 500Hz, and 2122Hz. They correspond to three time constants: 3180µs, 318µs, and 75µs. The RIAA equalization curve starts at its maximum gain at dc. At 50Hz, the gain starts to roll off at –20dB/decade until the breakpoint frequency 500Hz is reached. The gain remains relatively flat until the next breakpoint at 2122Hz. Again, the gain rolls off at –20dB/decade and continues to the high frequency.

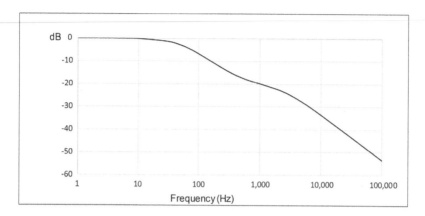

Figure 4.19 Standard RIAA equalization curve corresponding to three time constants: 3180µs, 318µs, and 75µs

Figure 4.20 shows a conventional MM phono amplifier employing op-amp and active RIAA equalization [5]. The reason that it is called active RIAA equalization is because the RIAA network is enclosed in the feedback loop. The gain for this phono amplifier is 35dB at 1kHz. The RIAA equalizing network is formed by components R3, R4, C3, and C4. The values shown in the circuit are not unique. Many combinations are possible, but they must satisfy the conditions for realizing the three breakpoint frequencies (f_1 = 50Hz, f_2 = 500Hz

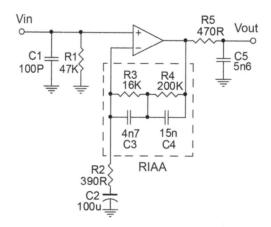

Figure 4.20 A conventional MM phono stage amplifier with active RIAA equalization network and 35dB gain

and $f_3 = 2122Hz$). For example, if the total resistance (R3 + R4 = 216kΩ) is considered rather high, which generates a considerable amount of thermal noise, smaller resistors can be used. However, C3 and C4 have to be increased so that the three breakpoint frequencies remain unchanged.

Figure 4.20 is a very popular MM phono stage amplifier design. It is good for setting a gain around 30–35dB at 1kHz. However, it will be a stretch for a 40dB gain as the low frequency gain is almost 60dB (including RIAA's –20dB attenuation). Imagine that an op-amp having a 120dB open loop gain has only 60dB left at the low frequency region to bring down the distortion. Thus, distortion will be compromised at low frequency for a 40dB gain.

Figure 4.21 shows an alternative approach to an MM phono stage amplifier [6]. It employs two op-amps and a passively equalized RIAA network. The first op-amp operates as a non-inverted amplifier with a voltage gain of 27.8dB, which is set by resistors R2 and R3. The second op-amp also operates as a non-inverted amplifier with a voltage gain of 32dB, which is set by resistors R6 and R7. A passive RIAA network is placed between the two amplifiers. Since the RIAA network has –20dB attenuation at 1kHz, the overall voltage gain for this MM phono stage amplifier is 40dB.

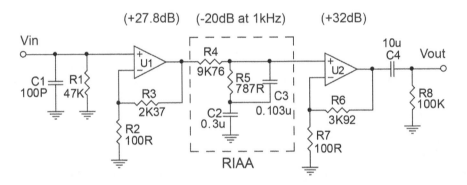

Figure 4.21 A passively equalized MM phono amplifier with 40dB gain

Figure 4.22 shows an MC phono stage amplifier with a passive RIAA equalization network and a 60dB gain. For any amplifier with a 60dB gain, noise is a primary concern. Low noise op-amp must be used. Additionally, it is well known that the signal-to-noise ratio (SNR) will be improved when the number of amplifiers is doubled. Therefore, a total of four op-amps (U1 × 4), which is in a non-inverted amplifier configuration, is paralleled in the first stage of the phono amplifier. The voltage gain of the first stage is set to 40dB by resistors R2 and R3.

The op-amp for the second stage, U2, is also a non-inverted amplifier with a gain of 40dB, which is set by resistors R6 and R7. U2 is also a low noise op-amp. However, it is not necessary to parallel an additional op-amp to the second stage. The noise level in the first stage is of primary importance because any noise from the first stage will be further amplified by the second stage producing greater noise. Thus, the effective way to reduce the noise is to reduce the noise produced from the first stage. In this example, four op-amps are used in parallel, improving SNR. Since the passive RIAA equalization network has a –20dB attenuation at 1kHz, the overall voltage gain of this MC phono stage amplifier is 60dB. More phono amplifier design examples are discussed in Chapter 10.

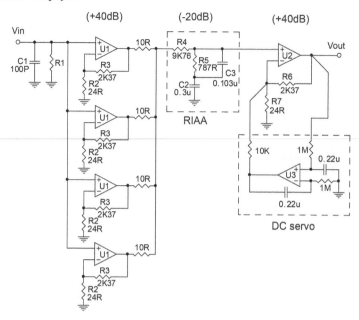

Figure 4.22 A passively equalized MC phono stage amplifier with a 60dB gain. R1 and C1 are a small resistor and capacitor that, when they are used correctly, extend the frequency response of the MC phono cartridge. A dc servo circuit is used to eliminate the output dc offset voltage

Tone control amplifier

An audio *tone control amplifier* allows the frequency response of the audio system to be altered to compensate for the loudspeaker or the listening room acoustic characteristics. Figure 4.23 shows an audio tone control amplifier that has a separate control for the bass and treble [7]. It is a variation of the classic design from Baxandall [8]. The input level of the source signal is first controlled by a volume control, VR1. The signal is then passed to a simple inverted amplifier with a gain of −2. Since the gain of the tone control circuit is also inverted, the output becomes non-inverted. The bass response can be boosted or cut by approximately ±20dB and the treble by approximately ±15dB. When the volume controls VR2 and VR3 are set to the middle, the frequency response becomes flat (Figure 4.24).

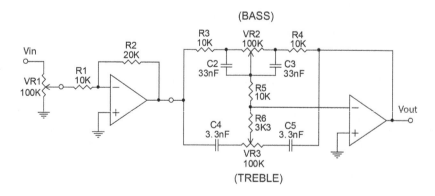

Figure 4.23 An audio tone control circuit that boosts and cuts the bass and treble

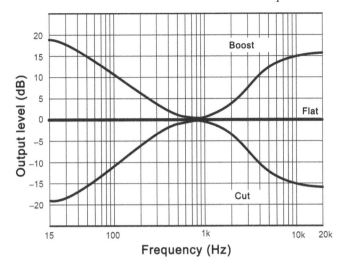

Figure 4.24 Frequency response of the audio tone control. (Courtesy of Texas Instruments)

Graphic equalizer

A *graphic equalizer* offers the user an array of tone controls within each intermediate frequency. When the frequency bands of the equalizer are separated by an octave, it is often called an *octave equalizer*. Figure 4.25(a) shows one section of a graphic equalizer [9,10]. The response of the individual section is adjusted by the variable resistor R2 and is shown in Figure 4.25(b). Note that each tone control performs as a narrow-band filter at a different center frequency f_o. If the component values are chosen such that

$$R3 \gg R1 \tag{4.40}$$
$$R3 = 10R2 \tag{4.41}$$
$$C1 = 10C2 \tag{4.42}$$

then the center frequency is given by

$$f_o = \frac{1}{20\pi R2 C2} \sqrt{2 + \frac{R2}{R1}} \tag{4.43}$$

and the voltage gain at the center frequency is

$$A_o = 1 + \frac{R2}{3R1} \tag{4.44}$$

A ten-section octave equalizer is shown in Figure 4.26. A switch at the output is provided so that the user can select either a "flat" or an "equalized" signal. In the input, a voltage buffer amplifier (U1) is used to buffer the source signal from the octave equalizer. Each section of the equalizer uses a set of different capacitors, C1 and C2, according to Table 4.1. The resistors are given by R1 = 10kΩ, R2 = 100kΩ, R3 = 1MΩ.

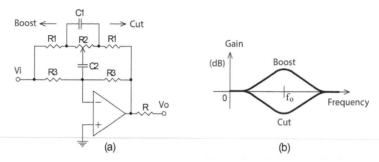

Figure 4.25 (a) One section of a graphic equalizer and (b) its frequency response

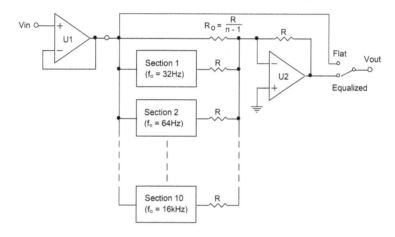

Figure 4.26 A ten-section octave equalizer. A switch allows the user to select a flat or an equalized signal

Table 4.1 The center frequencies and capacitances for C1 and C2 [10]

f_o (Hz)	C1	C2
32	0.18μF	0.018μF
64	0.1μF	0.01μF
125	47nF	4.7nF
250	22nF	2.2nF
500	12nF	1.2nF
1k	5.6nF	560pF
2k	2.7nF	270pF
4k	1.5nF	150pF
8k	680pF	68pF
16k	360pF	36pF

Output from each section is connected to the op-amp (U2) that works as a summing ampli-fier, via resistor R. Since there are ten sections, when summing up the total gain will become ten. In order to maintain the unity gain for the octave equalizer, a resistor Ro is added to the circuit. The resistor Ro is equal to R/(n–1), where n = number of sections. In this example, n = 10. If R = 10kΩ, Ro = 10kΩ/9 = 1.1kΩ. The function of the resistor Ro is as follows.

Each section of the equalizer produces an inverted output. The summing amplifier also produces an inverted output. As the signal is inverted twice, the output from the summing amplifier becomes non-inverted. If the input signal Vin is 1V, and each equalizer is set to flat, the summing amplifier's output becomes 10V. However, resistor Ro passes the input signal directly from the voltage buffer, which is non-inverted. When this signal passes through the summing amplifier, the output becomes inverted. For the same 1V input, the signal arriving from the resistor Ro produces –9V at the summing amplifier's output. Thus, the net output of the summing amplifier becomes 10V – 9V = 1V. Unity gain is retained.

4.5 DC servo applications

Many high-end audio amplifiers are implemented by discrete transistors. They include phono stage amplifiers, line stage amplifiers, and power amplifiers. Owing to the mismatch of a transistor's dc turn-on voltage (V_{BE} and V_{GS}), current gain and input bias current, an audio amplifier inevitably produces output dc offset voltage. For instance, if a line stage amplifier produces 0.05V output dc offset and it is further amplified by a dc power amplifier with a gain of 20, the power amplifier produces an output dc offset of 0.05×20 = 1V. This may not damage the voice coil of the loudspeaker immediately. However, the dc offset will push or pull the woofer away from its natural resting position. The result is a distorted and compressed sound. Thus, it is imperative to eliminate the output dc offset voltage from audio amplifier. In the following, examples of a dc servo circuit that can effectively eliminate the output dc offset are discussed.

DC servo for non-inverted audio amplifier

Figure 4.27(a) shows a non-inverted audio amplifier with a dc servo. Note that the audio amplifier's output is fed to the non-inverted input of the op-amp, V+. Capacitor C1 and R5, together with the op-amp, form a first order low-pass filter. The low-pass filter removes ac components from the audio amplifier's output. To remove ac components more effectively, capacitor C2 is added so that R4 and C2 also form another low-pass filter. Thus, the op-amp's output carries nearly no ac signal except an amplified dc offset voltage.

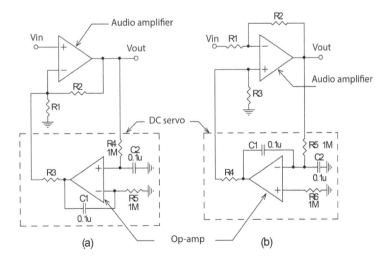

Figure 4.27 (a) A dc servo for a non-inverted audio amplifier, (b) a dc servo for an inverted audio amplifier

Assuming at the moment that the audio amplifier of Figure 4.27(a) has a positive output dc offset, Vos, when Vos passes through the non-inverted input of the op-amp, the op-amp produces a positive dc voltage k_1Vos, where k_1 is a positive multiplying constant. Therefore, k_1Vos, feeding to the audio amplifier's inverted input, is amplified by the audio amplifier and a negative output dc voltage is produced, say $-k_2$Vos, where k_2 is another positive multiplying constant. When this negative dc offset voltage is produced at the audio amplifier output, the net dc offset now becomes Vout = Vos – k_2Vos. The negative term ($-k_2$Vos), therefore, opposes the increase of output dc offset. In other words, the net output dc offset is reduced. The dc servo circuit continuous to produce a dc offset that counterbalances the audio amplifier's output dc offset until the inputs of the op-amp are equal, V+ = V-. Note that V- is connected to ground via resistor R5. Assuming that the op-amp's input bias current is negligible (by using JFET input op-amp), we have V- = 0. Therefore, the output of the audio amplifier will eventually rest at zero dc offset, Vout = V+ = V- = 0. Hence, the output dc offset is eliminated. Conversely, if the audio amplifier starts with a negative output dc offset voltage, the dc servo circuit will bring in a positive dc offset to counterbalance the dc offset. Again, the output dc offset will be eventually taken down to zero.

DC servo for inverted audio amplifier

Figure 4.27(b) shows an inverted audio amplifier employing a dc servo. Note that the output dc offset is passed through the inverted input of the op-amp. Then the op-amp's output is fed to the non-inverted input of the audio amplifier. As a consequence, an output dc offset is produced that counterbalances the audio amplifier's dc offset. Eventually, the audio amplifier's dc offset is eliminated by the dc servo. Note that resistor R6 is added to the op-amp's non-inverted input so as to minimize the dc offset due to the op-amp's input bias current by choosing R6 = R5. This has been discussed in section 4.3.

DC servo for a discrete JFET buffer amplifier

Figure 4.28(a) shows a buffer amplifier formed by a two N-channel JFET, 2SK246. Q1 works as a source follower with unity voltage gain. Q2 is a current source that works as an active load for Q1. When equal source resistors (R2 = R3) are used, the dc quiescent voltage at the drain of transistor Q2 should be ideally at ground level, i.e., $V_A = 0$. However, due to the mismatch of this pair of JFETs, we find $V_A = 40$mV, which is the unwanted output dc offset voltage.

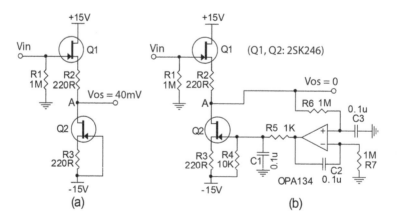

Figure 4.28 (a) A JFET buffer amplifier with output dc offset 40mV, (b) using a dc servo to eliminate the dc offset

As shown in Figure 4.28(b), a dc servo is now employed. Assume that at the moment there is a positive dc offset, Vos, present at the buffer amplifier's output. The dc offset Vos is first passed to the non-inverted input of the op-amp. At the op-amp's output, a positive dc voltage k_1Vos is produced, where k_1 is a positive multiplying factor. Note that a 10kΩ resistor, R4, is now connected between the gate of Q2 and a negative power supply of –15V. Q2 now works as a common source amplifier, which has a negative voltage gain. Therefore, at the drain of Q2, a negative dc voltage is produced, say –k_2Vos, where k_2 is a positive multiplying constant. The net dc offset at the output now becomes Vout = Vos – k_2Vos. The negative term (–k_2Vos), therefore, opposes the increase of the output dc offset. In other words, the net output dc offset is reduced. The dc servo circuit continues to provide a dc offset that counterbalances the audio amplifier's output dc offset until Vos = V+ = V- = 0. Hence, the output dc offset is eliminated.

A dc servo for DAC analog output

Figure 4.29(a) shows a typical digital-to-analog convertor (DAC) backend. The output from the DAC is passed to a I-to-V amplifier. Then a post filter is used to filter out any unwanted signals from the audio spectrum. Note that the DAC is often an analog and digital mixed signal

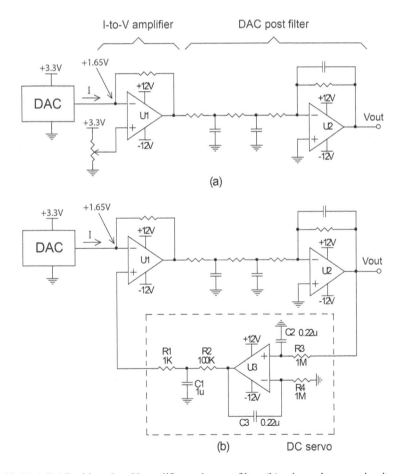

Figure 4.29 (a) A DAC with an I-to-V amplifier and a post filter, (b) using a dc servo circuit

integrated circuit that uses a single power supply. In this example, the power supply is +3.3V. However, a dual dc power supply (±12V) is used for the op-amps in the I-to-V amplifier and post filter. If the dc quiescent voltage at the DAC's output is biased to half of the power supply, +3.3V, then the inverted input of the op-amp U1 carries a dc +1.65V. Since the inverted and non-inverted inputs of an op-amp have to be maintained equal at all times, the non-inverted input (op-amp U1) has to be adjusted to +1.65V through a trimmer on a reference dc source. When the inverted and non-inverted inputs are maintained at the same potential, +1.65V, the output of the I-to-V amplifier rests at the ground level, 0V. As a consequence, the output of the post filter also rests at the ground level. However, the adjusted reference voltage +1.65V will drift over time due to temperature variation and component ageing. Eventually, the output from the post filter is deviated from the ground level and a dc offset voltage is, therefore, produced.

Figure 4.29(b) employs a dc servo to eliminate the need of a referenced dc source. Note that if a dc offset voltage is present at the post filter output, the dc offset voltage will be passed through op-amps U3, U1, and U2 before returning to the output again. Since the dc offset goes through the non-inverted input of op-amps U3 and U1, and the inverted input of U2, the dc offset voltage is inverted when reaching the output of the post filter. Therefore, the returning inverted dc offset voltage opposes the increase of the dc offset. Hence, the dc offset is eventually eliminated from the output of the post filter. When the output dc offset is eliminated, it can be seen that the inverted and non-inverted inputs of op-amp U1 are both equal to +1.65V.

4.6   DC regulated power supply applications

An op-amp is an important component in a dc regulated power supply. The op-amp compares the output (Vout), via a feedback resistor network, to a reference voltage (Vref), which is usually a low noise zener diode. Then the output voltage is stabilized by the op-amp to a specific level that is determined by the ratio of the feedback resistors. Figure 4.30 shows a *series type* dc-regulated power supply that contains an op-amp, a reference zener (Vref), a feedback resistor network (R2 and R3), and an output transistor Q1. Note that the transistor is in series between the unregulated dc power supply (Vin) and the regulated dc power supply (Vout). As a result, it is called a *series type* dc regulated power supply. The transistor Q1 is used to boost the output current.

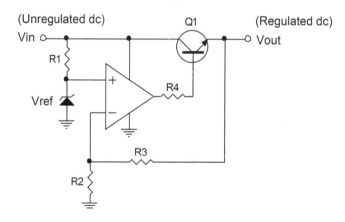

Figure 4.30 A series type dc regulated power supply

To determine the output, we examine the non-inverted (V+) and inverted (V-) input of the op-amp depicted in Figure 4.30:

$$V+ = Vref$$

and

$$V_- = \frac{R2}{R2 + R3} Vout$$

Since we must have V+ = V-, equating the above two equations and rearranging yields

$$Vout = \left(1 + \frac{R3}{R2}\right) Vref \tag{4.45}$$

Thus, for a given referenced voltage Vref, the output voltage is set by resistors R2 and R3. Note that resistor R1 sets up a proper dc biasing current for the reference zener. It is usually a few mA. Resistor R4 protects the op-amp by preventing the output transistor from drawing excessive current in case there is an output short-circuit. Figure 4.30 illustrates the basic structure of a typical series type dc regulated power supply. In a real application, more components are needed to ensure the circuit works properly and to reduce noise from the referenced zener and ripple from the unregulated dc supply.

A practical 15V dc regulated power supply is shown in Figure 4.31. Low noise voltage reference LM329 is used and it produces a 6.9V reference voltage. LM329's residual noise is filtered by a low-pass filter formed by R7 and C7. Resistors R5 and R6 form the feedback network that sets the desired output voltage. Eq. (4.45) is rewritten as

$$Vout = \left(1 + \frac{R5}{R6}\right) 6.9V \tag{4.46}$$

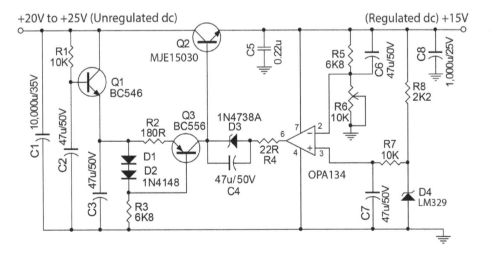

Figure 4.31 A 15V series type dc regulated power supply

R6 is a 10kΩ trimmer so that the output voltage can be precisely adjusted to 15V. Op-amp OPA134 is used in this example. However, many other op-amps can also be selected. The series transistor Q2 is a medium power MJE15030. Many medium power NPN transistors and N-channel MOSFETs are also suitable for use as the series transistor. Note that zener D3 is inserted between the base of transistor Q2 and the output of the op-amp. Zener D3 works as a level shifter that lifts the op-amp's output to a dc level that is approximately in the middle between the ground and 15V (i.e., 7.5V).

In the input side of the power supply, a large capacitor, C1, is used to reduce the ripple from the unregulated dc. Components R1, C2, and Q1 further reduce the ripple before supplying to transistor Q3. Transistor Q3 works as a current source that supplies a constant current to the op-amp. Without this current source, this dc regulated power supply will fail to start after switching on power. When the power is switched on, diodes D1 and D2 set up an emitter current for transistor Q3. The emitter current is equal to $[2V_F - V_{BE}(Q3)]/R2$, where V_F is the forward voltage of diode D1 and D2. If we take $V_F = 0.7V$ and $V_{BE}(Q3) = 0.7V$, the emitter current is equal to $[2 \times 0.7V - 0.7V]/180\Omega = 3.9mA$. After this emitter current has established, Q3 delivers a constant current 3.9mA to the zener diode D3 and the op-amp. This kick-starts the op-amp after switching on the power. Details about series type, as well as shunt type, dc regulated power supply designs are discussed in Chapter 14.

4.7   Composite op-amps

The maximum dc voltage rating for the high voltage op-amp LTC6090 is ±70V, while it is ±90V for OPA462. These two op-amps can deliver a signal with an output swing of over ±50V. However, the maximum dc power supply for most general purpose and audio op-amps is limited to ±18V. Inevitably, the maximum output swing is limited to less than ±18V. In addition, the maximum output current rating for most op-amps is limited to around 30mA except in special high current op-amps. For example, high current op-amp OPA548 can deliver up to 3A output current, and 10A for OPA541.

Fortunately, there are circuitries that can boost the output voltage swing and output current of an op-amp without the need to turn to high voltage or high current op-amps [11–12]. It takes an op-amp and several external components to form a *composite op-amp* that is capable of delivering an output voltage swing of over ±18V and an output current of over 30mA. A very low input noise composite op-amp can also be achieved. Here, we discuss a few examples.

Boosting output swing

Figure 4.32(a) shows a composite op-amp that boosts the output voltage swing. Resistors R4 to R7 and diodes D1 and D2 determine the dc bias current I_D, which is usually a few mA. If V_F is the forward voltage for the diode, D1 and R5 are chosen such that, after the $V_{BE}(Q1)$ voltage drop, +VS is equal to +15V. Similarly, –VS is equal to –15V. Therefore, the power supply to the op-amp is confined to ±15V, which is a safe power supply voltage for a 36V op-amp. When there is no input signal, ±VS stays at ±15V.

When an ac signal is present at the input, the output signal bootstraps the base of Q1 and Q2 so that the base varies according to the output waveform. However, since the power supply ±VS for the op-amp varies at the same time, the dc difference between them remains at 30V. If a rail-to-rail op-amp is used, and assuming $V_F \approx V_{BE}$, the maximum output is equal to

$$Vout(max) = \pm (VCC - V_F) \tag{4.47}$$

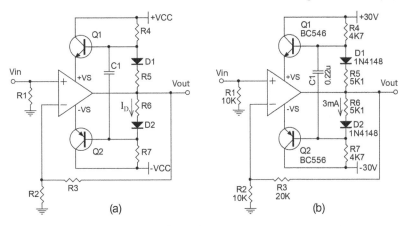

Figure 4.32 (a) A composite op-amp boosting output swing, (b) a composite op-amp producing a near to ±30V output swing for a rail-to-rail op-amp

Therefore, while the op-amp is operating safely at ±15V, the output swing can go up to ±(VCC – V_F) when a rail-to-rail op-amp is used.

An example is illustrated in Figure 4.32(b). The dual power supply is given as ±30V. R4 to R7 and diode D1 to D2 are chosen such that the bias current I_D is equal to 3mA. At no input signal, ±VS is equal to ±15V. R2 and R3 set the voltage gain of the amplifier, which is equal to (1 + R3/R2) = 3. Thus, if a rail-to-rail op-amp is used, the output swing can go up to ±(30V – 0.7V) = ±29.3V. On the other hand, if the op-amp is not rail-to-rail type, the output swing will be reduced to between ±26V and ±27V.

It should be cautioned that the input voltage at the non-inverted input of the op-amp must be within the device's common-mode input range. For example, the input common-mode range for OPA134 is ±13V and it is not a rail-to-rail output op-amp. The maximum output of the composite op-amp shown in Figure 4.32(b) is about 27V. The input signal at the non-inverted input of the op-amp is equal to (max. output/gain) = 27V/3 = 9V. It is still within the input common mode range for OPA134's 13V. However, if now the power supply for the composite op-amp is raised from ±30V to ±50V, the maximum output is close to 47V. Therefore, at the voltage gain of 3, the maximum input signal is equal to 47V/3 = 15.7V, which exceeds the input common mode range for OPA134. Either a different op-amp must be used or the composite op-amp's voltage gain has to be increased.

Boosting output current

A composite op-amp that can be used to boost the output current is shown in Figure 4.33(a). Note that a 33Ω resistor (R4 and R5) is connected in series between the dc power supply and the op-amp's power supply terminals. A voltage is created across the 33Ω resistor according to the amount of current supplied to the op-amp, I_S. If we ignore the small resistor R6 for the time being, the emitter-base of the transistor Q1 sees a voltage equal to ISR4. Assuming that V_{BE} is equal to 0.7V, when the op-amp supply current IS exceeds 0.7V/R4 = 0.7V/33Ω = 21mA, transistor Q1 is switched on, supplying additional output current. The small resistor R6 limits the emitter current for Q1 and compensates for the drop of VBE due to a temperature rise at the base-emitter junction to prevent thermal runaway. If the output current is less than 21mA, transistors Q1 and Q2 are turned off.

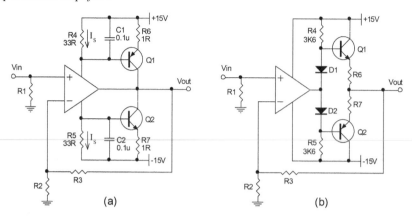

Figure 4.33 (a) A composite op-amp starts to boost output current when the output current exceeds 21mA, (b) a composite op-amp boosts the output current at all levels of output current

Figure 4.33(b) shows a different composite op-amp for boosting the output current. Diodes D1 and D2 create a V_{BE} spread to bias the push–pull output stage formed by Q1 and Q2. They are biased in class AB with a low biasing current to eliminate the so-called cross-over distortion. Since Q1 and Q2 are always turned on, they boost the output current at all output levels. If the output current exceeds 50mA, it is recommended to mount Q1 and Q2 in a heatsink.

If the composite op-amp depicted in Figures 4.32 and 4.33 are combined, we can form a composite op-amp that boosts the output voltage swing as well as the output current. Such a composite op-amp is shown in Figure 4.34.

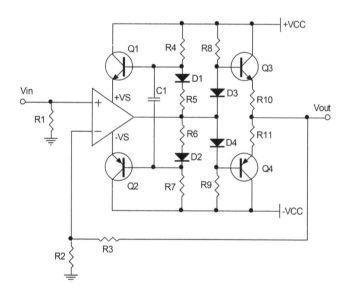

Figure 4.34 A composite op-amp boosting the output voltage swing and the output current

Low noise and gain boosting

A gain boosting composite amplifier is shown in Figure 4.35(a). It contains a differential amplifier directly coupling to an op-amp that forms a composite op-amp boosting voltage gain. The
gate of transistor Q1 is the non-inverted input, while the gate of transistor Q2 is the inverted
input. C1 and R4 form the frequency compensation network that stabilizes the composite amplifier at high frequency.

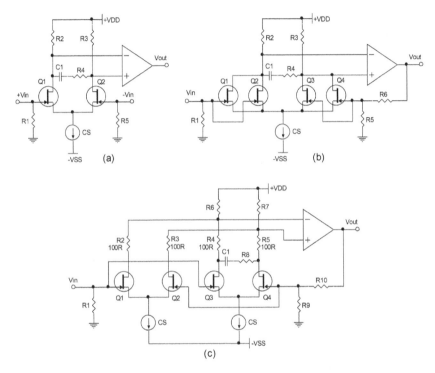

Figure 4.35 (a) A gain boosting composite amplifier, (b) a low noise and gain boosting composite amplifier, (c) a low noise and gain boosting composite amplifier with separate current sources

It is well known that when the number of amplifiers is doubled, the signal-to-noise ratio
(SNR) will be improved. Thus, Figure 4.35(b) offers an improvement in SNR over Figure
4.35(a). Figure 4.35(c) shows a composite amplifier with two differential amplifiers in parallel.
Since each differential amplifier has its own current source, the dc bias current on each transistor
is very close. It is it is expected that the composite amplifier shown in Figure 4.35(c) produces
slightly lower noise than the one in Figure 4.35(b). If more differential amplifiers are paralleled
to the input of the composite amplifier, the SNR can be further improved. Details of low noise
amplifier design is discussed in Chapters 9 and 10.

It is noted that the composite op-amp of Figure 4.35 relaxes the noise requirement for the op-
amp. In other words, by using low noise discrete transistors in the front, it allows most op-amps
to be used for forming a composite op-amp. As shown below, the noise as seen by the op-amp
can be reduced by a factor of 1/A, where A is the voltage gain of the input amplifier formed by
the discrete transistors.

The composite op-amps of Figure 4.35(b) and (c) contain a differential amplifier formed by discrete low noise JFETs, an op-amp. and a feedback network formed by two resistors. What a differential amplifier does is to amplify the difference between the two inputs. If the voltage gain is A1, the differential amplifier illustrated in Figure 4.36 contains a "mixer" or "adder" and a gain stage A1. Assuming the op-amp has a voltage gain of A2 and it sees noise N in the input, β is the feedback network formed by two resistors. From Figure 4.36, we have the following:

$$Vf = \beta Vout \tag{4.48}$$
$$Ve = Vin - Vf \tag{4.49}$$
$$Vout = A_1 A_2 Ve + A_2 N \tag{4.50}$$

Substituting Eqs. (4.48) and (4.49) for (4.50) and rearranging yields

$$Vout = \frac{A}{1 + A\beta}\left(Vin + \frac{N}{A_1}\right) = A_f\left(Vin + \frac{N}{A_1}\right) \tag{4.51}$$

where $A = A_1 A_2$ and

$$A_f = \frac{A}{1 + A\beta} \tag{4.52}$$

A_f is called the closed loop gain, i.e., the gain of the amplifier with feedback. More about feedback is discussed in Chapter 5. It is clear from Eq. (4.51) that the noise N is reduced by a factor of $1/A_1$. Thus, a composite op-amp reduces the noise in two ways. The first is improving the SNR by paralleling a number of discrete differential amplifiers in the input stage. Then the noise as seen by the op-amp is also reduced by means of negative feedback.

A list of representative op-amps is given in Table 4.2.

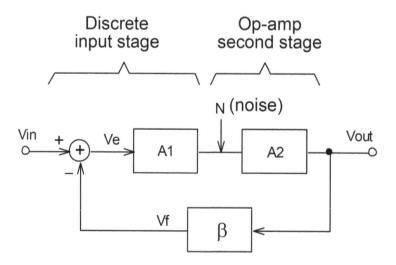

Figure 4.36 A composite op-amp with a low noise discrete input stage, an op-amp with noise and feedback

Table 4.2 A list of representative op-amps

Parts	Single	Dual	Quad	Total supply min (V)	Total supply max (V)	IQ/ch typ (mA)	VOS max (µV)	IB max (nA)	GBW typ (MHz)	Slew rate typ (V/µs)	En (nV√Hz) at 10Hz	Package
BJT input												
LM358A	•	•		-	32	1	3000	100	1	0.5	-	DIP-8, SO-8
LT6018	•			8	33	7.2	75	400	15	30	1.2	SO-8, DFN
LME49710	•	•		5	34	5	50	72	55	20	6.4	SO-8, TO-99
OPA189/2189	•	•		4.5	36	1.3	4	1	14	20	5.2	SO-8, SOT-23, VSSOP-8
OPAx192	•	•	•	4.5	36	1	50	5	10	20	30.0	SO-8, SOT-23, VSSOP-8, TSSOP-14
MAX4424x	•	•		2.7	36	0.5	5	1.25	5	3.8	9.0	SOT23–5, SO-8, SO-14, TSSOP-14
OPA1602/4		•	•	5	36	2.6	1000	200	35	20	5.2	SO-14, TSSOP-14
OPAx209	•	•	•	4.5	36	2	150	8	18	6.4	3.5	SOT-23, SO-8, VSSOP-8, TSSOP-14
OPAx228	•	•		5	36	3.7	100	10	33	11	3.4	DIP-8, SO-8
OPA1611/2	•	•		5	36	3.6	500	350	40	27	2.0	SO-8
AD797A	•			10	36	8.2	80	1500	110	20	1.8	DIP-8, SO-8
MAX9632	•			4.5	36	4	165	180	55	30	1.5	SO-8, TDFN
ADA4898-1/2	•	•		9	36	8.1	125	400	50	55	1.2	SO-8
NE5534A	•			10	40	4	5000	2000	10	13	7.0	DIP-8, SO-8
LT1126/7	•	•		8	44	2.6	100	30	65	11	3	DIP-8, SO-8, DIP-16, SO-16
LT1115	•			8	44	8.5	280	380	70	15	1	DIP-8, SO-16
ADA4522-x	•	•	•	4.5	55	0.9	5	0.5	2.7	1.7	6.0	SO-8, MSOP-8, SO-14, TSSOP-14
JFET input												
LF411A	•	•		-	44	1.8	2000	4	4	15	53	SO-8, DIP-8
OPAx134	•	•	•	5	36	4	3000	0.2	8	20	25	DIP-8, SO-8, DIP-14, SO-14
LME49880	•			10	34	14	10000	0.15	25	17	19	SO-8
ADA4627B	•			10	30	7	300	0.5	19	56	15	SO-8, LFCSP
OPA627	•			9	36	7	500	2	16	55	15	DIP-8, SO-8, TO-99
LT1792	•			9	40	4.2	800	0.8	4.3	3	8.2	DIP-8, SO-8
OPA827	•			8	36	4.8	150	5	22	28	7.3	SO-8, MSOP-8
AD743	•			9.6	36	8	1500	8.8	4.5	2.8	5.5	DIP-8, SO-16

(Continued)

Table 4.2 (Continued)

Parts	Single	Dual	Quad	Total supply min (V)	Total supply max (V)	IQ/ch typ (mA)	VOS max (µV)	IB max (nA)	GBW typ (MHz)	Slew rate typ (V/µs)	En (nV/√Hz) at 10Hz	Package
CMOS												
LMV791/2	•	•		2.5	5	0.95	1650	0.025	10	14	30	SOT-23, VSSOP-10
LTC6081/2		•	•	2.7	5.5	0.33	90	0.5	3.6	1	36	MSOP-8, SSOP-16, DFN-10, DFN-16
LMV751	•			2.7	5.5	0.5	1500	0.1	4.5	2.3	15	SOT-23
High voltage												
LTC2057HV	•			4.8	65	0.8	4	0.3	1.5	0.45	13	SO-8
ADA4700	•			10	100	1.7	2000	50	3.5	20	27	SO-8
OPA445	•			20	100	4.2	5000	20	2	10	35	SO-8, DIP-8
LTC6090	•			9.5	140	2.8	1600	0.8	12	21	100	SO-8, TSSOP-8
OPA462	•			12	180	4	4000	0.1	6.5	25	35 @1kHz	SO-8
Medium power												
OPA552	•			8	60	7	5000	0.1	12	24	50	DIP, DDPak
OPA547	•			8	60	10	5000	500	1	6	160	TO-220 (7)
OPA548	•			8	60	17	10000	500	1	10	200	TO-220 (11)
LM675	•			20	60	18	10000	2000	5.5	8	-	TO-220 (5)
OPA549	•			8	60	26	5000	500	0.9	9	120	TO-220 (11)
OPA541BM	•			20	80	20	1000	0.05	2	10	100	TO-3 (8), TO-220 (11)

Distinguished engineer/inventor

Robert Widlar [13] (November 30, 1937–February 27, 1991) was born in Cleveland, Ohio. He is considered the father of the analog monolithic integrated circuit. Widlar invented a number of basic building blocks for linear ICs, including the well-known Widlar current source and bandgap voltage reference. They are very popular building blocks that are still widely used today.

From 1964 to 1970, Widlar, together with David Talbert, created the first mass-produced operational amplifiers (μA702 and μA709), some of the earliest integrated voltage regulators (LM100 and LM105), and the first operational amplifier employing single capacitor frequency compensation (LM101).

The μA702 op-amp had limited common-mode range and weak output drive capabilities. The μA709, which was introduced in 1965, was an improved version over μA702. Widlar increased the μA709's voltage gain tenfold over that of the μA702, and improved output performance with a push–pull output stage. The μA709 became a technical and commercial success. It was Fairchild's revolutionary flagship product for several years.

Widlar worked for National Semiconductor in 1965 and retired in 1970. After being in retirement for several years, he returned to National Semiconductor (1974 to 1981) as a consultant. The notable products created by Widlar during this period were power amplifier LM12 and low voltage op-amp LM10. The LM10 is capable of operating at 1.1V supply voltage. There was no comparable low supply voltage op-amp in the industry for nearly ten years. The LM10 is still in production in the 21st century.

Widlar left National Semiconductor in 1981 and became one of the co-founders of Linear Technology. In 1984, he returned to National Semiconductor for the remainder of his life. Robert Widlar was inducted in the National Inventors Hall of Fame in 2009 and Electronic Design Hall of Fame in 2002.

4.8   Exercises

Ex 4.1   Figure 4.37 is a difference amplifier with high input impedance. Show that the voltage gain is given by Vout/(V1 – V2) = 1 + R1/R2.

Ex 4.2   There are two current-to-current amplifiers in Figure 4.38. Show that the current gain for Figure 4.38(a) is given by Iout/Iin = –R2/R1, while Iout/Iin = (1 + R2/R1) for Figure 4.38(b).

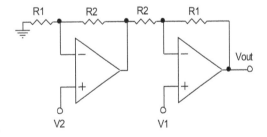

Figure 4.37 A difference amplifier

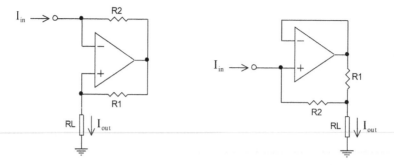

Figure 4.38 (a) An inverted current-to-current amplifier, (b) a non-inverted current-to-current amplifier

Ex 4.3　Figure 4.39 shows the output from a transducer bridge when, for example a stain gauge, is connected to a difference amplifier. Show that the output is given by

$$Vout = \frac{Vref}{4}\left(\frac{R2}{R1}\right)\left(\frac{\delta}{1+\delta/2}\right), \text{where } \delta = \frac{\Delta R}{R}$$

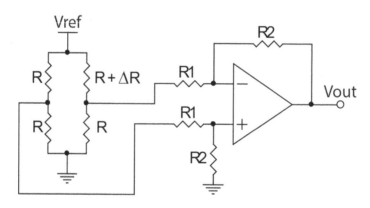

Figure 4.39 A transducer bridge is connected to a difference amplifier

Ex 4.4　Figure 4.40 is a voltage-to-current amplifier comprising two op-amps. Show that the output is given by

$$I_{out} = \frac{R2R4R6}{R1\{\ R3\left[R5R6+(R5+R6)RL\right]\ -\ R2R4RL\}}Vin \tag{4.48}$$

If the resistors are chosen such that $R1(R5 + R6) = R2R4$, the output current is simplified to

$$I_{out} = \frac{R2R4}{R1R3R5}Vin \tag{4.49}$$

If the resistors are further chosen such that R1 = R2 = R3 = R4, the output current is given by

$$I_{out} = \frac{Vin}{R5} \tag{4.50}$$

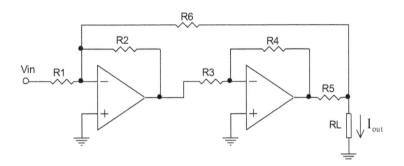

Figure 4.40 A voltage-to-current amplifier containing two op-amps

Ex 4.5 Figure 4.41 is a phono stage amplifier employing a low noise composite op-amp with active RIAA equalization. Design a dc servo to eliminate the output dc offset.

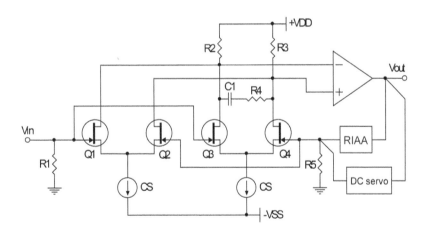

Figure 4.41 A phono stage amplifier employing a low noise composite op-amp

References

[1] Sedra, A. and Smith, K., *Microelectronic circuits*, Oxford, 7th edition, 2017.

[2] Gray, P. and Meyer, R., *Analysis and design of analog integrated circuits*, Wiley, 3rd edition, 1992.

[3] Franco, S., *Design with operational amplifiers and analog integrated circuits*, McGraw-Hill, 1988.

[4] Data sheet for OP176, Analog Devices.

[5] "High-performance audio applications of the LM833," application note AN-346, Texas Instruments.

[6] Jung, W., "Op amp applications," chapter 6, Analog Devices, 2002.

[7] Application note: AND8177/D, ON Semiconductor.

[8] Baxandall, P.J., "Negative-feedback tone control," pp. 402–405, Oct. 1952, Wireless World.

[9] Application note: Audio tone control using the TLC074 operational amplifier, Texas Instruments.

[10] Audio Handbook, National Semiconductor, 1976.

[11] King, G. and Watkins, T., "Bootstrapping your op-amp yields wide voltage swings," pp. 117–128, May 1999, EDN.

[12] Application note: AN87, Linear Technology.

[13] Widlar, R., Wikipedia.

[14] Jung, W., *Audio IC op amp application*, Howard Sams, 3rd edition, 1987.

[15] Jung W., *IC op amp cookbook*, Prentice Hall, 1986.

5 Feedback and frequency response

5.1 Introduction

Feedback has been mentioned in Chapter 4 where op-amp and its applications are discussed. Depending on the polarity of the feedback signal, feedback can be negative or positive. Positive feedback drives the circuit into oscillation and it should be avoided unless an oscillator circuit is your design goal. On the other hand, negative feedback results in reducing an amplifier's gain. The basic idea of feedback is to trade off gain for improvement to other properties, such as [1–5]:

- desensitizing the gain;
- modifying the input and output impedances;
- reducing nonlinearity, noise, and distortion;
- broadening the bandwidth.

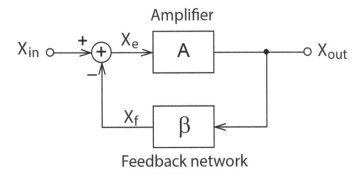

Figure 5.1 A signal-flow diagram for the general structure of a feedback amplifier. The quantity X can be either voltage or current

Figure 5.1 shows the general structure of a feedback amplifier. The quantity X can be either voltage or current. The basic amplifier has a unilateral nominal gain A, which is often called the *open loop gain*. The input to the amplifier X_e is related to the output X_{out} as

$$X_{out} = AX_e \tag{5.1}$$

DOI: 10.4324/9781003369462-5

The feedback network samples the output signal X_{out} and provides a feedback signal X_f that is related to the *feedback factor* β such that

$$X_f = \beta X_{out} \tag{5.2}$$

Let this not be confused with the same symbol (β) used for a bipolar transistor's current gain. To distinguish them, in this chapter we will use the symbol h_{fe} to represent a bipolar transistor's current gain.

When the feedback signal X_f is subtracted from the input signal X_{in}, the difference becomes the input to the basic amplifier. It is given by

$$X_e = X_{in} - X_f \tag{5.3}$$

where X_e is often called the *error signal*. Here, we have assumed that the feedback network does not have a loading effect on the input and output circuit of the amplifier. In other words, the feedback network does not have any impact on the amplifier's open loop gain A. However, in real life, the feedback network does create a loading effect on the amplifier's input and output circuit, depending on how the feedback network is configured. When we deal with real life amplifier circuits (to be discussed later), guidelines are given for taking into account the loading effect from the feedback network.

Assuming the feedback does not have a loading effect on the amplifier, the *closed loop gain* is defined by

$$A_f \equiv \frac{X_{out}}{X_{in}} \tag{5.4}$$

Substituting Eqs. (5.1) to (5.3) with (5.4) and rearranging yields

$$A_f = \frac{A}{1 + A\beta} \tag{5.5}$$

where the term $A\beta$ is called a *loop gain*. Note that open loop gain A and feedback factor β can be either negative or positive. However, the loop gain $A\beta$ must be a positive quantity if negative feedback is used. It can be easily seen that the feedback signal X_f has the same sign of the input signal X_{in}. As a result, Eq. (5.3) indicates that the error signal X_e is smaller than X_{in} for a negative feedback amplifier. Thus, the closed loop gain A_f must be smaller than the open loop gain. As indicated by Eq. (5.5), A_f is smaller than A by a factor of $(1 + A\beta)$, which is the *amount of feedback*. If loop gain $A\beta \gg 1$, then Eq. (5.5) can be simplified to

$$A_f \approx \frac{1}{\beta} \tag{5.6}$$

This is an interesting and useful result. When the loop gain is large, the gain of the amplifier is entirely determined by the feedback network. Since the feedback network is usually formed by passive components (i.e., resistors) that can be chosen to be as accurate as we want, as a result negative feedback brings accurate and stable gain to the amplifier. In other words, the closed loop gain (A_f) has very little dependence on the open loop gain (A) of the basic amplifier that is determined directly from the intrinsic properties of the amplifying devices, such as current gain, which may have wide manufacturing variations. This is especially true for audio amplifiers that contain many active and passive components.

- Open-loop gain $= A$

- Feedback factor $= \beta$

- Loop gain $=$

- Amount of feedback $= 1 + A\beta$

- Amount of feedback in dB $= 20\log(1 + A\beta)$

- Close-loop gain, $A_f = \dfrac{\text{open}-\text{loop gain}}{\text{amount of feedback}} = \dfrac{A}{1 + A\beta}$

- For large loop gain, $A\beta \gg 1$, $A_f \approx \dfrac{1}{\beta}$

Figure 5.2 Summary of the parameters and formulas for the feedback amplifier of Figure 5.1.

5.2 Effect of negative feedback on gain sensitivity

For most practical amplifiers, the amplifier open loop gain A has wide variation. The variation is dependent on temperature, the active device operation condition, transistor parameters such as h_{fe} for BJT and g_m for FET, and amplifying factor μ for vacuum tube triodes. As discussed in the previous section, the closed loop gain A_f reduces the variations in open loop gain A. This effect can be examined by differentiating Eq. (5.5),

$$\frac{dA_f}{dA} = \frac{(1 + A\beta) - A\beta}{(1 + A\beta)^2}$$

and this reduces to

$$\frac{dA_f}{dA} = \frac{1}{(1 + A\beta)^2}$$

If A is changed by δA, then A_f changes by δA_f such that

$$\delta A_f = \frac{\delta A}{(1 + A\beta)^2}$$

and dividing both sides of the above expression by A_f yields

$$\frac{\delta A_f}{A_f} = \frac{\delta A}{(1+A\beta)^2} \frac{1}{A_f} = \frac{\delta A}{(1+A\beta)^2} \frac{(1+A\beta)}{A}$$

The above expression is simplified to

$$\frac{\delta A_f}{A_f} = \frac{\dfrac{\delta A}{A}}{1+A\beta} \tag{5.7}$$

Therefore, the relative change in the amplifier gain with feedback is reduced by the factor of $(1+A\beta)$ compared to that without feedback. For example, an amplifier with an open loop gain $A = 1000$, and feedback factor $\beta = 0.1$ has a gain variation of 25% due to temperature, operational conditions, and current gain variation. The relative change in amplifier gain with feedback, using Eq. (5.7), becomes

$$\frac{\delta A_f}{A_f} = \frac{\dfrac{\delta A}{A}}{1+A\beta} = \frac{25\%}{1+1000\times0.1} = 0.247\%$$

Thus, this amplifier without feedback has a variation in open loop gain by 25%, but the variation in closed loop gain is reduced to only 0.247%. The improvement is 101 times. This is a trade-off in gain (reduced from 1000 to 100) for closed loop gain sensitivity reduction.

5.3   Effect of negative feedback on distortion

As seen from the analysis above, with the trade-off in gain, the gain sensitivity ($\delta A_f/A_f$) is reduced by a factor of $(1+A\beta)$ when feedback is applied. This may suggest that distortion

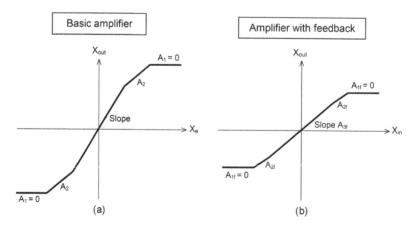

Figure 5.3 (a) The transfer function of X_{out}/X_e for the basic amplifier, (b) the transfer function of X_{out}/X_{in} for the amplifier with feedback

can also be reduced by the same factor. Assuming the slope of Figure 5.3(a) represents the input–output characteristics of the basic amplifier X_{out}/X_e, which is the gain of the basic amplifier, distortion is caused by the change in slope of the input–output transfer characteristics of the amplifier.

There are three different slopes in Figure 5.3(a) for the basic amplifier. Slope $A_1 = 0$ represents the amplifier in hard saturation and output is no longer dependent on the input. This is the region that we should avoid the amplifier operating in, as the output clips and creates very high distortion. Slopes A_2 and A_3 represent the non-linearity caused by the different stages of the amplifier. When feedback is now employed, the transfer function of the closed loop gain (X_{out}/X_{in}) is shown in Figure 5.3(b). As given by Eq. (5.5), the closed loop gain can be expressed in terms of the open loop gain as follows:

$$A_{1f} = \frac{A_1}{1 + A_1\beta} = 0, \text{ as } A_1 = 0 \qquad (5.8)$$

$$A_{2f} = \frac{A_2}{1 + A_2\beta} \qquad (5.9)$$

$$A_{3f} = \frac{A_3}{1 + A_3\beta} \qquad (5.10)$$

Eq. (5.8) suggests that feedback cannot improve the situation when the amplifier is in hard saturation. However, the slopes A_{2f} and A_{3f} suggest that the transfer characteristic of the feedback amplifier depicted in Figure 5.3(b) shows much improved linearity over the original basic amplifier of Figure 5.3(a). As a result, the feedback amplifier produces much lower distortion than the original basic amplifier. For example, a basic amplifier has an open loop gain of $A = 1,000$ and distortion $D = 2\%$. If feedback is applied, feedback factor $\beta = 0.049$, the closed loop gain becomes $A_f = A/(1 + A\beta) = 20$. The distortion of the feedback amplifier is reduced to $D_f = D/(1 + A\beta) = 0.04\%$. It is a great improvement in distortion reduction.

5.4 Effect of negative feedback on bandwidth

Consider a basic amplifier that has a high-frequency response dominated by a single pole. Assume that the frequency response starts to roll off at frequency f_H in –20dB/decade, as shown in Figure 5.4. The amplifier gain in complex frequency domain is given by

$$A(s) = \frac{A_o}{1 + s/\omega_H} \qquad (5.11)$$

where complex frequency $s = j\omega = j2\pi f$, $\omega_H = 2\pi f_H$, and A_o = amplifier mid-band gain.

According to Eq. (5.5), the closed loop gain of the amplifier becomes

$$A_f(s) = \frac{A(s)}{1 + A(s)\beta}$$

Substituting Eq. (5.11) for the above equation and simplifying yields

$$A_f(s) = \frac{A_o}{1 + A_o\beta + s/\omega_H} = \frac{\frac{A_o}{1 + A_o\beta}}{1 + s/\omega_H(1 + A_o\beta)} = \frac{\frac{A_o}{1 + A_o\beta}}{1 + s/\omega_{Hf}} \qquad (5.12)$$

$$\omega_{Hf} \equiv \omega_H(1 + A_o\beta) \qquad (5.13)$$

The frequency response of the feedback amplifier $A_f(s)$ starts to roll off at the frequency f_{Hf}, where $\omega_{Hf} = 2\pi f_{HL}$. In other words, f_H is increased to f_{Hf} by a factor of $(1 + A_o\beta)$.

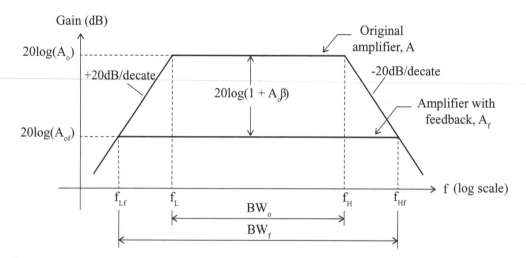

Figure 5.4 Frequency response of the original basic amplifier and the amplifier with feedback

Similarly, if the frequency response of the basic amplifier has a dominant low frequency pole at frequency ω_L, then the feedback amplifier will have a low frequency pole extended to ω_{Lf}:

$$\omega_{Lf} = \frac{\omega_L}{1 + A_o\beta} \tag{5.14}$$

The bandwidths of the original amplifier and the feedback amplifier are expressed as

$$BW_o = f_H - f_L \approx f_H \tag{5.15}$$

$$BW_f = f_{Hf} - f_{Lf} \approx f_{Hf} \tag{5.16}$$

as $f_H \gg f_L$ and $f_{Hf} \gg f_{Lf}$. From Eq. (5.13), we have

$$f_{Hf} = f_H(1 + A_o\beta)$$

Multiplying both sides of the above equation by A_o and rearranging yields

$$\frac{f_{Hf}}{1 + A_o\beta} A_o = f_H A_o$$

Substituting Eqs. (5.15) and (5.16) for the above equation and rearranging yields

$$(BW_f)(A_f) \approx (BW_o)(A_o) \tag{5.17}$$

As a result, Eq. (5.17) suggests that, while the gain of the feedback amplifier is reduced and bandwidth is increased, the *gain-bandwidth product* remains the same. For example, a basic amplifier has mid-band gain $A_o = 10,000$ and bandwidth $BW_o = 500Hz$. If feedback is applied and $A_f = 20$, the bandwidth of the feedback amplifier becomes

$$BW_f = \frac{(BW_o)(A_o)}{A_f} = \frac{(10,000)(500Hz)}{20} = 250kHz$$

Note that the frequency response of Figure 5.4 is drawn with ideal characteristics, where breaking points at the lower and upper corners are straight lines. In real life, the breaking points have rounded corners with continuous roll-off, as shown in Figure 5.5. At each corner, there is a drop of –3dB from the mid-band gain. Therefore, f_L is often called the lower –3dB frequency while f_H is the upper –3dB frequency. The bandwidth of an amplifier is equal to the difference between the two frequencies, $BW = f_H - f_L$, which is usually approximated to $BW \approx f_H$ for $f_H \gg f_L$. For example, if $f_H = 100kH$, $f_L = 10Hz$, $BW = 100kHz - 10Hz \approx 100kHz$.

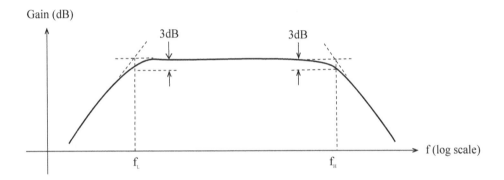

Figure 5.5 Frequency response showing a drop of –3dB at the breakpoint frequencies

5.5   Feedback configurations

Depending on the signal to be amplified (voltage or current) and the desired form of output (voltage or current), feedback amplifiers can be classified in four configurations. An amplifier from each configuration has a unique way of connecting the feedback network between the output and the input. A mixer or a junction of Figure 5.1 can be formed in two different ways, as shown in Figure 5.6. If the signal is in current form, a "series" configuration of Figure 5.6(b) is best to describe the current distribution. If the signal is in voltage form, a "shunt" configuration depicted in Figure 5.6(c) is best to describe the voltage distribution. It is expected that a series configuration increases the impedance Z, while a shunt configuration reduces the impedance Z. It may be also expected that the impedance Z is increased by a factor of $(1 + A\beta)$ or decreased by a factor of $1/(1 + A\beta)$, depending on the configuration.

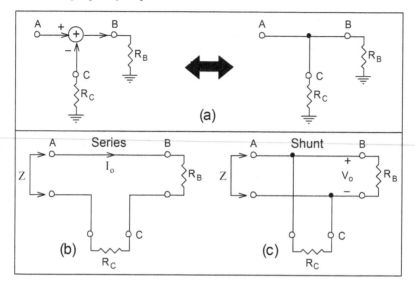

Figure 5.6 (a) A mixer or a junction formed by terminal A, B, and C, (b) a series configuration representing the mixer/junction, (c) a shunt configuration representing the mixer/junction

Similarly, the output junction of Figure 5.1 can also be formed by either a series or a shunt connection. Therefore, amplifiers can be classified in four configurations: (i) *series-shunt*, (ii) *shunt-shunt*, (iii) *series-series*, and (iv) *shunt-series*, as shown in Figure 5.7. Note that the name of the configuration is expressed in the form of XY, where X = input connection (series or shunt), Y = output connection (series or shunt). For example, series-shunt is referring to a series connection in input while shunt connection refers to output. Each configuration has different characteristics in terms of gain, input and output impedance.

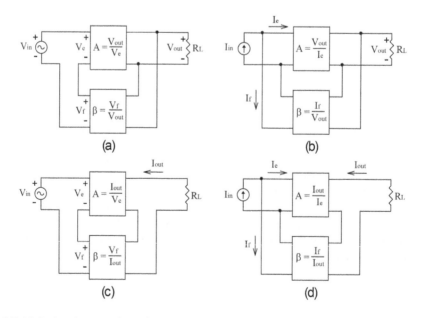

Figure 5.7 (a) Series-shunt configuration, (b) shunt-shunt configuration, (c) series-series configuration, (d) shunt-series configuration

5.6   Effect of negative feedback on input and output impedances

Series-shunt configuration (voltage amplifier)

Audio amplifiers such as the phono stage amplifier, line stage preamplifier, and power amplifier are voltage amplifiers. A voltage amplifier is basically a voltage-controlled voltage source. Thus, it is appropriate to use the Thévenin equivalent circuit to represent the voltage amplifier. As the output is in voltage, the feedback network samples the output voltage (V_{out}) and the feedback signal becomes V_f. Then the error signal, $V_e = V_{in} - V_f$, is passed to the basic amplifier. If we replace the signal symbol X with V, the signal flow diagram of Figure 5.1 can be represented by the ac equivalent circuit of Figure 5.8(a). Since the feedback network is in series connection with the input signal, while in shunt connection to the output, a voltage amplifier has the series-shunt feedback configuration.

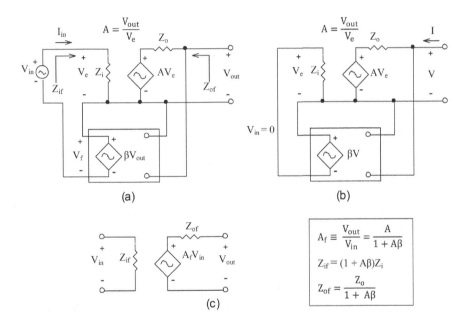

Figure 5.8 (a) An ac model for the series-shunt configuration, (b) input is shorted ($V_{in} = 0$) for determining the output impedance, (c) an ac equivalent circuit for the series-shunt feedback amplifier

For a good voltage amplifier, a high input impedance and low output impedance is needed. We will see that series-shunt feedback indeed increases the input impedance and reduces the output impedance of the voltage amplifier. Additionally, feedback improves the bandwidth and linearity of the amplifier, which has been already discussed in previous sections.

To determine the input impedance of the series-shunt amplifier in Figure 5.8(a), assuming that the feedback network does not have a loading effect on the input and output, we have the following

$$V_{out} = AV_e \qquad (5.18)$$

$$V_{in} = V_e + V_f = V_e + \beta V_{out} \qquad (5.19)$$

$$I_{in} = \frac{V_e}{Z_i} \qquad (5.20)$$

Substituting Eq. (5.18) with (5.19) and rearranging yields

$$V_{in} = (1 + A\beta) V_e \qquad (5.21)$$

By using Eqs. (5.20) and (5.21), the input impedance Z_{if} with feedback applied is given by

$$Z_{if} \equiv \frac{V_{in}}{I_{in}} = (1 + A\beta)Z_i \qquad (5.22)$$

Therefore, the input impedance with feedback is equal to Z_i multiplied by a factor of $(1 + A\beta)$.

To determine the output impedance, the input is rendered dead by shorting the input terminals so that $V_{in} = 0$, as shown in Figure 5.8(b). When a voltage V is applied to the output, we have

$$V_e + \beta V = 0 \qquad (5.23)$$

$$I = \frac{V - AV_e}{Z_o} \qquad (5.24)$$

Substituting Eq. (5.23) for (5.24) gives

$$I = \frac{V(1 + A\beta)}{Z_o}$$

The output impedance with feedback is given by

$$Z_{of} \equiv \frac{V}{I} = \frac{Z_o}{1 + A\beta} \qquad (5.25)$$

Thus, the output impedance with feedback is equal to Z_o multiplied by a factor of $1/(1 + A\beta)$. The increase in input impedance and decrease in output impedance makes the series-shunt feedback an excellent configuration for a voltage amplifier.

Example 5.1

Assume that the op-amp of Figure 5.9(a) has the following properties: $Z_i = 1M\Omega$, $Z_o = 10\Omega$, $A_o = 5 \times 10^5$, $f_H = 10Hz$ and $R1 = 1k\Omega$, $R2 = 9k\Omega$, $R3 = 47k\Omega$. First, the feedback factor is given by

$$\beta = \frac{R1}{R1 + R2} = \frac{1k\Omega}{1k\Omega + 9k\Omega} = 0.1$$

Assuming the voltage gain of the op-amp has a single pole expressed in the form

$$A(s) = \frac{A_o}{1 + s/\omega_H}$$

where $\omega_H = 2\pi f_H$. Using Eq. (5.12) and (5.13), the voltage gain with feedback applied is given by

$$A_f(s) = \frac{\dfrac{A_o}{1 + A_o\beta}}{1 + s/\omega_{Hf}}$$

where $\omega_{Hf} = (1 + A_o\beta) \omega_H$. Therefore, the mid-band voltage gain with feedback applied is given by

$$A_{of} = \frac{A_o}{1 + A_o\beta} = \frac{5 \times 10^5}{1 + 5 \times 10^5 \times 0.1} = \frac{500,000}{50,001} = 9.9998$$

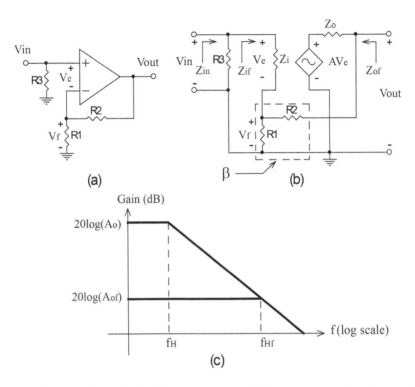

Figure 5.9 (a) An op-amp in series-shunt (non-inverted amplifier) configuration, (b) the ac model of the series-shunt feedback amplifier, (c) the frequency response of the op-amp and series-shunt feedback amplifier

As for large open loop gain A_o, the gain with feedback applied is approximately equal to $1/\beta = 1/0.1 = 10$. This is very close to the exact solution for A_{of}. We also have:

$$f_{Hf} = (1 + A_o\beta)f_H = (1 + 5 \times 10^5 \times 0.1) \times 10\text{Hz} = 500\text{kHz}$$

$$Z_{if} = (1 + A_o\beta)Z_i = (1 + 5 \times 10^5 \times 0.1) \times 1\text{M}\Omega = 50\text{G}\Omega$$

$$Z_{in} = R3 \| Z_{if} \approx R3 = 47\text{k}\Omega$$

$$Z_{of} = \frac{Z_o}{1 + A_o\beta} = \frac{10\Omega}{1 + 5 \times 10^5 \times 0.1} = 0.2\text{m}\Omega$$

Loading effect by feedback network

In the above analysis for determining input and output impedance of a feedback amplifier, we have assumed that the feedback network has no loading effect on the basic amplifier.

Therefore, the open loop gain is not affected. Example 5.1 shows that when we deal with an op-amp, which has very high open loop gain and relatively low output impedance, the loading effect caused by the feedback network can be neglected. However, when we deal with a discrete components amplifier circuit, the feedback network does create a loading effect that affects the open loop gain. As a result, the overall voltage gain, input and output impedance are also affected.

To permit accurate calculation of the open loop gain, the loading effect from the feedback network must be included in the input and output circuits [4]. The feedback network's input and output are either shorted to ground or an open circuit, according to whether the feedback network is connecting to a voltage or current source as shown in Table 5.1. The definition of the feedback network's input and output is given in Figure 5.10. The feedback network's input is where the network is connected to the amplifier's output circuit, while the feedback network's output is connected to the mixer or adder in the input circuit.

Table 5.1 Input and output circuit arrangement for taking into account the loading effect of the feedback network

	Feedback configuration			
	Series-shunt	*Shunt-shunt*	*Series-series*	*Shunt-series*
Input circuit loading arrangement	$V_{out} = 0$ Short feedback network's input to ground	$V_{out} = 0$ Short feedback network's input to ground	$I_{out} = 0$ Open feedback network's input	$I_{out} = 0$ Open feedback network's input
Output circuit loading arrangement	$I_f = 0$ Open feedback network's output	$V_f = 0$ Short feedback network's output to ground	$I_f = 0$ Open feedback network's output	$V_f = 0$ Short feedback network's output to ground

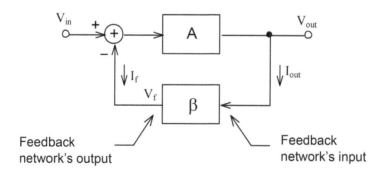

Figure 5.10 Feedback flow diagram defining the feedback network's input and output

An example of a general series-shunt feedback amplifier is shown in Figure 5.11. It is a two-stage amplifier with a nominal gain of A1 and A2 when no feedback is applied. The feedback network is formed by resistors R1 and R2. According to Table 5.1, first the feedback network's input (output stage A2) is shorted to ground. Therefore, R2 is in parallel with R1. Then the feedback network's output (input stage A1) is an open circuit. Therefore, R2 is in series with R1. Finally, the amplifier's open loop gain is determined under this new loading condition. The new gains now become A1′ and A2′. If A′ denotes the new open loop gain, we have A′ = A1′A2′, a product of the two amplifying stages. Then we have

$$\beta = \frac{R1}{R1+R2}, \quad A_f = \frac{A'}{1+A'\beta}, \quad Z_{if} = (1+A'\beta)Z_i, \quad Z_{of} = \frac{Z_o}{1+A'\beta}$$

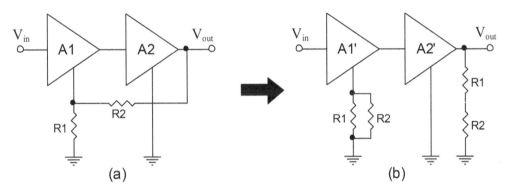

Figure 5.11 (a) A series-shunt feedback amplifier, (b) loading effect from the feedback network for the input and output circuit

Example 5.2

A two-stage amplifier, formed by a common emitter amplifier with a partially by-passed emitter resistor and a common source amplifier with a bypassed source resistor, is shown in Figure 5.12. The dc biasing condition for a common emitter amplifier can be found in Example 2.2, while Example 2.4 describes it for the common source amplifier. According to Table 5.1, the loading effect from the feedback network is given by paralleling resistor R10 to R5 for the common emitter amplifier. On the other hand, a resistor R5 is in series with R10 for the common source amplifier.

From Eq. (3.10) to (3.12):

$$Z_i = h_{fe}RE = h_{fe}(R5 \| R10) = 150(0.1k\Omega \| 2.2k\Omega) = 14.3k\Omega$$

$$A_{v1} = \frac{-RC}{r_e + RE} = \frac{-R3 \| R6 \| R7}{r_e + R5 \| R10} = \frac{-1.3k\Omega \| 3300k\Omega \| 1000k\Omega}{3.1\Omega + 0.1k\Omega \| 2.2k\Omega} = -13.1$$

For the common source amplifier, we have

$$Z_o = R8 \| (R5 + R10) = 1.2k\Omega \| (0.1k\Omega + 2.2k\Omega) = 788\Omega$$

From Eq. (3.37):

$$A_{v2} = -g_m RD = -g_m[R8 \| (R5 + R10)] = -(80 \times 10^{-3}S)(788\Omega) = -63$$

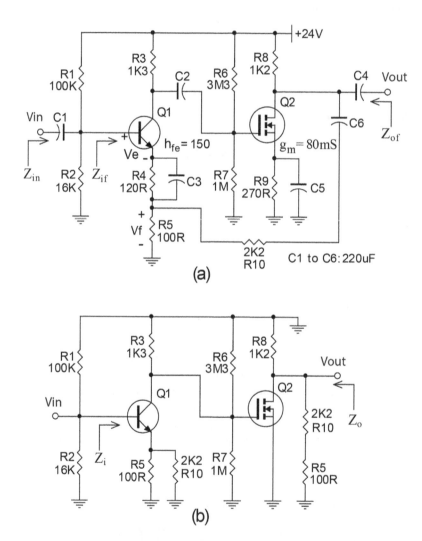

Figure 5.12 (a) A two-stage series-shunt feedback amplifier formed by a common emitter amplifier and a common source amplifier, (b) a simplified ac circuit for the amplifier without feedback but including the loading effect from the feedback network

Therefore, the overall open loop gain for the amplifier shown in Figure 5.12(b), which does not have the feedback but includes the loading effect from the feedback network, is given by

$$A = A_{v1}A_{v2} = (-13.1)(-63) = 825.3$$

Thus, the feedback factor, voltage gain, input and output impedance of the amplifier with feedback are given by

$$\beta = \frac{R5}{R5 + R10} = \frac{0.1k\Omega}{0.1k\Omega + 2.2k\Omega} = 0.0434$$

$$A_f = \frac{A}{1 + A\beta} = \frac{825.3}{1 + (825.3)(0.0434)} = 22.4$$

$$Z_{if} = (1 + A\beta)\, Zi = (1 + 825.3 \times 0.0434)(14.3k\Omega) = 526.5k\Omega$$

$$Z_{in} = R1 \parallel R2 \parallel Z_{if} = 100k\Omega \parallel 16k\Omega \parallel 526.5k\Omega = 13.4k\Omega$$

$$Z_{of} = \frac{Z_o}{1 + A\beta} = \frac{788\Omega}{1 + (825.3)(0.0434)} = 21.4\Omega$$

Note that Z_{of} is the output impedance of the amplifier with feedback applied. The input impedance Z_{in} is limited by the potential-divider biasing resistors R1 and R2. If high input impedance is required, JFET should be used for the input stage. It is also interesting to compare the closed loop gain approximated by $1/\beta = 1/0.0434 = 23$, which is indeed very close to the exact value of $A_f = 22.4$. The amount of feedback is $20\log(1+A\beta) = 31.3$dB.

Example 5.3

Figure 5.13 shows the Shigeru Wada amplifier, which is a variant of the Marantz 7 preamplifier. It is a popular vacuum tube line-stage amplifier. Details of the design are discussed in Chapter 8. Here, we aim to determine the overall voltage gain and input and output impedance of the Shigeru Wada amplifier. Assume that we have the following amplifying factor (μ) and plate impedance (r_p) for the triodes 12AX7 and 12AU7:

12AX7: $\mu = 100$, $r_p = 70k\Omega$; 12AU7: $\mu = 18$, $r_p = 8k\Omega$

Note that the Shigeru Wada amplifier is a three-stage amplifier in a series-shunt feedback configuration. The first two stages feature a common cathode amplifier using triode 12AX7. The output stage is a White cathode follower using triode 12AU7. Figure 5.13(b) shows the Shigeru Wada amplifier without feedback but including the loading effect from the feedback network. Let A_1, A_2 and A_3 denote the nominal gains of the three amplifying stages.

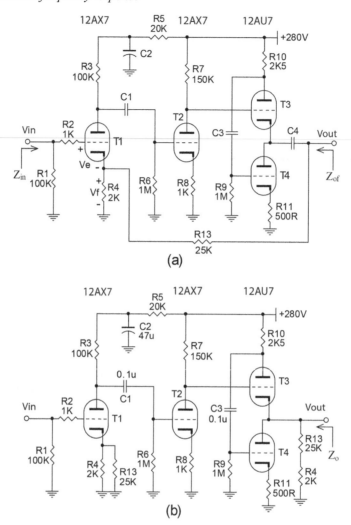

Figure 5.13 (a) The Shigeru Wada triode vacuum tube line stage preamplifier. Details can be found in Chapter 8, (b) the Shigeru Wada preamplifier without feedback but including the loading effect from the feedback network

Eq. (3.57):

$$A_1 = \frac{-\mu RP}{r_p + RP + (\mu+1)RK} = \frac{-\mu(R3 \| R6)}{r_p + (R3 \| R6) + (\mu+1)(R4 \| R13)}$$

$$= \frac{-(100)(100k\Omega \| 1000k\Omega)}{70k\Omega + (100k\Omega \| 1000k\Omega) + (100+1)(2k\Omega \| 25k\Omega)} = -26.1$$

$$A_2 = \frac{-\mu RP}{r_p + RP + (\mu+1)RK} = \frac{-\mu R7}{r_p + R7 + (\mu+1)R8}$$

$$= \frac{-(100)(150k\Omega)}{70k\Omega + 150k\Omega + (100+1)(1k\Omega)} = -46.7$$

Eq. (3.131):

$$A_3 \approx \frac{\mu}{\mu+1} = \frac{18}{18+1} = 0.947$$

Eq. (3.133):

$$Z_0 = \left[\frac{(r_p + RK)(r_p + RP)}{r_p + RP + (\mu+1)(r_p + RK) + \mu(\mu+1)RP} \right] \| (R4 + R13)$$

$$\approx \frac{(r_p + R11)(r_p + R10)}{r_p + R10 + (\mu+1)(r_p + R11) + \mu(\mu+1)R10}$$

$$= \frac{(8k\Omega + 0.5k\Omega)(8k\Omega + 2.5k\Omega)}{8k\Omega + 2.5k\Omega + (18+1)(8k\Omega + 0.5k\Omega) + 18(18+1)2.5k\Omega} = 87\Omega$$

The feedback factor, closed loop gain, input and output impedances are given by:

$$\beta = \frac{R4}{R4 + R13} = \frac{2k\Omega}{2k\Omega + 25k\Omega} = 0.074$$

$$A = A_1 A_2 A_3 = (-26.1)(-46.7)(0.947) = 1154.3$$

$$A_f = \frac{A}{1 + A\beta} = \frac{1154.3}{1 + (1154.3)(0.074)} = 13.3 \text{ (compared to } 1/\beta = 13.5)$$

$$Z_{of} = \frac{Z_0}{1 + A\beta} = \frac{87\Omega}{1 + (1154.3)(0.074)} = 1.0\Omega$$

The output impedance is just 1Ω, which is a respectable figure for a tube amplifier. The input impedance is equal to $R1 = 100k\Omega$. The amount of feedback $= 20\log(1 + A\beta) = 38.7$dB.

Shunt-shunt configuration (Trans-resistance amplifier)

Figure 5.14 shows the ac model for the shunt-shunt feedback configuration. We have the following:

$$V_{out} = AI_e \tag{5.26}$$

$$I_e = I_{in} - I_f \tag{5.27}$$

$$I_f = \beta V_{out} \tag{5.28}$$

Eliminating I_e and I_f from the Eqs. (5.26) to (5.28) yields

$$A_f \equiv \frac{V_{out}}{I_{in}} = \frac{A}{1 + A\beta} \tag{5.29}$$

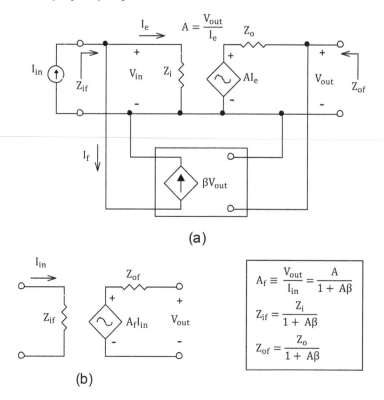

(b)

Figure 5.14 (a) An ac model for the shunt-shunt configuration, (b) the ac equivalent circuit for the shunt-shunt feedback amplifier

Here is the ideal feedback equation again. It is noted that A_f and A have the dimension of resistance, while having conductance for β.

To determine the input impedance with feedback, we substitute Eq. (5.26) with Eq. (5.29) and rearrange to get:

$$I_e = \frac{I_{in}}{1+ A\beta} \tag{5.30}$$

Since the input impedance of the basic amplifier and with feedback are given by,

$$Z_i \equiv \frac{V_{in}}{I_e} \tag{5.31}$$

$$Z_{if} \equiv \frac{V_{in}}{I_{in}} \tag{5.32}$$

Eq. (5.32) can be rearranged to

$$Z_{if} = \frac{V_{in}}{I_e} \frac{I_e}{I_{in}} = \frac{Z_i}{1+ A\beta} \tag{5.33}$$

To determine the output impedance, the input current source (I_{in}) is open circuit. A test voltage(V) and test current (I) are fed to the output as shown in Figure 5.15. From the output circuit, we have

$$V = AI_e + IZ_o \tag{5.34}$$

$$I_e = -\beta V \tag{5.35}$$

Eliminating I_e from Eqs. (5.34) and (5.35) yields

$$V(1 + A\beta) = IZ_o \tag{5.36}$$

Therefore, the output impedance with feedback is given by

$$Z_{of} \equiv \frac{V}{I} = \frac{Z_o}{1 + A\beta} \tag{5.37}$$

Thus, for shunt-shunt feedback configuration, the input and output impedances are reduced by a factor of $1/(1 + A\beta)$.

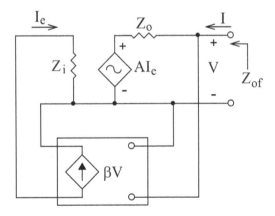

Figure 5.15 The input current source (I_{in}) is open circuit for determining the output impedance

Example 5.4

Figure 5.16(a) is an inverted amplifier. Assuming an ideal op-amp is used, the input voltage source can be replaced by a current source as shown in Figure 5.16(b). Since the op-amp is ideal, we do not need to determine the loading effect by the feedback network. Thus, we have

$$\beta = \frac{I_f}{V_{out}} = \left(\frac{V_- - V_{out}}{R2} \right) \frac{1}{V_{out}} = \frac{-1}{R2} \quad (\text{as } V_- = V_+ = 0) \tag{5.38}$$

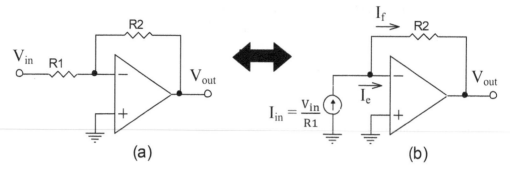

Figure 5.16 (a) A shunt-shunt configuration of an inverted amplifier using op-amp, (b) an ac model of the inverted amplifier with a current source I_{in}

The gain with feedback is given by

$$A_f \equiv \frac{V_{out}}{I_{in}} = \frac{A}{1+A\beta} = \frac{1}{\beta} = -R2 \tag{5.39}$$

as $A = \infty$ for the gain of an ideal op-amp. Note that A_f is trans-resistance. If we want to obtain the usual voltage gain of the inverted amplifier in Figure 5.16(a), we have

$$A_v \equiv \frac{V_{out}}{V_{in}} = \frac{V_{out}}{I_{in}} \frac{I_{in}}{V_{in}} = (-R2)\left(\frac{1}{R1}\right) = \frac{-R2}{R1} \tag{5.40}$$

This is expected from an inverted amplifier. It shows the consistency between the feedback analysis and the conventional op-amp analysis developed in Chapter 4.

Example 5.5

Figure 5.17(a) is a simple shunt-shunt feedback amplifier. Assume that the amplifier has the following data: $R1 = 33k\Omega$, $R2 = 4.7k\Omega$, $R_s = 5k\Omega$, transistor current gain $h_{fe} = 100$, and collector dc biasing current $I_C = 1mA$. According to Table 5.1, the loading effect from the feedback resistor becomes one R1 loading the base while another R1 loads the collector of the transistor Q1 as shown in Figure 5.17(b).

From Figure 5.17(b), we determine the Z_i, Z_o and open loop gain A as follows.

$$A \equiv \frac{V_{out}}{I_{in}} = \frac{-I_c(R1 \| R2)}{I_{in}} = \frac{-h_{fe}I_b(R1 \| R2)}{I_{in}} \tag{5.41}$$

From Figure 5.17(c), the base current is given by,

$$I_b = \frac{(RS \| R1)}{(RS \| R1) + h_{fe}r_e} I_{in} \tag{5.42}$$

where $h_{fe}r_e$ is the input impedance looking into the base of the transistor Q1, and $r_e = 26mV/I_C = 26mV/1mA = 26\Omega$. Substituting Eq. (5.42) for (5.41) and rearranging yields

$$A = \frac{-h_{fe}(R1 \| R2)(RS \| R1)}{(RS \| R1) + h_{fe}r_e}$$

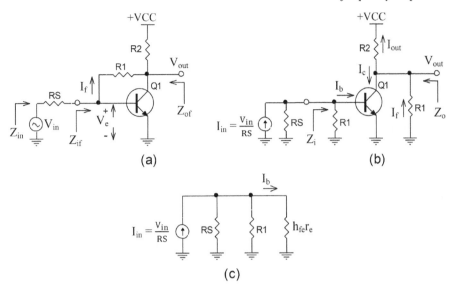

Figure 5.17 (a) A shunt-shunt feedback amplifier, (b) the amplifier without feedback but including the loading effect from the feedback resistor R1, (c) the ac model of the input circuit

where

$$R1 \| R2 = \frac{(33k\Omega)(4.7k\Omega)}{33k\Omega + 4.7k\Omega} = 4.1k\Omega$$

$$RS \| R1 = \frac{(5k\Omega)(33k\Omega)}{5k\Omega + 33k\Omega} = 4.3k\Omega$$

Therefore, we have

$$A = \frac{-(100)(4.1k\Omega)(4.3k\Omega)}{(4.3k\Omega)+(100)(0.026k\Omega)} = -255.5k\Omega \tag{5.43}$$

$$\beta = \frac{I_f}{V_{out}} = \frac{-1}{R1} = \frac{-1}{33k\Omega} = -0.03mA / V \tag{5.44}$$

$$Z_i = (R1) \| (h_{fe}r_e) = \frac{(33k\Omega)(100 \times 0.026k\Omega)}{33k\Omega + 100 \times 0.026k\Omega} = 2.4k\Omega \tag{5.45}$$

$$Z_o \approx R1 \| R2 = 4.1k\Omega \tag{5.46}$$

Thus, the feedback amplifier has the following properties:

$$A_f \equiv \frac{V_{out}}{I_{in}} = \frac{A}{1+ A\beta} = \frac{-255.5k\Omega}{1+(-255.5k\Omega)(-0.03mA / V)} = -29.5k\Omega$$

$$Z_{if} = \frac{Z_i}{1+ A\beta} = \frac{2.4k\Omega}{1+(-255.5k\Omega)(-0.03mA / V)} = 0.28k\Omega$$

$$Z_{of} = \frac{Z_o}{1+ A\beta} = \frac{4.1k\Omega}{1+(-255.5k\Omega)(-0.03mA\,/\,V)} = 0.47k\Omega$$

Expressed in terms of conventional voltage gain with feedback,

$$A_{vf} = \frac{V_{out}}{V_{in}} = \frac{V_{out}}{I_{in}RS} = \frac{A_f}{RS} = \frac{-29.5k\Omega}{5k\Omega} = -5.9$$

The input impedance looking from the voltage source is given by

$$Z_{in} = RS + Z_{if} = 5k\Omega + 0.28k\Omega = 5.28k\Omega$$

Series-series configuration (transconductance amplifier)

Figure 5.18 shows the series-series feedback configuration. The basic amplifier has the nominal gain $A = I_{out}/V_e$, which is transconductance, and feedback factor $\beta = V_f/I_{out}$, which is transresistance. The feedback network samples output current I_{out} and feeds back a proportional voltage V_f in series with the input. If $R_L \ll Z_o$, it can be shown that

$$A_f \equiv \frac{I_{out}}{V_{in}} = \frac{A}{1+ A\beta} \tag{5.47}$$

$$Z_{if} = (1+A\beta)Z_i \tag{5.48}$$

$$Z_{of} = (1+A\beta)Z_o \tag{5.49}$$

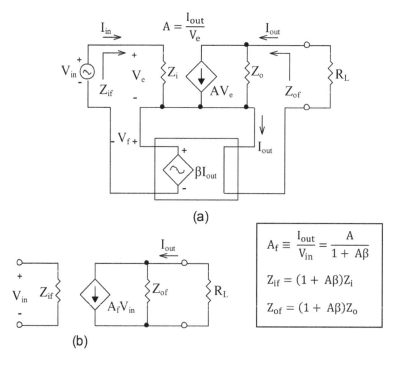

(a)

(b)

Figure 5.18 (a) An ac model for the series-series configuration, (b) an ac equivalent circuit for the series-series feedback amplifier

The series-series feedback configuration produces a transconductance amplifier with a stabilized gain A_f, high input impedance Z_{if} and high output impedance Z_{of}.

Example 5.6

A series-series configuration formed by a common emitter amplifier is shown in Figure 5.19(a). Note that the feedback resistor RE is not connected to the load resistor RC. The ac equivalent circuit is drawn in Figure 5.19(b) without the feedback resistor RE. However, the intrinsic collector-emitter resistor r_o is included in this example.

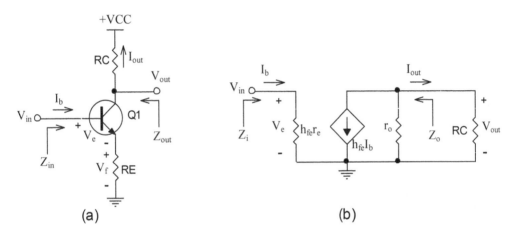

(a)   (b)

Figure 5.19 (a) A series-series configuration formed by a common emitter amplifier, (b) the ac equivalent circuit for the amplifier without feedback resistor RE

From Figure 5.19(a):

$$\beta \equiv \frac{V_f}{I_{out}} = -RE \tag{5.50}$$

From Figure 5.19(b):

$$A \equiv \frac{I_{out}}{V_e} \approx \frac{-h_{fe}I_b}{(I_b)(h_{fe}r_e)} = \frac{-1}{r_e} \quad \text{(for } r_o \gg RC) \tag{5.51}$$

Therefore, we have

$$1 + A\beta = 1 + \left(\frac{-1}{r_e}\right)(-RE) = \frac{r_e + RE}{r_e} \tag{5.52}$$

$$A_f = \frac{A}{1 + A\beta} = \frac{\left(\dfrac{-1}{r_e}\right)}{\left(\dfrac{r_e + RE}{r_e}\right)} = \frac{-1}{r_e + RE} \tag{5.53}$$

$$Z_i = h_{fe}r_e$$

$$Z_{if} = (1 + A\beta)Z_i = \left(\frac{r_e + RE}{r_e}\right)(h_{fe}r_e) = h_{fe}(r_e + RE) \approx h_{fe}RE \qquad (5.54)$$

$$Z_o = r_o$$

$$Z_{of} = (1 + A\beta)Z_o = \left(\frac{r_e + RE}{r_e}\right)(r_o)$$

$$Z_{out} = RC \| Z_{of} \approx RC \text{ (for } Z_{of} \gg RC) \qquad (5.55)$$

The input impedance Z_{in} and output impedance Z_{out} are consistent with a common emitter amplifier using the conventional ac analysis discussed in Chapter 3. Finally, the conventional voltage gain is given by

$$A_{vf} \equiv \frac{V_{out}}{V_{in}} = \frac{I_{out}RC}{V_{in}} = (A_f)(RC) = \frac{-RC}{r_e + RE} \approx \frac{-RC}{RE} \qquad (5.56)$$

which is consistent with a common emitter amplifier with an un-bypassed emitter resistor.

Shunt-series configuration (current amplifier)

Figure 5.20 shows a shunt-series feedback configuration. The basic amplifier has a current gain of $A = I_{out}/I_e$, and the feedback factor $\beta = I_f/I_{out}$. The feedback network samples output current I_{out} and feeds back a current I_f in series with the input. If $R_L \ll Z_o$, it can be shown that

$$A_f \equiv \frac{I_{out}}{I_{in}} = \frac{A}{1 + A\beta} \qquad (5.57)$$

$$Z_{if} = \frac{Z_i}{1 + A\beta} \qquad (5.58)$$

$$Z_{of} = (1 + A\beta)Z_o \qquad (5.59)$$

The shunt-series feedback configuration is a current amplifier with a stabilized gain A_f, low input impedance Z_{if} and high output impedance Z_{of}.

Example 5.7

Figure 5.21(a) shows a two-stage amplifier containing bipolar transistors Q1 and Q2. Each transistor forms a common emitter amplifier. The feedback resistor R3 is connected between the base of Q1 and the emitter of Q2, forming a shunt-series feedback amplifier configuration. From Table 5.1, the loading effect to the first and second stage of the amplifier without feedback is shown in Figure 5.21(b).

Assuming the two-stage amplifier has the following components: RS = 5kΩ, R1 = 3.3kΩ, R2 = 2.2kΩ, R3 = 1.8kΩ, R4 = 0.1kΩ, both transistors have identical current gain $h_{fe1} = h_{fe2} = 100$,

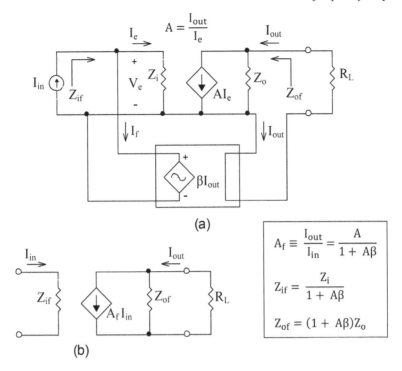

$$A = \frac{I_{out}}{I_e}$$

$$A_f \equiv \frac{I_{out}}{I_{in}} = \frac{A}{1 + A\beta}$$

$$Z_{if} = \frac{Z_i}{1 + A\beta}$$

$$Z_{of} = (1 + A\beta)Z_o$$

(a)

(b)

Figure 5.20 (a) An ac model for the shunt-series configuration, (b) the ac equivalent circuit for the shunt-series feedback amplifier

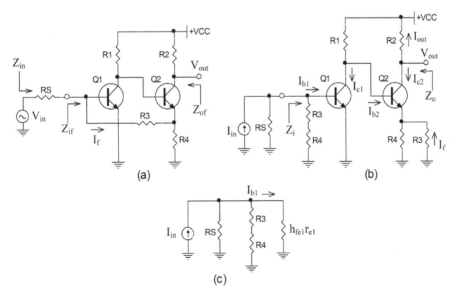

(a)

(b)

(c)

Figure 5.21 (a) A shunt-series feedback amplifier, (b) the amplifier without feedback but including the loading effect from the feedback resistor R3, (c) the ac model of the input circuit

dc collector biasing current 1mA, so that $r_{e1} = r_{e2} = 26\Omega$. Assume the nominal gains for the two amplifying stages are denoted by A_1 and A_2. From Figure 5.21(b), we have:

$$A_1 = \frac{I_{c1}}{I_{b1}} = h_{fe1} = 100 \tag{5.60}$$

$$A_2 = \frac{I_{c2}}{I_{b2}} = h_{fe2} = 100 \tag{5.61}$$

$$A \equiv \frac{I_{out}}{I_{in}} = \frac{-I_{c2}}{I_{in}} = \frac{-I_{c2}}{I_{b2}}\frac{I_{b2}}{I_{c1}}\frac{I_{c1}}{I_{b1}}\frac{I_{b1}}{I_{in}} = (-h_{fe2})\left(\frac{I_{b2}}{I_{c1}}\right)(h_{fe1})\left(\frac{I_{b1}}{I_{in}}\right) \tag{5.62}$$

The term I_{b1}/I_{in} in Eq. (5.62) can be determined from Figure 5.21(c), where $h_{fe1}r_{e1}$ is the input impedance looking into the base of Q1:

$$\frac{I_{b1}}{I_{in}} = \frac{RS\,\|\,(R3+R4)}{h_{fe1}r_{e1} + RS\,\|\,(R3+R4)} = \frac{(5k\Omega)\,\|\,(1.8k\Omega + 0.1k\Omega)}{(100)(0.026k\Omega) + (5k\Omega)\,\|\,(1.8k\Omega + 0.1k\Omega)} = 0.35 \tag{5.63}$$

Similarly, the term I_{b2}/I_{c1} in Eq. (5.62) is given by:

$$\frac{I_{b2}}{I_{c1}} = \frac{-R1}{h_{fe2}(R3\,\|\,R4)+R1} = \frac{-3.3k\Omega}{(100)(1.8k\Omega\,\|\,0.1k\Omega)+3.3k\Omega} = -0.26 \tag{5.64}$$

where $h_{fe2}(R3\|R4)$ is the input impedance looking into the base of Q2. Substituting Eqs. (5.60), (5.61), (5.63) and (5.64) with (5.62) yields

$$A = (-h_{fe2})\left(\frac{I_{b2}}{I_{c1}}\right)(h_{fe1})\left(\frac{I_{b1}}{I_{in}}\right) = (-100)(-0.26)(100)(0.35) = 910 \tag{5.65}$$

The feedback factor can be determined by examining the emitter current (I_{e2}) of Q2 where I_e2 is approximated by the collector current I_{c2}, which is equal to $-I_{out}$. Then we have

$$\beta = \frac{I_f}{I_{out}} \approx \frac{I_f}{I_{e2}} = \frac{R4}{R3+R4} = \frac{0.1k\Omega}{2.2k\Omega + 0.1k\Omega} = 0.0435 \tag{5.66}$$

$$A_f \equiv \frac{A}{1+A\beta} = \frac{910}{1+(910)(0.0435)} = 22.4$$

$$Zi = (R3+R4)\,\|\,(h_{fe1}r_{e1}) = (1.8k\Omega + 0.1k\Omega)\,\|\,(100)(0.026k\Omega) = 1.1k\Omega$$

$$Z_{if} = \frac{Z_i}{1+A\beta} = \frac{1.1k\Omega}{1+(910)(0.0435)} = 0.027k\Omega$$

$$Z_{in} = RS + Z_{if} = 5k\Omega + 0.027k\Omega \approx 5k\Omega$$

$$Z_o \approx R2 = 2.2k\Omega$$

$$Z_{of} = (1+A\beta)Z_o = [1+(910)(0.0435)]2.2k\Omega = 89.3k\Omega$$

The conventional voltage gain of the shunt-series feedback amplifier of Figure 5.21(a) is given by

$$A_{vf} \equiv \frac{V_{out}}{V_{in}} = \frac{I_{out}R2}{I_{in}RS} = A_f\left(\frac{R2}{RS}\right) = (22.4)\left(\frac{2.2k\Omega}{5k\Omega}\right) = 9.9$$

Table 5.2 shows a summary of parameters and formulas for the four feedback configurations while Figure 5.22 shows representative circuits for each configuration by means of op-amps and discrete components.

Table 5.2 Summary of parameters and formulas for the four feedback configurations

	Series-Shunt	*Shunt-Shunt*	*Series-Series*	*Shunt-Series*
Configuration	Figure 5.7	Figure 5.13	Figure 5.17	Figure 5.19
Amplifier type	Voltage	Trans-resistance	Trans-conductance	Current
Gain, A (no feedback)	$\dfrac{V_{out}}{V_e}$	$\dfrac{V_{out}}{I_e}$	$\dfrac{I_{out}}{V_e}$	$\dfrac{I_{out}}{I_e}$
Feedback factor, β	$\dfrac{V_f}{V_{out}}$	$\dfrac{I_f}{V_{out}}$	$\dfrac{V_f}{I_{out}}$	$\dfrac{I_f}{I_{out}}$
Gain, A_f	$\dfrac{V_{out}}{V_{in}} = \dfrac{A}{1+A\beta}$	$\dfrac{V_{out}}{I_{in}} = \dfrac{A}{1+A\beta}$	$\dfrac{I_{out}}{V_{in}} = \dfrac{A}{1+A\beta}$	$\dfrac{I_{out}}{I_{in}} = \dfrac{A}{1+A\beta}$
Input impedance, Z_{if}	$Z_i(1+A\beta)$	$\dfrac{Z_i}{1+A\beta}$	$Z_i(1+A\beta)$	$\dfrac{Z_i}{1+A\beta}$
	(increased)	(decreased)	(increased)	(decreased)
Output impedance, Z_{of}	$\dfrac{Z_o}{1+A\beta}$	$\dfrac{Z_o}{1+A\beta}$	$Z_o(1+A\beta)$	$Z_o(1+A\beta)$
	(decreased)	(decreased)	(increased)	(increased)

5.7   Miller effect capacitor

An ac analysis for single-stage amplifiers in the mid-band frequency was discussed in Chapter 3. The voltage gain is flat in the mid-band frequency. However, when the amplifier operates in a high frequency region, the impedance of the intrinsic capacitors of the amplifying device will roll off, limiting the bandwidth of the amplifier.

Figure 5.23 shows the high frequency hybrid-pi model for BJT. r_{bb} is the base spread resistor; r_π is the base-emitter resistor equal to $h_{fe}r_e$, and r_o is the collector-emitter resistor. These are the same parameters used in the hybrid-pi model in Figure 3.2. For the high frequency hybrid-pi model, there are two additional capacitors. C_π is the intrinsic base-emitter capacitor while C_μ is the intrinsic base-collector capacitor. When a BJT device is operated as a common emitter amplifier, the input is fed to the base while output is taken from the collector. Therefore, C_μ is connected across the input and output of a common emitter amplifier. Capacitor C_μ is similar to the capacitor placed between the input and output of the op-amp in the integrator circuit of Figure 4.5. At high frequency, the impedance of C_μ decreases and, therefore, limits the band-width of the amplifier. This bandwidth limiting effect is equal to placing a so-called *Miller effect capacitor* C_M, which is much larger than C_μ, at the input of the amplifier. In the following, we are going to see how the Miller effect capacitor C_M is determined.

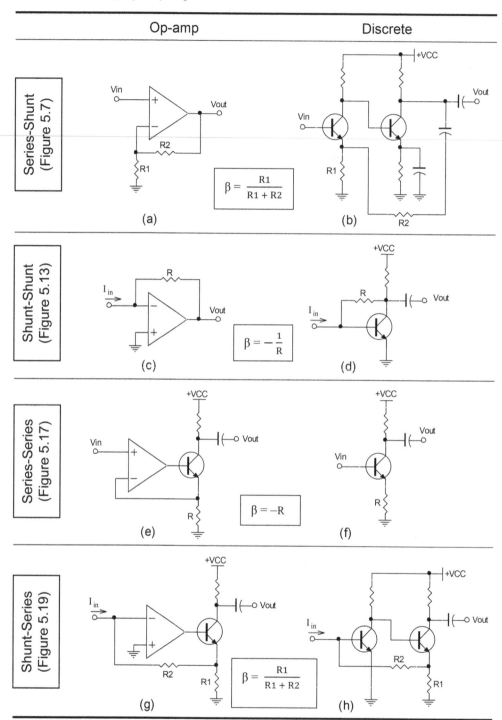

Figure 5.22 Representative circuits for the four feedback configurations in terms of op-amps and discrete components

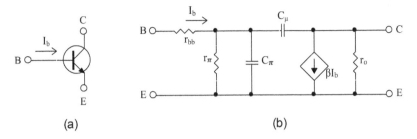

Figure 5.23 (a) An NPN transistor, (b) its high frequency hybrid-pi ac model

Figure 5.24 shows a capacitor, C, connected across an amplifying stage, which has voltage gain A_v and input impedance Z_i. From the input side, we have

$$I_{in} = I_1 + I_2 \tag{5.67}$$

where

$$I_{in} = \frac{V_{in}}{Z_{in}}, \ I_2 = \frac{V_{in}}{Z_i} \tag{5.68}$$

$$I_1 = \frac{V_{in} - V_{out}}{Z_C} = \frac{V_{in} - A_v V_{in}}{Z_C} = \frac{(1 - A_v)V_{in}}{Z_C} \tag{5.69}$$

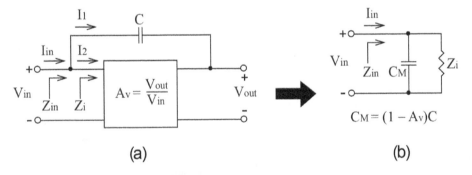

Figure 5.24 (a) A capacitor is connected across the input and output of an amplifier, (b) an equivalent circuit for the input side having the Miller effect capacitor C_M

Substituting Eqs. (5.68) and (5.69) for (5.67) yields

$$\frac{V_{in}}{Z_{in}} = \frac{(1 - A_v)V_{in}}{Z_C} + \frac{V_{in}}{Z_i}$$

and simplifying to

$$\frac{1}{Z_{in}} = \frac{1}{Z_C/(1 - A_v)} + \frac{1}{Z_i} = \frac{1}{Z_M} + \frac{1}{Z_i} \tag{5.70}$$

where

$$Z_M \equiv \frac{Z_C}{1 - A_v} \tag{5.71}$$

Since Z_C is the impedance of capacitor C, we have

$$Z_C = \frac{1}{sC}$$

If Z_M is the impedance of capacitor C_M, Eq. (5.71) can be written as

$$Z_M \equiv \frac{1}{sC_M} = \left(\frac{1}{sC}\right)\left(\frac{1}{1-A_v}\right)$$

Thus, the Miller effect capacitor is given by

$$C_M = (1-A_v)C \tag{5.72}$$

Eq. (5.70) can be viewed as Z_{in} is equal to C_M in parallel with Z_i, as shown in Figure 5.24(b). If the gain A_v is high, C_M becomes a large capacitor shunting the input of the amplifier. Thus, it limits the bandwidth of the amplifier.

On the other hand, a Miller effect capacitor will also appear in the output, which may also affect the high frequency response of the amplifier. To find the output Miller effect capacitance C_{Mo}, consider Figure 5.25(a). From the output side, we have

$$I_{out} = I_1 + I_2 \tag{5.73}$$

where

$$I_{out} = \frac{V_{out}}{Z_{out}}, \; I_2 = \frac{V_{out}}{Z_o} \tag{5.74}$$

$$I_1 = \frac{V_{out} - V_{in}}{Z_C} = \frac{V_{out} - V_{out}/A_v}{Z_C} = \frac{(1 - 1/A_v)V_{out}}{Z_C} \tag{5.75}$$

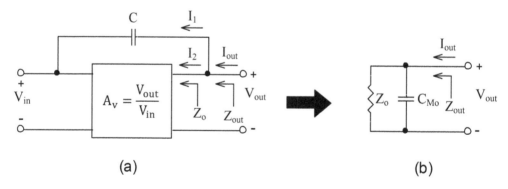

(a)   (b)

Figure 5.25 (a) A capacitor is connected across the input and output of an amplifier, (b) an equivalent circuit for the output side having the Miller effect capacitor C_{Mo}

Substituting Eqs. (5.74) and (5.75) for (5.73) yields

$$\frac{V_{out}}{Z_{out}} = \frac{(1 - 1/A_v)V_{out}}{Z_C} + \frac{V_{out}}{Z_o}$$

and simplifying to

$$\frac{1}{Z_{out}} = \frac{1 - 1/A_v}{Z_C} + \frac{1}{Z_o} = \frac{1}{Z_{Mo}} + \frac{1}{Z_o} \tag{5.76}$$

where

$$Z_{Mo} \equiv Z_C/(1 - 1/A_v) \tag{5.77}$$

Thus, if $Z_C = 1/sC$, $Z_{Mo} = 1/sC_{Mo}$, the output Miller effect capacitor is given by

$$C_{Mo} = (1 - 1/A_v)C \approx C \text{ (for } A_v \gg 1) \tag{5.78}$$

Eq. (5.76) can be viewed as Z_{out} is equal to C_{Mo} in parallel to Z_o, as shown in Figure 5.26(b). In other words, the Miller effect capacitor C_{Mo} is present in the output side. Since C_{Mo} is approximately equal to C, the output Miller effect capacitor is much smaller than the capacitor C_M presenting in the input. Thus, C_{Mo} does not have too much impact on the bandwidth.

If we combine the results, the Miller effect capacitors C_M and C_{Mo} are shown in Figure 5.26(b). In general, $C_M \gg C_{Mo}$, C_M is the Miller effect capacitor that sets the limit to the bandwidth of the amplifier.

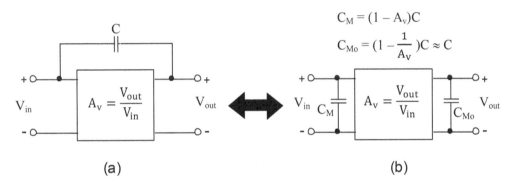

Figure 5.26 (a) A capacitor is connected across the input and output of an amplifier, (b) an equivalent circuit for the same amplifier with C removed but having Miller effect capacitors C_M and C_{Mo}

Example 5.8

The voltage gain for the common emitter amplifier of Figure 5.27 is given by $A_v = -RC/r_e$, where $r_e = 26\text{mV}/I_C$. If $C_\mu = 3\text{pF}$, $RC = 2.2\text{k}\Omega$, $I_E = 1\text{mA}$, we have $A_v = -2.2\text{k}\Omega/26\Omega = -85$. From Eqs. (5.72) and (5.78), we have $C_M = [1 - (-85)]3\text{pF} = 258\text{pF}$ and $C_{Mo} = [1 - (-1/85)]3\text{pF} = 3.03\text{pF}$.

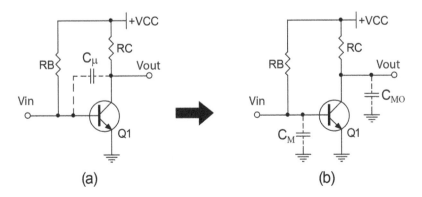

Figure 5.27 (a) A fixed bias common emitter amplifier with base-collector intrinsic capacitor C_μ, (b) an equivalent circuit with Miller effect capacitor C_M and C_{Mo}

5.8   Amplifier's frequency response

Figure 5.28(a) shows a typical common emitter amplifier, which has dc blocking capacitors C_i and C_o, emitter bypassed capacitor C_E, and intrinsic base-collector capacitor C_μ. The frequency response of the common emitter amplifier is shown in Figure 5.28(b) with the lower –3dB frequency f_L and the higher –3dB frequency f_H. It turns out that f_L is determined by C_i, C_o and C_E, while f_H is determined by C_μ.

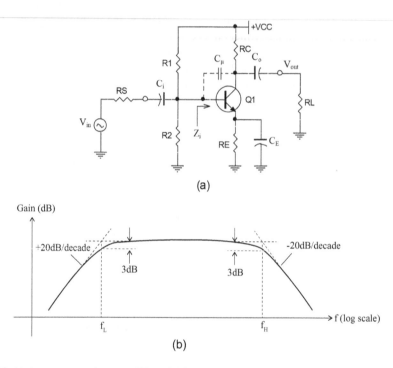

(a)

(b)

Figure 5.28 (a) A common emitter amplifier with input and output dc blocking capacitor, C_i and C_o, emitter bypassed capacitor C_E, and base-collector intrinsic capacitor C_μ, (b) frequency response of the amplifier showing the –3dB frequencies f_L and f_H

Low frequency response

Figure 5.29 shows a simple RC high pass filter and its frequency response. The impedance of the capacitor is $Z_C = 1/sC = 1/j\omega C$, which is inversely proportional to the frequency $\omega = 2\pi f$. At very high frequency, Z_C becomes very small. Therefore, V_2 is nearly equal to V_1 at high frequency. Conversely, Z_C becomes large at low frequency. As a result, Z_C and R forming a potential divider and making V_2 lower than V_1 at low frequency. f_L is given by

$$f_L = \frac{1}{2\pi RC} \text{ or } C = \frac{1}{2\pi f_L R} \tag{5.79}$$

For an audio amplifier with bandwidth 20Hz to 20kHz, f_L has to be 20Hz or lower. In practice, since R is the impedance associated with the amplifier, capacitor C must be chosen such that

$$f_L < 20Hz \tag{5.80}$$

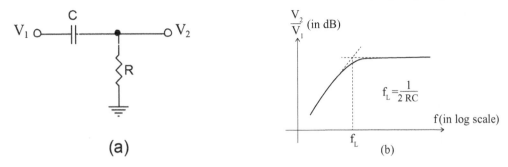

Figure 5.29 (a) A high pass filter formed by an RC network, (b) frequency response of the RC network

In Figure 5.28(a), the capacitor C_i and the input impedance of the common emitter amplifier forms an RC high pass filter. The lower $-3dB$ frequency is given by

$$f_L = \frac{1}{2\pi Z_{in} C_i} \tag{5.81}$$

where

$$Z_{in} = R1 \| R2 \| Z_i = R1 \| R2 \| (h_{fe} r_e) \tag{5.82}$$

Similarly, the capacitor C_o and resistor RL also form an RC high pass filter. The lower $-3dB$ frequency is given by

$$f_L = \frac{1}{2\pi R_L C_o} \tag{5.83}$$

On the other hand, the emitter bypassed capacitor C_E also produces a lower $-3dB$ frequency. The voltage gain of the common emitter amplifier can be expressed as,

$$A_V = \frac{-RC}{r_e + Z_E} \tag{5.84}$$

where

$$Z_E = RE \| (1/sC_E) \tag{5.85}$$

At high frequency, RE is bypassed by the capacitor C_E so that $Z_E \approx 0$. Therefore, the voltage gain $A_v = -RC/r_e$ is high at high frequency. At low frequency, Z_E is a high impedance so that the voltage gain is reduced to $A_v = -RC/(r_e + RE)$. Thus, the frequency response for the voltage gain A_v behaves like a high pass filter. The lower $-3dB$ frequency can be shown to be:

$$f_L = \frac{1}{2\pi REC_E} \tag{5.86}$$

In the above cases, it is a common practice to choose capacitors C_i, C_o and C_E such that f_L is much lower than 20Hz, forcing f_L outside of the audio frequency band.

High frequency response

Figure 5.30 shows a simple RC low pass filter and its frequency response. The impedance of the capacitor is $Z_C = 1/sC = 1/j\omega C$, which is inversely proportional to the frequency $\omega = 2\pi f$. At very high frequency, Z_C is very small. Therefore, V_2 drops to a very small value at high frequency. Conversely, Z_C is large at low frequency. As a result, V_2 becomes nearly equal to V_1. At the –3dB frequency f_H, we have

$$f_H = \frac{1}{2\pi RC} \text{ or } C = \frac{1}{2\pi f_H R} \tag{5.87}$$

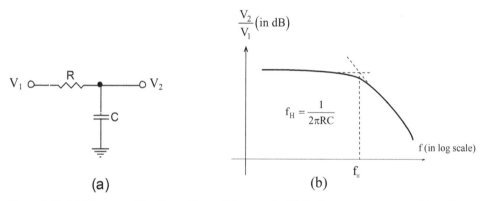

(a)   (b)

Figure 5.30 (a) A low pass filter formed by an RC network, (b) the frequency response of the RC network

For an audio amplifier with a desired bandwidth of 20Hz to 20kHz, f_H has to be at least 20kHz. In practice, f_H is limited by the Miller effect capacitor $C_M = (1 - A_v)C_\mu$ shunting the input. In order to push f_H higher than the desired 20kHz for a single stage amplifier, we need to choose a transistor with low C_μ and, perhaps, reduce the voltage gain A_v. For multi-stage amplifiers, the use of negative feedback will easily extend the bandwidth well beyond 20kHz, as well as improving the input impedance, output impedance, and distortion.

The high frequency ac equivalent circuit for the common emitter amplifier of Figure 5.28(a) is shown in Figure 5.31(a). The effect of the base-collector intrinsic capacitor C_μ now becomes a Miller effect capacitor C_M placed in the input side, and C_{Mo} in the output side. The input side is simplified in Figure 5.31(b) while the Thévenin equivalent circuit is shown in Figure 5.31(c). The Thévenin impedance is given by

$$\begin{aligned} R_{th} &= [(RS \| RB) + r_{bb'}] \| (r_\pi) = [(RS \| R1 \| R2) + r_{bb'}] \| (h_{fe}r_e) \\ &\approx (RS \| R1 \| R2) \| (h_{fe}r_e) \end{aligned} \tag{5.88}$$

and

$$C_{in} = C_\pi + C_M = C_\pi + (1 - A_v) C_\mu \tag{5.89}$$

Thus, the higher –3dB frequency is given by

$$f_H = \frac{1}{2\pi R_{th} C_{in}} \tag{5.90}$$

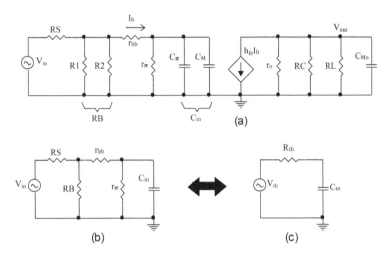

Figure 5.31 (a) A high frequency ac equivalent circuit for the common emitter amplifier of Figure 5.27 (a), (b) a simplified model for the input side, (c) the Thévenin equivalent circuit for (b)

Note that the output Miller effect capacitor C_{Mo} also limits the high frequency response of the amplifier. But it only happens at a much higher frequency. In other words, the Miller effect capacitor C_M shunts the input at a much lower frequency before C_{Mo} starts to roll off. The bandwidth of the amplifier is, therefore, limited by f_H, which is governed by Eq. (5.90).

As discussed above, the common emitter amplifier in Figure 5.32(a) creates a large Miller effect capacitor C_M. This is because the intrinsic base-collector capacitor C_μ appears to be right across the base input and collector output. Thus, C_M limits the bandwidth of the common emitter amplifier. However, the intrinsic capacitor C_μ does not significantly affect the emitter follower of Figure 5.32(b) and the common base amplifier of Figure 5.32(c), as the capacitor is not connected across the input and output. As a result, emitter follower and common base amplifiers have wider bandwidth than a common emitter amplifier.

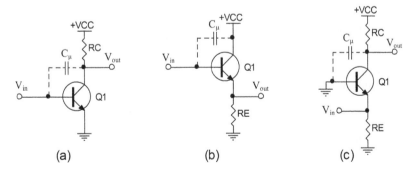

Figure 5.32 (a) A simplified common emitter amplifier, (b) a simplified emitter follower, (c) a simplified common base amplifier

Since a common base amplifier has a very low input impedance, it is rare for it to work alone. A common base amplifier is often paired with another device to form a compound amplifier such as cascode and emitter coupled amplifiers shown in Figure 5.33. Note that for the cascode amplifier in Figure 5.33(a), the input signal first feeds to the common emitter amplifier formed by Q1. The load resistor seen by the collector of Q1 is the input impedance of the common base amplifier formed by Q2, which is approximately equal to r_{e2}. Therefore, the gain of the common emitter amplifier is equal to $A_v = -r_{e2}/r_{e1} \approx -1$. The Miller effect capacitor for Q1 becomes $C_M = (1 - A_v) C_\mu = 2C_\mu$, which is low. On the other hand, Q2 is working in a common base amplifier configuration, which has a low Miller effect capacitor. As a result, the cascode amplifier has a very wide bandwidth.

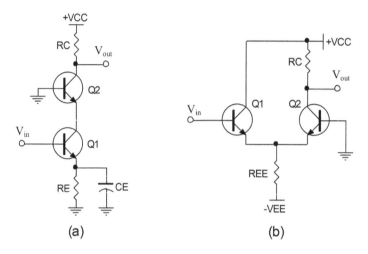

Figure 5.33 (a) A simplified cascode amplifier, (b) a simplified emitter coupled amplifier

The emitter coupled compound amplifier of Figure 5.33(b) is formed by an emitter follower Q1 directly coupled to a common base amplifier Q2. Since both emitter follower and common base amplifier have wide bandwidths, as a result the emitter coupled compound amplifier also has an excellent high frequency response. Details of the ac properties of the cascode and emitter coupled amplifiers are discussed in Chapter 3.

Generally speaking, the bandwidth for emitter follower, common base, cascode and emitter coupled amplifiers is at least an order of magnitude greater than the common emitter amplifier. For example, if a BJT device produces a 50kHz bandwidth in the common emitter configuration, the aforementioned amplifying configurations may produce a bandwidth of 500kHz or even higher. This also applies to FETs and vacuum tubes.

5.9   Stability, phase and gain margins

When negative feedback becomes positive, the feedback amplifier will break into oscillation in some frequency ranges. It becomes useless as an amplifier. Therefore, it is important to understand the conditions for establishing negative feedback and preventing positive feedback. First, let us rewrite Eq. (5.5) in the following form,

$$|A_f| = \left| \frac{A}{1 + A\beta} \right| \tag{5.91}$$

For a negative feedback amplifier, the amount of feedback must be

$$|1 + A\beta| > 1 \text{ so that } |A_f| < |A| \tag{5.92}$$

The gain with negative feedback $|A_f|$ is always smaller than the open loop gain $|A|$. However, for an amplifier with positive feedback, the amount of feedback becomes

$$|1 + A\beta| < 1 \text{ so that } |A_f| > |A| \tag{5.93}$$

In other words, the positive feedback gain $|A_f|$ is larger than the open loop gain $|A|$. This is a clear distinction between amplifiers with negative and positive feedback. If now the amount of feedback is reduced to zero, $(1 + A\beta) \rightarrow 0$, Eq. (5.91) suggests that $|A_f| \rightarrow \infty$. Clearly, this is the most favorable condition for positive feedback, such that

$$1 + A\beta = 0$$

or

$$A\beta = -1 \tag{5.94}$$

To interpret Eq. (5.94), let us examine the feedback amplifier in Figure 5.34. In this amplifier, no input is applied, $V_{in} = 0$. Assuming that, due to some transient disturbance, a signal V_{out} appears at the output, a portion of this output signal, $-\beta V_{out}$, is fed back to the input of the amplifier and appears to the output as an amplified transient disturbance $-A\beta V_{out}$. If this term is just equal to V_{out}, then the spurious output has regenerated itself. This is an amplifier with positive feedback. Therefore, the condition for positive feedback can be viewed as

$$-A\beta V_{out} = V_{out}$$

and simplifying to

$$A\beta = -1$$

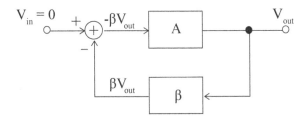

Figure 5.34 A feedback system without input signal

This is the same condition stated in Eq. (5.94). It should be noted that the loop gain $A\beta$ is a positive quantity for a negative feedback amplifier. However, the open loop gain A is frequency dependent, which is affected by the frequency dependent components – intrinsic capacitances of the amplifying devices. When the phase of the open loop gain reaches 180°,

A becomes "inverted" and, hence, Aβ may become negative. Thus, it turns an amplifier with negative feedback into one with positive feedback. In order to avoid positive feedback and oscillation happening to an amplifier, the loop gain should be less than unity when the phase reaches 180°:

$$|A\beta| < 1 \tag{5.95}$$

Figure 5.35 shows the frequency response of the loop gain Aβ including amplitude and phase. Both gain and phase margins are illustrated in the figure. The *gain margin* (GM) is defined as the value of |Aβ| in dB, when the phase angle of Aβ is 180°, with respect to 0dB (unity gain). Thus, at phase angle 180° we have GM ≡ 0 − 20 log(|Aβ|). For example, if loop gain at 180° is −15dB, then GM = 0 − (−15dB) = 15dB. The gain margin can be viewed as the amount of loop gain that can be increased or decreased without making the feedback amplifier unstable.

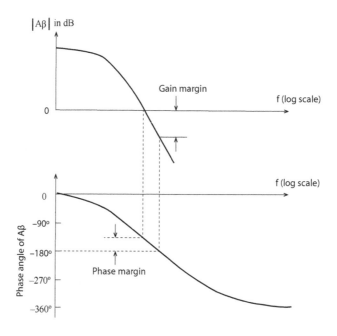

Figure 5.35 Frequency response of the loop gain Aβ illustrating the gain margin and phase margin

The *phase margin* is defined as 180° minus the angle of Aβ at which |Aβ| is 0dB (unity). The gain and phase margins give an indication of the feedback amplifier's stability. For a feedback amplifier with good stability, the gain margin is at least 10dB while the phase margin is at least 50°.

Exercises

Ex 5.1   Figure 5.36 is a cathode coupled amplifier (adapted from Figure 3.36[c]) with series-shunt feedback. If triodes T1 and T2 are identical with the same amplifying factor μ and plate impedance r_p, show that the loop gain $A\beta$, voltage gain with feedback A_f, and hence output impedance Z_{of} are given by

$$A\beta = \frac{\mu(RP)(R2)}{2r_p(RP+R1+R2)+RP(R1+R2)}$$

$$A_f = \frac{\mu(RP)(R1+R2)}{2r_p(RP+R1+R2)+RP[R1+R2(\mu+1)]}$$

$$Z_{of} = \frac{Z_o}{1+A\beta}, \text{ where } Z_o = r_p \| RP \| (R1+R2)$$

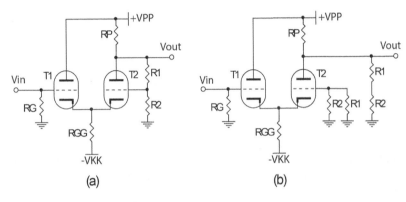

(a)   (b)

Figure 5.36 (a) A cathode coupled amplifier with series-shunt feedback, (b) the cathode coupled amplifier without feedback network but including the loading effect resistors

Ex 5.2 Figure 5.37 shows a shunt-shunt feedback amplifier formed by cascading three common source amplifiers. Assuming the transconductances for the JFETs are g_{m1}, g_{m2} and g_{m3}, respectively, show that the loop gain is given by

$$A\beta = \frac{g_{m1}g_{m2}g_{m3}R2R4(R1\|R8\|RS)(R6\|R8)}{R8(1+g_{m1}R3)(1+g_{m7}R7)}$$

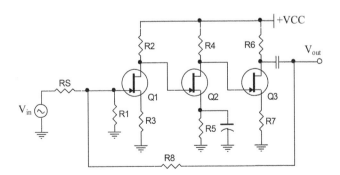

Figure 5.37 A shunt-shunt feedback amplifier formed by cascading three common source amplifiers

Ex 5.3 Determine the voltage gain of the triode vacuum tube amplifier in Figure 8.1. It is a variant of the Marantz line stage preamplifier, which has series-shunt feedback.

Ex 5.4 The voltage amplifier in Figure 5.38 is a series-shunt feedback amplifier. Assume that the transistors have same current gain $h_{fe} = 100$ and $V_{BE} = 0.7V$. Determine the voltage gain without feedback, loop gain $A\beta$, voltage gain with feedback A_f and output impedance Z_{of}.

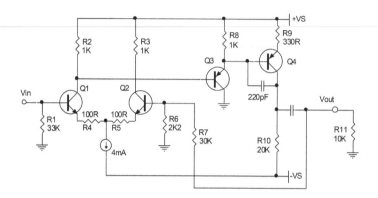

Figure 5.38 A voltage amplifier with series-shunt feedback

Ex 5.5 For the same amplifier shown in Figure 5.38, a 220pF capacitor is connected between the base and collector of transistor Q4. This creates a dominant pole to stabilize the feedback amplifier at high frequency. Assume the intrinsic base-collector capacitor is 5pF, which will, therefore, be added to the 220pF capacitor. Determine the Miller effect capacitor shunting the base of Q4, the impedance at the base of Q4. Then determine the higher –3dB frequency, f_H, without feedback. Finally, use the results from Ex. 5.4 to determine the higher –3db frequency with feedback, f_{Hf}, which is given by $(1 + A\beta) f_H$.

References

[1] Sedra, A. and Smith, K., *Microelectronic circuits*, Oxford, 7th edition, 2017.
[2] Boylestad, R. and Nashelsky, L., *Electronic devices and circuit theory*, Pearson, 10th edition, 2009.
[3] Gray, P. and Meyer, R., *Analysis and design of analog integrated circuits*, Wiley, 3rd edition, 1992.
[4] Millman, J., *Microelectronics, digital and analog circuits and systems*, McGraw-Hill, 1979.
[5] Gray, P.E., and Searle, C.L., *Electronic principles*, John Wiley and Sons, 1969.

6 Voltage buffer amplifiers

6.1 Introduction

In audio applications, we often come across situations where we need to connect source equipment (i.e., CD player) to receiving equipment (i.e., line-stage amplifier), or connect a line-stage amplifier to a power amplifier. We rarely encounter any problems in such situations. They work just fine. It is because the output impedance of the audio source equipment is low and the input impedance of the receiving equipment is high. Thus, most of the source signal is transferred into the receiving equipment. However, when the output impedance of the source and the input impedance of the receiving equipment are comparable, the signal drop across the source's output impedance will become significant.

Figure 6.1(a) shows a signal source, Vs, with impedance RS, connected to a resistive load, RL. RL can be the input impedance of the receiving equipment, or the input impedance of the following stage in an amplifier circuit. Vout can be easily determined by way of a voltage divider,

$$Vout = \frac{RL}{RS + RL} Vs \tag{6.1}$$

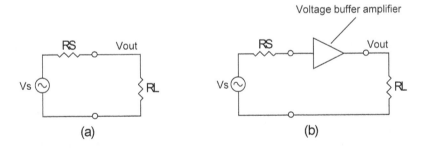

Figure 6.1 **(a)** A signal source Vs with impedance RS is connected to an RL in circuit, (b) a voltage buffer amplifier is inserted between the signal source and the load in circuit

It is clear that when RL ≫ RS, Vout is nearly equal to Vs. In other words, when RS is small and RL is large, most of the source signal is transferred to the load. But the problem comes when the condition RL ≫ RS is no longer valid. In order to maximize the source signal transferring

DOI: 10.4324/9781003369462-6

to the next stage, a voltage buffer amplifier is placed in between the source and RL as shown in Figure 6.1(b).

Figure 6.2(a) shows the ac equivalent circuit for Figure 6.1(b). The voltage buffer amplifier is modeled by a simple equivalent circuit with input impedance Zin, output impedance Zout and voltage gain Av. For an ideal voltage buffer amplifier, the voltage gain is unity; the input impedance is infinite and output impedance is zero. As a result, Vout is equal to Vs, as shown in

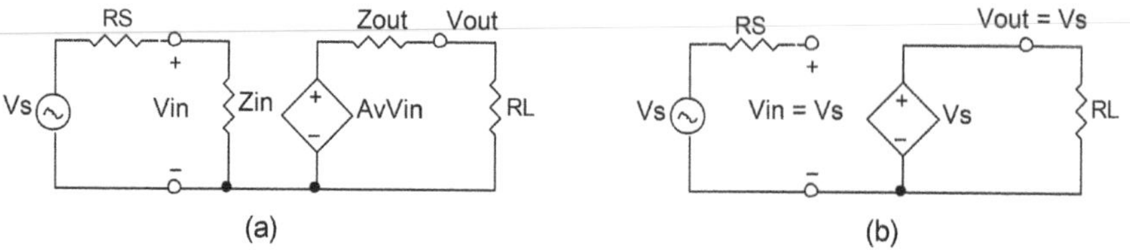

Figure 6.2 (a) The ac equivalent circuit for Figure 6.1(b). (b) The simplified ac equivalent circuit for the buffer amplifier having unity voltage gain (Av = 1), ideal infinite input impedance and zero output impedance

the simplified ac equivalent circuit in Figure 6.2(b). In other words, by placing an ideal voltage buffer amplifier between the source and the load, the source signal is transferred to the load.

In real life, the input impedance is not infinite and output impedance is not zero. However, a good voltage buffer amplifier should have the following properties:

• high input impedance,
• low output impedance,
• close to unity voltage gain,
• wide bandwidth.

All voltage buffer amplifiers discussed in this chapter have these properties. Specifically, the input impedance is greater than 100kΩ. Output impedance is lower than 200Ω for vacuum tube voltage buffer amplifiers and much lower for solid-state semiconductor voltage buffer amplifiers. Bandwidth is greater than several hundred kHz and many of the voltage buffer amplifiers may go up to MHz.

Another type of buffer amplifier is a current buffer amplifier. A current buffer amplifier is used in a system (or circuit) that transfers current from the first amplifier (or amplifying stage) with low output impedance to the second amplifier (or amplifying stage) with high impedance. The current buffer amplifier prevents the high input impedance of the second circuit from overloading the first circuit. In this book, we will only focus on the voltage buffer amplifier, which is simply referred to as a buffer amplifier in the remainder of this chapter.

We begin with a discussion of vacuum tube buffer amplifiers, followed by buffer amplifiers implemented by hybrid, op-amp and discrete components. A high current buffer amplifier, which is capable of delivering over 15A current, is also discussed. Finally, some practical applications for buffer amplifiers are illustrated.

6.2   Vacuum tube buffer amplifier

Figure 6.3 shows a simple buffer amplifier using a 6H30 triode in a cathode follower configuration. Vacuum tubes such as 12AU7, 6922, 6N1P, and ECC99 can also be considered. These tubes should work very well in this circuit for a plate current lower than 10mA. Owing to higher plate dissipation rating, ECC99 and 6H30 may work at a plate current higher than 10mA. Another advantage for ECC99 and 6H30 is that they have the lowest output impedance among these tubes.

First, for a vacuum tube to operate properly, we have to set up the desired plate current for the tube, as discussed in Chapter 2. When a vacuum tube is working in a class A operation, there is no grid current. Therefore, the buffer amplifier shown in Figure 6.3 has a dc bias current flowing from a power supply of +100V through the plate and cathode of the 6H30 and to the resistor R2. If we want to maximize output voltage swing, we must set the dc quiescent potential at point A, i.e., the cathode 6H30, to be halfway between the power supply +100V and –100V. When the dc quiescent potential at point A is set to ground level, the output can swing from 100V in the posi-

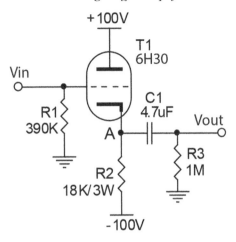

Figure 6.3 A cathode follower that requires a dual ±100V power supply

tive cycle, and –100V in the negative cycle. In both cycles, the output can swing with an amplitude of 100V. If we assume that the dc quiescent potential at point A is ground level, the plate and cathode current is given by [0 – (–100V)/R2] = 5.6mA. When a 6H30 operates at 5.6mA plate current, the grid-cathode voltage is found to be 5.5V. This makes dc quiescent at point A to be 5.5V above ground level. As a consequence, the maximum output voltage swing without clipping the waveform is 94.5V.

The input impedance is equal to R1. It is often called the grid leak resistor. It has sufficiently high input impedance (390kΩ) for a buffer amplifier. Since the grid is at ground level, this buffer amplifier can be directly connected to the source without the need of using a dc blocking capacitor. The gain of this buffer amplifier is given by Eq. (3.66) such that

$$A_v = \frac{\mu RK}{r_p + (\mu+1)RK} \tag{6.2}$$

where μ and r_p are the amplification factor and plate impedance of the triode, respectively. RK is the cathode resistor and in this example it is R2. When $(\mu+1)RK \gg r_p$, the above expression can be simplified to

$$A_v \approx \frac{\mu}{\mu+1} \tag{6.3}$$

Since the denominator is always greater than the numerator, it is obvious that the voltage gain for a cathode follower is always less than unity. When the amplifying factor is large, the voltage will approach unity. In this example, from the data sheet for 6H30, we have $\mu = 15$ and $r_p = 840\Omega$. Using RK = 18kΩ, both the exact and approximated expression for voltage gain gives $A_v = 0.93$. This is not an excellent figure but is good enough for a buffer amplifier. In comparison, the amplifying factors for ECC99 and 6922 are 22 and 33, respectively. These two tubes will produce a voltage gain slightly closer to unity. But when we look at the overall performance of 6H30 in terms of low output impedance, wide bandwidth, and high plate current rating, 6H30 is still an excellent choice for buffer amplifier application.

Figure 6.4 shows a buffer amplifier that is a variant of that shown in Figure 6.3, which only requires a single power supply. If the power supply requirement is reduced from dual to single, it reduces the size of the power supply as well as the requirement for the power transformer. Therefore, there is no surprise that Figure 6.4 is a more popular buffer amplifier than Figure 6.3. It is noted that even though the grid leak resistor R2 in Figure 6.4 has the same value, the input impedance is higher than that of Figure 6.3. This is because the grid leak resistor R2 is connected between the grid (input) and point B (near to the output). This creates a bootstrapping effect to R2 producing a greater effective input impedance.

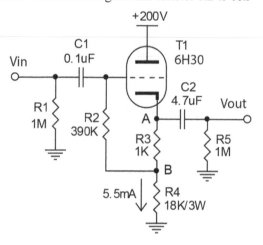

First, let us examine the dc quiescent condition of the buffer amplifier in Figure 6.4. 6H30 is a twin triode vacuum tube and has a maximum plate dissipation power of 4W per triode. In other words, if the plate cathode operates at 100V, the maximum plate current can be 4W/100V = 40mA. However, it is not recommended to operate a device at its maximum rating for power, current or voltage. In this example, 6H30's plate current is again set to around 5–6mA. When resistor R3 = 1kΩ and R4 = 18kΩ, the plate current is 5.5mA. Therefore, the dc quiescent potential at point A, the cathode of 6H30, is given by

Figure 6.4 A cathode follower with a single power supply

5.5mA × 19kΩ = 104V. It is close enough for the desired dc quiescent of 100V. The dc potential at the grid is lower than the cathode by a dc voltage drop across resistor R3. It is given by 5.5mA × R3 = 5.5V. The input impedance of this buffer amplifier may appear to be R2 + R4, but it is not. The reason is as follows.

In Figure 6.3, the input signal is fully across the grid leak resistor R1, which has one lead connected to the grid of 6H30 and the other lead connected to the ground. Clearly the input impedance is equal to R1. On the other hand, the grid leak resistor R2 of Figure 6.4 has one lead connected to the grid, and the other lead connected to point B (not ground), which carries a portion of output voltage. In other words, the input signal is not fully across resistor R2, which is therefore not the input impedance for the buffer amplifier. The amount of input signal across resistor R2 is determined as follows.

We first determine the voltage gain of the cathode follower. Since the same tube is used, the voltage gain is equal to $A_v = 0.93$, which is the same as in Figure 6.3. The output at point A is equal to Vout = 0.93Vin. The signal that appears at point B is easily calculated by way of a potential divider formed by R3 and R4. It is given by [R4/(R3 + R4)]×Vout = [18kΩ /(1kΩ + 18kΩ)]×0.93Vin = 0.881Vin. Therefore, the net signal across the grid leak resistor R2 is given as Vin − 0.881Vin = 0.119Vin. It presents an equivalent input impedance of R2/0.119 = 390kΩ/0.119 = 3.28MΩ. This is equal to the input impedance increased by eight-fold. Even though it is now necessary to use a dc blocking capacitor C1, it is just a small 0.1µF capacitor as the input impedance is very high.

It should be noted that in the above calculation for the signal appearing at point B, we have ignored the contribution from the input signal going to point B via resistor R2. Therefore, a more accurate model for calculating the signal presented at point B should be a sum of the signal coming from the input via resistor R2, and the signal coming from point A via resistor

R3. However, R2 is a large resistor and this makes the contribution from the input to point B insignificant compared to the signal from point A. If we use this accurate model to calculate the signal, the signal that appears at point B becomes 0.8813. It is very close to 0.881Vin that we have calculated earlier. The error is only 0.034%. Since the difference is so small, the equivalent input impedance remains at 3.28MΩ. The accurate model features as an exercise (Ex 6.1) at the end of the chapter.

From Eq. (3.62), the output impedance for a cathode follower is given by

$$Z_{out} = \frac{r_p RK}{r_p + (\mu + 1)RK}$$

where r_p = triode's plate impedance, RK = cathode resistor, μ = amplifying factor. If $(\mu+1)RK \gg r_p$, the above expression can be simplified to

$$Z_{out} \approx \frac{r_p}{\mu + 1}$$

In this example, $\mu = 15$, $r_p = 840\Omega$, the output impedance is approximately equal to 52Ω. When this buffer amplifier drives a 10kΩ load at a 4V peak-to-peak signal, the bandwidth is well over 1MHz. Even though it is not a perfect buffer amplifier yet, it behaves close to what we want from a buffer amplifier – high input impedance, low output impedance, and wide bandwidth.

From the expression for the voltage gain A_v, we know when the buffer amplifier has a large cathode resistor, RK, the gain will approach $\mu/(\mu+1)$. In other words, we will never achieve unity gain no matter how large a cathode resistor is unless the amplifying factor is approaching infinite. However, it should be noted that the linearity of the cathode follower is improved if a large cathode resistor is used. Therefore, it is always a design goal to make a large cathode resistor for a cathode follower.

A direct approach is to increase the power supply voltage. For instance, the supply voltage is increased from 200V to 250V as shown in Figure 6.5(b). If we keep the same plate-cathode voltage of 100V, the dc potential at the cathode is, therefore, 150V. If we keep the same plate current

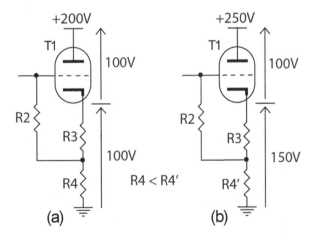

Figure 6.5 The circuit in (a) has a cathode dc quiescent voltage of 100V. In circuit (b), the cathode dc quiescent voltage increases to 150V so that a larger cathode resistor R4' can be used

for both Figure 6.5(a) and (b), we must have R4' > R4. A higher cathode resistor is therefore realized.

A second approach is to use a current source (CS), which has high impedance, as shown in Figure 6.6. There are several ways to implement the CS. We first discuss the use of vacuum tubes. The use of a solid-state semiconductor current source is discussed in the section on hybrid buffer amplifiers.

Since 6H30 is a twin triode vacuum tube, there are two triodes at our disposal. One triode can be used for the cathode follower. The second triode can be used for the current source as shown in Figure 6.7(a). Vacuum tube T2 and resistor R4 form the required current source. T2 is self-bias by the dc voltage across resistor R4. Since the cathode resistors R3 and R4 are identical, the power supply is equally split between the two vac-

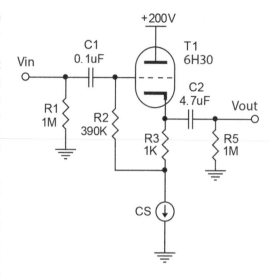

Figure 6.6 A current source is used to improve the linearity of a buffer amplifier

uum tubes, 100V for each. And the dc bias current also remains at 5.5mA, similar to the buffer amplifier in Figure 6.4(a).

From Eq. (3.52), it can be seen that the output impedance of the current source formed by vacuum tube T2 and the un-bypassed cathode resistor RK is given by

$$Z_o = r_p + (\mu + 1)\, RK, \text{ where } r_p = \text{plate impedance.}$$

In other words, the impedance of the current source is not just a simple plate impedance r_p, but it has increased by the term $(\mu + 1)RK$. If both the amplifying factor μ and the cathode un-bypassed resistor RK are large, the improvement can be significant. If 6H30 is used for the current source in Figure 6.7(a), the impedance is given as $Z_o = 0.84k\Omega + (15 +1) \times 1k\Omega = 16.8k\Omega$. This is not bad. However, it is not better than the buffer amplifier in Figure 6.4, where the R4 is a 18kΩ resistor. Nevertheless, the output impedance of the current source can be improved by using a triode vacuum tube with higher amplifying factor. Triodes 6922 and 12AT7, which have amplifying factor (μ) and plate impedance (r_p) higher than 6H30, are possible choices for the current source at plate current 5–6mA.

If we do not restrict ourselves to triodes, a pentode is a better choice as it has a higher plate impedance and amplifying factor. Figure 6.7(b) shows an arrangement where the triode cathode follower is sitting on top of a pentode current source. One popular pentode, which is still in production today for small signal amplification application, is EF86, or its variant EF806/6267. However, the maximum cathode current for EF86 is 6mA, making it not a good choice for this current source application. EF184/6EJ7 and EL83/6CK6 are better choices. However, production for these tubes ceased many years ago. The medium power pentode EL84, which is still production today, appears to be an appealing choice. The plate impedance for EL84 is 40kΩ. Maximum plate dissipation is 12W and maximum cathode current is 65mA. It has the right properties to work well as a current source. If EL84 is used, a higher voltage power supply is needed. If we allow plate-cathode voltage for tube T1 (6H30) to be 100V, and 150V for tube T2 (EL84), the power supply has to be 250V.

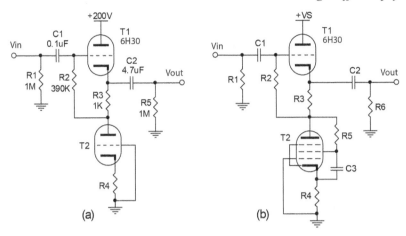

Figure 6.7 (a) A cathode follower using a triode current source, (b) a cathode follower using a pentode current source

6.3   Hybrid buffer amplifier

Figure 6.8(a) shows a hybrid buffer amplifier employing a vacuum tube cathode follower with an enhancement type MOSFET current source. Figure 6.8(b) shows the same cathode follower with a depletion type MOSFET current source. Let us first examine the dc quiescent condition of the hybrid buffer amplifier in Figure 6.8(a). The power supply is +100V and the MOSFET Q1 is sitting on a –45V power supply. In this arrangement, T1's grid is at ground level. Therefore, no input coupling capacitor is needed. The current source is formed by components R2, R3, R4, and Q1. Resistors R3 and R4 set up a quiescent dc at the gate of Q1 so that transistor Q1 delivers a constant drain current.

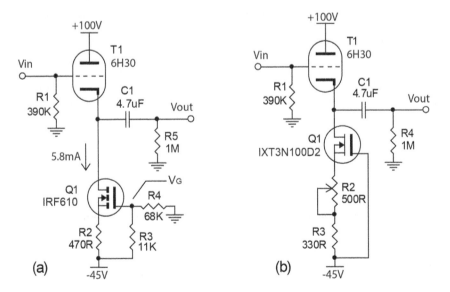

Figure 6.8 (a) A hybrid buffer amplifier employing a cathode follower and an enhancement mode MOSFET current source, (b) the same cathode follower using a depletion mode MOSFET current source

The resistance values for R2, R3, and R4 are determined as follows. Assume the gate dc quiescent potential for Q1 to be V_G as shown in Figure 6.9(a). Let $V_{GS(th)}$ be the gate-source threshold voltage of Q1. From Figure 6.9(b), we have

$$V_G = V_{GS(th)} + I_{bias} \times R2 - 45V \tag{6.4}$$

On the other hand, from Figure 6.9(c), we have

$$V_G = \frac{R4}{R3 + R4}(-45V) \tag{6.5}$$

Eliminating V_G from Eqs. (6.4) and (6.5) and simplifying yields

$$I_{bias} R2 = \frac{R3}{R3 + R4} 45V - V_{GS(th)} \tag{6.6}$$

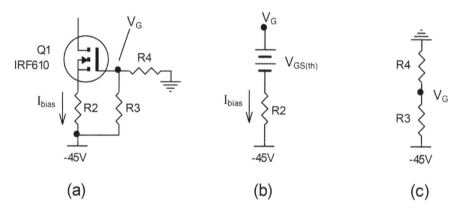

(a)   (b)   (c)

Figure 6.9 (a) The current source used in the buffer amplifier of Figure 6.8 (a), (b) a simplified model for determining the dc bias current for Q1 assuming a gate-source voltage $V_{GS(th)}$, (c) resistors R3 and R4 forming a potential divider to set up quiescent voltage at the gate, V_G

Here, we have some degree of freedom to determine the values for R2, R3, and R4. In order to set up a stable dc quiescent for Q1, we usually set up 2–3V across the source resistor R2. Let us choose a 2.5V across R2. Therefore, we have $I_{bias} \times R2 = 2.5V$. If we want to set $I_{bias} = 5.5mA$, we have R2 = 2.5V/5.5mA = 454Ω. We pick the nearest resistor, 470Ω.

Thus, Eq. (6.6) can be written as

$$2.5V = \frac{R3}{R3 + R4} 45V - V_{GS(th)}$$

From the data sheet for IRF610, the minimum $V_{GS(th)}$ is 2V and the maximum $V_{GS(th)}$ is 4V. Let us take a first approximation assuming the average $V_{GS(th)} = 3V$. The above expression can be written as

$$2.5V = \frac{R3}{R3 + R4} 45V - 3V \text{ and simplifying to}$$

$$R4 = 7.18 \times R3 \tag{6.7}$$

There are many possible combinations for selecting R3 and R4. Since there is no gate current for a MOSFET, we do not need to set a high dc bias current for the potential divider formed by R3 and R4. Let us say 0.5mA. Therefore, we must have 45V/(R3 + R4) = 0.5mA, or R3 + R4 = 90kΩ. Given this expression and Eq. (6.7), we obtain the first order approximation R3 = 11kΩ and R4 = 79kΩ. As we know, the first order approximation will only give us something to start with. Then we have to fine-tune the values for R2, R3 and R4 until the desired drain current of Q1 is achieved. The final values for the resistors are R2 = 470Ω, R3 = 11kΩ and R4 = 68kΩ, giving an output current of 5.8mA. The difference for R4 is due to the fact that the assumed $V_{GS(th)}$ of 3V is lower than the actual value, which is around 3.5–4V.

Figure 6.8(b) shows a hybrid buffer amplifier using a cathode follower and a depletion type MOSFET current source. The self-bias technique, which is often used for triodes and JFET, is also used for the Q1 depletion type MOSFET. When a dc bias current is established for Q1, the dc voltage across resistors R2 and R3 will set up a gate-source voltage. By varying the resistor R2, we can set up the desired dc current for Q1. It is noted that resistor R3 can be eliminated by using a 1kΩ variable resistor for R2. However, it is recommended to keep resistor R3 as it serves a dual purpose. The first is to determine the current by measuring the dc voltage across R3. The second is to limit the current when R2 is accidentally set to zero.

In addition to the two constant currents we have discussed in Figure 6.8(a) and (b), there are many other possible solutions. Figure 6.10 illustrates four options. Figure 6.10(a) is a simple current source formed by an NPN transistor, Q1. There are three diodes in series at the base of Q1. Resistor R2 is used to set up a proper bias current for the series diodes. For instance, the bias current of the diodes is given by $I_D = (VS - 3 \times V_F)/R2$. Usually, a few mA will be sufficient. V_F is the forward voltage of a diode. When the diodes are biased properly, they create $3 \times V_F$ forward voltage across the base of Q1 and resistor R1. Resistor R1 is used to set up the dc bias current for Q1, which becomes the output of the current source. Thus, the output current is given by Iout = $(3 \times V_F - V_{BE(Q1)})/R1$.

Figure 6.10(b) illustrates a popular current source formed by transistors Q1 and Q2 in a current mirror arrangement, often setting R1 equal to R2. When the collector current (I_{C2}) is set, the same amount of current will be reflected, like a mirror, to transistor Q1 as its collector current also. The collector current for Q1 is the desired output current for this current source. In other words, the output current (Iout) of transistor Q1 is simply determined by the dc bias current that we set for transistor Q2 and, hence, Iout = I_{C2}. Assuming the base current is negligible, the collector current for Q2 is given by $I_{C2} = [VS - V_{BE(Q2)}]/(R2 + R3)$. This current flows from the collector to the emitter. In contrast, in order to improve the stability of the current source, we usually set the voltage across the resistors R1 and R2 to be around 2V. Given this condition, we can easily find the value for R2. After R2 is determined, R3 can be determined by the expression for I_{C2}.

Figure 6.10(c) and (d) are current sources using enhancement type MOSFET. The resistors R1, R2, and R3 of Figure 6.10(c) can be determined in a similar way to the BJT version of Figure 6.10(b). Substituting a V_{BE} for a $V_{GS(th)}$, we can determine the resistors easily.

Figure 6.10(d) is an improved MOSFET Wilson current source [14]. A common Wilson current source used is one without transistor Q4. When Q4 is absent from Figure 6.10(d), the drain-source voltage (V_{DS}) on Q2 is greater than on Q1. For large threshold voltage $V_{GS(th)}$, this will cause a drain current mismatch between Q1 and Q2 due to the finite output impedance of the transistor. This can be overcome by adding transistor Q4 to the circuit. The Wilson current source has a higher output impedance than the simple current mirror. This is true for

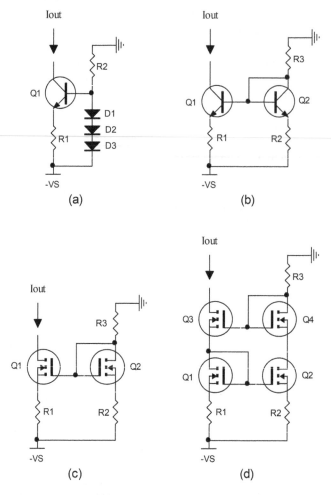

Figure 6.10 Some commonly used current sources. For convenience, it is often setting resistor R1 = R2
for current sources (b) to (d)

both BJT and MOSFET. This makes the current source of Figure 6.10(d) a better choice than
Figure 6.10(c), especially when it is used in an analog integrated circuit. Although it adds cost
and PCB space for the improved MOSFET Wilson current source by using discrete MOS-
FETs, I am pleased to use it for a cathode follower, a cathode coupled amplifier and a differential amplifier.

Figure 6.11 shows a buffer amplifier containing a cathode follower and a current source from
Figure 6.10(d). The output voltage swing is slightly over ±40V before the signal of the negative
cycle clips. If higher output voltage is needed, it is necessary to increase the negative power
supply. But when it is used as buffer amplifier in an audio line stage amplifier application, an
output swing of ±40V should be more than sufficient.

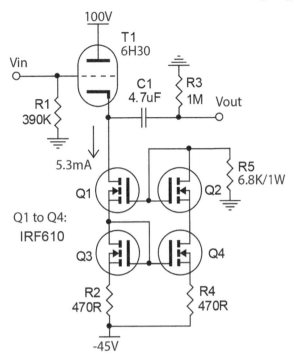

Figure 6.11 A hybrid buffer amplifier using a vacuum tube cathode follower and MOSFET current source

A note for the MOSFET device. There are several suppliers for IRF610 to IRF640, including Vishay, International Rectifier (now Infineon Technologies) and others. I have found that the same part from different suppliers produces noticeable differences in sound. General speaking, Vishay is my preferred supplier for IRF610 to IRF640.

6.4 Op-amp buffer amplifier

Op-amp provides the simplest form of buffer amplifier in terms of component count and PCB space. An op-amp buffer amplifier can be adapted easily, offering a flexible solution for unbalanced or balanced input and output. When we need a simple solution for voltage buffering purposes, we often consider the op-amp buffer amplifier first.

Figure 6.12(a) shows the simplest buffer amplifier configured by an op-amp. Unless otherwise stated in the data sheet, most op-amps are unity gain stable and can be operated as buffer amplifiers by connecting the inverted input to the output. This is a 100% feedback that returns all output to its negative input. Therefore, the buffer amplifier has unity gain, high input impedance, and low output impedance.

If we are looking for a buffer amplifier to handle a balanced input signal, Figure 6.12(b) is the answer. This is a simple difference amplifier. When all resistors are equal, as shown in the

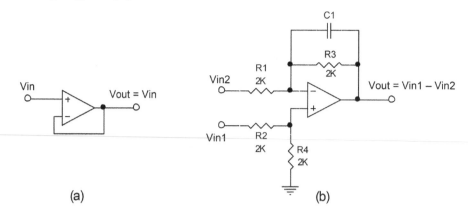

Figure 6.12 (a) An unbalanced input output buffer amplifier, (b) a balanced input–unbalanced output buffer amplifier

diagram, the voltage gain is unity. The compensation capacitor C1 is an optional component. It is used to ensure stability for the buffer amplifier at high frequency. C1 is usually around 10pF to 100pF. The balanced input impedance for the buffer amplifier of Figure 6.12(b) is R1 + R2 = 4kΩ. The input impedance is not very high, but it should be sufficient if the output impedance of the source is low. If high input impedance is needed, we can consider the instrumentation amplifier configuration of Figure 6.13.

The instrument amplifier of Figure 6.13 produces a voltage gain equal to $[1 + 2(R1/R2)]$. Unless we push the resistor R2 to an infinite high, the gain is always greater than unity. For

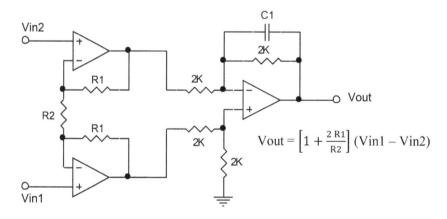

Figure 6.13 An instrumentation amplifier is used as buffer amplifier

instance, if we set R2 = 2×R1, the gain is 2. If we set R2 = 10×R1, the gain is 1.2. Therefore, the instrumentation amplifier will always provide some extra gain. It may not be a bad thing when the source signal is weak. The instrumentation amplifier has a very high input impedance, which is equal to the input impedance of the op-amp.

Figure 6.14 shows two buffer amplifiers that take an unbalanced input and produce a balanced output. The buffer amplifier in Figure 6.14(a) is the simpler solution. The op-amp on the top works as an inverted amplifier with unity gain, Vo1 = –Vin. The op-amp on the bottom works as a buffer amplifier with unity gain, Vo2 = Vin. Therefore, we have Vo2 – Vo1 = 2Vin. Thus, the differential output of this buffer amplifier has a gain of 2. Note that the top inverted amplifier and the bottom buffer amplifier are working independently. There is no interaction between the two amplifiers. The input impedance for the top inverted amplifier is 2kΩ while the bottom buffer impedance is very high. The difference in input impedance is likely to cause differences in phase shift to the two outputs at high frequency.

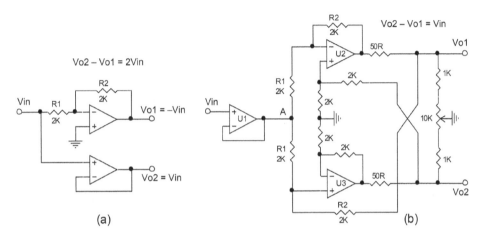

Figure 6.14 Buffer amplifiers that produce balanced outputs. The buffer amplifier shown in circuit (b) gives better balanced output than the one in circuit (a)

Figure 6.14(b) illustrates an improved version of Figure 6.14(a). Three op-amps (U1 to U3) are needed in this buffer amplifier arrangement [5]. Op-amp U3 produces the non-inverted output while U2 produces inverted output. Since the outputs are cross-coupled to the input of the op-amp that handles the opposite phase, the outputs are very well balanced. The gain of this amplifier circuit is determined by R2/R1. When we set R1 = R2, the gain is unity. Op-amp OP176 was originally used in this application with dual ±18V power supply. If output is limited to 14Vp-p, it can drive a differential load of 249Ω with a full-power bandwidth of 190kHz. The input impedance looking at point A is two R1s in parallel, i.e., 1kΩ. This is not a high impedance. Therefore, a simple buffer amplifier, formed by op-amp U1, is placed at the input.

The op-amp buffer amplifier that we have not yet discussed is a balanced input and balanced output buffer amplifier, as shown in Figure 6.15. The buffer amplifier featured in Figure 6.15(a) is the simplest form that we can achieve. It is easy to use and it will satisfy the need in most situation. However, the op-amps U1 and U2 are working independently to handle the signals from Vin1 and Vin2, respectively. As a result, this amplifier has a poor common mode rejection ratio (CMRR).

Figure 6.15(b) is an improved version of that shown in Figure 6.15(a). It is taken from the instrumentation amplifier of Figure 6.13. This is the simplest balanced input balanced output amplifier that we can achieve with such a low number of components. It has a gain of [1 + 2(R1/R2)].

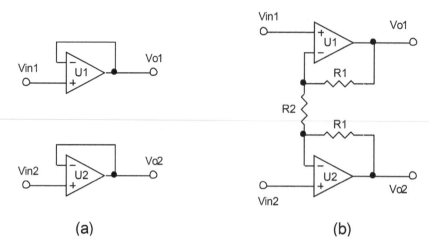

Figure 6.15 Balanced input and balanced output amplifiers. Two separate buffer amplifiers are used in circuit (a). An instrumentation amplifier is used in circuit (b)

As before, the gain is always greater than unity. When we choose resistor R2 ≫ R1, the gain will approach unity. If input impedance is not a concerning factor, the circuit in Figure 6.16 will provide a unity gain balanced input–balanced output buffer amplifier [6]. The input impedance is equal to 2×R1 = 4kΩ. The voltage gain is given by R2/R1. When we set R1 = R2, the voltage gain is equal to unity.

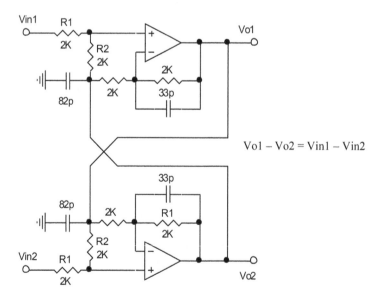

Figure 6.16 A balanced input and balanced output buffer amplifier with unity gain

6.5 Discrete buffer amplifier

If we do not restrict ourselves to just using vacuum tubes and op-amps, we open up a broader landscape of buffer amplifiers. When we move to discrete solid-state transistors there are many possibilities, ranging from low power buffer amplifiers (lower than 10mA output current) to high power buffer amplifiers (over 15A output current). The transistors can be a mix of JFET, BJT, and MOSFET. When a dual voltage power supply is used, the input and output can be biased at dc ground level. Thus, input and output coupling capacitors are eliminated.

Figure 6.17 shows several discrete buffer amplifiers employing JFET. Figure 6.17(a) is a basic buffer amplifier using two N-channel JFET, 2SK246. Transistor Q1 is working as a source

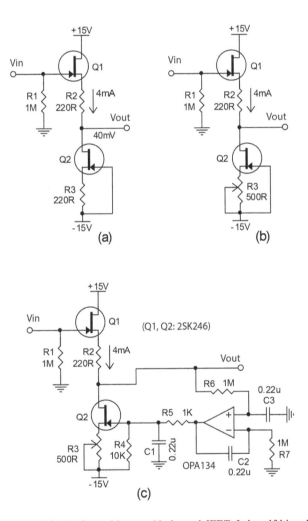

Figure 6.17 The buffer amplifier is formed by two N-channel JFET. It is self-biased in circuit (a) via a fixed resistor R3 and via a variable resistor R3 in circuit (b) so that the output dc offset can be adjusted to zero. (c) The buffer amplifier employs a dc servo circuit to eliminate dc offset at all times

follower, which is sitting on top of a current source formed by Q2. Transistor Q2 is self-biased by a dc voltage across resistor R3. When we set R3 = 220Ω, the drain current for both transistors is 4mA. When Q1's source resistor R2 is also set to 220Ω, and if transistors Q1 and Q2 are matched to the same I_{DSS}, the dc quiescent potential for Vout is nearly ground level, producing a very low output dc offset voltage. In practice, we can never get a pair of perfectly matched JFET and, therefore, in this example, there is a 40mV dc offset present at the output.

If the 220Ω resistor R3 is replaced by a 500Ω variable resistor, as shown in Figure 6.17(b), the output dc offset can be adjusted to zero. However, since V_{GS} and I_D are temperature dependent, the output dc offset is also temperature dependent. This is not desirable. Therefore, a dc servo circuit is needed to eliminate the output dc offset so that the output can be maintained at ground level at all times. Figure 6.17(c) shows how a dc servo is added to the buffer amplifier.

The output of the buffer amplifier contains two components – the ac signal coming from Vin and, the unwanted output dc offset voltage. What a dc servo does is to ignore the ac signal but amplifies the output dc offset voltage. When the amplified output dc offset voltage returns to the output, it is presented in the opposite polarity to the original output dc offset. As a consequence, the two dc offset voltages at the output with opposite polarities add up to zero. As a result, the output dc offset vanishes. In order to ensure that the amplified output dc offset is in the opposite polarity when reaching the output, we have to arrange the right path for it.

Note that from Figure 6.17(c), the path for the output dc offset starts from the buffer amplifier's output. Then it follows the op-amp, transistor Q2, and finally returns to the output. It is also important that the returning dc offset voltage must go through the gate of transistor Q2. If the input is to the gate and output is from the drain, Q2 operates as a common source amplifier for the returning dc-offset voltage. A common source amplifier is an inverted amplifier with negative gain. Then we have to select the correct input of the op-amp to amplify the dc offset voltage. The op-amp, which forms the dc servo, has two inputs to amplify the incoming dc offset from the buffer amplifier's output. Since the final amplified dc offset reaching the output must be opposite in polarity with respect to the original dc offset, and Q2 is an inverted amplifier, the op-amp must work as a non-inverted amplifier. Therefore, the buffer amplifier's output is fed to the op-amp's non-inverted input via resistor R6.

Since the dc servo circuit is only interested in the dc offset voltage, it will discard the ac signal. The ac signals are removed with the help of an RC filter formed by R6 and C3, R5, and C1, and a low-pass filter formed by the op-amp and capacitor C2. JFET 2SK246 is used for Q1 and Q2 in the buffer amplifiers of Figure 6.17. However, 2SK170 should work equally well with a change of source resistors R2 and R3 to suit a different $V_{GS(th)}$ and a different dc bias current. However, 2SK246 will be a better choice for the buffer amplifier in Figure 6.17(c) because 2SK246 has a higher $V_{GS(th)}$ than 2SK170. The $V_{GS(th)}$ for 2SK246 ranges from 0.7–6V while it is 0.2–1.5V for 2SK170. For small $V_{GS(th)}$, the dc potential at the op-amp's output in Figure 6.17(c) will be very close to the negative power supply rail (i.e., –15V). Op-amp OPA134 works very well in the dc servo circuit depicted in Figure 6.17(c) with 2SK246 for Q1 and Q2. However, OPA134 may or may not work when 2SK170 is used. If it does not work, then OPA134 has to be replaced by a rail-to-rail op-amp.

Depletion type MOSFET has electrical properties similar to JFET and, therefore, they can be used in a buffer amplifier as shown in Figure 6.18. MOSFET has a much higher drain-source voltage and drain current compared to JFET devices. For instance, the data sheet of 2SK246 states that the maximum drain-gate voltage is equal to 50V. However, the maximum drain-source voltage is not specified in the data sheet. But the two figures differ only by a V_{GS} voltage, which is just a few volts. Therefore, for convenience, we can assume that 2SK246 also has a maximum drain-source voltage of 50V. On the other hand, for example, the depletion type MOSFET DN2540N5, in a TO-220 package, has a maximum drain-source voltage of 400V and a maximum continuous drain current of 500mA. But 2SK246 can only handle drain current of less than 14mA. Therefore, there are great differences in terms of maximum drain-source voltage and drain current. In Figure 6.18 we have chosen DN2540N3, which is in the T0–92 package, and it can handle a maximum drain current of 120mA. This is a more suitable choice than DN2540N5 for this application. The buffer amplifiers from Figure 6.18(a) to (c) are biased to 6.2mA.

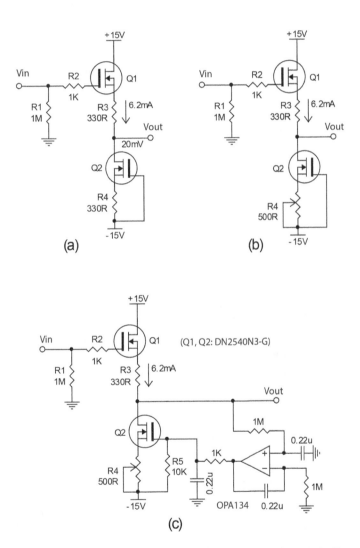

Figure 6.18 The buffer amplifier is formed by two depletion type MOSFET. The buffer amplifier is self-biased in circuit (a) via a fixed resistor R3 and (b) via a variable resistor R3 so that the output dc offset can be adjusted to zero. Buffer amplifier (c) employs a dc servo circuit to maintain zero dc offset at all times

In Figure 6.18(a), transistor Q2 is dc self-biased. It turns out that the output dc offset voltage is 20mV. It is even lower than the output dc offset from the JFET buffer amplifier of Figure 6.17(a), which has 40mV. This is not saying that a MOSFET buffer amplifier has lower output dc offset than a JFET buffer amplifier. Sometimes it may just happen that we have two closely matched MOSFET. When a variable resistor is used for R3, the output dc offset voltage of the buffer amplifier in Figure 6.18(b) can be adjusted to zero. If the dc servo is used as shown in Figure 6.18(c), we can ensure that the output dc offset is reduced to zero at all times.

We have seen that a pair of N-channel JFET and depletion type MOSFET can be easily used to implement buffer amplifier, which has the input and output at dc ground level. However, the shortcoming of such a simple buffer amplifier is that the output current is very limited. If we want to increase the output current, we can add one more stage using complementary NPN/PNP transistors, as shown in Figure 6.19. This buffer amplifier contains two stages. The first stage is a buffer amplifier formed by a pair of 2SK246s with a dc servo circuit similar to that shown in Figure 6.17(c). The second stage is formed by a pair of complementary NPN/PNPs (MJE15032 and MJE15033) arranged in a push–pull configuration. These devices are medium power transistors in the TO-220 package, which can be easily mounted on a heatsink. They are commonly used as driver transistors in an audio power amplifier application. A series of three diodes, D1 to D3, is inserted in between Q1 and Q2 so that Q3 and Q4 are biased to 40mA dc quiescent current. This buffer amplifier is suitable for applications where a few watts are needed to deliver to a load.

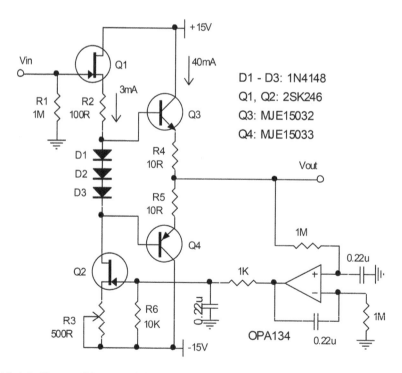

Figure 6.19 A buffer amplifier containing a pair of N-channel JFET and a complementary pair of NPN/PNP transistors

The buffer amplifiers discussed above have one thing in common. All of them are using one pair of N-channel JFETs or one pair of N-channel depletion type MOSFETs. If we now move to complementary N- and P-channel devices, we open up many possibilities for designing buffer amplifier. Before we move to the new buffer amplifier design, let us review the basic push–pull amplifiers in Figure 6.20.

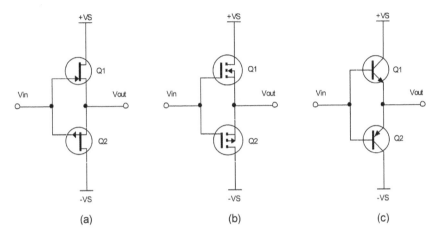

Figure 6.20 A push–pull buffer amplifier containing complementary JFET in circuit (a), complementary MOSFET in circuit (b), and complementary BJT in circuit (c)

If complementary MOSFET and BJT are used to form the buffer amplifier in Figure 6.20(b) and (c), they produce cross-over distortion. When the signal from Vin is less than the $V_{GS(th)}$ of a MOSFET, or V_{BE} of a BJT, the transistor is turned off. Therefore, there is no output until the input is greater than $V_{GS(th)}$ or V_{BE}. See the output in Figure 6.21. MOSFET and BJT devices must be biased to class A or class AB in order to eliminate the cross-over distortion. Even though the JFET buffer amplifier of Figure 6.20(a) does not produce cross-over distortion, there is a lack of control for setting the JFET to the desired dc quiescent current. Therefore, there is a need to develop a circuit that allows us to control the dc bias current of the complementary transistors (JFET, MOSFET and BJT) in Figure 6.20 before these buffer amplifiers can be of practical use. Figure 6.22 depicts a circuit that allows us to set the dc bias current for the complementary transistors.

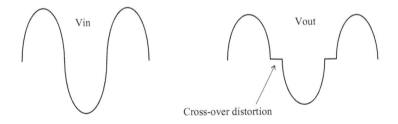

Figure 6.21 (a) Sinusoidal signal for Vin, (b) cross-over distortion appearing in Vout

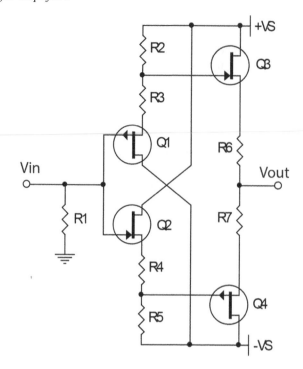

Figure 6.22 A buffer amplifier containing two pairs of complementary N- and P-channel JFETs. This configuration is also known as a "diamond buffer"

   In Figure 6.22, two pairs of complementary N- and P-channel JFET are employed to form a buffer amplifier [7, 8, 9, 10]. This configuration is also known as a *diamond buffer*, which first appeared in National Semiconductor's LH0002 buffer. If high input impedance is needed, it is recommended to use JFET or MOSFET devices for Q1 and Q2. Transistors Q1 and Q2 are configured into two separate source followers while Q3 and Q4 are arranged in a push–pull configuration. When the transistors Q1 and Q2 are biased properly, the dc voltage across R3 plus the V_{GS} of Q1 is sufficiently high to turn on Q3. Similarly, Q4 is turned on by the dc voltage across R4 plus the V_{GS} of Q2. This is a buffer amplifier with voltage gain close to unity and high input impedance. There is one shortcoming though. The output dc offset voltage can be substantial, as it is harder now to match two pairs of complementary JFET. We will address this later, by using the dc servo technique, showing how the output dc offset can be eliminated.

   Before we discuss using a dc servo circuit to eliminate the output dc offset, let us consider how we can improve the buffer amplifier featured in Figure 6.22. As we know, the current source has high impedance that improves linearity and reduces distortion to the source follower. Therefore, one important improvement is to use the current source to replace resistors R2 and R5 in Figure 6.22. The modified buffer amplifier is shown in Figure 6.23.

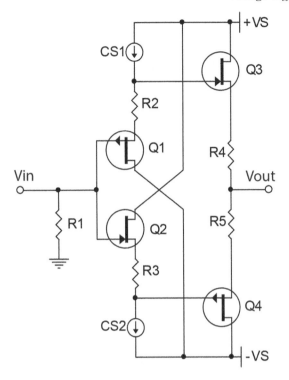

Figure 6.23 A diamond buffer amplifier containing two current sources and two pairs of complementary N- and P-channel JFETs.

Fixed-bias diamond buffer amplifier

There are two approaches that we can employ to implement the current sources CS1 and CS2. Figure 6.24 shows the first approach. Note that two complementary current sources are formed by transistors Q5 to Q8. In this arrangement, a constant current is first set up by Q5 and Q7. Then an equal amount of constant current is followed by Q6 and Q8. The constant current delivered by Q6 is used to bias the source follower Q1, while the constant current from Q8 is used to bias source follower Q2. Again, due to the mismatch of JFETs Q1 to Q4, a dc offset voltage is present at the output. A dc servo circuit is needed for the buffer amplifier to eliminate the output dc offset.

Figure 6.25 shows an all-MOSFET buffer amplifier with a dc servo circuit. Comparing to Figure 6.24, note that JFETs Q1 to Q4, which form the buffer amplifier, are now replaced by MOSFET. If JFET is preferred, it can be kept for Q1 and Q2. However, it is recommended to use MOSFET for Q3 and Q4 because it has greater current handling capability than JFET.

Assume that we want the dc bias current for the source follower formed by Q1 and Q2 to be 4.5mA and the output complementary push–pull source follower Q3 and Q4 to be 40mA. If higher or lower dc bias current is desired, we can easily change the components to meet the

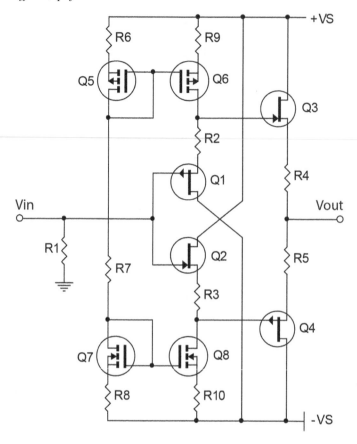

Figure 6.24 A fixed-bias diamond buffer amplifier containing two current sources formed by MOSFET
Q5 to Q8 and two pairs of complementary N- and P-channel JFETs

requirement. Note that there are two current sources. The first current source, which supplies the dc bias current for Q1, is formed by transistors Q5 and Q6. The second current source, which supplies the dc bias current for Q2, is formed by transistors Q7 and Q8. These two current sources are connected via resistor R4. Therefore, after the dc bias current is set up for Q5 and Q7, the dc bias current for Q6 and Q8 will follow automatically. The dc bias current of transistor Q6 is a "mirror image" of that for Q5. Therefore, it is logical that we first set up the dc bias current for Q5, and also for Q7.

To determine the dc bias current for Q5 and Q7, we draw a dc equivalent circuit for components R3, Q5, R4, Q7, and R5, as shown in Figure 6.26. There are three resistors (R3, R4, and R5) that we need to determine. In order to establish a stable dc bias current for Q5 and Q7, we set the dc voltage across the resistors R3 and R5 to be around 1.5V, or up to 2V, without sacrificing too much output voltage swing for the buffer amplifier. In this example, we set 1.5V across resistor R3 and R5. If we want I_{bias} to be 4.5mA, the resistor can be easily calculated as $1.5V/4.5mA = 330\Omega$. Then resistor R4 is determined by KVL such that,

$$15V - 1.5V - V_{SG(th)\_Q5} - I_{bias} \times R4 - V_{GS(th)\_Q7} - 1.5V = -15V$$

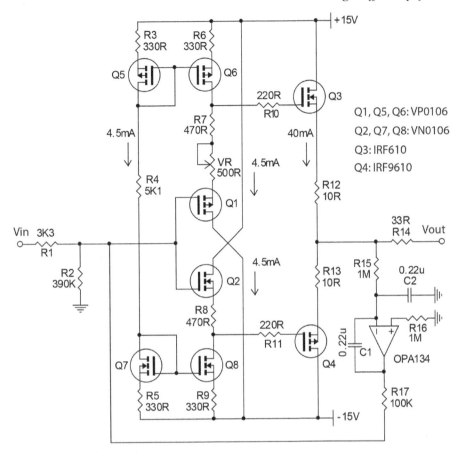

Figure 6.25 A fixed-bias diamond buffer amplifier containing two current sources formed by MOSFET Q5 to Q8 and two pairs of complementary N- and P-channel JFET in a source follower configuration. A dc servo is employed to eliminate the output dc offset voltage

After simplifying and rearranging, we have

$$R4 = [27 - V_{SG(th)\_Q5} - V_{GS(th)\_Q7}]/Ibias \tag{6.8}$$

Note that Q5 is a P-channel MOSFET, VP0106. From the data sheet we find that $V_{GS(th)}$ is ranging from –1.5V to –3.5V. We take an average of the values and assume $V_{GS(th)\_Q5} = -2.5V$. Since $V_{SG(th)}$ is equal to $-V_{GS(th)}$, we must have $V_{SG(th)\_Q5} = 2.5V$. On the other hand, Q7 is a N-channel MOSFET, VN0106. In a similar fashion we take an average of $V_{GS(th)}$ for Q7, which is ranging from 0.8V to 2.4V, so that $V_{GS(th)\_Q7} = 1.6V$. Solving Eq. (6.8) with $I_{bias} = 4.5mA$, we have R4 = [27V – 2.5V – 1.6V]/4.5mA = 5.088kΩ. Finally, we take the closest value 5.1kΩ for R4. Taking the calculated value for R3, R4, and R5, we find that the actual bias current is very close to 4.5mA that we wanted in the first place.

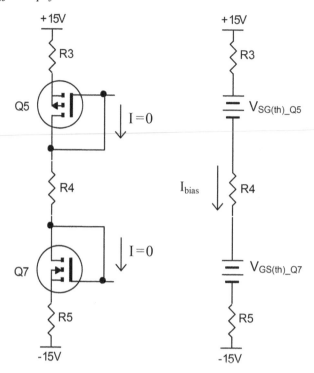

Figure 6.26 A dc equivalent circuit for determining dc bias current

After we have established the dc bias current for the current source, the dc voltages across R7, VR, R8, V_{GS_Q1}, and V_{GS_Q2} will be sufficient to establish the dc bias current for the push–pull output transistors Q3 and Q4. By adjusting the variable resistor VR, we can easily set a 40mA dc bias current for Q3 and Q4. Even though the supply voltage is not high, it is recommended to mount transistors Q3 and Q4 on a suitable heatsink.

A dc servo circuit, which is formed by OPA134, is included in the buffer amplifier to elimi-nate output dc offset. The dc servo circuit operates to amplify the dc offset voltage but discards the ac signal coming from Vout. The RC filter, R15 and C2, and the low pass filter formed by the op-amp and C1 removes the ac signal content. Since Vout is fed to the inverted input of the op-amp, the amplified dc offset voltage is in the opposite polarity to the original output dc offset. Then the amplified dc offset is fed to the input of the buffer amplifier. Because the buffer amplifier has a non-inverted gain of unity, when the amplified output dc offset voltage returns to the output, it remains in the opposite polarity to the original output dc offset. The two dc offset voltages at the output with opposite polarities, therefore, sum up to become a smaller dc offset. Gradually, the output dc offset is reduced to zero. More dc servo examples can be found in Chapter 4.

It should be noted that when the dc servo circuit operates, it will create a small dc offset at the input. In order to prevent the input source from affecting this small dc offset voltage, resis-tor R1 is placed in series to the input. In this buffer amplifier, we must have R1 ≥ 3.3kΩ. The input impedance of the buffer amplifier is equal to R1 + R2∥R17 = 83kΩ. In contrast, when the dc servo circuit operates, it has no impact on the dc bias current of the source followers Q1 and Q2. Therefore, we call it a fixed-bias diamond buffer amplifier so as to distinguish it from the floating-bias diamond buffer amplifier discussed below.

Floating-bias diamond buffer amplifier

The fixed-bias buffer amplifier shown in Figure 6.25 requires the dc servo to be directly connected to the input. Therefore, this reduces the input impedance. In Figure 6.25, the input impedance of the buffer amplifier is 83kΩ. It is not a bad impedance and it should work very well in most applications. If we put this buffer amplifier as the output stage of an audio line-stage amplifier or power amplifier, and global feedback is used, the dc servo will no longer need connecting to the input of the buffer amplifier. Then the high input impedance of the buffer amplifier can be preserved.

We now consider the second approach to implement the current source of Figure 6.23 in such a way that the dc servo no longer needs connecting directly to the input of the buffer amplifier. As a result, this new buffer amplifier always has a high input impedance. This buffer amplifier is shown in Figure 2.27. Transistors Q5 and Q6 work as current source for source followers Q1

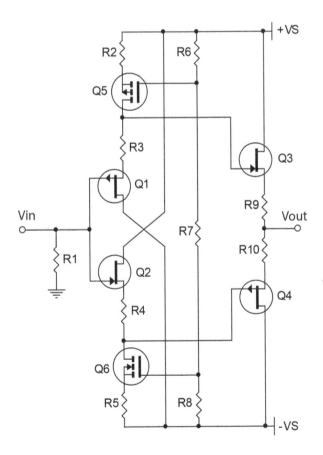

Figure 6.27 A diamond buffer amplifier employing two current sources formed by Q5 and Q6

and Q2, respectively. The dc bias current for Q5 and Q6 is set by the potential divider formed by R6, R7, and R8. When a sufficient dc voltage is applied to the gates of Q5 and Q6, they deliver constant current to Q1 and Q2.

It should be noted that the dc bias current for the potential divider formed by R6, R7, and R8 is determined by $(2 \times VS)/(R6 + R7 + R8)$. This dc bias current is set to create a sufficient dc voltage across R6 to turn on Q5 so that it creates a constant current to bias Q1. Similarly, a sufficient dc voltage across R8 turns on Q6 so that it creates a constant current to Q2. If we modify the potential divider and turn it into a folded potential divider, as shown in Figure 6.28, we can now adjust the variable resistor VR to alter the dc bias current for the folded potential divider. When the variable resistor is leaning toward +VS, there is lower dc voltage across R6 but higher dc voltage across R8. Therefore, Q5 has a lower dc bias current due to low gate-source voltage across R6, while Q6 has a higher dc bias current due to high gate-source dc voltage across R8. Since Q5 produces lower dc bias current supplying Q1, Q1 carries lower dc bias current also.

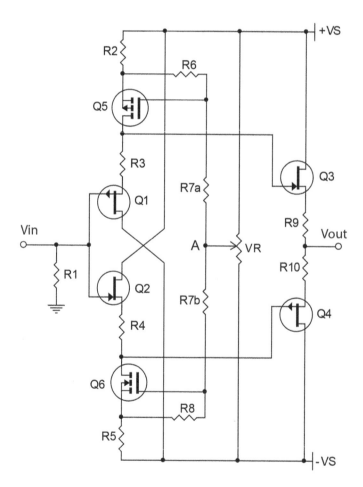

Figure 6.28 The diamond buffer amplifier in Figure 6.27 is modified by changing the fixed potential divider to a folded potential divider. A variable resistor VR is connected to power supply ±VS so that output dc offset can be adjusted to zero

However, Q6 produces higher dc bias current supplying Q2, therefore, Q2 must carry higher bias current. In contrast, if the variable resistor VR is leaning towards –VS, it creates an opposite effect to the dc bias current for Q1 and Q2. In other words, by adjusting the variable resistor VR, we can alter the dc bias current of the source follower formed by Q1 and Q2. As a result, the output dc offset voltage can be set to zero by adjusting the variable resistor VR.

Since the dc bias current for Q1 and Q2 can be varied by adjusting the variable resistor VR, the dc bias technique for the diamond buffer amplifier in Figure 6.28 is different from the fixed-bias diamond buffer amplifier discussed in the preceding section. In order to distinguish between the two buffer amplifiers, the buffer amplifier of Figure 6.28 is referred to as a floating-bias diamond buffer amplifier.

Note that by introducing a variable resistor VR, which is connected to the ±VS power supply, the output dc offset voltage can be adjusted to zero. However, we do not want to physically put up a variable resistor to manually cancel out the output dc offset. This is just to illustrate that it is feasible to do so by changing the dc quiescent potential at point A. Our goal is to employ a dc servo circuit to monitor the output dc offset and take it down to zero. Point A simply provides the best spot for a dc servo circuit to set control.

Figure 6.29 shows a floating-bias diamond buffer amplifier employing a dc servo circuit. The variable resistor VR of Figure 6.28 is now replaced by a dc servo circuit in Figure 6.29. JFET

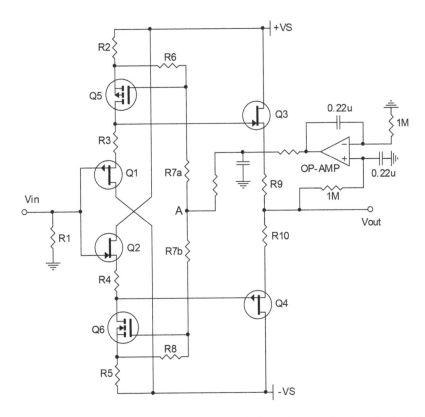

Figure 6.29 A floating-bias diamond buffer amplifier employing a dc servo circuit to eliminate the output dc offset voltage

transistors are kept for Q1–Q4. However, MOSFET can also be used for Q1–Q4, leading to an all-MOSFET buffer amplifier [11].

Figure 6.30 shows an all-MOSFET floating-bias diamond buffer. First, a folded potential divider is formed by resistors R3, R9a, R9b, R10, R11, R12b, R12a, and R8. The total sum of the resistance of the potential divider ranges from 26.24kΩ to 28.24kΩ depending the final resting position of the variable resistors R9b and R12b. If we assume the dc quiescent potential at point A is equal to zero, dc bias current I5 is equal to I6 and they can be roughly determined as 48V divided by the total resistance of the potential divider. This gives I5 and I6 equal to 1.7mA to 1.8mA. However, since point A is rarely resting at ground level in the actual circuit, the dc bias currents I5 and I6 are not equal. Also, the real life dc bias current for I5 and I6 is not exactly equal to 1.7–1.8mA. Nevertheless, this gives us a rough estimation of what dc bias currents I5 and I6 would be.

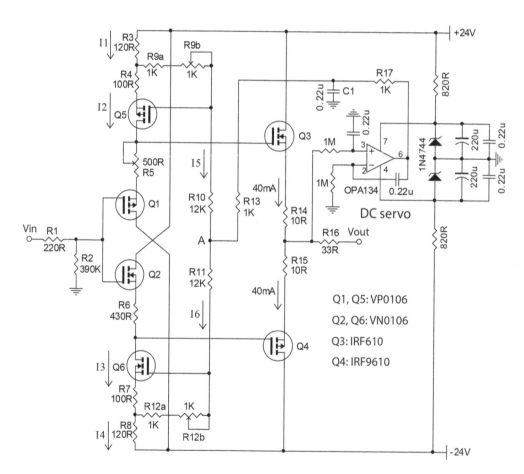

Figure 6.30 An all-MOSFET floating-bias diamond buffer amplifier with a dc servo circuit

In this buffer amplifier, the dc bias current is I2 for transistors Q1 and Q5, and I3 for transistors Q2, while Q6 is set to 5.5mA. It is an appropriate dc bias current for the chosen MOSFET. The sum of I2 and I5 (I1 = I2 + I5) is, therefore, equal to 7.2–7.3mA. Similarly, we also have

I4 = I3 + I6 = 7.2mA to7.3mA. In order to make sure we are getting the desired dc bias current, we adjust the variable resistor R9b in such a way that the dc voltage across R3 is equal to I1 × R3 = 0.86V to 0.88V. Similarly, we adjust R12b so that the dc voltage across R8 is equal to 0.86–0.88V.

On the other hand, resistor R4 is inserted between the source of Q5 and R3, and resistor R7 between source of Q6 and R8. These source resistors serve two purposes. The first is to limit the source current. The second is to provide an additional dc voltage drop so that the source is about 1.5V from the power supply rail. This additional dc voltage drop is equal to I2 × R4 = 5.5mA × 100Ω = 0.55V, and the same for I3 × R7.

In order to understand how the source resistor helps to limit the source current, let us consider Figure 6.31. The circuit of Figure 6.31(a) is taken from Figure 6.29, where a source resistor is not used. If dc bias currents I3, I4 and I6 remain constant throughout the operation, there is nothing to worry about. However, when transistor Q6 is heated up, the V_{GS} drops, the voltage across R8 (i.e., I6 × R8) will push the transistor to conduct more and, therefore, cause the transistor to run at a higher dc bias current. Then, the transistor will be further heated up and V_{GS} is further reduced. It is like a thermal run-away problem similar to that of a BJT transistor. In principle, a large resistor R5 could prevent such a thermal run-away from happening. If we now insert an additional resistor, R7, at the source of Q6, as shown in Figure 6.31(b), the additional dc voltage across R7 works more effectively to compensate for the decrease of V_{GS} due to temperature rise. Therefore, this greatly stabilizes the transistor against temperature change. A 100Ω resistor for R7 will be sufficient in this buffer amplifier.

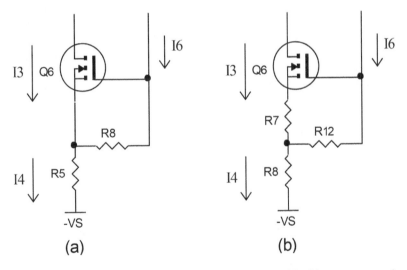

(a)   (b)

Figure 6.31 Circuit (a) shows the dc bias current of Q6 from Figure 6.29 without a source resistor. Circuit (b) shows the dc bias current of Q6 from Figure 6.30 with a source resistor, R7

Again, a dc servo circuit is employed to eliminate output dc offset voltage. The output dc offset is first fed to the non-inverted input of the op-amp. After the dc offset is amplified by the op-amp, while the ac signal content is discarded, it is then fed to point A. Transistors Q5 and Q6 are operating in a common source amplifier, which is an inverted amplifier having a negative gain, for the dc offset signal coming from point A. Therefore, the dc offset, after being amplified by Q5 and Q6, has an opposite polarity to that of the original dc offset. Transistors Q3 and

Q4 are operating as a source follower, which is a non-inverted amplifier. Therefore, when the amplified dc offset returns to the output, it has an opposite polarity. As a result, the two dc offset voltages presenting at the output with opposite polarities sum up to become a smaller dc offset. Gradually, the output dc offset is reduced to zero.

When the dc servo takes down the output dc offset voltage, the dc quiescent potential at point A may shift away from the ideal dc ground level. Depending on the condition of mismatching among the transistors, the dc quiescent potential at point A may be resting above or below the dc ground level. As a consequence, the dc bias currents I2 and I3 are not equal, so neither are I5 and I6. Note that the output of this dc servo is designed not to be connected to the buffer amplifier's input, because it may load down the buffer amplifier's input impedance. Thus, the input impedance of this buffer amplifier is high. Hence, the floating-bias diamond buffer's input impedance can be higher than that of the fixed-bias diamond buffer, which may require the output from the dc servo to be connected to the buffer amplifier's input. Finally, transistors Q3 and Q4 are running at 40mA dc bias current. They must be mounted on heatsinks with a suitable size.

6.6   High current buffer amplifier

The fixed-bias and floating-bias type buffer amplifiers can be modified by paralleling power transistors to boost the output current. In this section, we discuss how a floating-bias type buffer amplifier can be turned into a high current buffer amplifier [11]. This buffer amplifier is capable of delivering over 15A peak current to a load. If a higher output current is needed, it can be easily achieved by paralleling more power transistors to the output.

Figure 6.32(a) is a high current floating-bias diamond buffer amplifier containing MOSFET drivers and BJT output power devices. Pre-driver transistors Q1 to Q4 are MOSFET IRF610 and IRF9610, while driver transistors Q5 and Q6 are IRF630 and IRF9630. They are in the TO-220 package that allows them to be easily mounted on heatsinks. The output contains three pairs of complementary power BJTs, TTC5520 and TTA1943. These power BJTs are rated at 15A maximum collector current and 150W power dissipation. When three pairs of TTC5520 and TTA1943 are used to deliver a 15A peak current to a load, each pair of transistors will only be required to deliver 5A. This should not create much difficulty for the power transistors.

The folded potential divider that sets the dc bias current for Q1–Q4 is formed by resistors R3, R4, VR3, R9, R10, VR4, R7, and R8. These resistors are chosen in such a way that the dc bias current for the potential divider network is slightly below 2mA. The variable resistors VR3 and VR4 are used to set the dc bias current for Q1–Q4. The dc bias current for resistors R3 and R8 is a sum of the two dc bias currents and, therefore, becomes 12mA, as the gate of a MOSFET does not draw any current. By adjusting the variable resistor VR3 so that the dc voltage across R3 is equal to 1.2V, we set up a 10mA dc bias current for Q1 and Q2. Similarly, VR4 sets the dc voltage across R8 equal to 1.2V, so that Q3 and Q4 are biased to 10mA. Two zener diodes connecting in a reversed fashion are connected across the gate-source of Q2 and Q3. This prevents the input of the MOSFET from encountering an unexpected high dc or unwanted transient that may exceed $V_{GS(max)}$.

The buffer amplifier of Figure 6.30 only requires one variable resistor R5 to set the dc bias current for the push–pull output stage. But the buffer amplifier in Figure 6.32(a) employs two variable resistors VR1 and VR2 for setting the dc bias current to Q5 and Q6. The advantage of having two variable resistors is that point A, the mid-point of the potential divider, can be adjusted to a dc quiescent potential close to the ground level. In practice, the dc quiescent of point A can be above or below ground, depending on the conditions determined by the dc servo circuit. When the dc potential at point A is close to ground level, this implies that there is very

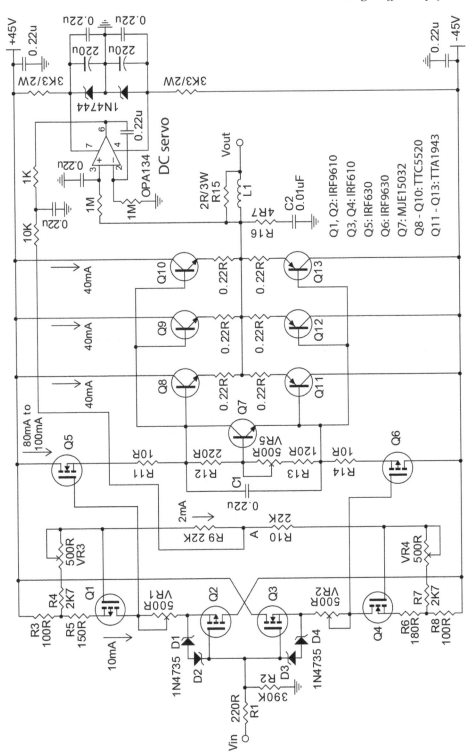

Figure 6.32 (a) A floating-bias diamond buffer amplifier capable of delivering over 15A output current

little current sourcing from, or sinking to, the op-amp of the dc servo circuit. Therefore, the same amount of dc bias current is running through the potential divider. As a result, the dc bias current for the source followers Q2 and Q3 will be nearly the same. When the dc quiescent potential at point A is farther away from the ground level, the source follower Q2 and Q3 will be running at different dc bias currents. However, if the difference is not too big, it does not affect the operation of the buffer amplifier. In summary, the use of two variable resistors will allow us to set the dc bias current for the complementary source followers Q5 and Q6, and, at the time, they minimize the difference in dc bias current between the input source followers Q2 and Q3.

The complementary source followers Q5 and Q6 operate as drivers that provide low impedance and sufficient current drive to the base of the output transistors Q8–Q13. Therefore, transistors Q5 and Q6 are set to a high dc bias current around 80mA to 100mA. By adjusting variable resistors VR1 and VR2, we can set up any desired dc bias current for Q5 and Q6. The bias current can be easily determined by measuring the dc voltage across resistor R11 and R14. If 100mA dc bias current is needed, the voltage across R11 and R14 should be 1V.

Transistor Q7 works as a V_{BE} multiplier. More details of V_{BE} multipliers are discussed in Chapter 12. The function of a V_{BE} multiplier is to set up a dc voltage spread between the bases of the output NPN and PNP power transistors so that they are biased in class AB with a small emitter current. In this buffer amplifier, the emitter current is set to 40mA. A small 0.22Ω resistor is placed at the emitter on each power transistor. This 0.22Ω emitter resistor works as a degeneration resistor that ensures the output current can be evenly distributed among the transistors; for example, if it happens that transistor Q8 starts delivering more emitter current than the transistors Q9 and Q10. This may be due to Q8's V_{BE} being lower than the other two transistors. The increased emitter current will increase a dc voltage across the 0.22Ω emitter resistor and, as a consequence, it raises the dc potential at the base of Q8. The dc potential at the base of Q8 becomes momentarily higher than the base of the other two transistors, Q9 and Q10. The higher base potential will oppose a further increase in emitter current in Q8. On the other hand, at a lower base dc potential than Q8, transistors Q9 and Q10 are now in favor to increase emitter current. Hence, the small emitter resistor at each power transistor plays an important role to ensure the output current is evenly distributed among output power transistors.

Additionally, the emitter resistor also prevents the thermal run-away problem from happening to the power transistor. The V_{BE} of bipolar transistor is usually around 0.6V–0.7V and it is temperature dependent. It is well known that V_{BE} has a negative temperature coefficient ($-2mV/°C$) such that higher the temperature rises, lower the V_{BE} will be. Thus, if, for instance, transistor Q8 carries a high emitter current and it causes the transistor to heat up. Due to a rise in temperature, the V_{BE} decreases. As V_{BE} is reduced, Q8 is in favor to carry more emitter current. If Q8 carries more emitter current, more heat dissipation is generated in the transistor and causes a further rise in temperature. The V_{BE} is further reduced and causes the transistor to carry more emitter current. The transistor is effectively being trapped in a positive feedback loop that eventually leads the transistor to self-destruct. If a small emitter resistor is in place, an increase in emitter current will cause an increase in the dc voltage across the emitter resistor. As a consequence, the dc potential of the transistor's base is also increased. The increased base dc potential will oppose any further increase in emitter current. Therefore, this prevents the transistor from experiencing the thermal run-away problem. In order to avoid excessive power loss in the emitter resistor, it is usually around 0.1Ω to 0.33Ω for a power BJT. Since a power MOSFET has a higher $V_{GS(th)}$, the source resistor is usually 0.33Ω or greater.

Input impedance of this buffer amplifier is $390k\Omega$. A simple output filter (Zobel network) formed by R15, R16, L1, and C2 is employed to improve the stability of the buffer amplifier for driving inductive load. Since the power dissipation from the transistors is high, transistors

Q5–Q13 must be mounted on heatsinks. Transistor Q7 should be mounted on the heatsink close to the power transistors to compensate for the V_{BE} change due to a rise in temperature. Power dissipations in Transistors Q1–Q4 are relatively small. However, they still have to be mounted on small individual heatsinks.

To implement a buffer amplifier for a fully balanced power amplifier, two single buffer amplifiers shown in Figure 6.32(a) are used. One single buffer amplifier is used for handling the output signal for the positive phase, while another buffer amplifier is employed for the negative phase. A total of six pairs of output power transistors are mounted on the surface of the heatsink. The entire buffer amplifier is shown in Figure 6.32(b). Four PCB supporters are provided for mounting the PCB of the combined input stage and voltage amplifying stage so that a complete fully balanced power amplifier can be built on top of a heatsink.

Figure 6.32(b) A balanced buffer amplifier formed by using the two floating-bias diamond buffers of Figure 6.32(a). Four PCB supporters are given for mounting the PCB of the combined input stage and voltage amplifying stage

Above we have discussed a high current floating-bias type buffer amplifier. However, a high current fixed-bias type buffer amplifier can also be realized in a similar fashion. As output power is high, it requires several pairs of power transistors in parallel at the output and a V_{BE} multiplier to set up the bias current.

6.7 Buffer amplifier applications

A buffer amplifier is commonly used as a line driver in a low level signal application. However, many of the buffer amplifiers discussed above are useful for low to high level audio applications. For instance, by adding a potentiometer, we can turn a vacuum tube buffer amplifier into a line-stage amplifier. Most modern CD players produce output higher than $4V_{p-p}$. If the source is from a CD player, even though a buffer amplifier has unity gain, $4V_{p-p}$ will be sufficient to drive most audio power amplifiers into rated output.

However, the most important application for a buffer amplifier is to work as the output stage in audio amplifiers. Audio line-stage amplifiers and power amplifiers are particularly suitable for a discrete buffer amplifier to fit in. Figure 6.33(a) shows a common configuration of a solid-state

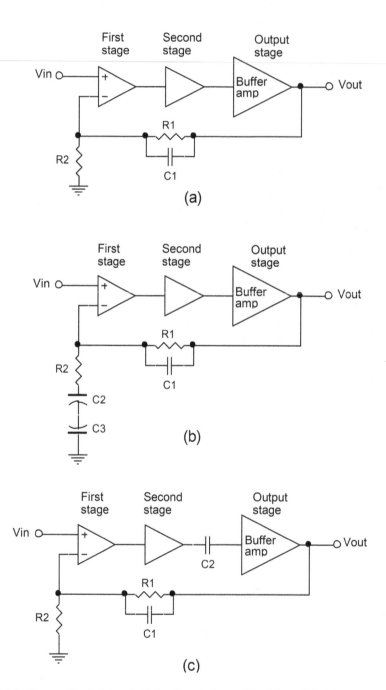

Figure 6.33 A buffer amplifier is integrated into a line-stage amplifier (a), a solid-state power amplifier (b), and a hybrid power amplifier (c)

semiconductor line-stage amplifier that has a flat voltage gain from dc to over 100kHz. The signal gain is low (20dB or lower) and the amplifier's output is within a few volts. The output dc offset voltage can be easily reduced to zero by employing a dc servo circuit.

Figure 6.33(b) shows a similar configuration but it limits the dc gain. This structure is commonly used for a solid-state power amplifier. Since the output dc offset is always a concern for power amplifiers, which have a signal gain of over 20dB, capacitors C2 and C3 are used to limit the dc gain to unity.

Figure 6.33(c) shows the structure of a hybrid power amplifier. Vacuum tubes are used for voltage amplification and solid-state semiconductor devices for current amplification. A dc blocking capacitor C2 is used to isolate the dc from the second stage, which is a vacuum tube amplifier operating at high dc supply voltage. Since the floating-bias type buffer amplifier does not require the dc servo output to be connected to the buffer amplifier's input, it is a better choice than the fixed-bias type buffer amplifier for using in the output stage of hybrid power amplifier. Below, we discuss briefly how a buffer amplifier is integrated into a line-stage amplifier and power amplifier.

Line-stage amplifier

The first example is an audio line-stage amplifier as shown in Figure 6.34. It contains a fixed-bias type buffer amplifier. The input stage of the line-stage amplifier employs a pair of complementary (N-channel and P-channel) JFET cascode differential amplifiers. The cascode differential amplifier on the top contains N-channel JFETs (Q1–Q4) of which the dc bias current is set by current source CS2. The cascode differential amplifier on the bottom contains P-channel JFETs (Q5–Q8), of which the dc bias current is set by the current source CS1. The second stage employs a pair of complementary MOSFET cascode amplifiers arranged in such a way that the drains of Q10 and Q11 are facing each other. The cascode amplifier on the top contains two P-channel MOSFETs (Q9 and Q10) while the bottom cascode amplifier contains two N-channel MOSFETs (Q11 and Q12). The two cascode amplifiers operate in such a way that the drain of Q11 works as an active load for Q10, while the drain of Q10 works as an active load for Q11. The output from the second stage is directly coupled to the output stage, which contains a fixed-bias type buffer amplifier formed by transistors Q13–Q20.

Note that a dc servo circuit is used to eliminate the output dc offset. The buffer amplifier of Figure 6.25 suggests that the output of the dc servo circuit is returned to the input of the buffer amplifier, i.e., the gates of Q15 and Q16 in Figure 6.34. However, this will create a loading effect to the drain of Q10 and Q11 so that the voltage gain of the second stage is reduced. In order to avoid this, the output of the dc servo circuit is now returned to the input of the line-stage amplifier. The input impedance is reduced but it is still sufficiently high for a line-stage amplifier. However, the open-loop gain of the amplifier remains high and it helps to reduce the distortion and broaden the bandwidth of the line-stage amplifier by negative feedback. Details of solid-state line-stage amplifier design are discussed in Chapter 7.

Power amplifier

The second example is an audio power amplifier as shown in Figure 6.35 that contains a fixed-bias type buffer amplifier. The input stage is a cascode differential amplifier formed by transistors Q1–Q4. The dc bias current of the cascode differential amplifier is set by a current source formed by transistors Q5 and Q6. The second stage contains an emitter follower formed by transistor Q7 that is then followed by a cascode amplifier formed by transistors Q8 and Q9. Transistors Q10 and Q11 work as active loads for Q8 and Q9. The output from the second stage is directly coupled to the output stage, which contains a high current fixed-bias type buffer amplifier.

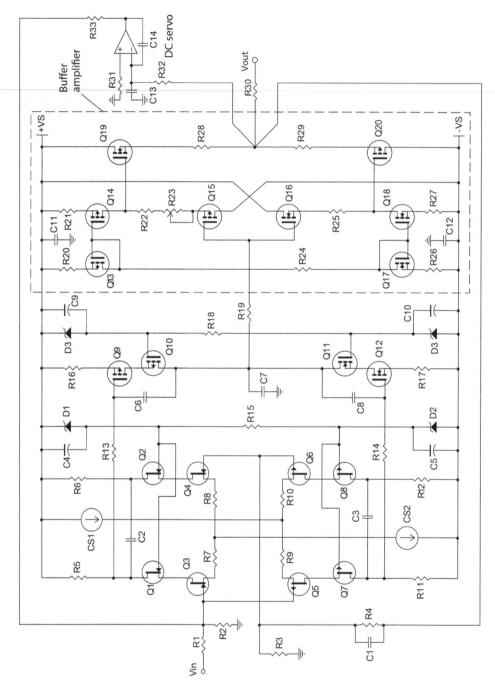

Figure 6.34 An audio line-stage amplifier employing a fixed-bias type buffer amplifier for the output stage

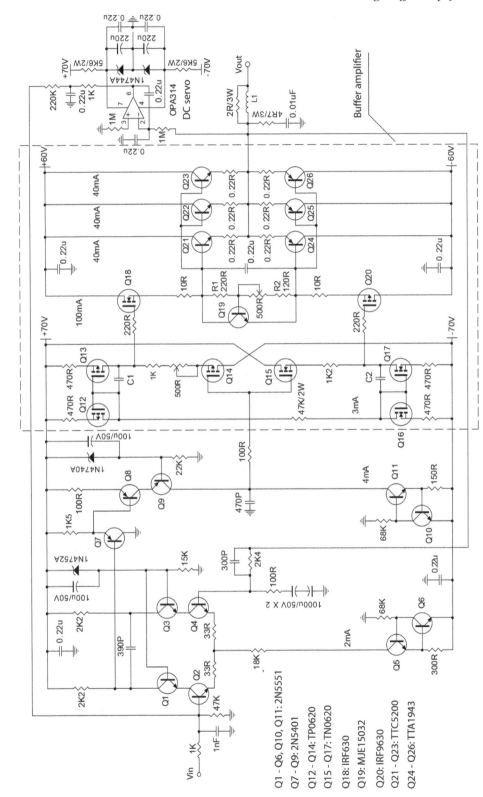

Figure 6.35 An audio power amplifier employing a fixed-bias type buffer amplifier for the output stage

To ensure high output current, the buffer amplifier employs three pairs of complementary BJT power transistors, Q21–Q26. In order to allow the output power transistors to swing near to ±60V, a second power supply ±70V is required. A dual ±70V dc power supply is now used for the first and second stage voltage amplifier. It is not common to use two separate dual dc power supplies for power amplifiers because this requires an addition winding from the power transformer and additional dc regulated power supply. Nonetheless, the power amplifier of Figure 6.35 can be easily modified to have just one set of dual power supply of ±60V. This example shows the flexibility of the buffer amplifier to adapt two sets of dual dc power supplies to maximize output swing and output power. Additionally, at the expense of a separate dc regulated power for the input stage and second stage, it improves the overall sonic performance of the entire power amplifier. Details of solid-state power amplifier designs are discussed in Chapter 12.

Hybrid power amplifier

The last example features using a buffer amplifier in a hybrid audio power amplifier, as shown in Figure 6.36. The input stage is a differential amplifier formed by triode 12AX7. The output from the differential amplifier is directly coupled to the second stage, which is a common cathode amplifier formed by triode 6922. The output of the second stage is a capacitor coupled to the output stage, which is a floating-bias type buffer amplifier similar to that shown in Figure 6.32(a).

A 100pF capacitor in series with a 330R resistor is connected to the 100kΩ plate resistor on the differential amplifier. A 200pF capacitor is connected in parallel to the 2.4kΩ feedback resistor. They work together to stabilize the power amplifier at high frequency. A 1N4005 diode in series with a 1N4740 zener diode is connected across the grid and cathode of the 6922 vacuum tube. This is to limit the dc voltage across the 6922 tube's grid-cathode during power up because a high grid-cathode dc voltage may damage a vacuum tube. After the tubes are warmed up, the 1N4005 diode and 1N4740 zener diode will be turned off by the inherent low grid-cathode dc quiescent voltage across 6922. The power-up process here is referring to the process for a vacuum tube amplifier going from a cold condition to fully warmed up, after which it starts to operate.

However, note that during the power-up process, which usually lasts for 30 seconds or so, the triode 6922's plate dc potential will fluctuate when the filament is warming up. At the end of the process, triode 6922 rests at the steady state plate dc potential, i.e., 207V in this example. This fluctuation of plate voltage appears to be a transient signal with large amplitude to the buffer amplifier during power up. If the buffer amplifier, a solid-state amplifier which is turned on instantly after switching on the power, is allowed to amplify this transient signal, the buffer amplifier will be driven into clipping. This is because the buffer amplifier is operating at a much lower dc power supply than the vacuum tubes. Therefore, a relay is needed to limit the transient signal during power up. In the first 30 seconds or so when the vacuum tubes start warming up, the input of the buffer amplifier is bypassed to the ground via a 220Ω resistor, which therefore limits the amplitude of the transient signal feeding to the buffer amplifier. After the vacuum tubes are completely warmed up, the relay disconnects the 220Ω resistor so that full signal can be feeding to the buffer amplifier. Details of hybrid audio power amplifier designs are discussed in Chapter 13.

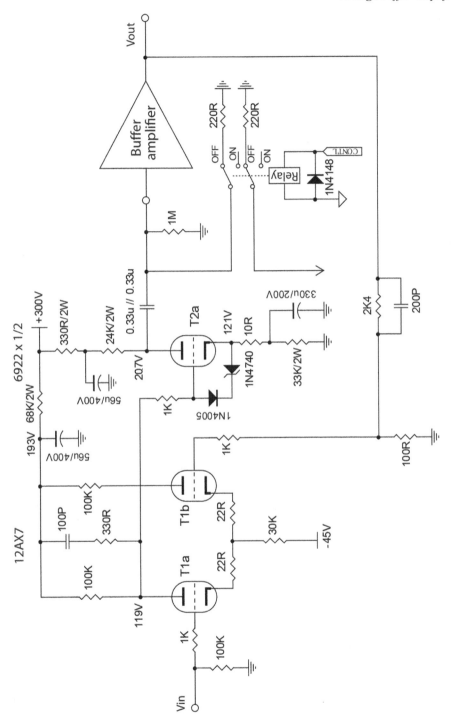

Figure 6.36 An audio hybrid power amplifier employing vacuum tubes for voltage amplification and a floating-bias buffer amplifier for the output stage

6.8 Exercises

Ex 6.1 If the resistors R2, R3, and R4 from Figure 6.4 are redrawn here, as shown in Figure 6.37, by using the method of superposition show that the signal at point B is given by 0.8813Vin.

Ex 6.2 Figure 6.38(a) is often called a supply-independent current source. The output current is set by resistor R1 in such a way that the dc voltage across R1 is equal to $V_{BE(Q2)}$, and it is independent of the power supply voltage. Therefore, it is a so-called supply-independent current source. However, this circuit is not completely supply independent because the base-emitter voltage of Q2 will change slightly with the power supply voltage. If the power supply – VEE is -35V, and the desired output current is 6mA, determine the values for resistors R1 and R2. We can assume that V_{BE} for Q1 and Q2 is 0.65V, the collector current for Q2 is 1mA, and the base current is negligible.

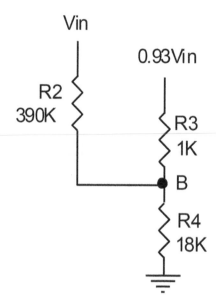

Figure 6.37 Cathode resistors for Figure 6.4

Ex 6.3 Figure 6.38(b) is a Wilson current source that has high output impedance compared to a simple current source shown in Figure 6.10(b). Additionally, the base current error is greatly reduced so that the output current Iout is very close to the reference current Iref. Assume the current gain of the transistors Q1, Q2 and Q3 are identical and equal to β. Show the following expression for Iout in terms of β and Iref,

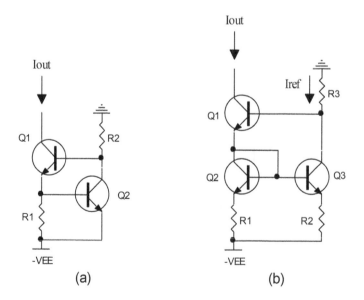

Figure 6.38 Circuit (a) is a supply-independent current source. Circuit (b) is a Wilson current source

$$I_{out} = I_{ref}\left(1 - \frac{2}{\beta^2 + 2\beta + 2}\right).$$

In order to simplify the calculation, we can assume R1 = R2 = 0 in the above calculation.

Ex 6.4 If the Wilson current source of Figure 6.38(b) is used to replace the MOSFET current source of Figure 6.11, determine the resistors R1, R2, and R3 if the desired output current Iout is 5mA. In order to simplify the calculation, we can assume R1 = R2, V_{BE} = 0.65V, and the dc voltage across R1 is 1.5V.

Ex 6.5 Figure 6.39 is a buffer amplifier employing a 6H30 triode vacuum tube. A dc servo circuit is used to eliminate the output dc offset voltage. Therefore, the output is biased at dc ground level. If the desired dc quiescent plate current is 6mA, complete the design by determining the unknown resistors. In this circuit, high voltage op-amp must be used. Note that in principle, if 6H30 is replaced by power triodes and employs a higher power supply voltage, this circuit may form the basis for the output stage of an OTL vacuum tube power amplifier.

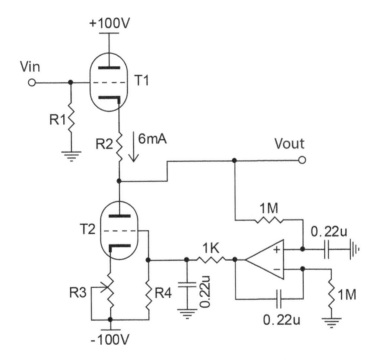

Figure 6.39 A voltage buffer amplifier using a triode vacuum tube. No input and output dc block capacitors are needed

References

[1] Valley and Wallman, "Vacuum tube amplifiers," M.I.T. radiation laboratory series. Reprinted by Audio Amateur Press, 2000.

[2] Ryder, J.D., *Engineering electronics with industrial applications and control*, McGraw-Hill, 1957.

[3] Jones, M., *Valve amplifiers*, 3rd edition, Newnes, 2004.

[4] Gray, P.R. and Meyer R.G., *Analysis and design of analog integrated circuit*, 3rd edition, Wiley, 1992.

[5] Data sheet for OP176, Analog Devices Inc.

[6] Data sheet for LTC6244, Linear Technology Corporation.

[7] "Applications of wide-band buffer amplifiers," National Semiconductor application note 227.

[8] Smith et al., "Wideband buffer amplifier with high slew rate," US patent 5049653.

[9] Lehmann, K., "Unity gain amplifier with high slew rate and high bandwidth," US patent 5177451.

[10] Bowers, D., "Rapid slewing unity gain buffer amplifier with boosted parasitic capacitance charging," US patent 5323122.

[11] Lam, C.M.J., "Buffer amplifier," US patent 8854138 B2.

7 Line-stage amplifiers − solid-state

7.1 Introduction

An audio lines-stage amplifier (also known as a *preamplifier*) is one of the key components in an audio system. It allows the user to switch between input sources, to control the input source signal level, and to amplify the signal before feeding to the audio power amplifier. The voltage gain of a line-stage amplifier often ranges from 5 to 10. Even though there is no absolute standard for it, a gain close to 10 is very common for a solid-state line-stage amplifier. In addition to the voltage gain, wide bandwidth (> 100kHz), low distortion (< 0.01%), and high signal-to-noise ratio (> 90dB) are also required for a good line-stage amplifier. Most amplifiers discussed in this chapter can easily meet these requirements.

Figure 7.1 shows a simple line-stage amplifier using op-amp OPA627. This device is a low noise Bi-FET op-amp, which has a JFET input stage. It is a popular audio op-amp that can be found in many audio applications, such as line-stage amplifiers, buffer amplifiers, active filter, I-to-V amplifiers in DAC, etc. It is a rather expensive device and the unit price today (in 2023) is $16.1 for the purchase of 1,000pcs of OPA627AU in SOIC-8 package. The unit price for another popular Bi-FET op-amp, OPA134UA, is just $1.278@kpcs. Thus, the price for a single OPA627AU will buy you 12pcs of OPA134UA. Refer to Table 9.3 in Chapter 9 for a list of popular low noise audio op-amps.

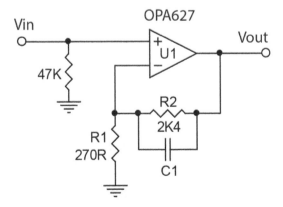

Figure 7.1 A simple line-stage amplifier with voltage gain of 10

DOI: 10.4324/9781003369462-7

The amplifier shown in Figure 7.1 is a non-inverted amplifier with a voltage gain equal to $(1 + R2/R1)$. Given the resistor values, the voltage gain is 9.9. A small compensation capacitor, C1, may be used to stabilize the amplifier at high frequency.

Even though many good sounding audio op-amps are available today, discrete transistor amplifiers are still preferred by many audio enthusiasts and designers. A discrete transistor amplifier will allow the designer to choose the most suitable electronic components – transistors, resistors, capacitors, diodes, and current sources, while setting optimum bias currents for each of the transistors. Therefore, a well deigned discrete transistor amplifier may offer better overall performance than a single op-amp. Before we move on to discussing the design details of a discrete transistor amplifier, let us walk through several basic line-stage amplifier examples.

Figure 7.2 is perhaps the simplest line-stage amplifier, employing just two transistors. Both transistors, Q1 and Q2, are JFET. Transistor Q1 operates as common source amplifier with a bypassed source resistor to maximize the voltage gain. Transistor Q2 operates as a source follower providing low output impedance. Feedback is provided by resistor R4 and, therefore, R1 and R4 set the overall voltage gain. A voltage gain of 10 can be achieved. Since the open-loop gain is provided by transistor Q1 only, it is not high. Therefore, the bandwidth is not very wide either. But it is sufficient for a line-stage amplifier.

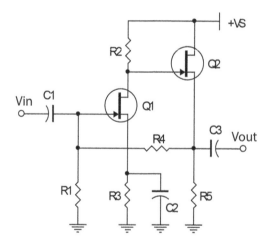

Figure 7.2 A basic two-transistor line-stage amplifier

Figure 7.3 shows an improved amplifier employing three transistors. Note that even though JFETs are illustrated, BJTs should work equally well. Transistors Q1–Q3 are dc coupled to extend the low frequency response. Since Q1 and Q2 are common source amplifiers, this line-stage amplifier produces a much higher open-loop gain than Figure 7.2. Source resistor R6 is bypassed to maximize the voltage gain while the un-bypassed source resistor R3 allows feedback to be fed from the output. Since the open-loop gain is high, when feedback is applied the bandwidth will be much more extended than is shown in Figure 7.2. The distortion level is also lower. Since a single voltage supply is used, input and output coupling capacitors are both required in Figure 7.2 while only an output capacitor is needed in Figure 7.3.

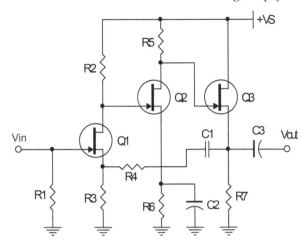

Figure 7.3 A three-transistor line-stage amplifier

When a dual voltage supply is used, the input and output can be biased at dc ground level, as shown in Figure 7.4. As a result, the input and output coupling capacitors can be removed. Figure 7.4(a) is a well-known amplifier popularized by John Curl in the 1970s. Normally, we would need to have a current source to set the tail current for a differential amplifier. Since there are two differential amplifiers in Figure 7.4(a), the circuit should have used two current sources – one for the differential amplifiers Q1 and Q2, and the second one for the differential amplifiers Q3 and Q4. This circuit cleverly uses the dc voltage across resistors R5 and R6 for self-bias (also called depletion mode bias) of the JFETs. This saves quite a few components that otherwise would be needed to implement two current sources. The original circuit does not have the resistor R6. It is added in this circuit so that the bias current can be easily set.

JFET transistors Q1–Q4 form a complementary differential amplifier. The outputs are amplified by the second stage, which contains two complementary BJTs in a folded common emitter push–pull amplifier configuration. Feedback is determined by resistors R9 and R10. Small capacitors C1–C3 may be required to stabilize the amplifier at high frequency.

The amplifier shown in Figure 7.4(a) is very popular among electronic DIY enthusiasts. As a result, there are several variants generated from the original design. Figure 7.4(b) shows one of the variants, which has a number of changes compared to the original design. The first change is to use JFET, Q5, and Q6 for the second stage. Since JFET has high input impedance, Q5 and Q6 will not create a loading effect to the input stage. The second change is an addition of an emitter follower push–pull output stage. The output stage boosts the output current as well as lowering the output impedance. The last change is the addition of a dc servo circuit to reduce the output dc offset voltage.

Note that the structure of the amplifier in Figure 7.4(b) consists of three stages: an input stage, a voltage amplifier second stage, and a unity gain output stage. In general, the input stage is often a differential amplifier. The voltage amplifier second stage contributes to a high open-loop gain for the entire amplifier. The output stage is a voltage buffer amplifier that produces unity gain and low output impedance. Various arrangements of these stages for balanced and unbalanced input and output line-stage amplifiers are discussed below.

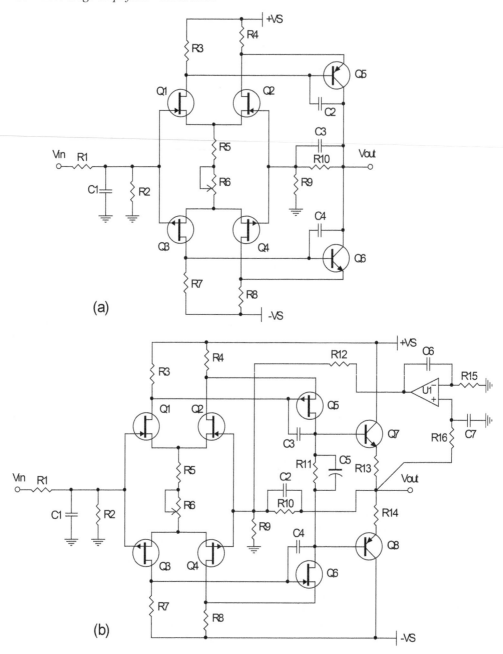

Figure 7.4 Circuit (a) is a two-stage amplifier popularized by John Curl. Circuit (b) is a modified version that contains three stages and a dc servo circuit

7.2 Unbalanced input–unbalanced output amplifiers

Figure 7.5 represents a basic unbalanced input–unbalanced output line-stage amplifier that contains three amplifying stages, including a differential amplifier input stage, a voltage amplifier second stage, and a unity gain output stage. Low noise JFET and BJT are popular choices for the input stage. The dc quiescent current of the differential amplifier is set up by the current source CS1. Since JFET has a high input impedance and triode-like behavior, JFET is a popular choice for the input stage. The source signal is feeding to the differential amplifier via transistor Q1, while the feedback signal from the output is feeding to transistor Q2. The difference between the input and the feedback signal is amplified and directly coupled to the second stage, which is a common emitter amplifier.

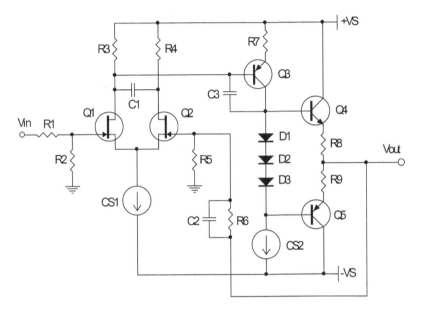

Figure 7.5 A basic unbalanced input–unbalanced output amplifier contains three amplifying stages: a differential amplifier input stage, a voltage gain second stage, and a unity gain output stage

A current source (CS2) with high output impedance is used in the second stage as an active load for Q3 so that the second stage produces a very high voltage gain. A string of diodes is used to set up a dc voltage spread to bias the output transistors Q4 and Q5 into class AB operation. Alternatively, a V_{BE} multiplier, or even a variable resistor, can be used to replace the diodes. Capacitors C1–C3 may be required to stabilize the amplifier at high frequency.

By improving each of the three stages, this basic amplifier can evolve into a high-performance amplifier. For instance, the input stage of Figure 7.5 can be improved by using a cascode differential amplifier, which has wide bandwidth and low distortion. Similarly, a cascode amplifier can be used for the second stage. The output stage can be changed to a double emitter follower (Darlington amplifier) for boosting output current. After all these changes are put together, the result is the line-stage amplifier shown in Figure 7.6.

The input stage is a cascode differential amplifier using low noise JFET 2SK170. A potential divider network is formed by resistors R4 and R6 setting up a 12V dc potential at the gate of transistors Q1 and Q2. Since the $V_{GS(th)}$ is small (around 0.4V) for 2SK170 at 2mA, the V_{DS} of

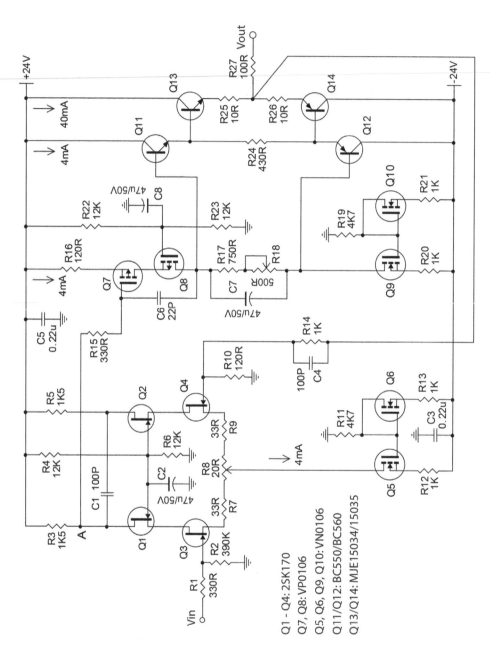

Figure 7.6 A high performance unbalanced input–unbalanced output line-stage amplifier

the transistors Q1, Q2, Q3, and Q4 are around 10–11V. Small degeneration resistors R7 and R9 (33Ω) are added to the source of the transistors Q3 and Q4. R8 is a small variable resistor that helps to balance the bias current at Q1/Q3 and Q2/Q4 so that the amplifier's output dc offset can be reduced to zero.

The differential amplifier's tail current is determined by the current source formed by Q5/Q6. Regarding the current source, there are many choices. It can be implemented by different arrangements and transistors such as BJT, JFET and MOSFET. Figure 7.7 reveals several common current sources. Figure 7.7(a) is the current source used in the amplifier of Figure 7.6, which sets up a 4mA tail current for the differential amplifier. Alternatively, NPN transistors are used in Figure 7.7(b) with the same configuration, which is often called a "current mirror." Figure 7.7(c) is a simple current source formed by a N-channel JFET. In choosing a JFET for the current source application, we have to make sure that the I_{DSS} of the JFET is greater than the required current. In this example, I_{DSS} must be greater than 4mA. Figure 7.7(d) is a current source employing an N-channel depletion type MOSFET. Since MOSFET's I_{DSS} is much greater than JFET, which is true for both enhancement and depletion type MOSFET, MOSFET is often the preferred choice for high current source applications.

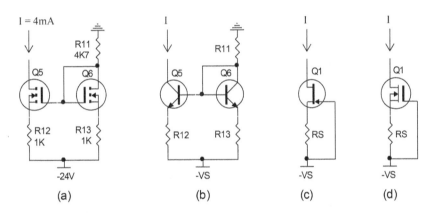

Figure 7.7 Circuit (a) is a current source employing an N-channel enhancement type MOSFET. Circuit (b) is a current source employing NPN BJT. Circuit (c) is a current source employing N-channel JFET. Circuit (d) is a current source employing N-channel depletion type MOSFET

The output from the input stage, at point A, is directly coupled to the second stage, which is a cascode amplifier formed by enhancement type MOSFET Q7 and Q8. Both transistors are biased at 4mA quiescent current determined by the current source formed by MOSFET, Q9, and Q10, as well as resistor R16. This current source is identical to the current source used in the differential amplifier of the input stage.

MOSFET is not the only choice for transistor Q7 and Q8. JFET and BJT should also work equally well. However, if JFET or BJT is used, R16 has to be changed accordingly. Figure 7.8 shows a simplified circuit of the first and second stage. The circuit in Figure 7.8(a) shows the same MOSFET that is used in Figure 7.6. The circuit in Figure 7.8(b) shows that JFET is used for Q7. Since the $V_{GS(th)}$ for 2SJ74 is different from VP0106, this affects the source resistor. The source resistor R16 can be calculated by tracking the dc potentials from power supply to point A.

$$24V - 4mA \times R16 + V_{GS(th)} = 21V, \text{ where } V_{GS(th)} \approx -0.47V \text{ for 2SJ74}$$

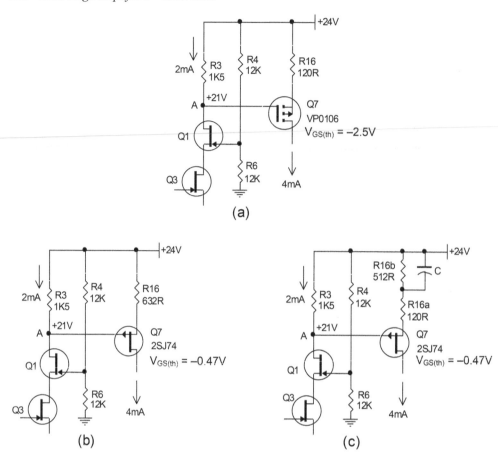

Figure 7.8 Circuit (a) shows that MOSFET (VP0106) is used for Q7 with source resistor 120 Ω. Circuit (b) shows that JFET (2SJ74) is used for Q7 and the source resistor is increased to 632Ω. Circuit (c) shows that the same JFET is used for Q7 and the source resistor is partially bypassed

Therefore, we find R16 = 632 Ω. This R16 is much higher than the 120Ω used for MOSFET Q7. A large source resistor R16 is not a bad thing, as it provides local feedback to the cascode amplifier formed by Q7 and Q8. However, the voltage gain is reduced. If we want to get a high voltage gain from this cascode amplifier so as to produce a sufficiently high open-loop gain for the entire amplifier, the resistor value for R16 must be kept low. Figure 7.8(c) shows that this can be achieved by using a partially bypassed resistor for R16. As a result, same 120Ω for R16 can be kept. In a similar fashion, if BJT is chosen for Q7, partially bypassed R16 can also be used.

The push–pull output stage of the amplifier in Figure 7.6 employs a double emitter follower formed by Q11–Q14. They are dc biased by the dc voltage spread created by resistors R17 and R18. The emitter follower formed by Q11 and Q12 is biased at 4mA, whereas the emitter follower formed by Q13 and Q14 is biased at 40mA so that the amplifier is able to drive more demanding loads. Due to heat dissipation, transistors Q13 and Q14 must be mounted on suitable heatsinks.

The overall feedback is set by resistors R10 and R14. The voltage gain of the amplifier is given by

$$A_v \equiv \frac{Vout}{Vin} = 1 + \frac{R14}{R10} = 1 + \frac{1000\Omega}{120\Omega} = 9.3$$

Any desired voltage gain can be easily set by a change of resistors R10 and R14. On the other hand, small frequency compensation capacitors, C1, C4, and C6, are required in this circuit. If JFET or BJT is used to replace the MOSFET of Q7, these compensation capacitors may be required to change also. Since MOSFET, JFET, and BJT have different sonic characteristics, it is worth trying different types of transistors for Q7/Q8, as well as the current sources formed by Q5/Q6, and Q9/Q10, and judge for yourself which type of transistor offers the better sonic performance.

Variation 1

After having discussed how the design of the amplifier in Figure 7.6 is achieved, we now examine where the amplifier can be improved. In so doing, this will lead to a number of variations to the original design. Here, the first variation is the change of the input stage from a single cascode differential amplifier to complementary cascode differential amplifiers, as shown in Figure 7.9.

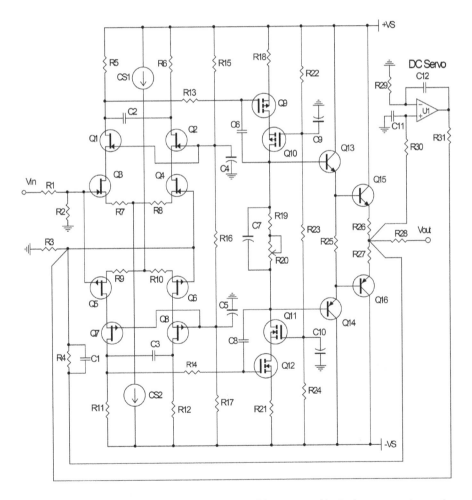

Figure 7.9 Variation 1: Complementary differential amplifiers are used in the input stage. A complementary cascode push–pull amplifier is used in the second stage

The differential amplifier on the top is formed by N-channel JFETs (Q1–Q4) while the differential amplifier on the bottom is formed by P-channel JFETs (Q5–Q8). The complementary differential amplifiers are connected to form one single input stage. Since two differential amplifiers are connected in parallel, signal-to-noise ratio of the entire amplifier is improved. Additionally, the complementary differential amplifiers will tend to reduce the total harmonic distortion of the entire amplifier – notably a reduction in second harmonic distortion.

The outputs of the input stage are directly coupled to the second stage, which now consists of cascode push–pull amplifiers formed by Q9/Q10 and Q11/Q12. Note that the drain of Q10 is operating as the loading resistor for the cascode amplifier formed by Q11/Q12. Likewise, the drain of Q11 becomes the loading resistor for the cascode amplifier formed by Q9/Q10. It should be noted that, in the original amplifier of Figure 7.6, the current source previously formed by Q9/Q10 is no longer needed. It is now replaced by the cascode amplifier formed by Q11/Q12 in Figure 7.9.

In the original amplifier of Figure 7.6, there is only one current source needed to bias the differential amplifier formed by Q1–Q4. That current source is formed by MOSFET Q5/Q6. Now it is represented by current source CS2 in Figure 7.9. In a similar fashion, a new additional current source, CS1, may be formed by employing P-channel MOSFETs, PNP transistors, a P-channel JFET or a P-channel depletion type MOSFET, as shown in Figure 7.10. Note that a P-channel enhancement type MOSFET (VP0106, which is a complementary device to VN0106), can be considered for the current source of Figure 7.10(a). To implement the current sources (b) and (c), there are many suitable devices. However, there is a problem with current source (d). We cannot find any P-channel depletion type MOSFET available in the market today. There are working principles that predict the electrical properties of a P-channel depletion type MOSFET. Therefore, it must be a device that can be produced. But perhaps its unavailability is due to commercial reasons, i.e., limited application and demand, or technical difficulties in manufacturing, that do not encourage any manufacturer to offer a P-channel depletion type MOSFET for commercial use. If and when such devices become available, Figure 7.10(d) shows how a current source can be developed with a P-channel depletion type MOSFET.

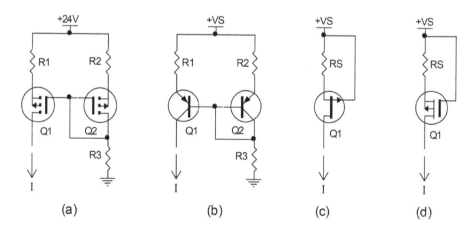

Figure 7.10 Alternative current sources for CS1 of Figure 7.9. Circuit (a) is a current source employing a P-channel MOSFET. Circuit (b) is a current source employing a PNP transistor. Circuit (c) is a current source employing a P-channel JFET. Circuit (d) is a current source employing a P-channel depletion type MOSFET. However, the P-channel depletion type MOSFET is not yet commercially available

The second stage of the amplifier in Figure 7.9 is a pair of complementary cascode amplifiers formed by Q9/Q10 and Q11/Q12. Assuming that the two current sources in the input stage are set to equal tail currents for the differential amplifiers, the dc voltage across load resistors R5 and R11 is therefore equal. However, the $V_{GS(th)}$ of Q9 (PMOS) and $V_{GS(th)}$ of Q12 (NMOS) are different. For instance, let us assume VP0106 and VN0106 are used for Q9/Q10 and Q11/Q12, respectively. We find that $V_{GS(th)}$ of VP0106 is equal to -2.5V (typ.) while 1.6V (typ.) for VN0106. If equal dc voltage is developed across R5 and R11, the source resistors for biasing Q9 and Q12, R18 and R21, have to be different. Otherwise, the transistors will fight for domination in setting the dc bias current. Since $|V_{GS(th)}|$ of VP0106 (Q9) is greater than that of VN0106 (Q12), if equal dc bias current is to be set for the second stage, we must have R21 > R18. Thus, these two resistors are not equal. Alternatively, if we prefer setting R21 = R18, then the current of CS1 must be lower than CS2. Even though these two approaches may affect the voltage gain of the amplifier, the impact should be very small. Both approaches should work just fine.

The output stage remains unchanged compared to the original version. However, a dc servo circuit is added. The variable resistor R8 of Figure 7.6, which is to reduce the output dc offset voltage, is not necessary for Figure 7.9. A dc servo circuit is more effective in eliminating the output dc offset.

Variation 2

The second variation employs an improved output stage, as shown in Figure 7.11. The output stage is a fixed-bias diamond buffer formed by transistors Q13–Q20. The dc bias current for this stage is set by current source Q13/Q14 and Q17/Q18, which are independent from the preceding second stage. Therefore, this output stage provides extra dc stability compared to Variation 1 shown in Figure 7.9.

If we re-examine the dc quiescent condition of Figure 7.9, it can be seen that the quiescent current of the complementary cascode push–pull amplifier (Q9/Q10 and Q11/Q12) in the second stage is dependent on the input stage. And the quiescent condition of the output stage is dependent on the second stage, which in turn is dependent on the input stage. Therefore, any variation of the quiescent condition of the input stage will affect the subsequent stages. In an extreme scenario, if the input stage is accidentally overdriven by a large signal, the transistors on the input stage will be driven into saturation. As a consequence, the quiescent conditions for the second stage and output stage become unpredictable. The feedback loop between the input stage and output stage may become temporarily out of control. Even though a dc servo circuit is in place, since the input stage is temporarily not working, a substantial dc voltage may appear in the output. When the abnormal signal is removed, it will still take considerable time for the input stage to stabilize before the entire amplifier is back to normal operation again.

It is true that such an extreme condition will rarely happen. Nevertheless, this scenario reveals that the dc stability of the amplifier in Figure 7.9 is very much dependent upon the input stage. On the other hand, the fixed-bias type diamond buffer output stage depicted in Figure 7.11 will always maintain its dc bias current regardless of the conditions in the preceding stages. This helps the entire amplifier to make a fast recovery from an overloaded input stage, if it ever happens. This makes the Variation 2 amplifier a better choice over the Variation 1 if stability of the amplifier is a primary concern. We know that the chance of driving the input stage of a line-stage amplifier into saturation is not common. However, we cannot claim the same thing for the audio power amplifier, which deals with large signals. Therefore, a fixed-bias diamond buffer may appear to be an attractive output stage for the audio power amplifier. Details of the fixed-bias diamond buffer are discussed in Chapter 6.

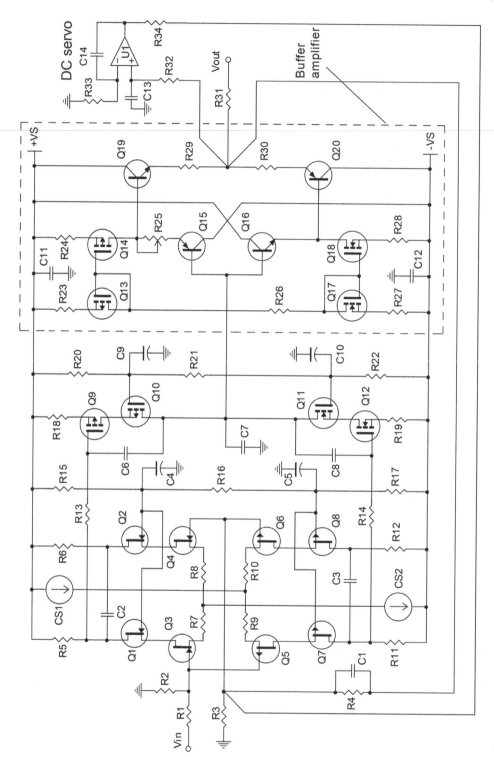

Figure 7.11 Variation 2: Input and second stages are identical to Variation 1 of Figure 7.9. Output stage employs a fixed-bias type diamond buffer

The dc bias current of the fixed-bias diamond buffer of Figure 7.11 is determined by current source formed by Q13/Q14 and Q17/Q18. MOSFETs are used in this example. However, it should be noted that BJT will work equally well. Therefore, the diamond buffer amplifier can employ mixed MOSFET/BJT as in this example, all MOSFET, or all BJT. It is worth exploring these devices to see how they affect the sonic performance of the amplifier.

Variation 3

In this design, a different input stage is employed. The input stage used in Variations 1 and 2 is now redrawn in Figure 7.12(a). It employs complementary cascode differential amplifiers. The differential amplifier on the top is formed by transistors Q1–Q4 and the tail current is set by current source CS2. The bottom differential amplifier is formed by Q5–Q8 and the tail current is set by current CS1. The gates of Q1, Q2, Q7, and Q8 are connected to a potential divider formed by resistor network R13–R15. Identical N-channel JFETs are often used for Q1–Q4. For instance, a popular choice is 2SK170, while 2SJ74 P-channel JFET is used for Q5–Q8. Alternatively, equivalent devices such as LSK170 or LSK389 (dual version) can be used for Q1–Q4. In other words, four identical N-channel and four identical P-channel JFETs are used to construct the complementary cascode differential amplifier of Figure 7.12(a).

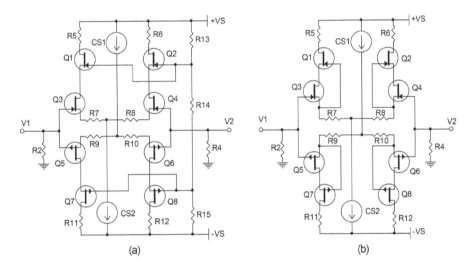

Figure 7.12 Circuit (a) is an input stage using complementary cascode differential amplifiers in which the gate of transistors Q1, Q2, Q7, and Q8 are dc biased via a potential divider network. Compound cascode amplifiers are used in circuit (b) that eliminate the need of a potential divider network

Alternatively, JFET with high $V_{GS(th)}$ can be used for Q1 and Q2, with low $V_{GS(th)}$ for Q3 and Q4. Given this choice of JFETs, we can simplify the complementary cascode differential amplifier as shown in Figure 7.12(b). Let us take a look at the compound transistor arrangement formed by Q1 and Q3. The V_{DS} of transistor Q3 is equal to the V_{GS} of Q1. In order for this configuration to work, there is one requirement – we must have $V_{GS(th)\_Q1} > V_{GS(th)\_Q3}$. For example, 2SK246 is used for Q1 and Q2, and 2SK170 is used for Q3 and Q4. Similarly, 2SJ103 is used for Q7 and Q8, and 2SJ74 is used for Q5 and Q6. This compound cascode amplifier configuration was popularized by Erno Borbely [1]. The primary advantage of this compound JFET cascode amplifier is that no external dc voltage is needed to bias the gate of Q1, Q2, Q7, and Q8. This saves us the components needed to implement the potential divider (R13–R15 of Figure 7.12[a]) as well as simplifying the PCB layout design work. The compound cascode amplifier operates like a single JFET with three terminals but it produces lower distortion and a wider bandwidth than a single JFET.

Another circuit technique that we can use to further simplify the complementary cascode differential amplifier is shown in Figure 7.13(a). This input stage is a complementary differential amplifier popularized by John Curl. The dc bias arrangement for this circuit has been discussed in Section 2.9 for special JFET configurations (see Figure 2.45). The complementary differential amplifier takes advantage of the dc voltage across resistor R6 so that self-bias can be established for the JFETs. In a similar fashion, self-bias can also be applied to the complementary cascode differential amplifier of Figure 7.13(b). In comparing with the input stage of Figure 7.12(a), the input stage of Figure 7.13(b) has eliminated two current sources, CS1 and CS2, and a potential divider network R13–R15.

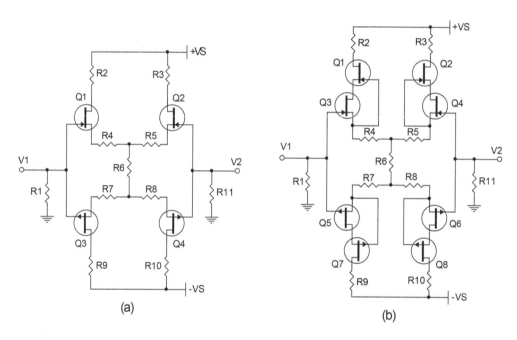

Figure 7.13 Circuit (a) is an input stage using complementary differential amplifiers that takes advantage of the dc voltage across resistor R6 so that a depletion mode bias is established for the JFETs. A compound cascode amplifier configuration is used in circuit (b) with similar depletion mode bias for the JFETs

When the input stage of Figure 7.13(b) is applied to the amplifier of Figure 7.11, this leads to the Variation 3 amplifier as shown in Figure 7.14. No current sources or potential divider network are needed to bias the complementary cascode differential amplifier of the input stage. The second stage and the output stage remain unchanged as of the Variation 2 amplifier.

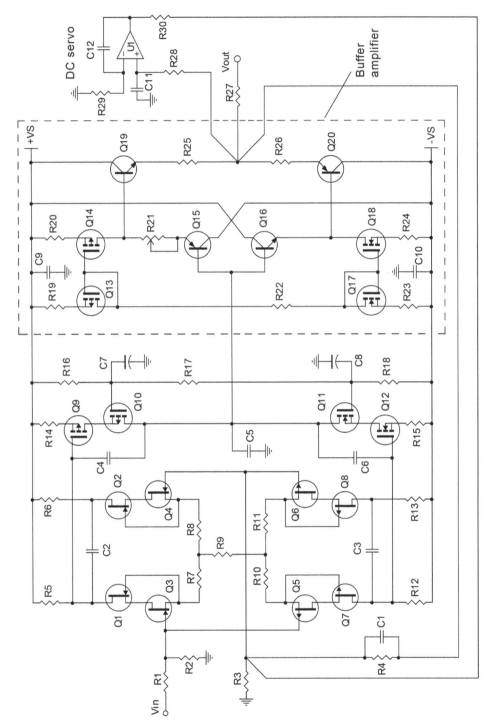

Figure 7.14 Variation 3: Input stage employs complementary differential amplifiers with a compound cascode amplifier configuration and using depletion mode bias for the JFETs

Variation 4

Above, we have discussed line-stage amplifiers with variations ranging from input stage to output stage. The Variation 1 amplifier employs a complementary cascode differential amplifier for the input stage. And the second stage has a complementary cascode amplifier. The Variation 2 amplifier is built upon the Variation 1 amplifier with a diamond buffer used for the output stage. The Variation 3 amplifier has simplified the input stage by using a compound cascode amplifier configuration. Self-bias is established for the JFETs of the differential amplifier input stage so that two current sources and a potential divider resistor network are eliminated. The Variation 3 amplifier can be further simplified by only employing half of the input stage and a simple output stage. This leads to the Variation 4 amplifier, which was popularized by Erno Borbely [2]. In this design, JFET is used for Q5 and Q8 instead of MOSFET as in previous variations. The output stage is a MOSFET push–pull amplifier. Feedback is cleverly applied to the source resistor R5 between two compound cascode amplifiers. This means the Variation 4 amplifier has the lowest components count among the four variations (Figure 7.15).

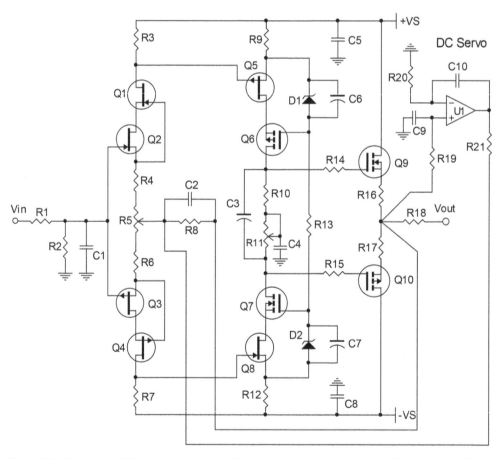

Figure 7.15 Variation 4: The input stage uses half of the complementary cascode differential amplifier of Figure 7.13(b). The output stage is a simple MOSFET push–pull amplifier.

7.3 Unbalanced input–balanced output amplifiers

Assume that we have a power amplifier that only takes balanced input. But the source component (e.g., CD player) has only unbalanced output. Therefore, we may need an unbalanced input–balanced output line-stage amplifier. This does not seem to be very common though. However, many old CD players only offer unbalanced output. In this scenario, an unbalanced input–balanced output line-stage amplifier may be a good choice.

Figure 7.16 shows an amplifier employing three op-amps (U1–U3) in an unbalanced input–balanced output configuration. It was developed to be used for a line driver application [3]. However, it can be adapted for an audio line-stage amplifier application. Op-amp U3 produces the positive phase signal while U2 produces the negative phase signal. Since the outputs are cross-coupled to the input of the op-amp that handles the opposite phase, the outputs are very well balanced. The gain of this amplifier circuit is given by R2/R1. When we set R2 = 10×R1, the gain is 10. Op-amp OP176 was originally used in this application with a dual ±18V power supply. If output is limited to 14Vp-p, it can drive a differential load of 249Ω with a full-power bandwidth of 190kHz. The input impedance looking into point A is two R1s in parallel, i.e., 1kΩ. This is not a high impedance. Therefore, a simple buffer amplifier, formed by op-amp U1, is placed at the input.

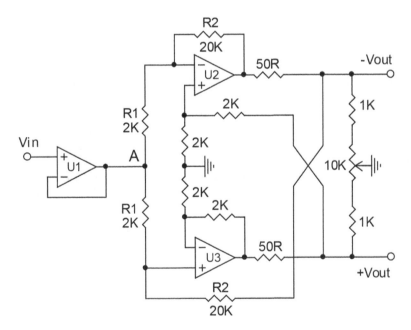

Figure 7.16 An unbalanced input–balanced output amplifier contains three op-amps. The gain of this amplifier is given by R2/R1

A better solution is to use a fully balanced amplifier that has both unbalanced and balanced input, and it produces unbalanced and balanced output. Details of fully balanced line-stage amplifier design are discussed in Section 7.5.

7.4 Balanced input–unbalanced output amplifiers

We can take advantage of the well-regarded instrumentation amplifier configuration for a balanced input–unbalanced output application. The voltage gain for Figure 7.17 is given by

$$A_v \equiv \frac{Vout}{V2 - V1} = \left(1 + 2\frac{R1}{R2}\right)\left(\frac{R4}{R3}\right) \qquad (7.1)$$

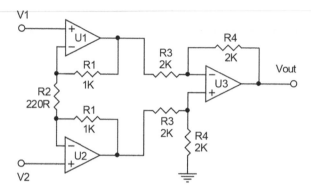

Figure 7.17 This is an instrumentation amplifier that takes balanced input and delivers unbalanced output

Given R1 = 1kΩ, R2 = 220Ω, R3 = R4 = 2kΩ, the voltage gain is 10. The input impedance looking into V1 and V2 is very high. It is a very stable and low noise amplifier. However, this op-amp circuit is not very attractive for implementation by discrete components because two identical discrete amplifiers must be used to replace U1 and U2. And a difference amplifier is required for converting balanced output into unbalanced output. The components count is high.

Note that the difference amplifier itself is already a balanced input to unbalanced output amplifier. Thus, it can be used for the desired balanced input–unbalanced output application so that the number of op-amps is reduced from three to just one. Now let us examine the difference amplifier shown in Figure 7.18. When the voltage gain is set to 10 while keeping a reasonable value for the input resistors R1 and R2, we choose R1 = R2 = 2kΩ and, R3 = R4 = 20kΩ.

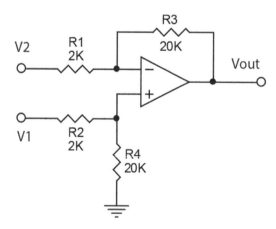

Figure 7.18 A difference amplifier with a voltage gain of 10

As discussed in Chapter 4, the input impedance for a difference amplifier is equal to R1 + R2. In this example, the input impedance becomes 4kΩ. This is not a very high input impedance for a line-stage amplifier. Let us say that if the desired input impedance is 20kΩ, so that R1 = R2 = 10kΩ, the feedback resistor R3 becomes 200kΩ for the same voltage gain of 10. However, 200kΩ is a rather high resistance that may produce a considerable amount of thermal noise. As a result, the difference amplifier does not appear to be a very attractive choice for realizing a balanced input–unbalanced output line-stage amplifier.

Figure 7.19 reveals a different type of inverted amplifier that offers high input impedance while keeping a low feedback resistor. For illustration purposes, assume that the conventional inverted amplifier of Figure 7.19(a) is required to have input impedance 10kΩ and voltage gain –100. Thus, the feedback resistor R2 is an enormous 1MΩ resistor. The noise generated by the 1MΩ resistor is likely to limit the usefulness of this circuit. But when the inverted amplifier of Figure 7.19(b) is used for the same input impedance and voltage gain, the sum of the feedback resistors is just around 70kΩ. As discussed in Chapter 4, it is called the *Tee-feedback network* (TFN) inverted amplifier. Clearly the TFN inverted amplifier is a great improvement over the conventional inverted amplifier of Figure 7.19(a). Given this TFN inverted amplifier, we can explore how to turn it into a balanced input–unbalanced output amplifier that meets the general input impedance requirement without using a high feedback resistor.

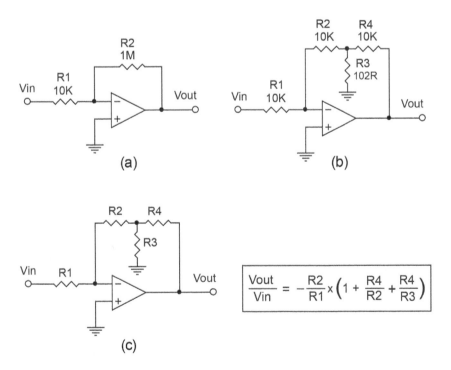

$$\frac{Vout}{Vin} = -\frac{R2}{R1} \times \left(1 + \frac{R4}{R2} + \frac{R4}{R3}\right)$$

Figure 7.19 Circuit (a) is a conventional inverted amplifier with a voltage gain of 100, input impedance 10kΩ and feedback resistor 1MΩ. Circuit (b) is the TFN inverted amplifier having the same voltage gain and input impedance, but the sum of the resistors in the Tee-feedback network is much lower than 1MΩ. Circuit (c) shows the expression for determining the voltage gain of a TFN inverted amplifier

By adding a potential divider resistor network at the non-inverted input, as shown in Figure 7.20, this leads to a *TFN difference amplifier*. The differential voltage gain for this amplifier is 9.7 and its differential input impedance is 20kΩ, which is more than sufficient for a line-stage amplifier in most practical audio applications. A small resistor, R7, is used in the circuit for reducing the common-mode error to a minimum. If we apply this TFN difference amplifier configuration to the unbalanced input–unbalanced output amplifiers discussed in the preceding section, they can be turned into balanced input–unbalanced output amplifiers. As an example, if the TFN difference amplifier configuration of Figure 7.20 is applied to the amplifier illustrated in Figure 7.6, we obtain a balanced input–unbalanced output line-stage amplifier, as shown in Figure 7.21.

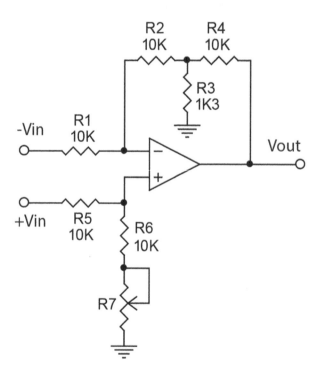

Figure 7.20 A TFN difference amplifier configuration that can be used for implementing a balanced input–unbalanced output line-stage amplifier. The differential input impedance is 20kΩ and the differential voltage gain is 9.7

Except for the use of the Tee-feedback resistor network, an input resistor divider (R1, R2, and R3), and a dc servo circuit, the rest of the amplifier is identical to the unbalanced input–unbalanced output amplifier of Figure 7.6. Note that when the two inputs of the differential amplifier in the input stage open up to amplify the incoming signals, it is necessary to have a dc servo circuit to bring down the output dc offset. In order to avoid the output of the dc servo circuit loading down the inverted input impedance, a large resistor R30 (150kΩ) is used.

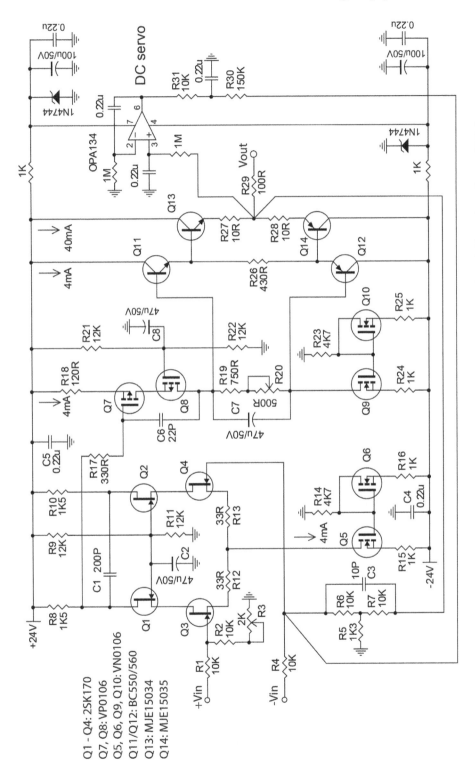

Figure 7.21 A balanced input–unbalanced output amplifier that has a voltage gain of 9.7 and a differential input impedance of 20kΩ

Since this is a balanced input amplifier, it is important to have a low common-mode gain. Even though the common-mode signal appearing at the output is not truly zero, it should be kept very low. A true fully balanced amplifier is one with balanced input and balanced output. As a common-mode signal can be effectively cancelled out at the output, a fully balanced amplifier has very high common-mode rejection ration (CMRR).

Note that the input stage of the amplifier in Figure 7.21 has a variable resistor, R3. It is used to reduce the common-mode gain for this amplifier. Figure 7.22 shows how this is done. The inputs of the amplifier of Figure 7.21 are shorted and connected to the same signal source, say a 1kHz sinusoidal wave. Then the output of the amplifier is observed by an oscilloscope. The output from the amplifier in this situation reflects the common-mode signal or error. For an ideal balanced input–unbalanced output amplifier, the common-mode error is zero, so that we see only a flat line in the oscilloscope. But, in real life, the common-mode error is not zero. Set the 1kHz signal to 5Vpp, or an even higher amplitude, and the oscilloscope vertical scale is set to 10mV/div. We should see a small signal appearing in the amplifier's output. This represents the common-mode error. We adjust variable resistor R3 so that that common-mode error is as small as possible. Even though there is no proven correlation between the common-mode gain and the sound quality of a balanced input–unbalanced output amplifier, it is always good practice to aim to design one with the lowest possible common-mode gain.

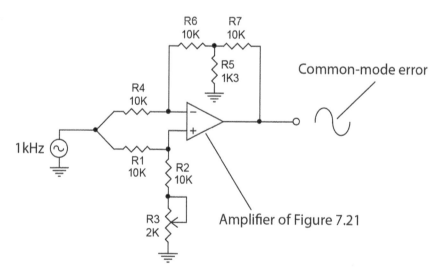

Figure 7.22 This circuit shows how to reduce the common-mode gain. When the non-inverted and inverted inputs are shorted together and connected to a common signal source with a 1kHz sinusoidal wave, adjust resistor R3 until the common-mode error is the smallest

Variation 1

Figure 7.23 shows a balanced input–unbalanced output amplifier [4] with a voltage gain of 10. The amplifier consists of only two stages, the input and output stages, and no global feedback. The voltage gain is determined by resistor R14.

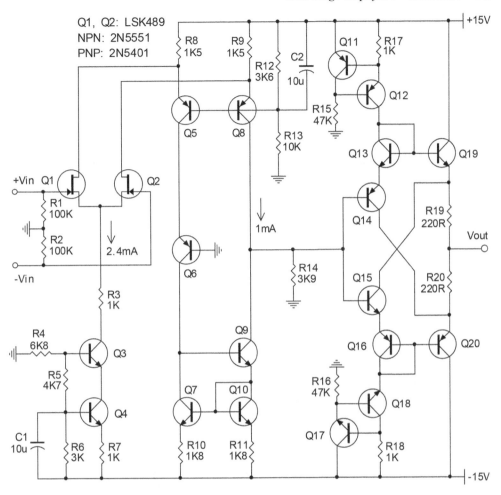

Figure 7.23 A balanced input–unbalanced output amplifier with a voltage gain of 10

The input stage is a folded cascode amplifier that comprises transistors Q1–Q10. Q1 and Q2 comprise a low noise dual JFET transistor, LSK489, and its single equivalent is LSK189. Transistors Q1 and Q2 form the input differential amplifier with tail currents set by a low noise cascode current source formed by Q3 and Q4. The outputs of Q1 and Q2 are directly coupled to Q5 and Q8, forming a cascode configuration. A Wilson current mirror, formed by Q7, Q9, and Q10, operates as an active load for Q8. Transistor Q6 works as common base amplifier that helps to reduce the V_{CE} of Q5 while keeping the collector of Q5 close to dc ground level.

The output stage is a *folded diamond buffer* that comprises transistors Q11–Q20. Q14, Q15, Q19, and Q20 are the four key transistors in an emitter follower configuration. Two current sources are employed to bias the output stage. Q11/Q12 is a current source on the top to bias Q14 and Q19, while Q17/Q18 is the other current source in the bottom to bias Q15 and Q20. For a regular diamond buffer, the collector of Q14 and Q15 is connected to the negative and positive power supply, respectively. In this folded diamond buffer configuration, the collectors are connected to the emitter of Q19 and Q20. As a result, V_{CE} of Q14 and Q15 is reduced and so is the heat dissipation from the transistor. This makes a folded diamond buffer an attractive output stage for a power amplifier where the power supply voltage is very high.

7.5   Balanced input–balanced output amplifiers

The simplest form for a balanced input–balanced output amplifier is shown in Figure 7.24. U1 and U2 are audio op-amps connected in an instrumentation amplifier configuration without a difference amplifier that converts the outputs into unbalanced output. The voltage gain is given by

$$A_v = 1 + \frac{2R1}{R2} \tag{7.2}$$

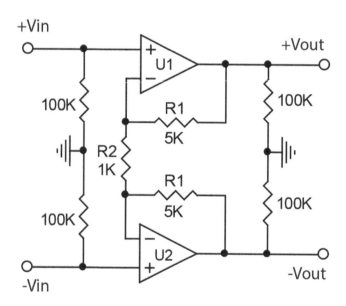

Figure 7.24 A fully amplifier is implemented by two op-amps in an instrumentation amplifier configuration without the output difference amplifier

Given the values for R1 and R2 in the circuit, the voltage gain is 11. This is a simple circuit for a fully balanced amplification. If the op-amps U1 and U2 are replaced by a discrete transistor amplifier, similar to those discussed in the preceding sections, we can turn the circuit into a discrete fully balanced amplifier. It can be easily seen that the components count for a discrete fully balanced amplifier that is double that of a discrete unbalanced input–unbalanced output amplifier.

A simple discrete fully balanced amplifier is shown in Figure 7.25. This amplifier contains only an input stage and output stage [5]. The input stage is a differential amplifier with compound JFET in a cascode arrangement. The outputs of the input stage (Q1 and Q2) are directly coupled to the source follower output stage for inverted and non-inverted outputs. The source follower Q8 is sitting on top of Q9, which works as an active load as well as a common source amplifier that amplifies the signal passing from the 0.22µF capacitor. It can be seen that the signal coming from the source of Q8, and the signal coming from the drain of Q9, have the same phase. Thus, the output stage produces an approximated push–pull output. This arrangement is a JFET equivalent to the White cathode follower using a triode. As a result, the output impedance is low. Note that no feedback is used in this amplifier.

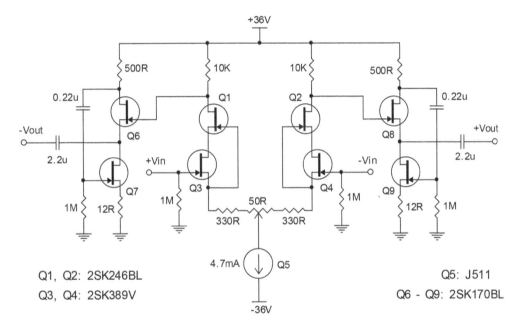

Figure 7.25 A discrete balanced input–balanced output amplifier with no feedback

Even without employing global feedback, the fully balanced discrete amplifier of Figure 7.25 should produce low distortion. But, on the other hand, if we want to set the voltage gain accurately, extending the bandwidth and reducing the distortion, feedback is something that we need to consider. Since the two inputs of the differential amplifier are already occupied by the input source, if we want to apply feedback, where will the feedback be connected without reducing the input impedance? If we also want to remove the 2.2uF output coupling capacitors, we need to have a more sophisticated output stage so that the output is resting at dc ground level. Let us see how these issues are addressed in the following two examples.

Example 1

A fully balanced (balanced input–balanced output) line-stage amplifier is shown in Figure 7.26. The structure of this amplifier is similar to an unbalanced input–unbalanced output amplifier

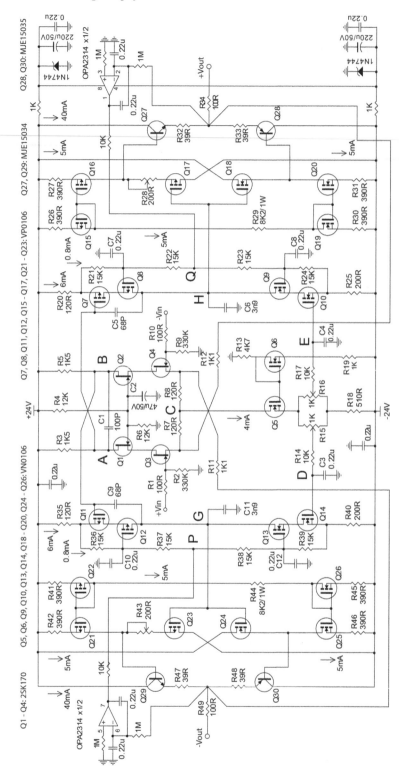

Figure 7.26 A fully balanced amplifier (Example 1) – cascode differential amplifier input stage

except that two second stages and two output stages are now needed. They are working with just one input stage, which is a cascode differential amplifier in this example. The structure of having two second stages and two output stages is something expected, as this amplifier has to produce two outputs. But this amplifier only has one input stage, making it different from the fully balanced amplifier of Figure 7.24, which has two input stages (i.e., two op-amps). Therefore, the feedback arrangement is also different.

The feedback from the inverted output is handled by resistor R11, and the non-inverted output by resistor R12. The feedback resistors are connected to the source degeneration resistors R7 and R8. Alternatively, the feedback resistors R11 and R12 can be connected to the gates of Q3 and Q4. Then the voltage gain will be determined by the ratio of the resistors R11/R2 and R12/R9. Imagine if we want input impedance to be 50kΩ for each input, the feedback resistors R11 and R12 will be 500kΩ for voltage gain of 10. Such a high resistor will generate considerable thermal noise that limits the usefulness of this amplifier. But when the feedback resistors are now connected as shown in Figure 7.26, they are just a small resistor of 1.1kΩ. The difference is obvious. Given this feedback arrangement in Figure 7.26, the voltage gain of the amplifier is given by R11/R8 = R12/R7 = 9.1. The tail current for the differential amplifier is set to 4mA by the current source formed by Q5 and Q6.

There are two second stages in the amplifier – one on the left formed by Q11–Q14, and the other one on the right formed by Q7–Q10. Since they are identical, it will suffice to discuss the one on the right-hand side. The second stage is a cascode amplifier formed by Q7 and Q8. The output from the input stage is taken from point A and amplified by the second stage so as to provide a sufficiently high open-loop gain for the entire amplifier. Transistors Q9 and Q10 form an active load for transistor Q8. The gate of Q10 is connected to a variable resistor, R16, that helps to set the bias current for the second stage. Furthermore, in order to effectively reduce the output dc offset, the potential divider network (R21–R24) is folded to the source of Q7 and Q10. When the dc servo's output is fed to point Q of the potential divider, the output dc offset is eliminated. Let us see how the dc servo works below.

Two different ways of connecting the potential divider network R21–R24 are shown in Figure 7.27. The first way is to connect the potential divider network to the power supply directly as depicted in Figure 7.27(a). When a dc servo sends its output to point Q, this will either pull up or push down the dc potential at point Q. As a consequence, the dc quiescent points of the transistors Q7 to Q10 are shifted so that the dc potential at point H is also shifted. When the shifting of the dc potential at point H is sufficiently high, the output dc offset is eventually reduced to zero. Unfortunately, Figure 7.27(a) fails to reduce the output dc offset. When we turn to Figure 7.27(b), where a folded potential divider is used, the currents flowing in the potential divider, I_1 and I_2, will also flow into the source resistors R20 and R25, respectively. Therefore, when the dc servo shifts the dc potential at point Q, this not only affects the dc potential at gate of Q8 and Q9, but the V_{GS} of Q7 and Q10 are also directly affected. Hence, when the dc servo shifts the dc potential at point Q, this creates a greater shift in the dc quiescent conditions of transistors Q7–Q10. The dc servo circuit continues shifting the dc potential at point Q until the dc potential at point H reaches a level such that the output dc offset disappears. It can be seen that this dc servo output arrangement is different from the one often used in amplifiers where the dc servo's output is fed directly to the input stage.

Figure 7.28 shows the conventional way of connecting the dc servo's output to an amplifier. The output of the dc servo is fed to one of the differential amplifier's inputs. Since the dc servo output signal is then amplified by the input and second stage, which has a high combined open-loop gain, the output of the dc servo is very small, just a fraction of a volt. However, the dc servo arrangement in Figure 7.26 is different. The output of the dc servo feeding to point Q is

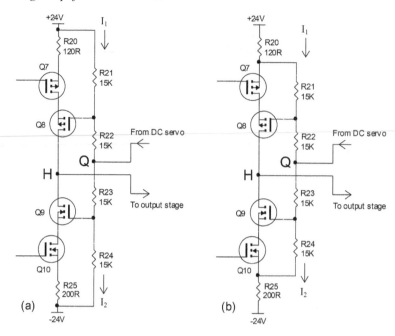

Figure 7.27 Circuit (a) shows the potential divider of the second stage is connected directly to the power
supply. Circuit (b) shows the potential divider is folded to the source of transistors Q7 and Q10

not amplified by the input and second stages. Without the help of the input and second stage, the
dc servo circuit has to work alone, shifting the dc potential at point Q either up or down, until
the dc quiescent conditions of transistors Q7–Q10 are shifted favorably towards a zero output
dc offset. In order for a sufficient shift of the dc quiescent condition to happen, the output of
the dc servo must be high. It is not like the case in Figure 7.28 where it is only a fraction of a
volt. In Figure 7.26, the output of the dc servo can be ±3V or even higher. Fortunately, we can
take advantage of the variable resistors R15 and R16 adjusting the output dc offset to zero while
keeping the dc potential at points P and Q less than ±2V.

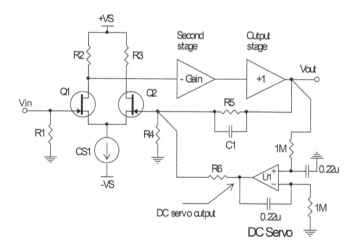

Figure 7.28 A simplified circuit shows the output of a dc servo feeding to the input stage of a conventional
amplifier

The output stage for Figure 7.26 is a fixed-bias diamond buffer. It is noted that the diamond buffer uses a mix of MOSFET and BJT. But there is no restriction to the selection of transistors. They can be all MOSFET, all BJT, or a mix of both, as in this example. Since the output transistors Q27 to Q30 are biased at 40mA, they must be mounted on a heatsink of a suitable size.

Example 2: Diamond balanced amplifier

In Example 1, illustrated by Figure 7.26, cascode amplifiers (Q7/Q8 and Q11/Q12) are used for the second stage. Since the active loads (Q9/Q10 and Q13/Q14) are also in a cascode configuration, if we modify the input stage to a complementary cascode differential amplifier, as shown in Figure 7.29(a), the second stage and output stage can be kept unchanged. But the active load has a second role working as an amplifier.

There are two current sources in the input stage of Figure 7.29(a). The current source at the top (Q9/Q10) sets a tail current (2.8mA) for the differential amplifier at the bottom (Q5–Q8). The current source at the bottom (Q11/Q12) sets a tail current (4mA) for the differential amplifier at the top (Q1–Q4), which is identical to the current source in the earlier example of Figure 7.26. Here, two different tail currents (2.8mA and 4mA) are used for the differential amplifiers because the MOSFETs on the second stage cascode amplifiers have different $|V_{GS(th)}|$. Q17 is a P-channel MOSFET, VP0106, which has $V_{GS(th)} = -2.5V$ (typ). Q20 is an N-channel MOSFET, VN0106, which has a $V_{GS(th)}$ of 1.6V (typ). Therefore, there is a conflict, and a fight for domination between the top cascode amplifier (Q17/Q18) and the bottom cascode amplifier (Q19/Q20), to control the bias current for the second stage if the two current sources in the input stage have the same tail current, say 4mA, giving the same voltage drop of 3V across R3 and R17. If we reduce the tail current of the top current source to 2.8mA so that the voltage drop across R17 is also reduced, this appears to be an appropriate amount for bias Q20, which has a lower $V_{GS(th)}$. As a result, the dc bias current for the second stage is peacefully set without the unwanted conflict.

The source resistor for Q17, R26, is now 100Ω, lower than the resistor used in the previous example, which is 120Ω. Lower the source resistor and the cascode amplifier (Q17/Q18) will produce higher voltage gain for the second stage. For a reduced source resistor, the bias current for the second stage is increased from 6mA to 8mA. It probably does not cause a big difference in the overall performance, whether source resistor R26 is 100Ω or 120Ω. Since it works very well in the circuit, 100Ω is used for R26. If JFET, which has low $V_{GS(th)}$, or BJT, which has low V_{BE}, is used for Q17 and Q20, the source resistors R26 and R31 + R32 can be partially bypassed in a similar way to that shown in Figure 7.8(c).

The variable resistor R31 helps to set the bias current for the second stage as well as reducing the output dc offset to zero. When adjusting R31, we have to observe the dc potential at point Q, which is the output from the dc servo circuit. Ideally, this should be close to dc ground level. However, the dc servo has to create a sufficiently high dc voltage at point Q in order that the dc quiescent conditions of Q17–Q20 are shifted in such a way that the output dc offset is zero. In the earlier example of Figure 7.26, the dc potential at point Q is around ±2V. For Figure 7.29(a), the dc potential at point Q can be as high as ±5V. Again, we can take advantage of the variable resistor R31 to bring down the dc potential at point Q. Similarly, adjust the variable resistor R47 to bring down the dc potential at point P.

The output stage employs a fixed-bias diamond buffer, which contains Q21–Q28 for the non-inverted output, and Q29–Q36 for the inverted output. In comparison to the previous example, BJTs are now used for the driver transistors Q25/Q26 and Q33/Q34 of Figure 7.29(a), while the use of MOSFET for current sources Q21/Q22, Q23/Q24, Q29/Q30, and Q31/Q32 remains unchanged. We found that the present mix of transistors works very well in the diamond

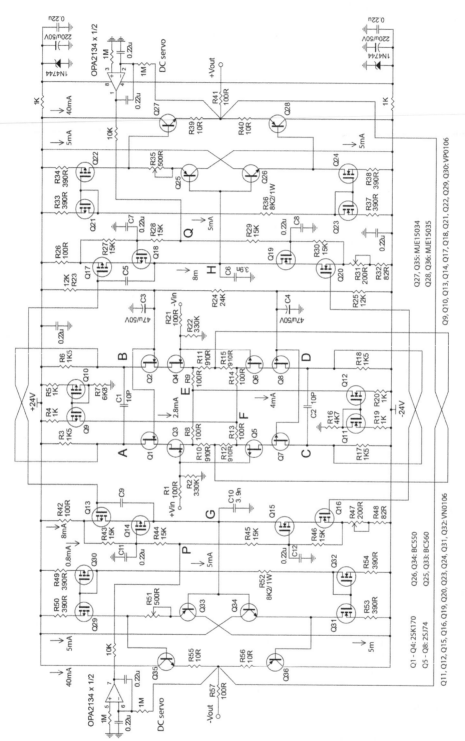

Figure 7.29(a) A fully balanced amplifier (Example 2): the *diamond balanced amplifier* employs a complementary cascode differential amplifier input stage

buffer, as well as for the entire amplifier, in terms of stability and sonic performance. However, one can easily modify the diamond buffer to all-MOSFET or all-BJT, if needed. Since the output transistors Q27/Q28 and Q35/Q36, are running at 40mA dc bias current, it is recommended to mount them in a heatsink. Finally, the voltage gain of the amplifier is determined by the ratio of R11/R9, which is 9.1 when R8 = R9 = R13 = R14 = 100Ω, and R10 = R11 = R2 = R15 = 910Ω.

Figure 7.29(b) shows two versions of a PCB layout for a two-channel (stereo) fully balanced line-stage amplifier in which each channel employs the same diamond balanced amplifier of Figure 7.29(a). The first version uses through-hole components (PCB size = 188×216mm), while the second version uses surface-mounted devices (PCB size = 188×112mm). The surface-mounted version has reduced the size of the PCB by 48%.

Figure 7.29(b) (a) Photo of a two-channel (stereo) line-stage amplifier using two diamond balanced amplifiers of Figure 7.29(a) with through-hole components (PCB dimension: 188×216mm), (b) an equivalent two-channel line-stage amplifier to (a) but using surface-mount devices (PCB dimension: 188×112mm) – the top layer, (c) the bottom layer where the output transistors are for mounting directly to the chassis

7.6 Potentiometer for volume control

A potentiometer is an integral part of an audio line-stage amplifier. It is a variable resistor that samples the input signal before feeding it to the amplifier. When a potentiometer is turned to one end, no incoming signal is feeding to the amplifier. When a potentiometer is turned to the end of the opposite direction, the maximum level of incoming signal is feeding to the amplifier. Therefore, by turning the potentiometer, we can select the right level of incoming signal such that, after being amplified by the line-stage amplifier and power amplifier, the sound level coming out of the loudspeakers is right for us. There are several popular types of potentiometer available in the market today including rotary, motorized, step attenuator, and digital potentiometers. They are discussed below.

Rotary potentiometer

A rotary potentiometer is perhaps the most commonly used potentiometer for volume control in a line-stage amplifier. It is a simple and cost-effective solution. It works to sample the input signal, as shown in Figure 7.30. This arrangement is often called a series type volume control in that the source signal is connected right across the potentiometer. If we turn a 50kΩ potentiometer to a position such that the center tap is positioned at resistance R_{adj} above the ground, as shown in Figure 7.30(b), the reminder of the resistance above the center tap is therefore equal to $50kΩ – R_{adj}$. The level of signal feeding to the amplifier is given by

$$V_{adj} = \frac{R_{adj}}{50kΩ} V_{in}$$

(7.3)

When R_{adj} is adjusted between 0 to 50kΩ, V_{adj} varies from 0 to V_{in}.

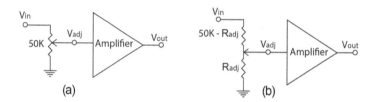

Figure 7.30 A rotary potentiometer is used in a series type volume control. Circuit (a) shows a 50kΩ rotary potentiometer being used. Circuit (b) shows the center tap of the potentiometer positioned at a resistance R_{adj} above the ground

Rotary potentiometers can also be used for shunt type volume control, as shown in Figure 7.31. In this arrangement, a fixed resistor, R1, must be added in series with the potentiometer. The center tap of the potentiometer is shorted to one of the other two taps. In this arrangement, the potentiometer is not in the signal path. The potentiometer simply works to attenuate the incoming signal. Therefore, theoretically speaking, the potentiometer does not bring any coloration to the signal. In comparison, since the potentiometer in a series type volume control is right on the signal path, it will certainly bring coloration to the signal. It should be noted that as there is also a fixed resistor R1 used in shunt type volume control, it is right on the signal path in the exact same way that the series type potentiometer is. However, a fixed resistor is highly regarded as more reliable and producing a better sound than a rotary potentiometer, in which the film resistive surface is constantly touched by a wiper with metallic contact.

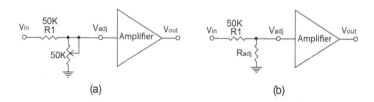

Figure 7.31 A rotary potentiometer used in shunt type volume control. Circuit (a) shows a resistor, R1, in series with a 50kΩ rotary potentiometer. Circuit (b) shows the center tap of the potentiometer positioned at a resistance R_{adj} above the ground

Figure 7.31(b) shows that the potentiometer is positioned at a resistance R_{adj} above the ground. Therefore, the signal V_{adj} is given by

$$V_{adj} = \frac{R_{adj}}{50k\Omega + R_{adj}} V_{in} \tag{7.4}$$

When we turn the potentiometer from one end to the other end, R_{adj} is between 0 and 50kΩ. Therefore, V_{adj} will become 0 to 0.5V_{in}. In other words, there is a 50% reduction of the incoming signal. Note that if the series resistor R1 is greater than the potentiometer, reduction will be greater than 50%. This appears to be a penalty for the shunt type volume control. If we use a smaller resistance R1, we can minimize the incoming signal reduction. But it will be difficult to set V_{adj} for a low listening level when R1 is too small. However, for Figure 7.31 even if there is a 50% reduction of V_{in}, as long as the gain of the amplifier is greater than 2, the output of the amplifier V_{out} is still greater than V_{in}.

For an amplifier with balanced inputs, again both series type and shunt type volume controls can be used, as shown in Figure 7.32. It can be seen that a dual potentiometer is required for series type volume control, while a single potentiometer can be used in shunt type volume control. Since there are two resistors, R1 and R2, used in the shunt type volume control, the maximum signal passing to the amplifier is only 1/3 of the input signal in this example. Therefore, the voltage gain of the amplifier has to be greater than 3 in order to get $V_{out} > V_{in}$.

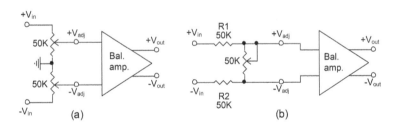

Figure 7.32 Circuit (a) shows a series type volume control used in a fully balanced amplifier. Circuit (b) shows a shunt type volume control used in a fully balanced amplifier

There is one problem associated with a rotary potentiometer. Since the potentiometer has a wiper with metallic contact that is constantly touching the resistive film surface, that surface is inevitably deteriorating over time. As a result, it will create some kind of cracking noise at the output of the power amplifier. The cracking noise will be more prominent when there is a small dc presented at the potentiometer, which may be a dc output offset coming from the source, or coming from the line-stage amplifier's input stage.

Motorized potentiometer

If we take the advantage of using an infrared remote control and microcontroller, we can use a motorized potentiometer for volume control. A H-bridge motor driver is often employed to control the motor, as shown in Figure 7.33. Since the potentiometer itself is identical to the rotary potentiometer discussed in the preceding section, a motorized potentiometer can be used to implement series type and shunt type volume controls.

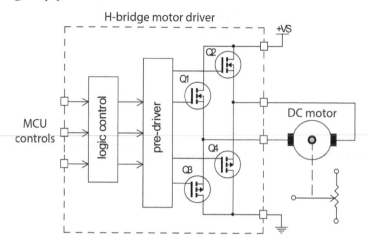

Figure 7.33 An H-bridge motor driver for controlling a motorized potentiometer

Figure 7.34(a) shows a simplified discrete motor control circuit for a motorized potentiometer. This circuit only requires two signals to control the motor, +5V (HIGH) and 0V (LOW). A total of four transistors is used with 2 NPN and 2 PNP. The resistors are selected such that a transistor is either in cut-off or saturation mode when responding to a HIGH or LOW control signal. When Control 1 is LOW and Control 2 is HIGH, transistors Q2 and Q3 are cut off while Q1 and Q4 are in saturation, as shown in Figure 7.34(b). The dc voltage applied to the motor is V_{AB}, which is equal to 5V – 2×Vsat, where Vsat is the $V_{CE(sat)}$ of the transistor.

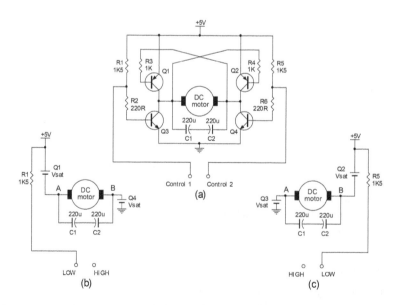

Figure 7.34 A simplified discrete version for controlling a motorized potentiometer in circuit (a). Circuit (b) shows the motor rotated in one direction by LOW–HIGH control signals. Circuit (c) shows the motor rotated in the opposite direction by HIGH–LOW control signals

The V_{AB} drives the dc motor to rotate in one direction. When the control signals are reversed, becoming HGH and LOW as shown in Figure 7.34(c), V_{AB} becomes $- (5V - 2 \times Vsat)$, which is the opposite of the case in Figure 7.34(b). Therefore, the dc motor rotates in the opposite direction.

Most $V_{CE(sat)}$ of BJTs vary from 0.1–0.4V. Therefore, in order to maximize the dc voltage across the motor (V_{AB}), it is recommended to select transistor with a low $V_{CE(sat)}$ and current gain greater than 150 for Q1–Q4.

Stepped attenuator potentiometer

A stepped attenuator potentiometer employs a series of discrete resistors connected to a multi-throw switch. The popular number of throws is 24 and 48. For making a stereo 2-channel potentiometer, a double-pole 24-throw or a double-pole 48-throw switch is needed. For smooth transition between steps, the switch must be a make-before-break type.

Figure 7.35 shows a 24-step attenuator potentiometer for one channel. It is operating as a series type volume control. Twenty-four resistors are used for a total of 24 steps. The input signal is connected to R24, which is the highest value resistor. Step #1 will get the highest attenuation so that the output Vadj is the lowest. It is common to set an even 2dB attenuation step size so that the highest attenuation is 46dB for the configuration shown in Figure 7.35. It can be seen that the highest attenuation is not 48dB because resistor step 24 is directly connected to the input source without attenuation.

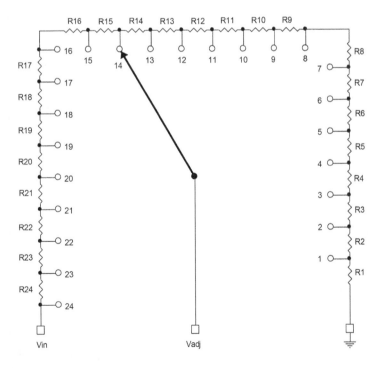

Figure 7.35 A 24-step attenuator potentiometer

Table 7.1 Resistors for a 24-step attenuator potentiometer are computed for implementing resistances of 1kΩ, 10kΩ, 5kΩ, and 50kΩ in a step of –2dB attenuation

Tap		Exact values (Ω)	(dB)	E96 values (Ω)	(dB)	E96 values (Ω)	(dB)	Tap		Exact values (Ω)	(dB)	E96 values (Ω)	(dB)	E96 values (Ω)	(dB)
24	R24	205.672	0	205	0	2050	0	24	R24	1,028.4	0	1,020	0	10,200	0
23	R23	163.371	-2	162	-2.01	1620	-2.01	23	R23	816.85	-2	820	-1.98	8,200	-1.98
22	R22	129.770	-4	130	-4.00	1300	-4.00	22	R22	648.85	-4	649	-3.99	6,490	-3.99
21	R21	103.080	-6	102	-6.02	1020	-6.02	21	R21	515.4	-6	511	-5.99	5,110	-5.99
20	R20	81.879	-8	80.6	-8.01	806	-8.01	20	R20	409.4	-8	412	-7.98	4,120	-7.98
19	R19	65.039	-10	64.9	-9.99	649	-9.99	19	R19	325.2	-10	324	-9.99	3,240	-9.99
18	R18	51.662	-12	51.1	-12.00	511	-12.00	18	R18	258.31	-12	261	-11.98	2,610	-11.98
17	R17	41.037	-14	41.2	-13.98	412	-13.98	17	R17	205.18	-14	205	-14.00	2,050	-14.00
16	R16	32.597	-16	32.4	-16.00	324	-16.00	16	R16	162.98	-16	162	-16.00	1,620	-16.00
15	R15	25.893	-18	25.5	-18.00	255	-18.00	15	R15	129.46	-18	130	-17.99	1,300	-17.99
14	R14	20.567	-20	20.5	-19.98	205	-19.98	14	R14	102.84	-20	102	-20.00	1,020	-20.00
13	R13	16.337	-22	16.2	-21.98	162	-21.98	13	R13	81.685	-22	82	-21.99	820	-21.99
12	R12	12.977	-24	13	-23.97	130	-23.97	12	R12	64.885	-24	64.9	-24.00	649	-24.00
11	R11	10.308	-26	10.2	-25.98	102	-25.98	11	R11	51.54	-26	51.1	-26.00	511	-26.00
10	R10	8.188	-28	8.25	-27.96	82.5	-27.96	10	R10	40.94	-28	41.2	-27.98	412	-27.98
9	R9	6.504	-30	6.49	-29.98	64.9	-29.98	9	R9	32.52	-30	32.4	-29.99	324	-29.99
8	R8	5.166	-32	5.11	-31.98	51.1	-31.98	8	R8	25.831	-32	26.1	-31.99	261	-31.99
7	R7	4.104	-34	4.12	-33.96	41.2	-33.96	7	R7	20.518	-34	20.5	-34.01	205	-34.01
6	R6	3.260	-36	3.24	-35.97	32.4	-35.97	6	R6	16.298	-36	16.2	-36.02	162	-36.02
5	R5	2.589	-38	2.61	-37.96	26.1	-37.96	5	R5	12.946	-38	13	-38.01	130	-38.01
4	R4	2.057	-40	2.05	-39.99	20.5	-39.99	4	R4	10.284	-40	10.2	-40.02	102	-40.02
3	R3	1.634	-42	1.62	-41.99	16.2	-41.99	3	R3	8.1685	-42	8.2	-42.02	82	-42.02
2	R2	1.298	-44	1.3	-43.98	13	-43.98	2	R2	6.4885	-44	6.49	-44.03	64.9	-44.03
1	R1	5.012	-46	4.99	-45.99	49.9	-45.99	1	R1	25.059	-46	24.9	-46.04	249	-46.04
Total =		1,000 Ω		994.4 Ω		9,943.8 Ω		Total =		5000 Ω		4,993 Ω		49,932 Ω	

Resistor values are computed for a 24-step attenuator potentiometer with 2dB steps, as shown in Table 7.1. The resistors are computed for implementing a 1kΩ, 5kΩ, 10kΩ, and 50kΩ potentiometer. These are the common resistances often used in stepped attenuator potentiometers. If some other resistance is required, say 20kΩ, we can multiply the column of the exact value of 1kΩ by 20 to get the exact values. Then look up a E96 series resistor table to select the closest resistors. When E96 series resistors are used, the error for attenuation can be kept within ±0.2dB.

Resistor values for a 48-step attenuator potentiometer with 1dB steps are computed and shown in Table 7.2. The resistors are computed for 10kΩ and 50kΩ potentiometers. For other resistances, again we can multiply a factor to the exact values. Then look up a E96 series resistor table to select the closest resistors.

Table 7.2 Resistors for a 48-step attenuator potentiometer are computed for implementing resistances of 10kΩ and 50kΩ in a step of −1dB attenuation

Tap		Exact values		E96 values		Tap		E96 values		E96 values	
		(Ω)	(dB)	(Ω)	(dB)			(Ω)	(dB)	(Ω)	(dB)
48	R48	1087.5	0	1100	0	24	R24	68.616	-24	68.1	-24.01
47	R47	969.23	-1	976	-1.01	23	R23	61.154	-25	61.9	-25.00
46	R46	863.82	-2	866	-2.02	22	R22	54.504	-26	54.9	-26.02
45	R45	769.88	-3	768	-3.03	21	R21	48.576	-27	48.7	-27.03
44	R44	686.16	-4	681	-4.03	20	R20	43.294	-28	43.2	-28.03
43	R43	611.54	-5	604	-5.02	19	R19	38.586	-29	38.3	-29.03
42	R42	545.04	-6	549	-6.01	18	R18	34.389	-30	34	-30.03
41	R41	485.76	-7	487	-7.02	17	R17	30.650	-31	30.9	-31.02
40	R40	432.94	-8	432	-8.02	16	R16	27.317	-32	27.4	-32.03
39	R39	385.86	-9	383	-9.02	15	R15	24.346	-33	24.3	-33.04
38	R38	343.89	-10	340	-10.02	14	R14	21.698	-34	21.5	-34.04
37	R37	306.50	-11	309	-11.01	13	R13	19.339	-35	19.1	-35.03
36	R36	273.17	-12	274	-12.02	12	R12	17.236	-36	17.4	-36.02
35	R35	243.46	-13	243	-13.02	11	R11	15.361	-37	15.4	-37.04
34	R34	216.98	-14	215	-14.02	10	R10	13.691	-38	13.7	-38.04
33	R33	193.39	-15	191	-15.02	9	R9	12.202	-39	12.1	-39.05
32	R32	172.36	-16	174	-16.00	8	R8	10.875	-40	10.7	-40.05
31	R31	153.61	-17	154	-17.02	7	R7	9.692	-41	9.76	-41.03
30	R30	136.91	-18	137	-18.02	6	R6	8.638	-42	8.66	-42.05
29	R29	122.02	-19	121	-19.02	5	R5	7.699	-43	7.68	-43.05
28	R28	108.75	-20	107	-20.02	4	R4	6.862	-44	6.81	-44.06
27	R27	96.92	-21	97.6	-21.00	3	R3	6.115	-45	6.19	-45.06
26	R26	86.38	-22	86.6	-22.01	2	R2	5.450	-46	5.49	-46.08
25	R25	76.99	-23	76.8	-23.01	1	R1	44.668	-47	44.2	-47.09
								Total = 10,000 Ω		**10,002 Ω**	

Tap		Exact values		E96 values		Tap		Exact values		E96 values	
		(Ω)	(dB)	(Ω)	(dB)			(Ω)	(dB)	(Ω)	(dB)
48	R48	5437.5	0	5490	0	24	R24	343.08	-24	340	-24.05
47	R47	4846.1	-1	4870	-1.009	23	R23	305.77	-25	301	-25.05
46	R46	4319.1	-2	4320	-2.015	22	R22	272.52	-26	274	-26.04
45	R45	3849.4	-3	3830	-3.016	21	R21	242.88	-27	243	-27.04

(*Continued*)

Table 7.2 (Continued)

Tap		Exact values (Ω)	(dB)	E96 values (Ω)	(dB)	Tap		Exact values (Ω)	(dB)	E96 values (Ω)	(dB)
44	R44	3430.8	-4	3400	-4.011	20	R20	216.47	-28	215	-28.05
43	R43	3057.7	-5	3090	-5.002	19	R19	192.93	-29	191	-29.05
42	R42	2725.2	-6	2740	-6.013	18	R18	171.95	-30	169	-30.04
41	R41	2428.8	-7	2430	-7.019	17	R17	153.25	-31	154	-31.03
40	R40	2164.7	-8	2150	-8.021	16	R16	136.58	-32	137	-32.04
39	R39	1929.3	-9	1910	-9.015	15	R15	121.73	-33	121	-33.04
38	R38	1719.5	-10	1740	-10.01	14	R14	108.49	-34	107	-34.04
37	R37	1532.5	-11	1540	-11.02	13	R13	96.693	-35	97.6	-35.03
36	R36	1365.8	-12	1370	-12.02	12	R12	86.178	-36	86.6	-36.04
35	R35	1217.3	-13	1210	-13.03	11	R11	76.806	-37	76.8	-37.05
34	R34	1084.9	-14	1070	-14.03	10	R10	68.453	-38	68.1	-38.06
33	R33	966.93	-15	976	-15.01	9	R9	61.009	-39	60.4	-39.06
32	R32	861.78	-16	866	-16.02	8	R8	54.375	-40	54.9	-40.05
31	R31	768.06	-17	768	-17.03	7	R7	48.461	-41	48.7	-41.07
30	R30	684.53	-18	681	-18.03	6	R6	43.191	-42	43.2	-42.08
29	R29	610.09	-19	604	-19.03	5	R5	38.494	-43	38.3	-43.09
28	R28	543.75	-20	549	-20.03	4	R4	34.308	-44	34	-44.09
27	R27	484.61	-21	487	-21.04	3	R3	30.577	-45	30.1	-45.09
26	R26	431.91	-22	432	-22.05	2	R2	27.252	-46	27.4	-46.09
25	R25	384.94	-23	383	-23.05	1	R1	223.34	-47	221	-47.1
								Total = 50,000 Ω		50,045 Ω	

A stepped attenuator can be used for both series type volume control and shunt type volume control, as shown in Figure 7.36. In the shunt type volume control configuration, a series resistor,

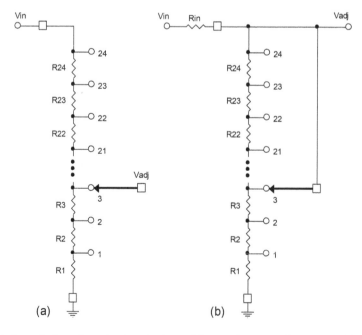

Figure 7.36 A stepped attenuator potentiometer is used for a series type volume control in circuit (a) and a shunt type volume control in circuit (b)

Rin, is needed. This resistor must have a high resistance so that it creates sufficient attenuation. 50kΩ is a good start for Rin pairing with a 50kΩ potentiometer. Even though any power rating can be used for Rin, as there is no power dissipation in the resistor, it is recommended to use a power rating higher than 0.5W. A 1W resistor does often produce sound better than a 0.5W when it is used for Rin. The power rating for the rest of the resistors can be 0.25W to 0.5W.

Digitally controlled potentiometer

With the advance of semiconductor technologies, there are many high performance digitally controlled potentiometers available on the market. Some of them work as simple variable resistors, but others have a built-in op-amp that makes the device itself a complete line-stage amplifier with volume control. The popular number of steps are 64, 128, and 256. Nowadays, some digital potentiometers offer as many as 1024 steps.

Figure 7.37 shows the structure of a simplified digitally controlled potentiometer. The potentiometer illustrated in Figure 7.37(a) works like a simple rotary potentiometer with three taps. In other words, digitally controlled potentiometers can be used to replace rotary potentiometers, shown in Figures 7.30–7.32. However, it should be noted that there is an inherited wiper resistance in a digitally controlled potentiometer, known as a wiper resistor, RW, as shown in Figure 7.37(b). The value of RW varies from 50Ω to 400Ω. Typically, RW is around 100Ω. If a digitally controlled potentiometer is used for series type volume

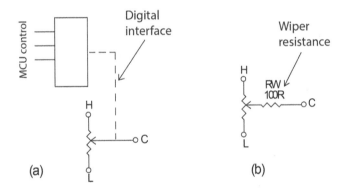

Figure 7.37 Circuit (a) shows a simplified diagram of a digitally controlled potentiometer. Circuit (b) shows an inherited wiper resistance, RW

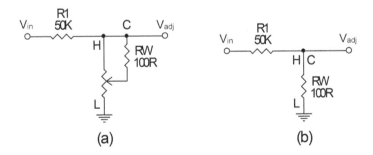

Figure 7.38 Circuit (a) shows a digitally controlled potentiometer used in a shunt type volume control. Circuit (b) shows the equivalent circuit when the potentiometer has reached the lowest resistance value

controls, the presence of RW is not a problem, as it is just another series resistor. But when a digitally controlled potentiometer is used for shunt type volume controls, as shown Figure 7.38, the wiper resistance will play a role when the wiper has reached the lowest resistance at position L.

When the wiper has reached the lowest resistance at position L, we would expect that $V_{adj} = 0$ so that no signal can get through. But RW is now present as a shunt resistor in parallel to the potentiometer, as shown in Figure 7.38(b). For Figure 7.38(a), since the resistance of the potentiometer is much higher than the wiper resistance, we can ignore RW in this situation. For Figure 7.38(b), the wiper resistor 100Ω appears to be shunting the input. For example, if it is a 20kΩ potentiometer, 20kΩ ∥ 100Ω ≈ 100 Ω. As a result, there is a small amount of signal getting through to the amplifier. Thus, when a digitally controlled potentiometer is used for shunt type volume control, this is something we need to overcome.

Figure 7.39 shows two digitally controlled potentiometers used in a shunt type volume control for a balanced amplifier. It should be noted that the configuration of Figure 7.32(b) can also be realized by a digitally controlled potentiometer in which only one potentiometer is needed. But when a digitally controlled potentiometer is used, I feel more comfortable using two potentiometers by grounding both tap "L", as shown in Figure 7.39(a). Since many digitally controlled potentiometers come in a dual package, using two potentiometers per channel should not be too difficult to implement and would not add too much cost. Now let us consider the case when the potentiometers reach the lowest resistance. Owing to the wiper resistance, RW is shunting on each of the inputs as shown in Figure 7.39(b). In order to reduce the shunting resistance RW, we can take advantage of using switched relays.

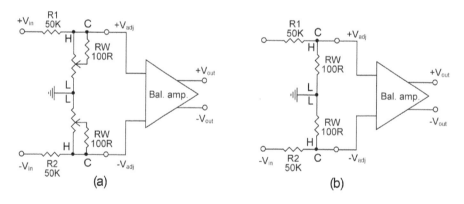

Figure 7.39 Dual digitally controlled potentiometers are used for balanced shunt type volume control in circuit (a). Circuit (b) shows a wiper resistor, RW, is shunting to ground as the potentiometer has reached its lowest resistance

Switched relays are used to reduce the shunting resistance caused by the wiper resistance RW in Figure 7.40. It is shown that a total of eight relays with different shunting resistances is used. Depending upon the actual wiper resistance, if it is low (RW < 100Ω), 4–5 relays are enough. But for a high wiper resistance (RW > 100Ω), 6–8 relays should be used to ensure smooth transition. In this design, only one relay is switched on at a time. When a relay is not switched on, the resistor associated with that relay does not affect the wiper resistor RW.

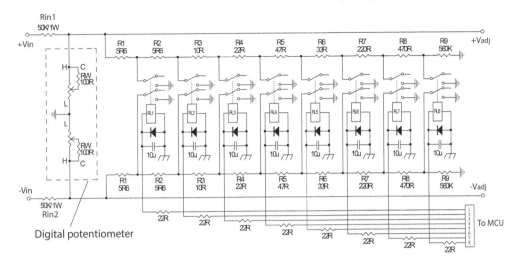

Figure 7.40 Switched relays are added to a balanced shunt type volume control employing a digitally controlled potentiometer for reducing the wiper resistance when the potentiometer has reached its lowest resistance

It starts with relay RL1 for the lowest shunting resistance that represents the lowest signal passing to the amplifier. This is then followed by RL2, RL3, and up to RL8. After RL8, the digitally controlled potentiometer itself will take over and no further relay is switched on. When RL1 is switched on, the effective shunting resistance will be $R_{eff} = RW \| R1 = 5.3\Omega$. In this calculation, we ignore the resistance of the potentiometer itself, which is often a high value $20k\Omega$ or larger resistor that has little loading effect to the small shunting resistor. For RL2 and RL3, the R_{eff} is 10Ω and 17.5Ω, respectively. These are shown in Figure 7.41. Therefore, a steady increase of the R_{eff} occurs when relays are switching from RL1 to RL8. Note that in each relay control line, there is a 22Ω resistor and $10uF$ capacitor. This is to create a small delay to switching on the relay so as to avoid unpleasant transient noise.

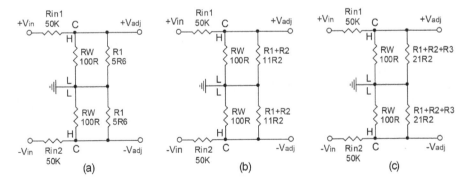

Figure 7.41 When the potentiometer is turned to its lowest resistance with its wiper resistor RW shunting at the input, circuit (a) is showing relay RL1 is switched on, connecting R1 to ground. Circuit (b) is showing relay RL2 is switched on, connecting (R1 + R2) to ground. Circuit (c) is showing relay RL3 is switched on, connecting (R1 + R2 + R3) to ground

7.7   Exercises

Ex 7.1   Figure 7.42 is a two-stage amplifier. It can be considered as a JFET equivalent to the vacuum tube amplifier of Ex 8.2 in Chapter 8. The input stage is a cascode amplifier. The second stage is a source follower with a bootstrapping active load. Assume voltage across resistor R4, V_{DS} of Q1 and V_{DS} of Q2 sharing the power supply equally, i.e., $V_B = 15V$, $V_A = 30V$. If 2SK170 is used for Q1–Q4, design the amplifier with a voltage gain of –20 and input impedance of 100KΩ. The dc quiescent current for the input stage and second stage are 2mA, and $V_C = 15V$. If necessary, drain resistor R4 and source resistor R5 can be partially bypassed.

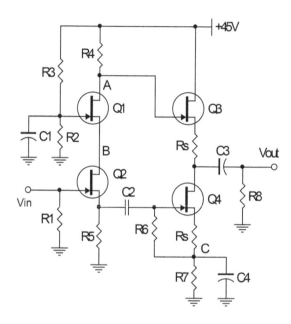

Figure 7.42 A two-stage amplifier for Ex 7.1

Ex 7.2   Complete the design of the unbalanced input–unbalanced output amplifier in Figure 7.9. Suggest using 2SJ74 for Q5–Q8. Other transistors follow Figure 7.6.

Ex 7.3   Complete the design of the unbalanced input–unbalanced output amplifier in Figure 7.11. Suggest using 2SJ74 for Q5–Q8. Other transistors in the input stage and second stage follow Figure 7.6. For the output stage, use VP0106 for P-MOS Q13 and Q14, and VN0106 for N-MOS Q17 and Q18. Transistor Q15 is BC560 while BC550 is for Q16. The dc bias current in Q15 and Q16 is 5mA. For output transistors Q19 and Q20, choose any parts that you prefer. The dc bias current for Q19 and Q20 is 40mA.

Ex 7.4   Complete the design of the unbalanced input–unbalanced output amplifier in Figure 7.15. In the input stage, suggest using 2SK246 or its equivalent for Q1, 2SK170 for Q2, 2SJ74 for Q3, and 2SJ103 or its equivalent for Q4. In the second stage, change JFET Q5 to MOSFET so that Q5 and Q6 are VP0106. Also change JFET Q8 to MOSFET so that Q7 and Q8 are VN0106. Change the output stage to the fixed-bias diamond buffer used in Ex 7.3.

Ex 7.5 The circuit of Figure 7.43 is a Tee-feedback network difference amplifier. Determine the resistors for a differential input impedance of 15kΩ and a voltage gain of 10.

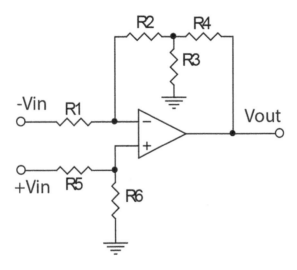

Figure 7.43 A Tee-feedback network difference amplifier for Ex 7.5

Ex 7.6 Use the difference amplifier in Ex 7.5 and apply it to Ex 7.3. Thus, change the amplifier of Figure 7.11 from unbalanced input to balanced input with voltage gain 10 and differential input impedance 30kΩ.

Ex 7.7 Table 7.2 shows the calculation for a 48-step attenuator with –1dB attenuation step implementing a 10kΩ and a 50kΩ potentiometer. Now compute 48 exact resistor values with a –1.5dB attenuation step for a 50kΩ potentiometer.

References

[1] Borbely, E., "JFET: the new frontier, part 1," Audio Electronics, May 1999.
[2] Borbely, E., "The all-FET line amp," AudioXpress, May 2002.
[3] Data sheet for OP176, Analog Devices Inc.
[4] Cordell, B., "Low noise dual monolithic JFET," LSK489 application note.
[5] Borbely, E., "JFET: the new frontier, part 2," Audio Electronics, June 1999.
[6] Self, D., "Small signal audio design," Focal Press, 2nd edition, 2015.
[7] Jung, W.G., *Audio IC op amp application*, Howard Sams, 3rd edition, 1987.

8 Line-stage amplifiers – vacuum tubes

8.1 Introduction

As discussed in Chapter 7, solid-state semiconductor line-stage amplifiers often consist of three stages: input, second, and output stage. Whether the amplifier is to amplify unbalanced or balanced input, a differential amplifier is often employed for the input stage. Given the availability of complementary JFET, BJT, and MOSFET devices, a designer has the flexibility to employ a complementary differential amplifier for the input stage, a complementary voltage amplifier for the second stage, and a complementary push–pull output stage. In designing solid-state semiconductor line-stage amplifiers, we are very fortunate in having many devices available to choose from. However, the vacuum tube amplifier is very different.

First, we do not have both N-type and P-type vacuum tubes. In an analogy to the JFET transistor, the triode vacuum tube is like an N-type (or N-channel) JFET. Therefore, designing a vacuum tube amplifier is similar to designing a solid-state amplifier using only N-type devices. It is not impossible, but it is something that most circuit designers are not very keen on. No one wants to turn an otherwise simple design into a difficult one.

Owing to a lack of complementary vacuum tubes, designing a vacuum tube amplifier is different from a solid-state amplifier. Fortunately, most vacuum tubes used in line-stage amplifier application are triodes, which have high linearity and low distortion. As a result, employing only a few pieces of triodes will be sufficient to give us a highly musical line-stage amplifier. A list of commonly used audio triodes is given in Table 8.1.

In the following section, we discuss two classic vacuum tube line-stage amplifiers that comprise three stages. Then, in the subsequent sections, we will discuss various line-stage amplifier designs for unbalanced and balanced input and output.

Three-stage amplifier (a variant of Marantz 7)

A three-stage configuration is employed by the amplifier illustrated in Figure 8.1. The first stage is not a differential amplifier that we often find in the solid-state amplifier design. It is just a simple common cathode amplifier formed by triode T1 that provides amplification to the input signal as well as feedback from the output. The difference between the input and feedback is amplified and capacitor coupled to the second stage formed by triode T2. The second stage is also configured in a common cathode arrangement so as to build up the necessary open-loop gain for the entire amplifier.

DOI: 10.4324/9781003369462-8

Table 8.1 A list of commonly used triodes for line-stage amplifier applications

| Tube | | Typical value | | Maximum value | | Heater voltage/current | | | |
	Amplifying factor, μ	Transcon- ductance, g_m (mA/V)	Plate resistance, r_p (kΩ)	Plate voltage (V)	Plate dissipation (W)	6.3V	Current (A)	12.6V	Current (A)
ECC81/ 12AT7	60	4 to 5.5	15 to 11	300	2.5	●	0.3	●	0.15
ECC82/ 12AU7	20	2.2 to 3.1	7.7 to 6.5	300	2.75	●	0.3	●	0.15
ECC83/ 12AX7	100	1.2 to 1.6	80 to 62	330	1.2	●	0.3	●	0.15
ECC88/ 6922	33	11.5 to 12.5	2.6 to 2.9	250	1.5	●	0.3		
ECC99	22	9.5	2.3	400	5	●	0.8	●	0.4
6H30	15±3	18±5	0.52 to 1.3	250	4	●	0.83		
6N1P	35±8	3.5 to 5.5	4.9 to 12	300	2.2	●	0.6		
6SL7	70	1.6	44	300	1	●	0.3		
6SN7	20	2.6 to 3	7.7 to 6.7	450	5	●	0.6		
12AY7	40	1.75	22.8	300	1.5	●	0.3	●	0.15
12BH7	16.5	3.1	5.3	300	3.5	●	0.6	●	0.3

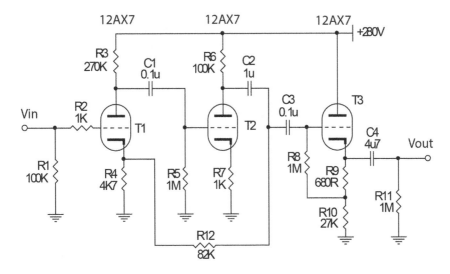

Figure 8.1 A three-stage design popularized by the Marantz 7 preamplifier

Since the output impedance of a common cathode amplifier is rather high, it is not desirable to drive a power amplifier directly. Therefore, an output stage with low output impedance is needed. Naturally, a cathode follower, which is formed by triode T3, is used for the output stage. The output impedance for a cathode follower is approximately equal to $1/g_m$ (see Eq. [3.63]). For 12AX7, if we take g_m to be 1.2mA/V, the output impedance is 833Ω. It is not very small but it is a respectable output impedance for a vacuum tube line-stage amplifier. If the feedback resistor R12 is applied to the output of the cathode follower so that the feedback loop also encloses the output from the cathode follower, it becomes a series-shunt feedback configuration. Then the feedback could greatly reduce the output impedance.

However, in this circuit, feedback is applied between the input stage and the second stage via resistor R12. The overall voltage gain is approximately equal to $1 + R12/R4 = +18.4$. The positive sign reminds us that it is a non-inverted amplifier. Additionally, feedback will effectively help with extending the bandwidth. This is a rather simple amplifier that contains only three triodes, four capacitors, and 12 resistors – a total of 19 components. The component count is a lot fewer than is needed for most discrete sold-state line-stage amplifiers. In comparison, the solid-state line-stage amplifier in Figure 7.6 employs a total of 49 components. As a matter of fact, some vacuum tube amplifiers, which will be discussed in the next section, contain even fewer than ten components. Nevertheless, this simple three-stage amplifier offers remarkable performance in such a simple design. It was popularized by the *Marantz 7* amplifier and followed by many variants. Figure 8.1 is one of the Marantz 7 variants using slightly different components and arrangement.

Shigeru Wada amplifier

One of the better-known Marantz 7 variants was developed by *Shigeru Wada* as shown in Figure 8.2. Comparing with Figure 8.1, there are several improvements provided by Figure 8.2. First, the feedback is now taken from the output stage. Since the output is enclosed by the feedback loop, output impedance is greatly reduced by negative feedback. As discussed in Example 5.3, the output impedance is only 1Ω. Second, the output stage is a White cathode follower formed by

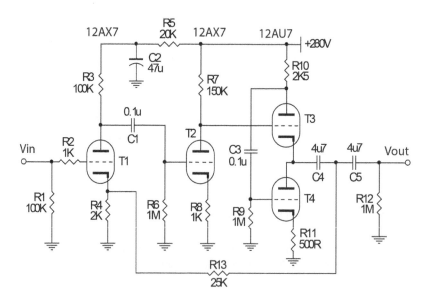

Figure 8.2 This is the Shigeru Wada amplifier, which is a variant of Marantz 7 preamplifier

triodes T3 and T4. The White cathode follower has a lower output impedance than a normal cathode follower (see Chapter 3 for details). Third, 12AU7 is used for the output stage. Since transconductance, g_m, of 12AU7 is higher than 12AX7, this also helps to reduce output impedance. Finally, an R-C filter formed by R5 and C2 is used to remove ripples from the power supply that may, otherwise, get into the input stage.

The amplifier's voltage gain is approximately equal to $1 + R13/R4 = 13.5$. Given the improvements discussed above, it is not surprising to see that the Shigeru Wada amplifier in Figure 8.2 produces lower output impedance, lower distortion, and broader bandwidth than the amplifier shown in Figure 8.1.

8.2 Unbalanced input–unbalanced output amplifiers

In the previous section, we illustrated two classic three-stage line-stage amplifiers. However, a vacuum tube amplifier does not have to be confined to three-stages and to using feedback. We are going to discuss in this section that many practical line-stage vacuum tube amplifiers can be just a single-stage or two-stage design with or without employing feedback.

For a single-stage design, we may consider the basic single tube amplifier configurations such as (i) common cathode, (ii) common grid, and (iii) cathode follower. A common cathode amplifier offers a desirable voltage gain, which is needed for a line-stage amplifier. However, the high output impedance will limit its use to only drive power amplifiers with very high input impedance. For example, vacuum tube power amplifiers or solid-state semiconductor power amplifiers with a JFET input stage are probably fine. Even so, we would still prefer the output impedance of a line-stage amplifier to go down to a lower level that, unfortunately, a single tube common cathode amplifier configuration has little hope of achieving.

For a common grid amplifier configuration, it also provides a desirable voltage gain. The voltage gain is non-inverted versus inverted gain for the common cathode amplifier. However, the input impedance of a common grid amplifier is just too low for practical use. It must pair with a second tube to form a compound amplifier so that the compound amplifier can offer sufficiently high input impedance, voltage gain, and low output impedance. A cascode compound amplifier may seem to be a possible solution. However, the output impedance of a cascode amplifier is still considered high. But if a cathode follower output stage is added, it becomes a possible solution for an audio line-stage amplifier application. However, it is a two-stage design.

The cathode follower configuration has the desirable high input impedance and low output impedance. However, the voltage gain for a cathode follower is close to unity. Except in a situation where the source signal amplitude is so high that no further voltage gain is needed by a line-stage amplifier, a cathode follower may find a niche in such an application.

If we rule out a single tube amplifier, there are several compound tube amplifier configurations that can be considered for single-stage amplifier application including SRPP, half-mu, and cathode-coupled amplifier. When a cathode follower second stage is added, it will transform a compound tube amplifier to a two-stage amplifier that retains the properties of the compound amplifier while having a low output impedance. This makes a two-stage amplifier more appealing than a single-stage compound tube amplifier for audio line-stage application. Single-stage and two-stage amplifiers for unbalanced input–unbalanced output applications are discussed in the following.

Shunt regulated push–pull (SRPP) amplifier

Figure 8.3 shows an SRPP amplifier. It also has other names, such as *totem-pole amplifier* [1], *series-balanced amplifier* [2] and *bootstrap amplifier* [3], etc. It is a compound tube amplifier that comprises two triodes in a single-stage arrangement. The triode on the bottom, T2, operates in a common cathode configuration. Triodes T1 and T2 work together, producing an approximated push–pull output. Since the output is taken at the cathode of T1 (i.e., point A), the output impedance is low. Given medium voltage gain and high input impedance, SRPP amplifiers are often found in line-stage amplifiers and even in power amplifiers.

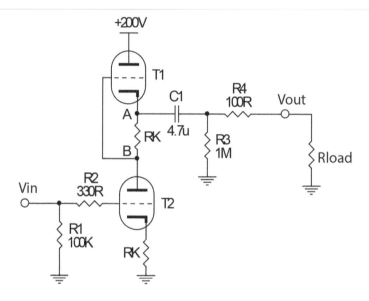

Figure 8.3 This is an SRPP amplifier with two equal cathode resistors, RK

In the 1950s and 1960s, an SRPP amplifier was often used in color TV systems, where linearity of operation is of primary importance. The SRPP was also very popular for driving 75Ω cables where high output current and low load impedance are of great concern. When an SRPP amplifier is designed to drive a load with optimized output current, a plate resistor (RP, not shown in Figure 8.3) is inserted between the power supply and the plate of triode T1. There is a relationship between the cathode resistor RK, load resistor R_{load}, and the vacuum tube's inherent plate impedance r_p when the output current for a given load resistor R_{load} is optimized [4–6]. However, in an audio line-stage amplifier application, optimizing the output current for a particular output load is not needed. We often use an SRPP amplifier to provide a certain voltage gain and a relatively low output impedance driving a high impedance load. Therefore, it is not necessary to insert a plate resistor (RP) in T1's plate.

The operation of an SRPP amplifier is given as follows. The input signal is first amplified by T2, which is a common cathode amplifier. The signal appearing at point B is no different to the output signal (Vout) if we ignore the resistor RK between point A and B for the time being. Therefore, the signal at pint B is Vout. The grid of T1 picks up the signal from point B and delivers it to its cathode (point A). Since T1 is operating in a cathode follower configuration with unity voltage gain, T1 delivers an identical signal (Vout) to point A. If we view the output of T1 as pushing down to point A, then the output from T2 is pulling up to point A, where point A is the output of this SRPP amplifier. This is an approximated push–pull operation that we can obtain from an SRPP amplifier without using a transformer. However, we can never get a perfect push–pull operation out of this simple circuit because the voltage gain of a cathode follower is always slightly less than unity.

The SRPP amplifier works best when identical cathode resistors RK for T1 and T2 are used, as shown in Figure 8.3. When identical cathode resistors and closely matched triodes are used, the two triodes should have very close quiescent conditions – the same plate-cathode dc potential, and the same plate current. In this example, if a 200V power supply is used, the dc potential at point B will be very close to 100V.

When the SRPP amplifier drives a high impedance load, the voltage gain and output impedance can be approximated by the following expressions [4–6],

$$A_v \approx \frac{-\mu(r_p + \mu RK)}{2r_p + 2RK(\mu+1)} \tag{8.1}$$

$$Z_{out} \approx \frac{(r_p + 2RK)[r_p + RK(\mu+2)]}{2r_p + 2RK(\mu+2)} \tag{8.2}$$

where μ = amplification factor, r_p = plate impedance of the triode vacuum tube.

We have chosen five triodes (6H30, 12AU7, ECC99, 6922, and 6N1P) with low to medium amplifying factors for implementing the SRPP amplifier. The amplification factor varies from 15 to 35. For triodes 6H30 and ECC99, which have a higher plate dissipation rating, the plate current is set to around 7mA. For other tubes with lower plate dissipation ratings, the plate current is set to around 4mA to 5mA. Table 8.2 compares the calculated and measured values for voltage gain and output impedance for the five triodes.

Table 8.2 Voltage gain and output impedance of an SRPP amplifier for five selected triodes at no load condition

				Output impedance		Voltage gain	
Tube (T1 to T2)	*Tube Amplifying Factor*	*RK (Ω)*	*Plate Current*	*Calculated (Ω)*	*Measured (Ω)*	*Calculated*	*Measured*
6H30	15	820	6.8mA	1,237	1.1K	-7.1	-6.7
12AU7	20	620	4.2mA	3,846	3.7K	-9.7	-8.1
ECC99	22	430	6.5mA	1,588	1.4K	-10.6	-10.2
6922	33	620	4.2mA	1,940	1.8K	-16.1	-14.1
6N1P	35	180	5mA	5,180	4.5K	-17.3	-17.8

Note: measured output impedance does not include R4.

There is a general trend showing that the measured values are slightly lower than the calculated values. Since the electrical properties of vacuum tubes may vary a lot, it is not surprising to see that the measured values vary by more than 10% using triodes of the same type. However, if it is desired to increase the plate current, we can lower the cathode resistor, RK. The output impedance and voltage gain will be slightly affected. Even though the plate dissipation of 6H30 and ECC99 may suggest that a much higher plate current can be used, I recommend not exceeding 10mA so as to extend the useful life-span of the vacuum tube.

Half-mu amplifier

The SRPP and *half-mu amplifier* look almost identical. When the output is taken at point A, it is an SRPP amplifier. But when the output is moved to point B, it becomes the so-called half-mu amplifier (Figure 8.4). The reason for the amplifier having such a name is due to the fact that when an equal cathode resistor RK is used for T1 and T2, the voltage gain of this amplifier becomes $-\mu/2$, where μ is the amplification factor of the triode.

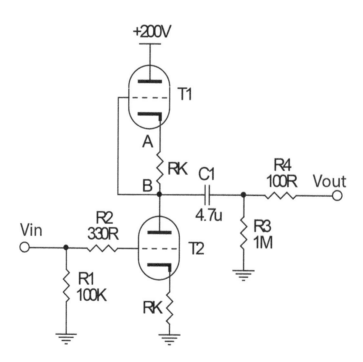

Figure 8.4 A half-mu amplifier with two equal cathode resistors RK

When an equal cathode resistor (RK) and closely matched triodes are used for T1 and T2, the dc potential at B is approximately equal to half of the supply voltage. In Figure 8.4, the dc potential at B is, therefore, around 100V. Because of the push–pull operation, the half-mu and

SRPP amplifiers are very effective in canceling out the variation in the vacuum tube's heater ac hum noise at the output when ac supply is used for the filament. The output impedance of the half-mu amplifier is given as follows [4–7],

$$Z_{out} = \frac{r_p + RK(\mu+1)}{2} \tag{8.3}$$

Table 8.3 compares the calculated and measured voltage gain and output impedance of the half-mu amplifier for the five triodes. The voltage gains between Table 8.2 and 8.3 are very close. In general, the voltage gain for a half-mu amplifier is slightly higher than that of an SRPP amplifier. However, the output impedance of a half-mu amplifier is much higher than that of an SRPP amplifier. The high output impedance makes a half-mu amplifier not the most desirable choice to drive a low impedance load directly. It requires an additional cathode follower output stage, forming a two-stage amplifier to reduce output impedance.

Table 8.3 Showing the voltage gain and output impedance of a half-mu amplifier for five different triodes at no load condition

Tube (T1 to T2)	Tube Amplifying Factor	RK (Ω)	Plate Current	Output impedance Calculated (Ω)	Measured (Ω)	Voltage gain Calculated	Measured
6H30	15	820	6.8mA	6,977	6.8K	-7.5	-7
12AU7	20	620	4.2mA	9,736	9.7K	-10	-8.5
ECC99	22	430	6.5mA	6,103	6.2K	-11	-10.4
6922	33	620	4.2mA	11,860	11.1K	-16.5	-14.4
6N1P	35	180	5mA	8,240	7.3K	-17.5	-18.3

Note: measured output impedance does not include R4.

Example 8.1

Determine the value of cathode resistor RK in the amplifier shown in Figure 8.5 so that the plate current for the triodes (T1 and T2 = ECC99) is 10mA.

First, as we know, when the two cathode resistors are identical, the plate-cathode dc voltage for both tubes in this circuit are very much the same. Therefore, the dc potential at point B is half of the power supply voltage, i.e., 100V. The plate-cathode dc voltage for each tube will be 100V – RK ×10mA. Since the dc potential drop across the cathode resistor RK is just a few volts, in simplifying the calculation we ignore this for the time being. Now, assuming the plate-cathode dc potential for each tube to be 100V, we examine the transfer characteristics of tube ECC99, as given in Figure 8.6.

In Figure 8.6, we locate the quiescent point Q for a plate-cathode voltage of 100V and a plate current of 10mA. The corresponding grid-cathode dc potential is –2.6V. Since there is no grid current in normal operation, the cathode current must be equal to the plate current. The cathode resistor RK is easily calculated to be RK = 2.6V/10mA = 260Ω. When RK = 260Ω is used in the circuit, the tube's plate current is measured to be 9mA. It is very rare that we can hit the correct

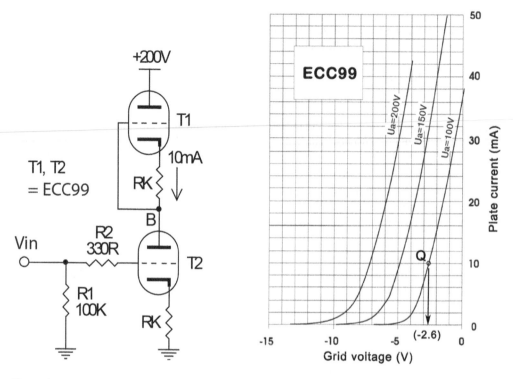

Figure 8.5   *Figure 8.6* Transfer characteristics of ECC99

value at the first attempt because the electrical properties of a vacuum tube can deviate from the data sheet by over 20%. Anyway, if we reduce the resistor to 230Ω, the measured plate current becomes 9.9mA, which is close enough to the desired value. In a similar way, we can determine the cathode resistors, which correspond to different triodes and desired plate currents, for the SRPP and half-mu amplifiers.

Aikido amplifier

As we have seen from Table 8.3, the output impedance of a half-mu amplifier is rather high. Even for 6H30 and ECC99 at a plate current near to 7mA, the output impedance is greater than 6kΩ. If we want to make use of the half-mu amplifier, we need to add a cathode follower output stage to reduce output impedance. To improve the performance of the cathode follower, an active load is added, as shown in Figure 8.7. This is called the *Aikido amplifier*, which was popularized by John Broskie [8].

The active load formed by the triode T4 is in a common cathode amplifier configuration. As a matter of fact, the active load serves two purposes. The first is to operate as an active load for the cathode follower formed by triode T3 so as to improve the linearity of the output stage. This configuration looks very similar to a White cathode follower. It produces an approximated push–pull operation for the amplifier to reduce noise coming from the power supply. To see how the noise cancellation works, let us examine the R-C network formed by C2, R4, and R5.

If noise e_n is present in the power supply, the noise will eventually get to the point C, the cathode of triode T3, say with amplitude $K_1 \cdot e_n$, where K_1 is a positive multiplying constant. Why

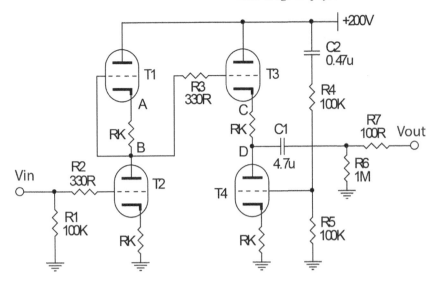

Figure 8.7 The Aikido amplifier is a two-stage amplifier comprising an input stage of a half-mu amplifier and an output stage of a compound cathode follower with an active load. Noise cancellation is introduced to the compound cathode follower

is K1 a positive constant, not negative? We have to look at the paths for the noise e_n getting to the point C. There are two paths. The first path comes down from tube T1 to point B. It is then amplified by the cathode follower (T3) with unity gain. The second path comes down from triode T3 directly. Since there is no device in these two paths to invert the noise e_n, the multiplying constant must be positive. In order for the noise at the output stage to be cancelled, we introduce a noise with opposite phase ($-K_2 \cdot e_n$) so that, ideally, the sum of the noises $(K_1 \cdot e_n) + (-K_2 \cdot e_n)$ is cancelled out at the output, if K_1 and K_2 are equal. At the very least, the noise will be reduced.

The way to invert the noise e_n is to send the noise via capacitor C2 to the common cathode amplifier formed by triode T4. Since a common cathode amplifier has an inverted voltage gain, the amplified noise coming from the plate of tube T4 is inverted, say $-K_2 \cdot e_n$, where K_2 is a positive multiplying constant. The total noise appearing at point D is the sum of the two noises, $(K_1 \cdot e_n) + (-K_2 \cdot e_n)$. The total noise is zero if $K_1 = K_2$. Since K_1 is very much determined by the tubes T1 and T2, there is not much we can do. However, we can control K_2 by varying the resistor ratio of R4 and R5. For example, if R4 is fixed and R5 is replaced by a variable resistor, we may get a wide range of K_2 so that the noise at the output can be taken down to a very low level. Even though two fixed resistors are now used for R4 and R5, the result is still better than it would be without using the noise cancellation arrangement.

Table 8.4 shows the measured voltage gain and output impedance of the Aikido amplifier for the five triodes. In comparing to Table 8.3, we find that the voltage gain of the Aikido amplifier is slightly lower than the half-mu amplifier. It should be obvious, because the Aikido amplifier has added a cathode follower output stage, which has a voltage gain lower than unity. On the other hand, the output impedance of Aikido amplifier is now much lower than the half-mu amplifier. For instance, if 6H30 is used, the output impedance for half-mu amplifier is 6.8kΩ while Aikido amplifier is only 0.84kΩ.

Table 8.4 This table shows the measured voltage gain and output impedance of an Aikido amplifier with five different triodes at no load condition

Tube (T1 to T4)	Tube Amplifying Factor	RK (Ω)	Input Stage Plate Current	Output Stage Plate Current	Voltage Gain	Output Impedance (Ω)
6H30	15	820	6.8mA	6.8mA	-6.8	840
12AU7	20	620	4.2mA	4.2mA	-8.2	1050
ECC99	22	430	6.5mA	6.5mA	-10.1	520
6922	33	620	4.2mA	4.2mA	-14	710
6N1P	35	180	5mA	5mA	-17.9	440

Note: measured output impedance does not include R7.

Another thing we can experiment with in the Aikido amplifier is to use different tubes for the half-mu amplifier and output stage. For example, if we prefer the sound of 12AU7, then this triode is used for the half-mu amplifier first stage. However, 6H30 or ECC99 is used for the output stage to ensure low output impedance.

Wing Chun amplifier

As far as voltage gain is concerned, the SRPP and half-mu amplifiers are almost the same. The half-mu amplifier is only slightly higher under no load condition. Therefore, when an SRPP amplifier is used for the input stage and the same cathode follower is used for the output stage, as shown in Figure 8.8, the voltage gain and output impedance are expected to be very similar to the Aikido amplifier. Here, let us call it the *Wing Chun* amplifier.

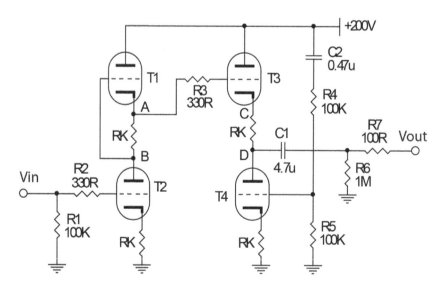

Figure 8.8 The *Wing Chun* amplifier is a two-stage amplifier comprising an input stage of an SRPP amplifier and an output stage of a compound cathode follower with an active load. Noise cancellation is introduced to the compound cathode follower

Table 8.5 shows the measured voltage gain and output impedance of the Wing Chun amplifier. It has an inverted voltage gain similar to the Aikido amplifier. However, as the Aikido amplifier employs a half-mu amplifier for the input stage, while the input stage of the Wing Chun amplifier uses an SRPP amplifier, this will lead to slightly different sound performance. Since the circuits of the two amplifiers are identical except the location of where the output of the input stage is taken, it is worth exploring the sound of these two amplifiers by moving from point A to point B, and vice versa, listening to the differences.

Table 8.5 Measured voltage gain and output impedance of the Wing Chun amplifier in Figure 8.8 for five different triodes at no load condition

Tube (T1 to T4)	Tube Amplifying Factor	RK (Ω)	Input Stage Plate Current	Output Stage Plate Current	Voltage Gain	Output Impedance (Ω)
6H30	15	820	6.8mA	6.8mA	-6.5	840
12AU7	20	620	4.2mA	4.2mA	-7.6	1050
ECC99	22	430	6.5mA	6.5mA	-9.7	520
6922	33	620	4.2mA	4.2mA	-13.8	710
6N1P	35	180	5mA	5mA	-17.4	440

Note: measured output impedance does not include R7.

Tai Chi amplifier

If an SRPP amplifier is used for the input stage, and a compound cathode follower output stage is allowed to work in push–pull operation, the result is an amplifier illustrated in Figure 8.9. Here let us call it *Tai Chi amplifier* because the way resistor R3 and capacitor C1 connecting to T1 and T2 looks like the pushing and pulling hand movement demonstrated in Tai Chi exercise. Beside the name, let us see how a Tai Chi amplifier works.

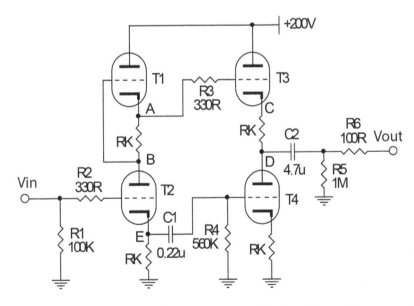

Figure 8.9 This is the *Tai Chi* amplifier. It is a two-stage amplifier comprising an input stage of an SRPP amplifier and an output stage of a compound cathode follower in push–pull operation

In the previous examples of the Aikido and Wing Chun amplifiers, triode T4 operates to invert the phase of the noise so that the push–pull operation at the output tends to cancel out common noise. However, it should be noted that triode T4 in the Aikido and the Wing Chun amplifier does not amplify the signal from the input Vin. On the other hand, triode T4 on the Tai Chi amplifier in Figure 8.9 operates in a common cathode amplifier configuration and it does amplify the input signal taken from the cathode follower formed by tube T2. Since a cathode follower is a non-inverted amplifier and a common cathode is an inverted amplifier, the signal from the output of tube T4 at point D is also inverted.

It is clear that triodes T1 and T2 have the same plate current, which comes from the same power supply. Therefore, at the cathode resistor RK of tube T2 (point E), noise is contained in a similar nature to the noise appearing at the cathode of tube T1 (point A). For example, if e_n represents the noise that appears at point E, the noise that appears at point A may be expressed as $K_1 \cdot e_n$, where K_1 is a positive multiplying constant. Since T3 operates as a cathode follower with a non-inverted voltage gain of unity, the noise that coming out from point C is also $K_1 \cdot e_n$. However, T4 operates as a common cathode amplifier that has an inverted gain. The input noise to the grid of T4 is e_n from point E. After being amplified by T4, the noise coming out from point D becomes $-K_2 \cdot e_n$, where K_2 is a positive multiplying constant with the negative sign indicating an inverted phase. Thus, point D produces a sum of $(K_1 \cdot e_n) + (-K_2 \cdot e_n)$. Ideally, if $K_1 = K_2$, the noises will be completely cancelled at the output. In practice, even though they are not completely cancelled, the noise at the output is greatly reduced.

Since no feedback is applied and there is no component loading down the voltage gain of the SRPP first stage, the voltage gain and output impedance of the Tai Chi amplifier are identical to that of the Wing Chun amplifier in Figure 8.8. Thus, the Tai Chi amplifier offers similar voltage gain and output impedance to the Aikido and Wing Chun amplifier. Refer to Table 8.5 for the choice of cathode resistors RK to pair with various triodes.

Cathode-coupled amplifier

All the amplifiers that we have discussed earlier in this section have one thing in common. They all have inverted voltage gain. In other words, the output is 180 degrees out of phase with respect to the input signal. If such a line-stage amplifier is used to drive a power amplifier also with an inverted voltage gain, the output from the power amplifier becomes non-inverted. Then there is nothing we have to worry about. However, due to the circuit design topology used for power amplifiers, most of them are the non-inverted design. Therefore, when an inverted line-stage amplifier is used to drive a non-inverted power amplifier, the output from the power amplifier is inverted. A solution to reverse the output polarities is to swap the loudspeaker cables connecting to the power amplifier. It appears to be a simple issue.

The Marantz 7 variance and Shigeru Wada amplifier discussed in Section 8.1 are non-inverted amplifiers. Both employ negative feedback in a three-stage amplifier configuration. On the other hand, all amplifiers discussed in this section so far have inverted gain, including SRPP, half-mu, Aikido, Wing Chun, and Tai Chi amplifiers. It is interesting to see if there is any single-stage or two-stage amplifier that produces non-inverted voltage gain. Yes, there are two examples shown in Figure 8.10.

In Figure 8.10(a), it is a cathode coupled amplifier. Triode T1 and resistor RKK work as cathode followers with unity gain for the input signal. Then the signal is coupled to the cathode of the tube T2, which is operating in a ground grid configuration. Note that the output

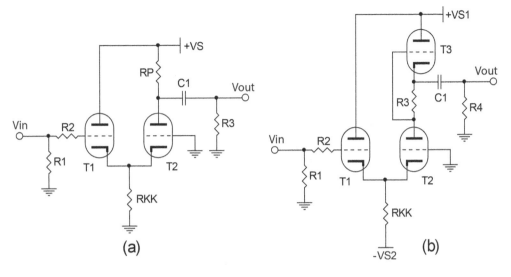

Figure 8.10 Circuit (a) is a cathode-coupled amplifier in which vacuum tube T1 operates as a cathode follower while T2 operates in a grounded grid arrangement. Circuit (b) employs vacuum tube T3 to produce an approximated push–pull output

impedance of the cathode follower T1 and the input impedance of the grounded grid amplifier T2 are low. In order for this arrangement to work well, it is suggested to employ a large RKK. In general, the voltage gain is given by Eq. (3.94), which is re-written as follows:

$$A_v \approx \frac{\mu RP}{2r_p + RP} \tag{8.4}$$

where

μ = amplification factor for triodes T1 and T2
r_p = plate impedance for triodes T1 and T2

Generally speaking, if triodes with low to medium amplifying factors are used, the voltage gain of a cathode-coupled amplifier is around 10 or less. The output impedance will be similar to a common grid amplifier.

On the other hand, since triode T2 has a plate resistor RP, the dc potential at the plate of T2 is usually designed to be half of the power supply voltage +VS. But the dc potential at the plate of T1 is the full supply voltage +VS. As triode T1 has a high plate voltage, T1 may carry a high plate current. Generally speaking, the plate current in T1 will be several times higher than the plate current in T2. For example, if the plate current in T2 is 3mA, the plate current in T1 can be as high as 10mA. Therefore, caution has to be exercised to ensure the total plate dissipation and individual plate dissipation do not exceed the maximum rating of the triode. The direct approach to bring equal plate current to the tubes is to add a plate resistor in triode T1 so that both tubes have an equal plate resistor RP.

Figure 8.10(b) shows a modified version of a cathode-coupled amplifier. The plate resistor RP is replaced by triode T3. This will produce an approximate push–pull output. In addition, a negative power supply (–VS) is used so as to increase the cathode resistor RKK. On the other hand, the cathode resistor RKK can be replaced by a current source with high impedance. This is discussed in the description of the CCPP amplifier below.

Cathode-coupled push–pull (CCPP) amplifier

An improved cathode-coupled amplifier is shown in Figure 8.11. This is a two-stage amplifier without using feedback. The input stage is a cathode-coupled amplifier formed by T1 and T2 having an equal plate resistor (15kΩ) and it employs a current source. The output stage is formed by T3 and T4 so that an approximated push–pull output is produced. Here, let us call this two-stage arrangement a *cathode-coupled push–pull amplifier* (CCPP). CCPP is a non-inverted amplifier with a voltage gain similar to a cathode-coupled amplifier.

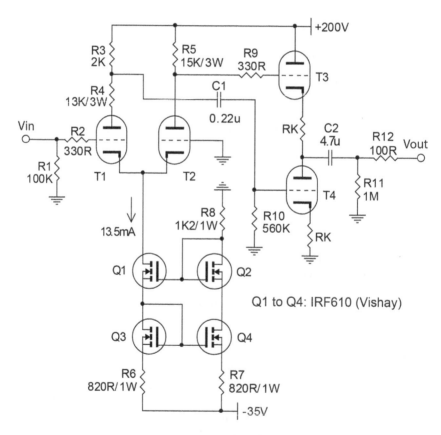

Figure 8.11 This is a CCPP (cathode-coupled push–pull) amplifier. Cathode resistor RKK is replaced by a current source formed by transistors Q1 to Q4

A current source, formed by transistors Q1–Q4, is employed to replace the cathode resistor RKK shown in Figure 8.10(a). The current source is a modified Wilson current source implemented by four MOSFETs (IRF610). The Wilson current is known for producing very high impedance. A high impedance current source is very useful, as it does not load down the cathodes of T1 and T2.

It should also be noted that resistors R3 and R4 form the plate load resistor for triode T1. Since the total resistance of R3 + R4 is equal to R5, 15kΩ, this will ensure that both tubes carry equal plate current. R3 is a small resistor that allows a small portion of the amplified input signal to be coupled to the tube T4. Since this small portion of amplified input signal is inverted in phase, while T4 forms a common cathode amplifier that has an inverted voltage gain, the output signal at the plate of T4 becomes non-inverted. The output of the CCPP amplifier will combine the signal pushing down from the cathode follower T3, and the signal pulling up from the common cathode amplifier T4, giving an approximated push–pull output.

For triode T1, if output is taken from the plate, T1 is a common cathode amplifier with high gain. But if output is taken from the cathode, T1 is a cathode follower with unity gain. Now let us take a look at the push–pull operation in the output stage. First, the "pushing" signal path starts from Vin, and is then followed by the T1 cathode follower, T2 ground grid amplifier, and T3 cathode follower. Therefore, in the "pushing" signal path, only T2 provides voltage gain, which is a common grid amplifier. Thus, the T1 × T2 × T3 combination produces a voltage gain equal to a common grid amplifier, which is similar to the gain of a common cathode amplifier. On the other hand, the "pulling" signal path starts from Vin, and is then followed by the T1 common cathode amplifier and the T4 common cathode amplifier. Clearly, the voltage gain of the T1 × T4 combination is a product of two common cathode amplifiers and, therefore, it is much greater than the T1 × T2 × T3 combination. Therefore, in order to equalize the push–pull signal to the same voltage gain, the signal is taken at a small resistor R3 from the plate of T1.

When R3 is set to 2kΩ, as shown in Figure 8.11, the CCPP amplifier performs very well. However, we cannot say whether 2kΩ is the optimized value or not for this CCPP amplifier with the given selection of triodes. I believe an optimum value for R3 may lie somewhere between 1kΩ and 5kΩ, depending on the triode being used. Therefore, it is recommended to try different values for R3 and evaluate the performance in terms of sound and output distortion. When changing the value for R3, we have to make sure the sum of R3 + R4 remains equal to R5.

We have tested the CCPP amplifier for two triodes, 6H30 and ECC99. First, when 6H30 is used, we have RK = 820Ω, voltage gain = 7 (non-inverted), and output impedance = 840Ω. When ECC99 is used, we have RK = 430Ω, voltage gain = 9.6 (non-inverted), and output impedance = 520Ω. The low output impedance is expected when a cathode follower output stage is employed. The performance and non-inverted voltage gain make the CCPP amplifier an attractive alternative to the audio line-stage amplifier application.

Cascading SRPP amplifier

The SRPP amplifier has an inverted gain. If we cascade two SRPP amplifiers, the composite amplifier produces non-inverted gain. Figure 8.12 shows a two-stage amplifier that comprises two cascaded SRPP amplifiers. Since the open-loop gain is more than we normally need for an audio line-stage amplifier application, feedback is used to lower the overall voltage gain while broadening the bandwidth and reducing distortion and output impedance.

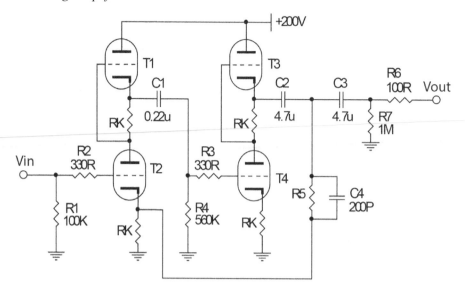

Figure 8.12 A two-stage amplifier formed by cascading two SRPP amplifiers. It is a non-inverted amplifier and the voltage gain is approximately equal to R5/RK

For illustration purpose, two identical SRPP amplifiers are used so that T1–T4 are identical triodes. Again, triodes with low to medium amplifying factors are employed in the test. They are 6H30, 12AU, ECC99, 6922, and 6N1P. As shown in Figure 8.12, the output of the first SRPP amplifier is capacitor coupled to the second SRPP amplifier. The cathode resistor RK for triode T4 is un-bypassed in this circuit. If needed, it can be bypassed by a capacitor producing a higher open-loop gain. Feedback is taken from the output and connected to the cathode of the tube T2, forming a series-shunt feedback configuration. When the open-loop gain is sufficiently high, the overall voltage gain of the composite amplifier is approximately equal to R5/RK. Measurements of the voltage gain and output impedance are summarized in Table 8.6. Voltage gain is ranging from 5 to 10.4.

Table 8.6 Measured voltage gain and output impedance of the two-stage amplifier in Figure 8.12 for five triodes at no load condition

Tube (T1 to T4)	Tube Amplifying Factor	RK (Ω)	R5 (Ω)	Input Stage Plate Current	Output Stage Plate Current	Voltage Gain	Output Impedance (Ω)
6H30	15	820	4K3	6.8mA	6.8mA	5	160
12AU7	20	620	5K1	4.2mA	4.2mA	7.1	550
ECC99	22	430	3K	6.5mA	6.5mA	6.9	160
6922	33	620	5K1	4.2mA	4.2mA	8.3	140
6N1P	35	180	2K	5mA	5mA	10.4	170

Note: measured output impedance does not include R6.

Another three-stage amplifier

In Section 8.1, we discussed two three-stage amplifiers, the Marantz 7 amplifier and the Shigeru Wada amplifier. Both have something in common, i.e., non-inverted voltage gain and the use of feedback. Even though some people may prefer a vacuum tube audio line-stage amplifier with no feedback, we cannot underestimate the usefulness of feedback, and the overall performance that a three-stage feedback amplifier can offer. Without applying feedback, the open-loop gain, as well as the noise and distortion, of a three-stage amplifier will be too high for practical use. Even if we choose low amplifying factor triodes such as 6H30 and ECC99 for Figures 8.1 and 8.2, the open-loop gain is easily over 50. However, when the design work is properly carried out, a three-stage feedback amplifier can offer remarkably high performance. The Marantz 7 and Shigeru Wada amplifiers are two good examples.

Figure 8.13 shows another three-stage vacuum tube amplifier. The input stage employs a differential amplifier, which is the most commonly used input stage in solid-state semiconductor amplifiers. The output from the first stage is capacitor coupled to the second stage, which is a common cathode amplifier. The product of the voltage gains from the input and second stages produces the necessary open-loop gain for the entire amplifier, as the output stage is a cathode follower of unity gain. Generally speaking, the structure of the three-stage amplifier in Figure 8.13 very much resembles a solid-state amplifier. This circuit configuration is not new. However, Figure 8.13 garners little popularity in vacuum tube amplifier designs.

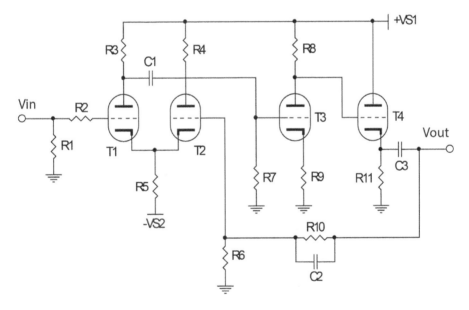

Figure 8.13 This is a three-stage non-inverted amplifier using a differential amplifier (long-tailed pair) input stage

One reason for not being a popular design is because it is not a necessity and there is no definite advantage to using a differential amplifier for the input stage, which employs two triodes. A simple common cathode amplifier for the input stage will do the job. This saves us one triode. Another reason is that a differential amplifier needs to have a negative power supply to get a high tail resistor R5. Therefore, the negative regulated power supply can also be saved when a common cathode amplifier is used for the input stage. To sum up, by simplifying the input stage from a differential amplifier to a common cathode amplifier, we save one triode and a negative dc regulated power supply.

8.3   Balanced input–unbalanced output amplifier

Differential amplifier with push–pull cathode follower

When dealing with a balanced input signal, we will naturally employ a differential amplifier for the input stage. Since the output is unbalanced, we find that a CCPP amplifier is very close to what we want for handling balanced input–unbalanced output signals. The CCPP amplifier in Figure 8.11 is modified in Figure 8.14.

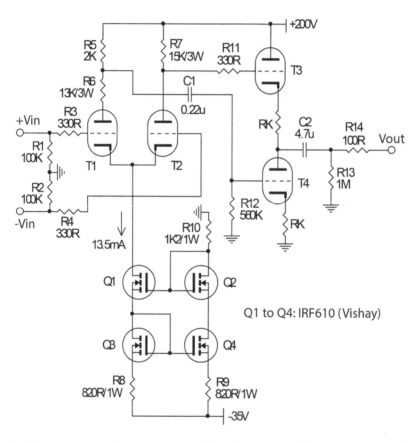

Figure 8.14 This is a two-stage balanced input–unbalanced output amplifier that comprises a differential amplifier input stage and a compound cathode follower output stage in push–pull operation

Instead of grounding the grid of triode T2, the grid is now used for handling the negative phase input signal, –Vin. Therefore, the differential amplifier in the input stage is ready to handle a balanced input signal. The differential amplifier's tail current is set by the modified Wilson current source formed by transistors Q1–Q4. Given the values for resistors R8–R10, the tail current is set to 13.5mA. Therefore, each of the tubes in the input stage carries a 6.7mA plate current. When a 15kΩ plate resistor is chosen for T1 and T2, the dc voltage across the resistor is about 100V, which is half of the positive power supply. Thus, the plate-cathode dc potential of both T1 and T2 is around 100V.

The output of the differential amplifier is taken from the plate of T2 and directly coupled to T3, which operates as a cathode follower with unity voltage gain and low output impedance. Triode T4 is operating in dual roles. First, T4 is functioning as an active load for the cathode follower formed by T3. The other role for T4 is to operate in a common cathode arrangement that amplifies a small portion of the signal taken from the plate of T1.

Note that triode T1 is a common cathode amplifier for the positive input signal +Vin. If the full output signal at the plate of T1 is taken and coupled to T4 of the output stage, which is also a common cathode amplifier, the combined gain will be too much for proper push–pull operation at the output stage. Let us examine the other half of the push–pull operation. The −Vin signal is amplified by T2 and T3. T2 is a common cathode amplifier, with the same voltage gain of T1. But T3 is a cathode follower of unity gain. As a result, the T2 × T3 combination has a lower voltage gain than the T1 × T4 combination. In order to equalize the two combined voltage gains, the output from T1 is taken from a small resistor R5. When we set R5 + R6 = R7, triodes T1 and T2 will be biased with equal plate current.

When R5 is set to 2kΩ, as shown in Figure 8.14, the amplifier performs very well for balanced input–unbalanced output operation. However, we cannot say whether 2kΩ is the optimized value or not for this two-stage amplifier arrangement with the given selection of triodes. I believe an optimum value for R5 may lie somewhere between 1kΩ to 5kΩ depending on the triode being used. Therefore, try different values for R5 and evaluate the performance in terms of sound and output distortion. When changing the value for R5, make sure the sum of R5 + R6 is equal to R7.

We have tested the balanced input–unbalanced output amplifier of Figure 8.14 for two triodes: 6H30 and ECC99. First, when 6H30 is used, we have RK = 820Ω, voltage gain = 7 (non-inverted), output impedance = 840Ω. When ECC99 is used, we have RK = 430Ω, voltage gain = 9.6 (non-inverted), output impedance = 520Ω. The low output impedance is expected when a cathode follower output stage is employed.

8.4   Balanced input–balanced output amplifiers

A balanced input–balanced output amplifier is often called a fully balanced amplifier. In the solid-state line-stage amplifier design, the number of transistors of a fully balanced amplifier is almost twice that of an unbalanced input–unbalanced output amplifier. Given one more input channel and one more output channel, a fully balanced amplifier inevitably requires the use of many components. However, it may not be the case for a vacuum tube fully balanced line-stage amplifier.

We have discussed unbalanced input-unbalanced output amplifiers in Section 8.2. Most of them contain two to four triodes. For fully balanced amplifiers, it is often found that four triodes are also sufficient. For some sophisticated fully balanced designs, the number of triodes may go up to six or even more. Four to six triodes is not a high number, but they can offer remarkable sonic performance in a relatively simple design. In this section, we are going to discuss several fully balanced amplifier designs that contain four to six triodes.

Differential amplifier with cathode follower

Figure 8.15 shows a two-stage fully balanced amplifier. The input stage employs a differential amplifier formed by triodes T1 and T2 with a current source, which is formed by MOSFETs Q1–Q4 setting the tail current. The current source is a modified Wilson current source that has a high output impedance. A high impedance current source is always desirable as it will help to improve the CMRR of the differential amplifier for dealing with a balanced input signal. The output of the input stage is directly coupled to the cathode follower second stage, which is formed by tubes T3 and T4.

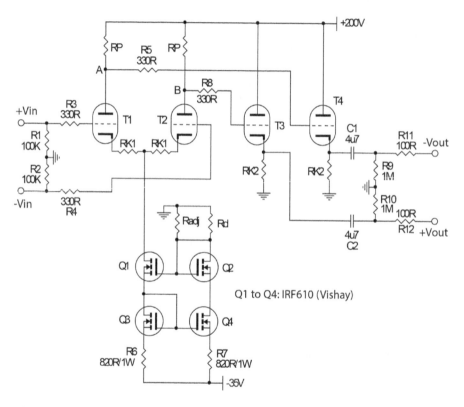

Figure 8.15 This is a two-stage fully balanced amplifier that comprises a differential amplifier input stage and a cathode follower output stage

This is a simple fully balanced amplifier design employing only four triodes. Since no feedback is used, it is important that closely matched triodes are used for T1 and T2. When closely matched triodes are used, the input stage produces the following desirable results. First, the plate current in the tubes T1 and T2 will be nearly identical, so that the tail current is equally shared by the two triodes. For example, if the current source sets up a 10mA tail current, the plate current for T1 and T2 is about 5mA each when the triodes are closely matched. As a consequence, the dc potentials at point A and B, i.e., the plate voltages, are also nearly equal. Since the output of the differential amplifier is directly coupled to the cathode follower output stage, triodes T3 and T4 also have same dc plate voltage and plate current.

The other benefit of using closely matched triodes for T1 and T2 is that the voltage gain of the amplifier will be likely to be the same. Without employing feedback to control the voltage gain, the amplifier has to rely on the vacuum tubes themselves in setting the voltage gain for the non-inverted and inverted outputs. On the other hand, if degeneration cathode resistor RK1 is added in the differential amplifier, this provides local feedback, improving linearity as well as setting the desired voltage gain of the differential amplifier.

For illustration purpose, triodes with low to medium amplifying factors are employed in the tests. They are 6H30, 12AU, ECC99, 6922, and 6N1P. The voltage gain and output impedance measurements are summarized in Table 8.7.

Table 8.7 Measured voltage gain and output impedance of the fully balanced amplifier in Figure 8.15 for five different triodes at no load condition

Tube (T1 to T4)	Tube Amplifying Factor	RP (kΩ) 3W	RK1 (kΩ)	RK2 (kΩ) 3W	RD (kΩ)	Radj (kΩ)	Input Stage Plate Current	Output Stage Plate Current	Voltage Gain	Output Impedance (Ω)
6H30	15	15	1	12	1.2	open	6.7mA × 2	8.6mA × 2	5.9	230
12AU7	20	24	1.2	15	2.4	open	4.2mA × 2	6.6mA × 2	6.9	740
ECC99	22	15	1	12	1.2	open	6.7mA × 2	8.6mA × 2	7.3	270
6922	33	24	1.8	15	2.4	open	4.2mA × 2	6.6mA × 2	8	330
6N1P	35	20	1.5	15	2	39	5mA × 2	6.6mA × 2	8.2	430

Note: measured output impedance does not include R11 and R12.

Differential amplifier with push–pull cathode follower

The fully balanced amplifier of Figure 8.15 is now modified by turning two simple cathode followers into two compound cathode followers, as shown in Figure 8.16. The new fully balanced amplifier has one compound cathode follower formed by tubes T3 and T5, and the other compound cathode follower formed by T4 and T6. Since the voltage gain of the cathode follower output stage remains unity, the voltage gain of the fully balanced amplifier in Figure 8.16 is identical to Figure 8.15. However, by introducing triodes T5 and T6, the compound cathode followers produce an approximated push–pull output.

Similar to the CCPP amplifier, triodes T5 and T6 are operating in dual roles. First, each triode is operating as an active load for the cathode follower right above it. The other role is to operate in a common cathode arrangement that amplifies a small portion of the signal taken from the plate of the differential amplifier in such a way that both compound cathode followers (T3/T5 and T4/T6) work in an approximated push–pull operation. Details of the push–pull operation in a compound cathode follower stage has been discussed in Section 8.3. Plate resistors should be chosen such that RP1 + RP2 = RP, where RP is the plate resistor used in Table 8.7 and Figure 8.15. By choosing a small RP1, a small portion of output from the differential amplifier is fed to the output stage for the push–pull operation. For example, if RP = 15kΩ, we may use RP1 = 2kΩ and RP2 = 13kΩ. It is suggested to vary RP1 from 1kΩ to 5kΩ and evaluate which resistor value offers better sound and measured performance. On the other hand, cathode resistors RK1 should follow the values in Table 8.7 if we want to obtain the same voltage gain. However, RK2 should follow the values for RK in Table 8.5 or 8.6.

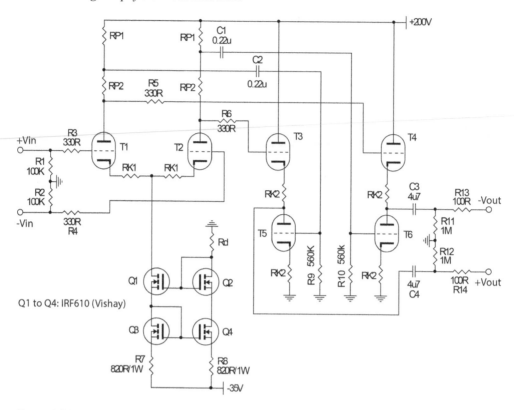

Figure 8.16 A two-stage fully balanced amplifier contains a differential amplifier input stage and compound cathode follower in a cross-coupled push–pull operation

SRPP differential amplifier

We have discussed two fully balanced amplifiers as illustrated in Figures 8.15 and 8.16. Both are
two-stage amplifiers. The first stage is a differential amplifier and the output stage is a simple or
compound cathode follower. If we are willing to compromise by having a slightly higher output
impedance, we can look for a single-stage amplifier, as shown in Figure 8.17(a). It combines
two SRPP amplifiers with the cathode resistors at the bottom connected. Here, let us call it an
SRPP differential amplifier. Two SRPP differential amplifiers are used to form a two-channel
(stereo) line-stage amplifier, as shown in Figure 8.17(b).

From the measured output impedance of various amplifiers discussed in preceding sections,
we know the output impedance of an SRPP amplifier is higher than that of a cathode follower
but lower than a common cathode amplifier. Therefore, by the same token, the output impedance
of an SRPP differential amplifier is higher than that of a simple common cathode differential
amplifier having a cathode follower output stage, but lower than a simple common cathode differential amplifier without a cathode follower output stage.

Triode 6H30 is chosen for the SRPP differential amplifier. The current source, formed by
Q1–Q4, sets up a tail current of 16.5mA. Therefore, each 6H30 has a dc plate current of 8.2mA.

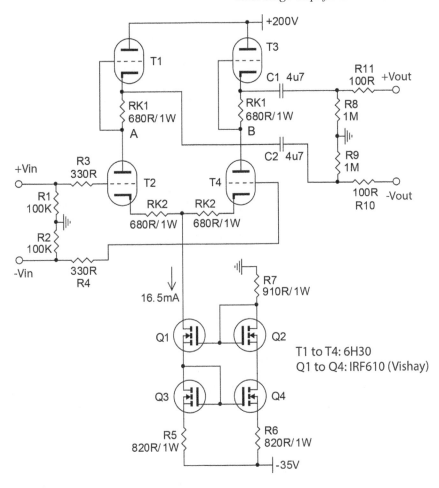

Figure 8.17(a) An SRPP differential amplifier

The dc potential at point A and B is about 100V, which is half of the dc power supply voltage. The voltage gain for this amplifier is 6.8 at no load condition. Output impedance is 1.9kΩ (not including resistors R11 and R12), which is a respectable value for a vacuum tube amplifier. If higher gain is required, the degeneration resistor RK2 can be reduced accordingly. Since no feedback is employed, we have to choose closely matched triodes for T1/T3 and T2/T4.

It is strongly suggested to carefully choose the right resistors for this amplifier and any other vacuum tube amplifiers. First, we have to distinguish whether a resistor is in the signal path or not. For resistors in the signal path, we recommend using audio grade metal film resistors with power rating of 1W or higher. For resistors not in the signal path, it will be sufficient to use ordinary metal film resistors with the appropriate power rating. In this SRPP differential amplifier, the resistors in the signal path include R3, R4, RK1, RK2, R10, and R11. They should be audio grade 1W metal film resistors. Even though resistors R3, R4, R10, and R11 do not carry

4.7µF x 4

IRF610 x 4

Q1 on heatsink

Figure 8.17(b) Photo of a two-channel (stereo) line-stage amplifier, which contains two amplifiers shown
in Figure 8.17 (a). (Dimension: L×W×H = 188×216×83mm)

dc current and, therefore, do not produce heat dissipation, use of 1W power rating resistor will
help to improve the sound of the amplifier. On the other hand, resistors R1, R2, R5, to R9 are not
in the signal path. In addition, resistors R1, R2, R8, and R9 do not carry any dc current. Using
ordinary 0.5W metal film resistors will be just fine for R1, R2, R8, and R9. However, resistor
R7 carries a dc current of 16.5mA and a power dissipation of 0.25W. It is recommended to use
a power rating at least three times the actual power dissipation in the resistor. Therefore, 1W
rating metal film resistors are chosen for R7, R6, and R5.

We may have somehow compromised the output impedance, but definitely not the sound
quality. I have developed several line-stage amplifiers at JE Audio. Line-stage amplifiers VL19,
VL20, and Reference 1 follow the same SRPP differential amplifier design except with different
biasing current, voltage gain, and output coupled capacitors. When good quality dc regulated
power supplies are used, vacuum tube amplifiers produce exceptionally good sound. As a matter
of fact, some of the vacuum tube amplifiers discussed in this chapter absolutely have no problem
competing with sound performance in any of the solid-state amplifiers discussed in Chapter 7,
including the diamond balanced amplifier that contains 36 transistors. In a vacuum tube audio
amplifier, perhaps it is true that *less is more*.

A note for the MOSFET device. There are several suppliers of IRF610–IRF640, including
Vishay, International Rectifier (now Infineon Technologies), and others. I have found that the
same part from different suppliers produces noticeable differences in sound. Generally speaking, Vishay is my preferred supplier for IRF610–IRF640.

Cascode differential amplifier with active load

Another single-stage fully balanced amplifier is shown in Figure 8.18. It looks similar to
the SRPP differential amplifier of Figure 8.17(a). However, a cascode differential amplifier

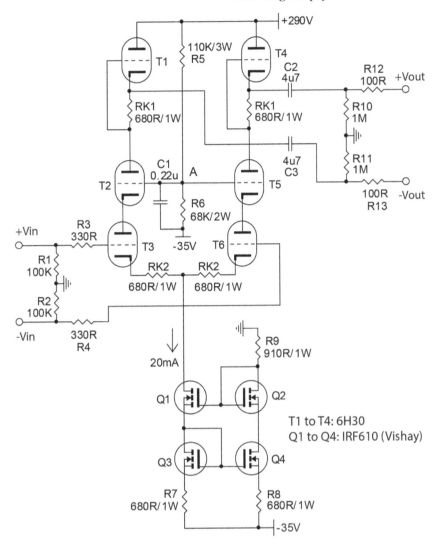

Figure 8.18 A single-stage fully balanced amplifier contains a cascode differential amplifier with an active load

configuration is used in Figure 8.18. Therefore, given the benefit of using a cascode differential amplifier configuration, the fully balanced amplifier of Figure 8.18 has higher voltage gain and wider bandwidth. This fully balanced amplifier contains a total of six triodes [9].

Transistors Q1–Q4 are in a modified Wilson current source arrangement, setting up 20mA tail current for the differential amplifier. This current is slightly higher than that of the SRPP differential amplifier. Resistors R5 and R6 form a potential divider network that sets up a 90V dc voltage at the grids of triode T2 and T5, i.e., point A. Given this dc potential at point A, the power supply voltage (290V) will be equally spread among triodes T1–T3, and T4–T6. When

6H30 is used in Figure 8.18, the amplifier has a voltage gain of 11.8 at no load and an output impedance of 3.3kΩ, which does not include output resistors R12 and R13. JE Audio's VL10.1 line-stage amplifier has followed a similar design, except with different voltage gain and tail current. The merits of this fully balanced amplifier are higher voltage gain and wider bandwidth than that illustrated in Figure 8.17(a). However, these advantages come with penalties – a higher output impedance and the need to use six triodes (6H30) in this amplifier.

8.5   Series and shunt type output muting arrangement

Muting is an essential function for an audio line-stage amplifier. Therefore, it is helpful to see how the output muting arrangement is implemented in a line-stage amplifier. Figure 8.19 shows two different muting arrangements for unbalanced output.

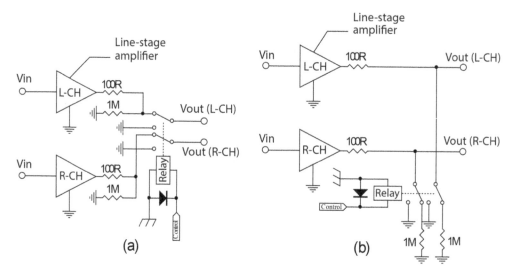

Figure 8.19 Circuit (a) is a series type output muting arrangement. Circuit (b) is a shunt type output muting arrangement. Outputs are unbalanced. Both L- and R-channels are shown

Figure 8.19(a) is a series type output muting arrangement. When the relay is inactive and the control signal = LOW, output is unaffected. But when the relay is activated and the control signal = HIGH, the amplifier's output is open so that no signal is coming out. And, at the same time, the output post is shorted to ground so as to *mute* the output of the amplifier. However, it should be noted that a relay contact is in series between the amplifier's output and the output post during normal operation when the relay is inactive. Because of this, we call it a series type output muting arrangement. For signal switching relay, the contact resistance is around 50mΩ. Good quality gold-plated contact has lower contact resistance. However, no matter how low the relay's contact resistance is, it is always in the signal path.

Figure 8.19(b) is a shunt type output muting arrangement. When the relay is inactive and the control signal = LOW, output is unaffected. But when the relay is activated and the control signal = HIGH, the output signal is shorted to ground so as to mute the amplifier. At this moment, the amplifier's output is shorted to ground via a 100Ω resistor. This 100Ω resistor prevents the amplifier from delivering excessive output current during the moment when the signal at the output is shorted to ground. Since the output is shunted to ground when the mute function is activated, it is called a shunt type output muting arrangement. Obviously, the relay's contact is not in the signal path and does not affect the sound quality.

Figure 8.20 shows the series type and shunt type output muting arrangement for line-stage balanced amplifier of one channel. The same arrangement can be applied to the other channel.

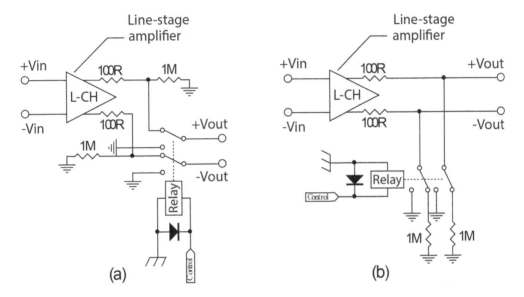

Figure 8.20 Circuit (a) is a series type output muting arrangement. Circuit (b) is a shunt type output muting arrangement. Outputs are balanced. Only the L-channel is shown

8.6 Exercises

Ex 8.1 Design a Tai Chi amplifier (shown in Figure 8.9) using 6H30 tubes. It is required that the plate current for the SRPP amplifier in the first stage is 6.8mA. The plate current for the compound cathode follower second stage is 10mA. Determine the cathode resistors.

Ex 8.2 Figure 8.21 is a two-stage amplifier. It can be considered as a vacuum tube equivalent to the JFET amplifier of Ex 7.1 in Chapter 7. The input stage is a cascode amplifier. The second stage is a compound cathode follower producing an approximated push–pull output. Assume dc voltage across resistor R5, plate-cathode voltage of T1 and plate-cathode voltage of T2 dividing the power supply equally, i.e., $V_B = 97V$, $V_A = 195V$. If ECC99 is used for T1–T4, complete the amplifier design for a voltage gain of -12. The plate current for the first stage and second stage are both around 6mA to 7mA, and $V_C = 90V$. If necessary, plate resistor R5 or cathode resistor R6 can be partially bypassed.

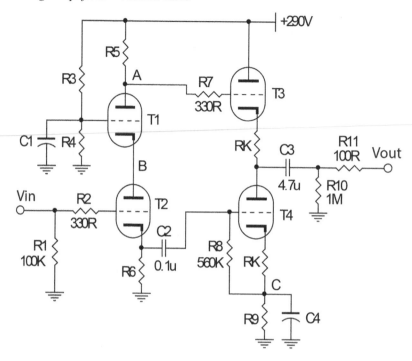

Figure 8.21 This is a two-stage line-stage amplifier. It can be considered as a vacuum tube equivalent of the JFET amplifier in Ex 7.1 of Chapter 7

Ex 8.3   Design a CCPP amplifier (shown in Figure 8.11) using 6SN7 for T1–T4. It is required that tail current for the cathode-coupled input stage is 12mA so that the plate current for T1 and T2 is 6mA each. Plate current for the triodes in the output stage is 8mA. Determine resistors R3–R8, and RK. What is the voltage gain and output impedance of this amplifier?

Ex 8.4   Design the cascaded SRPP amplifiers shown in Figure 8.12. Triode 6SL7 is used for T1 and T2 and 6SN7 for T3 and T4. The plate current for T1/T2 is 2mA and 7mA for T3/T4. If a voltage gain of 15 is required, determine the resistor values for RK and R5. Note that cathode resistors RK for T1/T2 and T3/T4 are different.

References

[1] Millman, J. and Taub, H., *Pulse and digital circuits*, McGraw-Hill, p. 100, 1956.
[2] Artzt, M., "Survey of dc amplifiers," Electronics, pp. 212–218, August 1945.
[3] Keen, A.W., "Bootstrap circuit technique," Electronic and Radio Engineer, pp. 345–354, September 1958.
[4] Blencowe, M., "The optimized SRPP amp (part 1)," Audio Xpress, pp. 13–19, 2010.
[5] Blencowe, M., "The optimized SRPP amp (part 2)," Audio Xpress, pp. 18–21, 2010.
[6] Blencowe, M., *Designing high-fidelity tube preamps*, Merlin Blencowe, 2016.
[7] Valley, G.E. and Wallman, H., *Vacuum tube amplifiers*, McGraw-Hill, pp. 456–464, 1948.

[8] Broskie, J., "Aikido amplifier," Tube CAD.

[9] Lam, C.M.J., "Single-stage balanced voltage amplifier," US patent 7482867.

[10] "Tube circuits for audio amplifiers," reprinted by Audio Amateur Press.

[11] Hedge, L.B., "The long-tailed cascade pair," Radio Electronics, pp. 40–42, October 1956.

[12] Jones, M., *Valve amplifiers*, Newnes, 3rd edition, 2003.

[13] Hood, L.H., *Valve & transistor audio amplifiers*, Newnes, reprinted 2001.

9 Noise in audio amplifiers

9.1 Introduction

Noise is a random signal and its instantaneous amplitude or phase cannot be predicted at any specific time. Therefore, noise is usually expressed in an average root-mean-square (rms) term. Noise can be generated either internally by the electronic devices (i.e., resistor, transistor, op-amp, and vacuum tube, etc.) or externally by electromagnetic interference. The external noise is often the dominant noise source. In the following, we discuss some common intrinsic noise generated by electronic devices [1–5]. By understanding the intrinsic noise of electronic devices, we may find clues for designing low noise amplifiers.

9.2 Thermal noise

Thermal noise is also called *Johnson noise*, named after John Johnson, who discovered it at Bell Labs in 1926. Thermal noise is generated by thermal agitation of electrons in a conductor. Unless the conductor is at absolute zero temperature, electrons are never at rest and they are always in motion. The average rms thermal noise can be determined by the expression

$$E_n = \sqrt{4kTRB} \tag{9.1}$$

where k = Boltzmann constant, 1.3806×10^{-23} J/°K, T = absolute temperature in °K, R = resistance in ohm, B = the noise measurement bandwidth in Hz.

For example, the noise generated by a 10kΩ resistor at 25°C (298°K) over the audio frequency band 20Hz to 20kHz is given by

$$E_n = \sqrt{4 \times \left(1.3806 \times 10^{-23}\right) \times 298 \times 10,000 \times \left(20,000 - 20\right)} = 1.81\mu V$$

For the same 10kΩ resistor, if the temperature is reduced from 25°C to 0°C, thermal noise is reduced to 1.74μV, a drop of 0.34dB. However, if we keep the temperature at 25°C but the resistance is reduced from 10kΩ to 1kΩ, thermal noise is reduced to 0.57μV, a drop of 10dB. Even though both lowering the temperature and the resistance can reduce thermal noise, it is clear that lowering resistance is a more effective way of doing so. Therefore, it is always good practice to use a low value resistor, especially for the input stage of an amplifier.

It should be noted that the power spectrum of thermal noise is constant and independent of frequency. Thermal noise is white noise, which will be discussed later.

DOI: 10.4324/9781003369462-9

9.3 Shot noise

Shot noise is also referred to as *quantum noise*. It is caused by random motion of charged carriers in a conductor. If we look at it closely, current flow is not a smooth and continuous event. Instead, it is a consequence of electrons (charged particles) that move under the influence of an external potential difference. When an electron comes across a barrier, the external potential difference will help in building sufficient energy for the electron to cross the barrier. When an electron has gained sufficient potential energy, it is abruptly transformed to kinetic energy, enabling it to cross the barrier. The aggregate effect of all of the electrons jumping across the barrier creates the so-called shot noise, which is described as the sound of lead shot pouring into a drum.

The average rms shot noise current is given as

$$I_n = \sqrt{2qI_{dc}B} \tag{9.2}$$

where q = electron charge, 1.6×10^{-19} Coulombs, I_{dc} = average dc current in Ampere, B = the noise measurement bandwidth in Hz.

The expression for the shot noise current indicates that it is independent of temperature. However, shot noise is always associated with dc current flow. It vanishes when the current is stopped. Shot noise is present in a conductor as well as a semiconductor when there is a dc current flow. The level of shot noise in metal conductors is very small, due to the relative size of potential barriers. However, due to high potential barriers, shot noise in a semiconductor is much higher than that of a metal conductor.

Note that the power spectrum of thermal noise is constant and independent of frequency. Thermal noise is white noise.

9.4 Flicker noise (1/f noise)

Flicker noise is also called *1/f noise* and *excess noise*. It is present in all semiconductor devices and many passive devices. In semiconductor devices, the flicker noise is related to imperfections in the crystalline structure of the semiconductor. Better device processing can reduce the flicker noise. Flicker noise is also found in carbon composite resistors, where it is referred to as excess noise because it appears to be additional to the thermal noise. It should be noted that the power spectrum of flicker noise is not constant and it is dependent on frequency. Flicker noise is pink noise.

Figure 9.1 shows the equivalent input noise voltage density of a typical op-amp. The flicker noise dominates at the low frequency range, usually below 100Hz. At this low frequency range, the noise is proportional to 1/f. The lower the frequency, the higher the noise is. At the frequency where the flicker noise meets the white noise in equal value, it is called *corner frequency*. If white noise and pink noise are displayed on an oscilloscope, the waveforms look very different, as shown in Figure 9.2. White noise appears to have a relatively uniform amplitude while the amplitude of the pink noise fluctuates widely.

9.5 Burst noise (popcorn noise)

Burst noise is also called popcorn noise. It is related to imperfections in semiconductor material and heavy ion implants. Burst noise is characterized by making a popping sound at a frequency below 100Hz when it is played through a loudspeaker. Burst noise can be reduced by clean device processing. Therefore, given a semiconductor device, the circuit designer does not have anything with which to control the burst noise.

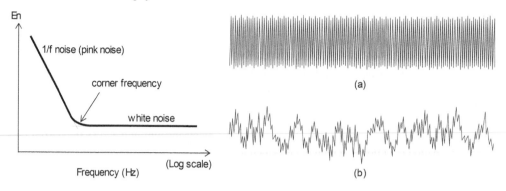

Figure 9.1 Equivalent input noise voltage density of a typical op-amp

Figure 9.2 Noise waveform is displayed on an oscilloscope. White noise with uniform amplitude is shown in (a). Pink noise (1/f noise) with non-uniform amplitude is shown in (b)

9.6   Avalanche noise

Avalanche noise in a semiconductor device is generated when a pn junction is operated in the reverse breakdown mode. For example, the base-emitter junction of a BJT and zener diode in reverse bias. However, it should be noted that when the reverse bias is less than 7V, the noise is shot noise for both the zener diode and BJT. Only when the reverse voltage is over 7V, as the semiconductor junction becomes thicker, is the breakdown mechanism avalanche. When the reverse voltage is high, the junction's depletion region is under a strong reverse electric field. Electrons have sufficient kinetic energy to collide with the atoms of the crystal lattice. As a consequence, additional electron-hole pairs are formed and more collisions occur, with avalanche multiplication. These collisions are random and produce a noise similar to shot noise, but in greater intensity. When a zener diode is used in an amplifying circuit, it is necessary to suppress the noise by connecting a large electrolyte capacitor (around 100μF) across the zener diode or insert a R-C low pass filter to remove the noise.

9.7   Adding noise sources

If there are independent multiple noises in a circuit, the noises are independent and uncorrelated. The total rms noise is the square root sum of the mean square values of the individual noise sources,

$$E_{n(total)} = \sqrt{E_{n1}^2 + E_{n2}^2 + E_{n3}^2 + \cdots} \tag{9.3}$$

For example, there are three independent noise sources in a circuit and each of them is $1\mu V$ (rms). The total noise is not $3\mu V$ but it is equal to

$$E_{n(total)} = \sqrt{1^2 + 1^2 + 1^2}\,\mu V = 1.73\mu V$$

The total noise is not triple the value of a single noise source, which one may expect. It may be an increase of $20 \times \log(3/1) = 9.5dB$. The actual increase is only $20 \times \log(1.73/1) = 4.7dB$. This is perhaps the most fortunate thing that we have when dealing with noise because total noise is not a linear sum of the individual noise sources.

Another important property is that the worst noise source in the circuit dominates the total noise. For example, there are three independent noise sources in a circuit. They are $1\mu V$, $2\mu V$, and $3\mu V$, respectively. The total noise becomes

$$E_{n(total)} = \sqrt{1^2 + 2^2 + 3^2}\,\mu V = 3.74\mu V$$

The total noise is not much higher than the worst noise source, $3\mu V$. The increase from $3\mu V$ to $3.74\mu V$ is only $20\times\log(3.74/3) = 1.9dB$. However, if we can reduce the worst noise source from $3\mu V$ to $1\mu V$, the total noise will become

$$E_{n(total)} = \sqrt{1^2 + 2^2 + 1^2}\,\mu V = 2.45\mu V$$

The total noise decreases from $3.74\mu V$ to $2.45\mu V$, i.e., $20\times\log(2.45/3.74) = -3.6dB$. An improvement of 3.6dB! In other words, when there are multiple noise sources present in a circuit, the most effect way to reduce the total noise is to bring down the highest noise source. Note that in an amplifier circuit, the input stage of the amplifier is often the place that creates the highest noise in the amplifier if it is not properly designed.

9.8   Equivalent input noise, noise unit, and S/N ratio

In modeling an amplifier or op-amp, it is often the practice to use a noiseless amplifier with an equivalent input noise voltage density (E_n) and input noise current density (I_n), as shown in Figure 9.3. Then the total noise at the output is the sum of the E_n and I_n that interacts with the source and multiplies with the gain of the amplifier. It should be noted that if the source impedance is zero, the source will not contribute noise to the output.

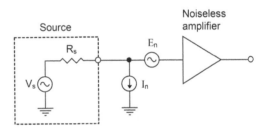

Figure 9.3 A noisy amplifier is modeled by a noiseless amplifier with an equivalent input noise voltage density En and input noise current density In

The equivalent input noise voltage density E_n is specified in $V/\sqrt{Hz}$ and the input noise current density I_n in $A/\sqrt{Hz}$. Figure 9.4 shows the equivalent input noise sources for op-amp OPA1611. The shape follows closely that shown in Figure 9.1. At frequencies below 40Hz, the 1/f noise dominates. Above 40Hz, white noise dominates.

At 1kHz, the voltage noise density is $1.1nV/\sqrt{Hz}$. For example, OPA1611 is configured to a voltage amplifier with a voltage gain of 40dB (i.e., 100). This amplifier is connected to a source signal with zero source impedance. If we assume that input noise voltage density is flat from 20Hz to 20kHz, then the equivalent noise at the input is approximately equal to $1.1nV\times\sqrt{20,000-20} = 155.5nV$. The noise at the output is equal to $155.5nV\times 100 = 15.5\mu V$.

In order to compare the merits of different amplifiers, we often use the term signal-to-noise ratio, which is defined as

$$SNR = 20\log\left(\frac{signal}{noise}\right) \tag{9.4}$$

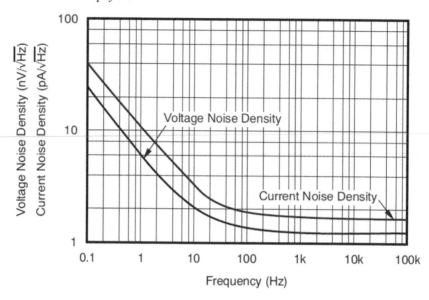

Figure 9.4 Input noise voltage density and input current noise density of OPA1611. (Courtesy of Texas Instruments)

In the above example, if the output signal is 1V(rms), the SNR is equal to $20\log(1V/15.5\mu V) = 96dB$. This is a very good SNR for an amplifier at 40dB gain. However, the actual SNR will be much lower., which is due to the fact that we have assumed the input noise voltage density is flat from 20Hz to 20kHz. But, due to the nature of 1/f noise, the noise level from 20Hz to 1kHz is higher than the assumed flat noise density $1.1nV/\sqrt{Hz}$ from 20Hz to 20kHz. In addition, the source impedance is not zero and it creates thermal noise. The input current noise will also interact with the source impedance and feedback resistors of the amplifier and, therefore, they contribute to an increase of total noise. Hence the SNR is much lower.

9.9   Noise in bipolar transistors (BJTs)

Figure 9.5 shows the noise equivalent model of a BJT with noise sources. The noise sources include thermal noise of the base spreading resistor r_{bb}, flicker noise of the base current (I_{n1}), shot noise of the base current (I_{n2}), and shot noise of the collector current (I_{n3}). They are shown in Figure 9.5(a). However, these individual noise sources are not very helpful for a circuit designer. Fortunately, these noise sources can be translated into two equivalent input noise components, input noise voltage density (E_n) and input noise current density (I_n). Semiconductor manufacturers publish E_n and I_n in data sheets for components that are classified as low noise transistors. But the value of the base spreading resistor r_{bb} is not often published separately. Since the noise contribution from the base spreading resistor is already absorbed into input noise voltage density E_n and input current noise density I_n, when these two equivalent noise densities are given, the output noise can be calculated for an amplifying circuit at a given bandwidth.

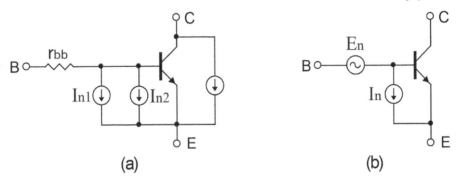

Figure 9.5 Circuit (a) is a noise equivalent model for a bipolar transistor that contains thermal noise of a base spreading resistor r_{bb}, flicker noise (I_{n1}) of the base current, shot noise (I_{n2}) of the base current, and shot noise (I_{n3}) of the collector current. Circuit (b) translates all noise sources into just two components: equivalent input noise voltage density E_n and equivalent input noise current density I_n

Since the base spreading resistor r_{bb} creates thermal noise and base current creates shot and flicker noise, a low noise BJT should ideally have a low base spreading resistor r_{bb}, and high current gain. For the same quiescent collector current, it is obvious that a high current gain lowers the base current, and, hence, a lower shot and flicker noise. For instance, in a pair of classic low noise BJTs produced by ROHM, 2SD786 (NPN) and 2SB737 (PNP), the base spreading resistor r_{bb} is 2–4Ω and current gain is 270–560. At a collector current of 1mA, the equivalent input noise voltage density is given as $E_n = 0.55V/\sqrt{Hz}$ at 10Hz while input noise current density is $I_n = 3.3pA/\sqrt{Hz}$. The equivalent input noise voltage density is not just one of the lowest from BJT devices, but it is also lower than JFET devices. Unfortunately, ROHM has ceased the production of 2SD786/2SB737. Even though they can still be purchased through some electronics supply specialists, the cost is high and the authenticity is uncertain.

Table 9.1 shows a list of representative low noise BJTs. Even though 2SD786/2SB737 are no longer available, some BJTs offer very close performance. On the other hand, it is well known that paralleling two transistors (BJT or JFET) together, the SNR is improved by 3dB. If the number of transistors is doubled again, from two to four, the SNR is further improved by 3dB. Thus, even without the lowest noise device, we can still develop low noise amplifiers by paralleling a sufficient number of transistors. However, it should be noted that the BJT's input noise current density I_n is in the order of $pA/\sqrt{Hz}$. For JFET, the input noise current density is in the order of $fA/\sqrt{Hz}$, which is much lower than that of BJT. Therefore, if we just look at the input noise current density, it may suggest that BJT is a good choice for a low impedance source, while JFET could be better for a high impedance source.

Table 9.1 shows the input noise voltage densities arranged in descending order. It should be noted that when we compare the input noise voltage density, we have to compare the value at the same frequency. For example, both that of transistors SSM2220 and THAT300 is stated to be $0.8nV/\sqrt{Hz}$. However, SSM2220 is measured at 10Hz while THAT300 is measured at 1000Hz. As the 1/f noise increases at low frequency, the input noise voltage density for THAT300 will

Table 9.1 A list of some representative low noise bipolar transistors

Parts	NPN	PNP	Single	Dual	Quad	V_{CEO} min (V)	f_T (MHz)	$h_{FE(min)}$	at I_C (mA)	r_{bb} (Ω)	E_n (nV/√Hz)	at I_C (mA)	at freq (Hz)	I_n (pA/√Hz)	at I_C (mA)	at freq (Hz)	Package
2N5551	•		•			160	150	80	1	–	–	–	–	–	–	–	TO-92
2N5401		•	•			150	150	50	1	–	–	–	–	–	–	–	TO-92
BC337	•		•			45	150	250	1	–	–	–	–	–	–	–	TO-92
BC327		•	•			45	150	250	1	–	–	–	–	–	–	–	TO-92
BC550	•		•			45	300	420	2	–	–	–	–	–	–	–	TO-92
BC560		•	•			45	150	420	2	–	–	–	–	–	–	–	TO-92
2SC2240	•		•			120	100	350	2	–	–	–	–	–	–	–	TO-92
2SA970		•	•			120	100	350	2	–	–	–	–	–	–	–	TO-92
ZTX851	•		•			60	130	100	10	–	–	–	–	–	–	–	TO-92
ZTX951		•	•			60	120	100	10	–	–	–	–	–	–	–	TO-92
BC850	•		•			45	300	420	2	–	–	–	–	–	–	–	SOT-23
BC860		•	•			45	150	420	2	–	–	–	–	–	–	–	SOT-23
FJV1845	•		•			120	110	600	1	–	–	–	–	–	–	–	SOT-23
FJV992		•	•			120	100	400	1	–	–	–	–	–	–	–	SOT-23
2SC3324	•		•			120	100	350	2	–	–	–	–	–	–	–	SOT-23
2SA1312		•	•			120	100	350	2	–	–	–	–	–	–	–	SOT-23
MAT14	•				•	40	300	300	1	–	2	1	10	–	–	–	SOIC-14
LM394	•			•		35	100	300	1	–	1.8	0.1	100	–	–	–	Obsolete
MAT02E	•			•		40	200	500	1	–	1.6	1	10	3	1	10	Obsolete
MAT12	•			•		40	200	300	1	–	1.6	1	10	–	–	–	TO-78
SSM2212	•			•		40	200	300	1	–	1.6	1	10	–	–	–	SOIC-8 LFCSP
THAT300B	•			•		36	350	300	1	32	1	1	1000	–	–	–	DIP-14 SOP-14
MAT03E		•		•		36	40	100	1	–	0.8	1	10	–	–	–	TO-78
SSM2220		•		•		36	190	80	1	–	0.8	1	10	–	–	–	DIP-8 SOIC-8
THAT300	•				•	36	350	60	1	32	0.8	1	1000	–	–	–	DIP-14 SOP-14
THAT320					•	36	325	50	1	25	0.75	1	1000	–	–	–	DIP-14 SOP-14
2SD786	•		•			40	100	270	10	4	0.55	10	10	–	10	10	Obsolete
2SB737		•	•			40	100	270	10	2	0.55	10	10	16	10	10	Obsolete

inevitably become greater than $0.8\text{nV}/\sqrt{\text{Hz}}$ at 10Hz. Therefore, THAT300 will produce a higher noise level than SSM2220 even though both of them are given with the same figure of input noise voltage density. On the other hand, the input noise current density is not usually stated in data sheets. But one thing we are certain of is that the BJT's input noise current density is much higher than that of the JFET. Therefore, leave the high source impedance application to JFET. In a low source impedance application, the input noise voltage density is more dominant than the input noise current density.

A comprehensive table of low noise BJTs can be found in Horowitz and Hill [6]. They do not just state the parameters taken from manufacturers' data sheets, they also measure the actual base spreading resistor and input noise voltage densities for most of the low noise BJTs. Since not many manufacturers publish such data, the table provides valuable insight to some BJTs that we might, otherwise, overlook for their noise behavior. In particular, the measured input noise voltage densities of transistor ZTX851 and ZTX951 are so impressive that they are actually similar to, or better than, 2SD786 and 2SB737 at collector current 10mA. ZTX851 and ZTX951 appear to be an excellent choice for a very low source impedance application. But high dc bias current (10mA) will inevitably increase the shot noise.

9.10   Noise in junction FET (JFETs)

Figure 9.6 shows an equivalent noise model of a JFET. The input noise voltage density E_n comes from two sources: (i) thermal noise of the channel resistance and (ii) flicker noise of the drain current. The input noise current density I_n comes from the shot noise of the gate current. In normal operation, the gate current for a JFET is very small and so is the shot noise. Generally speaking, input noise current density I_n for a JFET is in the order of $\text{fA}/\sqrt{\text{Hz}}$, which is much lower than for a BJT of the order of $\text{nA}/\sqrt{\text{Hz}}$. Therefore, JFET is a better choice than BJT for low noise application when the source impedance is high.

The thermal noise of the JFET's channel resistance can be expressed as $\sqrt{4kT(2/3g_m)}$ $\text{V}/\sqrt{\text{Hz}}$, where g_m is the transconductance of the JFET. As the thermal noise is inversely proportional to g_m, the thermal noise can be reduced by

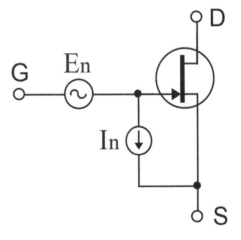

Figure 9.6 Noise equivalent model of JFET: input noise voltage density E_n and input noise current density I_n

increasing g_m. Paralleling a pair of JFETs will double the g_m and, therefore, reduce the thermal noise by a factor of $\sqrt{2}$. This translate to an improvement of $20\log(\sqrt{2}) = 3\text{dB}$. Generally speaking, a high number of JFETs can be paralleled in order to reduce the input noise voltage density. However, since the input capacitance of JFET also increases when paralleling JFETs, the bandwidth of the amplifier may be compromised.

Table 9.2 shows a list of representative low noise JFETs. If we compare Table 9.2 detailing JFETs to Table 9.1 listing BJTs, we find that in general the input noise voltage density of BJTs and JFETs is close. While this is the case, the BJTs are still slightly better unless special JFETs

Table 9.2 A list of some representative low noise JFETs

Parts	N-channel	P-channel	Single	Dual	$V_{GS(BR)}$ (V)	$V_{GS(off)}$ (V)	g_m (mS)	C_{iss} (pF)	C_{rss} (pF)	E_n (nV/√Hz)	at I_D (mA)	at freq (Hz)	Package
2SK208	•		•		50	0.4 to 5	1.2	8.2	2.6	-	-	-	SOT-23
2SK209	•		•		50	0.2 to 1.5	15	13	3	-	-	-	SOT-23
2SK2145	•	•		•	50	0.2 to 1.5	15	13	3	-	-	-	SOT-25
MMBFJ211	•		•		25	2.5 to 4.5	12	-	3	12	10	100	SOT-23
J211	•		•		25	2.5 to 4.5	12	-	3	12	10	100	TO-92
MMBFJ309L	•		•		25	1 to 4	18	5	2.5	10	10	100	SOT-23
MMBFJ202		•	•		40	0.8 to 4	>1	-	2	10	1	1000	SOT-23
MMBF5460		•	•		40	0.75 to 6	2	5	1	10	3	1000	SOT-23
2N2652	•		•		25	0.5 to 3.6	15 to 30	25	5	8	5	1000	TO-72
2N2654	•		•		25	0.75 to 5	20 to 40	25	5	8	5	1000	TO-72
J107	•		•		25	0.5 to 4.5	0.1	38	22	3	10	100	TO-92
2N2651	•		•		20	0.5 to 3.5	15 to 30	25	5	3	5	1000	TO-72
2N2653	•		•		20	0.75 to 5	20 to 40	25	5	3	5	1000	TO-72
LS846	•		•		60	0.5 to 3.5	1.5	8	3	3	0.5	1000	SOT-23, TO-92
LSJ289		•	•		50	1.5 to 5	1.5	8	3	2	2	1000	SOT-23, TO-92
LSJ689		•	•		50	1.5 to 5	1.5	8	3	2	2	1000	SO-8, TO-71
LSK489	•			•	60	1.5 to 3.5	1.5	8	3	1.8	2	1000	SO-8, TO-71
LSK189	•			•	60	0.5 to 3.5	1.5	8	3	1.8	2	1000	SOT-23, TO-92
2SK170	•		•		40	0.2 to 1.5	22	30	6	1	1	1000	TO-92
2SJ74		•	•		25	0.15 to 2	22	105	32	0.92	1	1000	TO-92
LSK170	•		•		40	0.2 to 2	10	20	5	0.9	2	1000	SOT-23, TO-92
LSJ74		•	•		25	0.15 to 2	22	105	32	0.9	2	1000	TO-92, SOT-89
LSK389B	•			•	40	0.15 to 2	20	25	5.5	0.9	2	1000	SO-8, TO-71
IF3601	•		•		20	0.35 to 2	750	300	200	0.3	5	100	TO-39
IF3602		•		•	20	0.35 to 3	750	300	200	0.3	5	100	TO-78

IF3601 and IF3602 are considered. Another thing is that since input noise current density I_n for a JFET is very small, this data is not often given in data sheets.

Example 9.1

In this example, illustrated in Figure 9.7, it is given that $g_m = 12\text{mS}$ and $En = 5\text{nV}/\sqrt{\text{Hz}}$ for the JFET. From Eq. (3.36), the voltage gain for a common source amplifier is given as

$$A_{vs} = \frac{g_m R2}{1+g_m R3} = \frac{12 \times 10^{-3} \times 10 \times 10^3}{1+12 \times 10^{-3} \times 330} = 24.2 \text{ (the negative sign is discarded)} \tag{9.5}$$

And from Eq. (3.47), the voltage gain for a common gate amplifier is given as

$$A_{vg} = g_m R2 = 12 \times 10^{-3} \times 10 \times 10^3 = 120 \tag{9.6}$$

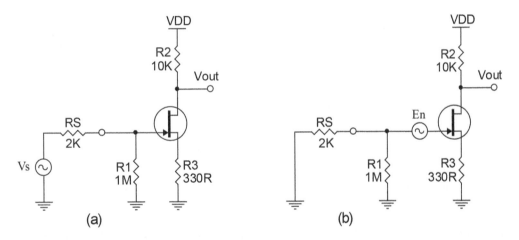

Figure 9.7 Circuit (a) is a simple common source amplifier in a self-bias configuration. In circuit (b), the JFET is replaced by its noise equivalent model with an input noise voltage density E_n. Since the input noise current density is small for a JFET, it is ignored in this example

The output noise of the amplifier can be expressed as follows.

$$E_{n(total)} = \sqrt{\left(E_{n1}^2 + E_{n2}^2 + E_{n3}^2\right) \cdot B} \tag{9.7}$$

where

$$E_{n1} = \sqrt{4kT(Rs \parallel R1)} \times \left(A_{vs}\right), \text{ thermal noise of resistor } Rs \parallel R1 \tag{9.8}$$

$$E_{n2} = \sqrt{4kTR3} \times \left(A_{vg}\right), \text{ thermal noise of resistor } R3 \tag{9.9}$$

$$E_{n3} = E_n A_{vs}, \text{ voltage noise of JFET} \tag{9.10}$$

Therefore, we have

$$E_{n1} = \sqrt{4 \times 1.38 \times 10^{-23} \times 295 \times (2k\Omega \parallel 1M\Omega)} \times (24.2)V / \sqrt{Hz} = 138nV / \sqrt{Hz}$$

$$E_{n2} = \sqrt{4 \times 1.38 \times 10^{-23} \times 295 \times 330} \times (120)V / \sqrt{Hz} = 278nV / \sqrt{Hz}$$

$$E_{n3} = 5 \times 24.2nV / \sqrt{Hz} = 121nV / \sqrt{Hz}$$

$$E_{n(total)} = \sqrt{\left(1.11 \times 10^{-13}\right) \cdot (20000 - 20)}V = 47.1\mu V \tag{9.11}$$

9.11   Noise in MOSFETs

Figure 9.8 shows a noise equivalent model of MOS-FET. Similar to that of JFET, the input noise voltage density E_n of a MOSFET comes from two sources: (i) thermal noise of the channel resistance and (ii) flicker noise of the drain current. On the other hand, since there is no gate current in MOSFET, the input noise current density can be ignored for the equivalent noise model.

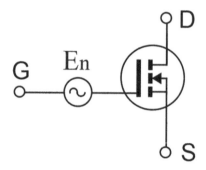

Figure 9.8 A noise equivalent model of MOSFET is represented by an input noise voltage density E_n

In discrete MOSFET devices, flicker noise is the dominant noise and it is much higher than JFET and BJT, as illustrated in Figure 9.9. The corner frequency for the flicker noise can be 10kHz or even higher, making MOSFET an undesirable device for low noise application at low frequency. However, MOSFETs perform better in monolithic op-amps, where they are processed in the form of CMOS. With a better control in integrated circuit fabrication, CMOS op-amps have an acceptable noise performance even though they are not the best in class among low noise op-amps (see Table 9.3).

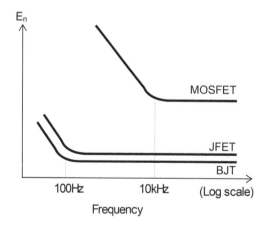

Figure 9.9 A quantitative comparison of input noise voltage density for discrete MOSFET, JFET and BJT devices. MOSFET has a much higher flicker noise and higher corner frequency, making MOSFET an undesirable device for low noise application

9.12 Noise in op-amps

Figure 9.10 shows an equivalent noise model for an op-amp. An input noise voltage density E_n is present at the non-inverted input. Input noise current density I_n is present at both of the inputs. As we know, the input noise voltage density is different for BJT, JFET, and MOSFET, op-amps with BJT input stage have the lowest E_n, followed by JFET input stage and CMOS op-amps. However, for input noise current density, the result is just the opposite. JFET input stage and CMOS op-amps have the lowest I_n, followed by BJT input stage op-amps.

Table 9.3 shows a list of representative low noise op-amps. The input noise voltage and current densities (E_n and I_n) are specified at 10Hz, unless otherwise stated. The BJT

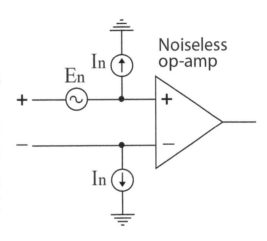

Figure 9.10 Equivalent noise model of an op-amp: input noise voltage density E_n and input noise current density I_n

op-amps have the lowest E_n at $1nV/\sqrt{Hz}$. Even though op-amps with JFET input stage have slightly higher E_n, they are still better than CMOS op-amps. For hand-held equipment where low quiescent current and power consumption is of prime importance, CMOS op-amps are the natural choice. However, the shortcoming for CMOS op-amps is that supply voltage is limited to around 5V. This greatly limits the signal voltage swing and their usage in audio application such as phono-stage and line-stage amplifier. For phono-stage amplifier application, op-amps with BJT and JFET input stage are the preferred choice.

Figure 9.11 shows the inverted and non-inverted amplifier, which are two common configurations in an op-amp application. Note that they produce different output noise [3–4]. The source is modeled by a source resistor RS in series with a source signal Vs. Figure 9.11(a) is a non-inverted amplifier configuration while Figure 9.11(b) shows an inverted amplifier configuration. When the input noise voltage and current densities are given, we can express the output noise in terms of the source resistor (RS) and feedback resistors (R1 and R2).

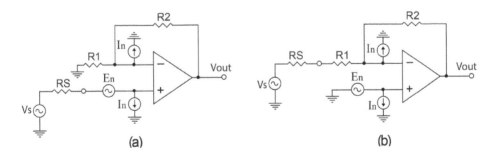

(a)　　　　　　　　　　　　(b)

Figure 9.11 Circuit (a) is an equivalent noise model for a non-inverted amplifier while circuit (b) illustrates an inverted amplifier

For the non-inverted amplifier of Figure 9.11(a), the output noise is expressed in the following:

$$E_{n(total)} = \sqrt{\left(E_{n1}^2 + E_{n2}^2 + E_{n3}^2 + E_{n4}^2 + E_{n5}^2 + E_{n6}^2\right) \cdot B} \tag{9.12}$$

Table 9.3 A list of some representative low noise op-amps: BJT, JFET input stage and CMOS

Parts	Single	Dual	Quad	Total supply min (V)	max (V)	I_q/ch typ (mA)	V_{os} max (μV)	GBW typ (MHz)	Slew rate typ (V/μs)	E_n (nV/√Hz) at 10Hz	I_n (pA/√Hz) at 10Hz	Package
BJT input												
NE5534A	•			10	40	4	5000	10	13	7.0	4.5	DIP-8, SO-8
LME49710	•			5	34	5	50	55	20	6.4	3.1	SO-8, TO-99
LME49720		•		5	34	5	700	55	20	6.4	3.1	DIP-8, SO-8
LME49743			•	8	34	2.5	1000	30	12	6.4	3.1	TSSOP-14
LME49723	•			5	34	3.35	1000	19	8	5.5	-	SO-8
OPAx1602/4		•	•	5	36	2.6	1000	35	20	5.2	4.4	SO-14, TSSOP-14
OPAx209	•	•	•	4.5	36	2	150	18	6.4	3.5	0.6	SOT-23, SO-8, VSSOP-8, TSSOP-14
AD8675/6	•	•		10	30	2.9	240	10	2.5	3.5	0.3	SO-8, MSOP-8
OPAx227	•	•		5	36	3.7	200	8	2.3	3.4	2	DIP-8, SO-8
OPAx228	•	•		5	36	3.7	100	33	11	3.4	2	DIP-8, SO-8
ADA4075-2		•		9	36	2.25	1200	6.5	12	3.4	1.9	SO-8
LT1124/5	•	•		8	44	2.3	140	12.5	4.5	3	1.3	DIP-8, SO-8, DIP-16, SO-16
LT1126/7		•		8	44	2.6	100	65	11	3	1.3	DIP-8, SO-8, DIP-16, SO-16
LT1007	•			10	30	2.6	60	8	2.5	2.8	1.5	DIP-8, SO-8
LT1037	•			8	44	2.6	60	60	15	2.8	1.5	DIP-8, SO-8
AD8671/2/4	•	•	•	8	36	3.5	125	10	4	2.3	1.4	SO-8, MSOP-8
OPA1611/2	•	•		5	36	3.6	500	40	27	2.0	3	SO-8
OPA211	•	•		4.5	36	3.6	125	45	27	2.0	3	SO-8, MSOP-8
ADA4004-1/2/4	•	•	•	10	30	2.2	300	12	2.7	1.9	3.9	SO-8, SOT-23, MSOP-8, SO-14
AD797A	•			10	36	8.2	80	110	20	1.8	2 at 1kHz	DIP-8, SO-8
MAX9632	•			4.5	36	4	165	55	30	1.5	23	SO-8, TDFN
LME49990	•			10	36	9	2000	110	22	1.4	4.2	Obsolete
AD8597/9	•	•		9	30	5.7	180	10	16	1.4	4.7	SO-8
ADA4898-1/2	•	•		9	36	8.1	125	50	55	1.2	4.6	SO-8
LT6018	•			8	33	7.2	95	15	30	1.2	17	SO-8, DFN

Device										Package
LT1115	•	8	44	8.5	200	70	15	1	4.7	DIP-8, SO-16
JFET input										
OPAx134	• •	5	36	4	3000	8	20	25	0.02	DIP-8, SO-8, DIP-14, SO-14
LME49880	• •	10	34	14	5000	25	17	19	0.073	SO-8
ADA4627B	•	10	30	7	300	19	56	15	–	SO-8, LFCSP
OPAx627	• •	9	36	7	500	16	55	15	0.0016 at 100Hz	DIP-8, SO-8, TO-99
OPAx141	• •	4.5	36	1.8	4300	10	20	12	0.0008 at 1kHz	DIP-8, SO-8, DIP-14, SO-14
LT1793	•	10	40	4.2	1600	4.2	3.4	11.5	0.001	DIP-8, SO-8
LT1792	•	9	40	4.2	800	4.3	3	8.2	0.01	DIP-8, SO-8
OPAx140	• •	4.5	36	1.8	220	11	20	8	0.0008 at 1kHz	SOT23–5, DIP-8, SO-8, DIP-14, SO-14
OPA1641-2-4	• •	4.5	36	1.8	3500	11	20	8	0.0008 at 1kHz	SO-8, VSSOP-8, SO-14, TSSOP-14
OPA828	•	8	36	5.5	350	45	150	7.5	0.0012 at 1kHz	SO-8
OPA827	•	8	36	4.8	150	22	28	7.3	0.002 at 1kHz	SO-8, MSOP-8
AD745	•	9.6	36	8	1500	20	12.5	5.5	0.032	SO-16
AD743	•	9.6	36	8	1500	4.5	2.8	5.5	0.027	DIP-8, SO-16
CMOS										
LTC6081/2	• •	2.7	5.5	0.33	1650	3.6	1	36	0.0005	MSOP-8, SSOP-16, DFN-10, DFN-16
LMV791/2	• •	2.5	5	0.95	90	10	14	30	0.01 at 1kHz	SOT-23, VSSOP-10
LMV751	•	2.7	5.5	0.5	1500	4.5	2.3	15	0.01 at 1kHz	SOT-23

where

$$E_{n1} = \sqrt{4kTRS}\left(1 + \frac{R2}{R1}\right), \text{ thermal noise of resistor RS} \qquad (9.13)$$

$$E_{n2} = \sqrt{4kTR1}\left(\frac{R2}{R1}\right), \text{ thermal noise of resistor R1} \qquad (9.14)$$

$$E_{n3} = \sqrt{4kTR2}, \text{ thermal noise of source resistor R2} \qquad (9.15)$$

$$E_{n4} = E_n\left(1 + \frac{R2}{R1}\right), \text{ voltage noise of op-amp En} \qquad (9.16)$$

$$E_{n5} = I_n RS\left(1 + \frac{R2}{R1}\right), \text{ current noise of source resistor RS} \qquad (9.17)$$

$$E_{n6} = I_n R2, \text{ current noise of resistor R2} \qquad (9.18)$$

k = Boltzmann constant, 1.3806×10^{-23} J/°K
T = absolute temperature in °K
B = noise measurement bandwidth in Hz

For the inverted amplifier configuration of Figure 9.11 (b), the output noise is expressed in the following,

$$E_{n(total)} = \sqrt{\left(E_{n1}^2 + E_{n2}^2 + E_{n3}^2 + E_{n4}^2 + E_{n5}^2\right) \cdot B} \qquad (9.19)$$

where

$$E_{n1} = \sqrt{4kTRS}\left(\frac{R2}{R1 + RS}\right), \text{ thermal noise of resistor RS} \qquad (9.20)$$

$$E_{n2} = \sqrt{4kTR1}\left(\frac{R2}{R1 + RS}\right), \text{ thermal noise of resistor R1} \qquad (9.21)$$

$$E_{n3} = \sqrt{4kTR2}, \text{ thermal noise of source resistor R2} \qquad (9.22)$$

$$E_{n4} = E_n\left(1 + \frac{R2}{R1 + RS}\right), \text{ voltage noise of op-amp En} \qquad (9.23)$$

$$E_{n5} = I_n R2, \text{ current noise of resistor R2} \qquad (9.24)$$

Example 9.2

Let us determine the output noise at room temperature for the non-inverted amplifier of Figure 9.11(a) with a voltage gain of 101. Assume RS = 600Ω, R1 = 20Ω, R2 = 2kΩ and OPA1611 is used. From the data sheet, we find that the input noise voltage density at 10Hz, 100Hz, and 1kHz is given as $2nV/\sqrt{Hz}$, $1.5nV/\sqrt{Hz}$ and $1.1nV/\sqrt{Hz}$, respectively. This clearly shows the 1/f noise increases at low frequency. Since the noise density is not flat from 20Hz to 20kHz, for accurate calculation we need to find the noise voltage and current densities at every frequency point. This is certainly not a delightful task to do. Therefore, here we take an approximation. We take the noise voltage density at 100Hz ($1.5nV/\sqrt{Hz}$) and assume this is flat from 20Hz to 20kHz. Similarly, we take noise current density at 100Hz ($1.9pA/\sqrt{Hz}$). Each of the noise sources is calculated as follows.

$$E_{n1} = \sqrt{4kTRS}\left(1+\frac{R2}{R1}\right) = \sqrt{4\times1.38\times10^{-23}\times295\times600}\times(101)\,V/\sqrt{Hz} = 316nV/\sqrt{Hz}$$

$$E_{n2} = \sqrt{4kTR1}\left(\frac{R2}{R1}\right) = \sqrt{4\times1.38x10^{-23}\times295\times20}\times(100)\,V/\sqrt{Hz} = 57nV/\sqrt{Hz}$$

$$E_{n3} = \sqrt{4kTR2} = \sqrt{4\times1.38\times10^{-23}\times295\times2000}\,V/\sqrt{Hz} = 5.7nV/\sqrt{Hz}$$

$$E_{n4} = E_n\left(1+\frac{R2}{R1}\right) = 1.5\times(101)nV/\sqrt{Hz} = 151\,nV/\sqrt{Hz}$$

$$E_{n5} = (I_n RS)\left(1+\frac{R2}{R1}\right) = \left(1.9\times10^{-12}\times600\right)\times(101)\,V/\sqrt{Hz} = 115nV/\sqrt{Hz}$$

$$E_{n6} = I_n R2 = 1.9\times10^{-12}\times2000\,V/\sqrt{Hz} = 3.8\,nV/\sqrt{Hz}$$

Therefore, the total output noise is given as

$$E_{n(total)} = \sqrt{\left(E_{n1}^2 + E_{n2}^2 + E_{n3}^2 + E_{n4}^2 + E_{n5}^2 + E_{n6}^2\right)\cdot B}$$

$$= \sqrt{\left(1.39\times10^{-13}\right)\times19980}\,V = 52.7\mu V,\text{ where } B = (20,000-20)Hz$$

For an output signal of 1V(rms), the signal-to-noise ratio is given as,

SNR = 20log(1V/52.7 μV) = 85.5dB.

It should be noted that the source resistor RS dominates the noise. The noise of the source resistor alone (E_{n1} and E_{n5}) already gives 47.5μV. In order to reduce the output noise, the source impedance must be kept low.

9.13   Noise in triodes

The noise equivalent model for triode is shown in Figure 9.12(a). The noise sources include (i) shot noise (I_{n1}) from the grid current, (ii) flicker noise (I_{n2}) from the plate current, and (iii) shot noise (I_{n3}) from the plate current. In a similar fashion to discrete transistors, the noise of the vacuum tube can be translated into input equivalent noise voltage (E_n) and current densities (I_n), as shown in Figure 9.12(b). Unfortunately, equivalent input noise densities were not given in vacuum tube data sheets. Most of the vacuum tube data sheets were published from 1940s to 1960s, when the equivalent input noise densities were not yet defined.

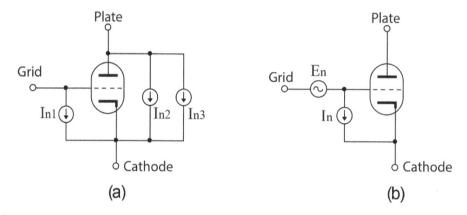

Figure 9.12 Circuit (a) is an equivalent noise model for a triode that contains shot noise (I_{n1}) from the grid current, flicker noise (I_{n2}) from the plate current, and shot noise (I_{n3}) from the plate current. Circuit (b) translates all the noise sources into just two components: input equivalent noise voltage density E_n and input equivalent noise current density I_n

However, the shot noise and flicker noise can be modeled for the triode and expressed in terms of the electrical properties of the tube. Therefore, the input equivalent noise density can be fairly accurately approximated. The square of input noise voltage density (i.e., $E_n{}^2$) for several triodes (12AT7/ECC81, 12AU7/ECC82, 12AX7/ECC83, 6922/ECC88 and triode connected pentode 6J52P) is calculated by Blencowe [5]. If we follow the same calculation as in Blencowe, we can obtain the input noise voltage density (E_n) as shown in Table 9.4. Since the grid current is negligible for a triode in normal operation, the input noise current density should be very low. It is expected to be in the same order of magnitude as a JFET in the fA/$\sqrt{\text{Hz}}$ range.

ECC88/6922 has one of the lower input noise voltage densities at 39nV/$\sqrt{\text{Hz}}$ among the vacuum tubes shown in Table 9.4. It is not surprising that 6922 is often the choice of triode for use in line-stage preamplifiers and phono amplifiers. The vacuum tube 6J52P is a Russian-made pentode that has electrical properties similar to a D3A pentode, which was obsolete for a long time. Even though a triode connected 6J52P gives a respectable input noise voltage density 16nV/$\sqrt{\text{Hz}}$ at 10Hz, it does not compare favorably to low noise JFETs, which can be as low as only a few nV/$\sqrt{\text{Hz}}$ while less than 1nV/$\sqrt{\text{Hz}}$ for low noise BJTs. But the reason that some designers choose vacuum tubes over semiconductor devices is because of the tube sound. Noise is important, but sound is also important for audio amplifiers.

Table 9.4 Input noise voltage densities are calculated following Blencowe [5] for some popular audio triodes at a plate current of 2mA

En (nV/√Hz), noise voltage density at 10Hz				
ECC81/ 12AT7	*ECC82/ 12AU7*	*ECC83/ 12AX7*	*ECC88/ 6922*	*6J52P (triode connected)*
100	95	45	39	16

9.14   Basics of low noise design

In designing low noise amplifier circuits, we have to keep the following in perspective:

- The first stage of the amplifier always dominates the noise and SNR. Make efforts to design the first stage with as little noise as possible.
- Paralleling the amplifying devices (transistor, op-amp, vacuum tube) will improve the SNR. It is well known that double the number of amplifying devices in parallel improves the SNR.
- Avoid using degenerating resistors on transistors and triodes in the first stage. For example, if a JFET/BJT differential amplifier configuration is employed for the first stage, do not use a source/emitter resistor.
- Keep the value of any series resistor between the source and the amplifier in the signal path as low as possible.
- Keep the value of a feedback resistor as low as possible. However, the value of the feedback resistor should not be so low that it creates a severe loading effect to the amplifying device (e.g., op-amp, transistor, etc.) when the amplifier is responding to a maximum allowable signal. See Example 9.3.
- Avoid using carbon film and carbon composite resistors. In addition to thermal noise, a carbon composite resistor has very high flicker noise (1/f noise), which is added to the thermal noise that a resistor already has.
- High quality, low noise regulated power supplies must be used.
- The power transformer must be placed far away from the first stage of the amplifier. It is best to have the power transformer properly shielded. And the power transformer is placed in the corner of the chassis.
- Amplifiers with fully balanced configuration are used. A fully balanced amplifier tends to cancel out common mode noise so that the SNR is better than that of an unbalanced amplifier.

Example 9.3

Figure 9.13(a) shows a non-inverted amplifier with a gain of 101. It is given that the op-amp has a maximum output current rating of 25mA and maximum output voltage of 12V(peak). In order to minimize thermal noise, very low resistors are used. They are given by R1 = 3Ω and R2 = 300Ω, so that the voltage gain is equal to $(1 + R2/R1) = 101$. It seems that the first amplifying stage may work just fine. However, we have to check the output current when the op-amp is producing the maximum output voltage of 12V.

Since the input impedance of the second stage is 20kΩ, it draws a current of 12V/20kΩ = 0.6mA. In addition, the feedback resistor network draws a current of 12V/(3Ω + 300Ω) = 39.6mA. Therefore, the op-amp has to deliver a total of 40.2mA in responding to a 12V output. However, this exceeds the maximum allowable current in the op-amp, which is 25mA. The output signal is severely clipped by the op-amp and produces a very high distortion.

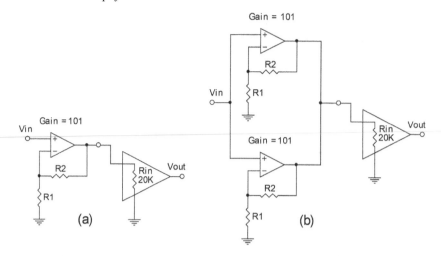

Figure 9.13 A low noise system contains two amplifying stages. Circuit (a) uses a single op-amp for the first stage while circuit (b) is using two op-amps in parallel. The second stage has an input impedance of 20KΩ

If we increase the resistors R1 and R2 by twofold (i.e., R1 = 6Ω, R2 = 600Ω), the output current is reduced to 20.4mA, which is now below the op-amp's maximum rating. The increase in resistor value will inevitably cause an increase of thermal noise. Therefore, we can employ two (or more) first stages in parallel to improve the overall SNR, as shown in Figure 9.13(b).

9.15   Reducing common mode noise

One of the major advantages of a fully balanced amplifier over an unbalanced amplifier is its ability to reduce common mode noise. Because of this, fully balanced audio equipment is the preferred equipment used in a professional audio studio when low noise is of primary concern. In the following, we will see how common mode noise is eliminated by a fully balanced amplifier.

Figure 9.14(a) shows a conventional amplifier in a noisy environment. There are two noise sources, n_1 and n_2, in the vicinity of the amplifier. The noise sources may come from hum noise generated by a nearby power transformer or induced noise picked up by the copper conductors under a magnetic field influenced by power lines. The amplifier has a voltage gain A, such that $A = Vo/Vi$. Since the input signal is now corrupted by noise n_1, while output is corrupted by noise n_2, we have

$$Vi = Vin + n_1 \qquad (9.25)$$

$$Vout = Vo + n_2 \qquad (9.26)$$

Thus we have

$$Vout = Vi\,A + n_2 = (Vin + n_1)A + n_2 = Vin\,A + (n_1A + n_2) \qquad (9.27)$$

The unwanted noise $(n_1A + n_2)$ is unfortunately inseparable from the output, as shown in the above expression. Since n_1 is amplified by a factor of A, the first term is the dominant noise in the output.

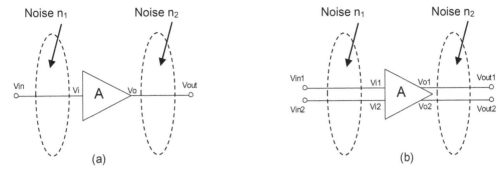

Figure 9.14 Circuit (a) is a conventional amplifier in the vicinity of noise source n_1 and n_2. Circuit (b) employs a fully balanced amplifier in the same environment

Figure 9.14(b) shows a fully balanced amplifier in the same noisy environment. The balanced amplifier has a voltage gain of A such that A = (Vo1 – Vo2)/(Vi1 – Vi2). Here we have the following.

$$Vi1 = Vin1 + n_1 \tag{9.28}$$

$$Vi2 = Vin2 + n_1 \tag{9.29}$$

$$Vout1 = Vo1 + n_2 \tag{9.30}$$

$$Vout2 = Vo2 + n_2 \tag{9.31}$$

Therefore, the differential output is given as

$$Vout1 - Vout2 = (Vo1 + n_2) - (Vo2 + n_2) = Vo1 - Vo2 \tag{9.32}$$

Note that noise n_2 is cancelled out in the above expression. Then Eq. (9.32) can be written as

$$\begin{aligned} Vout1 - Vout2 &= (Vi1 - Vi2)A = \{(Vin1 + n_1) - (Vin2 + n_1)\}A \\ &= (Vin1 - Vin2)A \end{aligned} \tag{9.33}$$

Again, noise n_1 is cancelled out and, therefore, the output is free from common mode noises. The term "common mode" noise is often used because n_1 is viewed as noise common to both inputs, Vi1 and Vi2, while n_2 is common to the outputs Vout1 and Vout2 of the balanced amplifier. In order to make the noise appear to be common to the inputs and outputs of the balanced amplifier so that the above noise cancellation scheme works best, the copper wires of the interconnecting cables must be closely tied together. Therefore, twisted cable or professional shielded balanced interconnect cable must be used. By using professional cable, we can run the cable very long before noise could become an issue. Fully balanced audio components are particularly useful for a professional recording studio as the microphones are often placed far away from the recording equipment.

9.16   Exercises

Ex 9.1 Determine the output noise from the inverted amplifier in Figure 9.11(b). Assuming the same op-amp and resistors in Example 9.2 are used, such that R1 = 20Ω, R2 = 2kΩ, RS = 600Ω, input equivalent noise densities are En = 1.5nV/√Hz) and In = (1.9pA/√Hz), which are flat from 10Hz to 20kHz. Then determine the SNR for an output of 1V(rms).

Ex 9.2 Figure 9.15(a) is a simplified common emitter amplifier having input equivalent noise densities E_n and I_n. Resistor RS represents the source impedance while RB represents the amplifier's Thévenin equivalent resistor for dc biasing the transistor. To evaluate the output noise, the common emitter amplifier is broken down into five cases, Figure 9.15(b) to (f), where only one noise source is given at a time. The output noise is written as

$$E_{n(total)} = \sqrt{\left(E_{nb}^2 + E_{nc}^2 + E_{nd}^2 + E_{ne}^2 + E_{nf}^2\right) \cdot B}$$

where E_{nb} to E_{nf} are the noise densities for Figure 9.15(b) to (f), respectively. Assume RS = 2kΩ, RB = 22kΩ, RC = 10kΩ, RE = 330Ω, En = 2nv/√Hz, In = 5pA/√Hz, B = 20kHz. Determine the output noise.

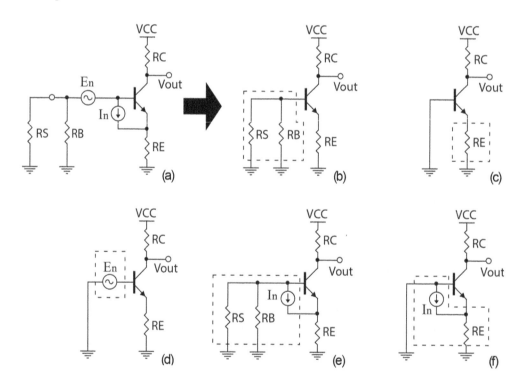

Figure 9.15 Circuit (a) is a simplified common emitter amplifier. Only one noise source is given in each circuit from (b) to (f)

References

[1] Wilamowsk, B.M. and Irwin, J.D., *Fundamentals of industrial electronics*, chapter 11, CRC Press, 2017.
[2] Mancini, R., "Op amps for everyone – design reference," chapter 10, Texas Instruments.
[3] Trump, B., "The signal," chapter 6, Texas Instruments.
[4] OPA1611 data sheet, Texas Instruments.
[5] Blencowe, M., *Designing high-fidelity tube preamps*, Merlin Blencowe, 2016.
[6] Horowitz, P. and Hill, W., *The art of electronics*, 3rd ed., Cambridge University Press, 2015.

10 Phono-stage amplifiers

10.1 Introduction

RIAA equalization is needed for phonograph record playback. The RIAA equalization is embedded in a phono-stage amplifier (or, simply, a phono amplifier) with sufficient voltage gain for moving magnet (MM) and moving coil (MC) cartridges. The gain for an MM phono amplifier is generally around 30 to 40dB (i.e., a gain of 56–100). For an MC phono amplifier, the gain is around 60–70dB (i.e., a gain of 1,000–3,162). Given such a high gain, the inherent noise of a phono amplifier is a primary concern. Therefore, a well-designed phono amplifier must have accurate RIAA equalization, high gain as well as low noise. The origins of some intrinsic noise from electronic devices and the basics of low noise design technique are discussed in Chapter 9. It is recommended to review Chapter 9 before moving on to this chapter. Various types of RIAA equalization and phono amplifier design examples are discussed below.

10.2 RIAA equalization

A phono amplifier performs three major tasks: high voltage gain, low noise, and accurate RIAA equalization. A gain of 40dB at 1kHz is generally required for a moving magnet (MM) phono amplifier. However, since a moving coil (MC) cartridge produces output –20 to –30dB lower than an MM cartridge, the gain for an MC phono amplifier has to be around 60dB–70dB.

Figure 10.1 shows the standard RIAA equalization curve from dc to 100kHz. The curve is normalized to 0dB at dc. This curve has three inflection points at 50Hz (f_1), 500Hz (f_2), and 2122Hz (f_3). They correspond to three time constants: 3180μs, 318μs, and 75μs. The RIAA equalization

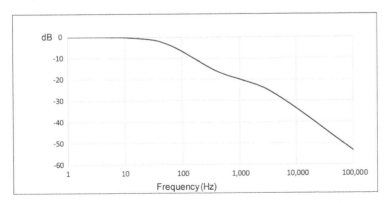

Figure 10.1 The standard RIAA equalization curve corresponds to three time constants: 3180μs, 318μs, and 75μs

DOI: 10.4324/9781003369462-10

curve starts at its maximum gain from dc. Above 50Hz, the gain starts to roll off at –20dB/decade until the 500Hz breakpoint frequency. The gain remains relative flat until the 2122Hz breakpoint frequency. Again, the gain rolls off at –20dB/decade and continues to high frequency.

If we examine the data of the RIAA equalization curve in Figure 10.1 closely, the amplitude at 1kHz is –19.9dB ($\approx$ 20 dB) below dc gain. At 20kHz, the amplitude is –39.5dB below dc gain or –19.6dB below the gain at 1kHz. When the gain of a phono amplifier is said to be 40dB, it is referred to as the gain at 1kHz. Therefore, the dc gain is equal to 40dB + 19.9dB = 59.9dB, while the gain at 20kHz is equal to 40dB – 19.6dB = 20.4dB. On the other hand, the three breakpoint frequencies are related to the time constant T, such that

$$f = \frac{1}{2\pi T} \tag{10.1}$$

Therefore, we have

$$f_1 = \frac{1}{2\pi \cdot 3180 \cdot 10^{-6}} = 50 \text{Hz} \tag{10.2}$$

$$f_2 = \frac{1}{2\pi \cdot 318 \cdot 10^{-6}} = 500 \text{Hz} \tag{10.3}$$

$$f_3 = \frac{1}{2\pi \cdot 75 \cdot 10^{-6}} = 2122 \text{Hz} \tag{10.4}$$

Types of RIAA equalization

There are several approaches [1–5] for implementing the RIAA equalization as shown in Figure 10.2. The conventional one op-amp RIAA equalization is shown in Figure 10.2(a). The

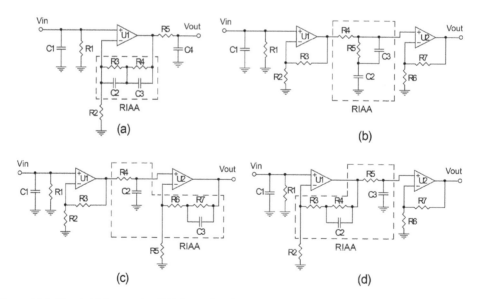

Figure 10.2 Circuit (a) is an active RIAA equalization. Circuit (b) is a passive RIAA equalization. Circuit (c) is a passive-active RIAA equalization. Circuit (d) is an active-passive equalization.

three frequency breakpoints are set by the network formed by R3, R4, C2, and C3. Resistor R2 determines the gain. Since this is a non-inverted amplifier configuration, the gain will never go below unity. Therefore, a low-pass filter formed by R5/C4 is added to the circuit so that the gain will continue to roll off at high frequency. This is a simple RIAA equalization using just one op-amp and it is a very popular phono amplifier design. It is good for setting a gain around 30dB–35dB at 1kHz. It will be a stretch for 40dB gain, as the dc gain becomes 60dB. Imagine that an op-amp having 120dB open-loop gain has only 60dB left at the dc to low frequency region to enforce the RIAA equalization and bring down the distortion. Distortion will be inevitably compromised at low frequency.

When the number of op-amps is increased to two, we have more flexibility to set up the gain while achieving high accuracy for RIAA equalization and low distortion. Figure 10.2(b) to (d) shows three different approaches. Figure 10.2(b) is a passively equalized RIAA phono amplifier. The equalizing network, which contains R4, R5, C2, and C3, is placed between the two op-amps. As the equalizing network has an attenuation of –20dB at 1kHz, the gain for the first amplifier and the second amplifier can be set to 30dB so that overall total 40dB gain is achieved. This arrangement can easily allow us to achieve a 40dB gain at 1kHz requirement for an MM phono amplifier. To improve the SNR, op-amps can be paralleled in the first amplifier. Figure 10.2(b) can be turned into a 60dB gain MC phono amplifier. Since the first amplifier provides a flat gain, the cartridge's high frequency gain signals get through the first amplifier before rolling off by the RIAA network. This arrangement has a compromise of overload margin.

Figure 10.2(c) and (d) are two semi-passive (or semi-active) RIAA equalizations. A passive-active RIAA equalization is shown in Figure 10.2(c). The 2122Hz roll-off (time constant 75 µs) is realized passively by the R4 and C2. Network R6, R7, and C3 are used to realize the other two time constants, 318 and 3180µs, forming an active equalization for the second amplifier. Since the cartridge's signals with a frequency lower than 2122Hz get through the first amplifier before rolling off in the second amplifier, this arrangement also has a compromise of overload margin.

Figure 10.2(d) is an active-passive RIAA equalization. It is a reverse arrangement of Figure 10.2(c). Network R3, R4, and C2 are used to realize the time constants, 318 and 3180µs, forming an active equalization for the first amplifier. The 2122Hz roll-off (time constant 75 µs) is realized passively by the R5 and C3 network. Again, this arrangement also has a compromise of overload margin.

For passive and semi-passive RIAA equalization, we have to pay attention when setting the gain for the first and second amplifier so as to maximize the overload margin. For example, when setting the gain for passive RIAA equalization in Figure 10.2(b), the first and second amplifier can be set to 30dB gain. Given the 20dB attenuation by the RIAA equalization, the MM phono amplifier achieves a gain of 40dB. On the other hand, since the output of the MC cartridge is around 20dB lower than that of an MM cartridge, the two amplifiers must overcome this. Again, if Figure 10.2(b) is used for realizing an MC phono amplifier of 60dB gain, the first and second amplifiers can be set to 40dB. And if an MC phono amplifier of 70dB gain is needed, both amplifiers can be set to 45dB. When a phono amplifier has a gain 60dB–70dB, noise could be an issue. Fortunately, we can parallel op-amps or transistors of a composite op-amp for the first amplifier. It is well known that when the number of amplifiers (or transistors in the input stage of an amplifier) is doubled, the SNR will be improved. When a phono amplifier is carefully designed, it should have accurate RIAA equalization, low noise, low distortion, and sufficient overload margin.

10.3   MM phono amplifiers

An MM cartridge is connected to a phono amplifier, as shown in Figure 10.3. The ac equivalent circuit of an MM cartridge can be modeled by means of a resistor Rc and an inductor Lc. The value for Rc and Lc varies between different phono cartridge manufacturers. Rc often ranges from 500Ω to 800Ω, while it is 300mH to 500mH for Lc. In the phono amplifier's side, capacitor CL and resistor RL shunt the input. RL is often a 47kΩ resistor that provides damping for the MM cartridge. Capacitor CL resonates with the inductance Lc and the correct C1 will extend the frequency response of the cartridge. CL includes the capacitance of the interconnecting cable and input stage of the phono amplifier. For best performance, follow the recommendations from the cartridge manufacturer. For MM phono amplifiers to be discussed in this section, a 100pF capacitor and 47kΩ resistor are often used in the input stage of the amplifier. However, note that emphasis is placed on the phono amplifier circuit design rather than optimizing a particular cartridge model. The active devices employed in the phono amplifiers include vacuum tube, op-amp, discrete transistor, and semi-discrete devices. RIAA equalization includes active, passive, and semi-active.

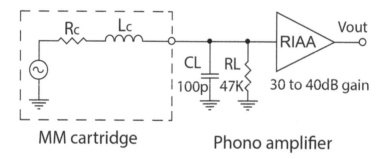

Figure 10.3 An MM cartridge connected to an MM phono amplifier, which has a gain of 30–40dB at 1kHz

Vacuum tube MM phono amplifier

Figure 10.4 shows a classic vacuum tube MM phono amplifier. It is a three-stage amplifier configuration, similar to that of the line stage amplifier of Figure 8.1 discussed in Chapter 8. Here, a conventional active RIAA equalization is used. The RIAA equalizing network contains two resistors (R1 and R2) and two capacitors (C1 and C2). A detailed discussion of how to determine the three breakpoint frequencies required by an RIAA equalization will be given later when we discuss the active RIAA MM phono amplifier design.

Resistor R0 determines the overall gain at 1kHz. Since vacuum tubes produce relatively higher noise than solid-state devices, the gain is usually set below 35dB. As shown in Figure 10.4, the cathode resistor in vacuum tube T1 is partially by-passed. This allows us to set the overall gain by feedback while maintaining a high voltage gain for the input stage. On the other hand, if we want to reduce noise, one more 12AX7 can be connected, paralleling to T1. This is perhaps the best we can do for this design. T3 is in a cathode follower configuration that provides a low output impedance.

Figure 10.5 shows a passive RIAA equalization using three 6SL7 twin triodes. The input stage is an SRPP amplifier, which has a relatively low output impedance and good current drive

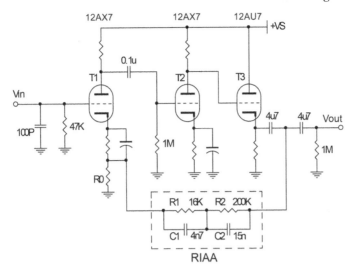

Figure 10.4 An MM phono amplifier formed by a three-stage vacuum tube configuration with active RIAA equalization

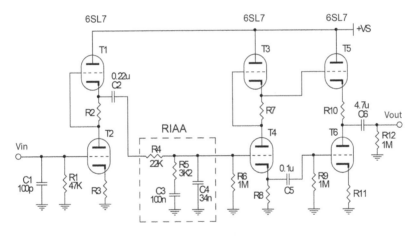

Figure 10.5 An MM phono amplifier formed by an SRPP amplifier for the first stage followed by a passive RIAA equalization and Tai Chi amplifier

capability. The gain of an SRPP amplifier is approximately equal to half of the tube's amplifying factor (μ). The second stage is a Tai Chi amplifier, which is discussed in Chapter 8. It also produces a gain similar to an SRPP amplifier. For 6SL7, μ is 70. Therefore, the gain of the SRPP and Tai Chi amplifiers is expected to be around 30 (29.5dB) each. The combined gain of the two amplifiers is about 59dB. After subtracting 20dB loss from the RIAA equalization network, the gain for this phono amplifier could be 39dB. In reality, the actual gain for this circuit will be less than 39dB but more than 35dB at 1kHz.

Early discrete MM phono amplifier

Figure 10.6 shows a three-transistor MM phono amplifier with active RIAA equalization, commonly found in the 1960s. As usual, C1 and R1 are chosen to meet the cartridge manufacturer's recommendation in order to provide the best damping to the cartridge's inductance and flatten the frequency response. Since BJT has low input impedance, in order to achieve a 47kΩ impedance at the input, R1 is often chosen to be greater than 47kΩ. When R1 is paralleled with the amplifier's input impedance, the effective input impedance will be close to the desired 47kΩ.

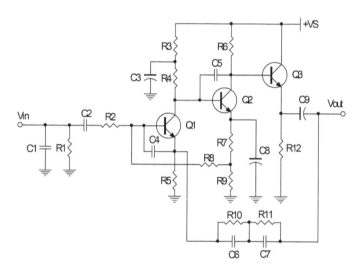

Figure 10.6 An MM phono amplifier formed by an SRPP amplifier for the first stage followed by a passive RIAA equalization and Tai Chi amplifier

Transistors Q1 and Q2 are arranged in a common emitter configuration. They provide the necessary open-loop gain for the entire phono amplifier. The emitter resistor for Q2 is formed by R7 and R9. They are bypassed by C8 so as to produce a high open-loop gain. The RIAA equalizing network comprises R10, R11, C6, and C7. The output stage is a simple emitter follower configuration. It provides a low output impedance and it also prevents the output from loading down the second stage. On the other hand, in addition to the ac feedback through the RIAA equalizing network, transistor Q1 also needs a dc bias network to properly set up a desired dc quiescent current. The dc bias network is formed by R7, R8, and R9. It should be noted that because R7 and R9 are bypassed by C8, no ac feedback can be established between Q1 and Q2 via resistor R8, which may otherwise reduce the open-loop gain of the amplifier.

Capacitors C4 and C5 are frequency compensation capacitors that stabilize the amplifier at high frequency. They do not affect the RIAA equalization, since there are only two transistors providing the open-loop gain, which is not very high. Thus, the distortion will be compromised. Op-amps, which typically have a 120dB open-loop gain, will greatly improve the performance.

Active RIAA MM phono amplifier

Figure 10.7 shows a conventional MM phono amplifier employing an op-amp and active RIAA equalization [3]. The gain is 35dB at 1kHz. Since the open-loop gain of an op-amp is typically around 120dB, this phono amplifier will definitely produce lower distortion than the three-transistor phono amplifier in Figure 10.6. The RIAA equalizing network is formed by components R3, R4, C3, and C4. Their values are not unique. Many combinations are possible, but they must satisfy the conditions for realizing the three breakpoint frequencies ($f_1 = 50$Hz, $f_2 = 500$Hz and $f_3 = 2122$Hz). For example, if we consider the resistance total (R3 + R4 = 216kΩ) rather high, which generates a considerable amount of thermal noise, smaller resistors can be used, but, at the same time, C3 and C4 have to be changed accordingly. In the following, we express the transfer function of this phono amplifier in terms of the components. Thus, resistors R3 and R4 and capacitors C3 and C4 can be chosen freely, as long as they satisfy the expression for the three breakpoint frequencies.

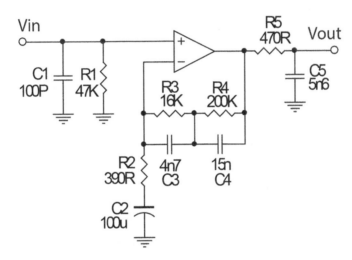

Figure 10.7 A conventional MM phono amplifier employing an op-amp and active RIAA equalization

In network theory, the impedance (or reactance) of a capacitor C is written as $Z_c = 1/sC$, where s is a complex variable. See Appendix B for an overview of complex variable and transfer functions. When the phono amplifier is simplified as shown in Figure 10.8, the transfer function is given as follows:

$$H(s) \equiv \frac{Vout(s)}{Vin(s)} = 1 + \frac{Z}{R2}$$

where $Z = \left(R3 \parallel \dfrac{1}{sC3}\right) + \left(R4 \parallel \dfrac{1}{sC4}\right)$.

After simplifying the expression for Z, we have

$$H(s) = 1 + \left(\frac{1}{R2}\right)\left(\frac{R3}{1+sR3C3} + \frac{R4}{1+sR4C4}\right)$$

$$= 1 + \frac{C3+C4}{R2C3C4} \cdot \frac{\left(\dfrac{R3+R4}{R3R4(C3+C4)} + s\right)}{\left(\dfrac{1}{R3C3}+s\right)\left(\dfrac{1}{R4C4}+s\right)} \qquad (10.5)$$

The zero of the numerator determines the second breakpoint frequency (f_2). For C3 = 4.7nF, C4 = 15nF, R3 = 16kΩ and R4 = 200kΩ, we have

$$|s| = |j\omega| = 2\pi f = \frac{R3+R4}{R3R4(C3+C4)}, \text{ leading to } f = 545\text{Hz}, \text{ (exact } f_2 = 500\text{Hz)}.$$

The first pole of the denominator determines the third breakpoint frequency (f_3),

$$|s| = |j\omega| = 2\pi f = \frac{1}{R3C3}, \text{ leading to } f = 2116\text{Hz}, \text{ (exact } f_3 = 2122\text{Hz)}.$$

The second pole of the denominator determines the first breakpoint frequency (f_1),

$$|s| = |j\omega| = 2\pi f = \frac{1}{R4C4}, \text{ leading to } f = 53\text{Hz}, \text{ (exact } f_1 = 50\text{Hz)}.$$

The three breakpoint frequencies are off from the exact values by 9%. Using 1% E-96 series resistors and 5% tolerance capacitors, connecting the components in series and parallel to obtain a practical component value closer to the calculated value may help to improve the RIAA equalization accuracy. On the other hand, since this is a non-inverted amplifier configuration, the gain will never go below unity. Therefore, a low-pass filter formed by R5/C5 is added to the circuit so that the gain will continue to roll off at 60kHz. Figure 10.7 is a simple RIAA equalization using just one op-amp and it is also a very popular phono amplifier design. It is good for setting a gain around 30dB–35dB at 1kHz. It will be a stretch for 40dB gain, as the gain at dc is 60dB. Imagine that an op-amp having a 120dB open-loop gain has only 60dB left at the low frequency region to bring down the distortion. Thus, distortion will be compromised at low frequency.

Given the transfer function of Eq. (10.5), the amplitude of the phono amplifier can be determined analytically. It can be shown that the amplitude of the phono amplifier of Figure 10.8 is given as

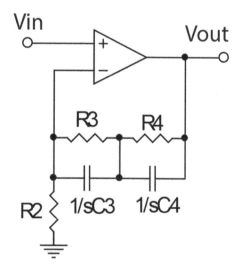

Figure 10.8 This is a simplified circuit for calculating the transfer function of the MM phono amplifier

$$A(\omega) \equiv |H(j\omega)| = 1 + \left(\frac{C3 + C4}{R2C3C4}\right)\frac{\sqrt{a^2 + \omega^2}}{\sqrt{(b^2 + \omega^2)(c^2 + \omega^2)}} \tag{10.6}$$

where

$$a = \frac{R3 + R4}{R3R4(C3 + C4)}, \quad b = \frac{1}{R3C3}, \quad c = \frac{1}{R4C4}, \quad \omega = 2\pi f \tag{10.7}$$

A plot of $A(\omega)$ versus frequency gives the phono amplifier's frequency response, which is identical to Figure 10.1 except that the three breaking frequencies are slightly off from the exact values.

Semi-passive RIAA MM phono amplifier

Figure 10.9 is a semi-passive RIAA equalizing MM phono amplifier [3]. It contains two non-inverted amplifiers. The overall gain is 36dB at 1kHz. The first amplifier is implemented by op-amp U1 realizing two breakpoint frequencies, f_1 and f_2. The third breakpoint frequency f_3 is realized by a passive network formed by R5 and C5. The second amplifier is implemented by op-amp U2, producing a flat gain of 10dB. A dc servo circuit is not used in Figure 10.9(a). Instead, it employs C3 to reduce the dc gain of the first amplifier to unity and a dc blocking capacitor C4 to block the first amplifier's output dc offset voltage. Alternatively, a dc servo circuit is used in Figure 10.9(b) to eliminate output dc offset. Thus, capacitors C3 and C4 are removed.

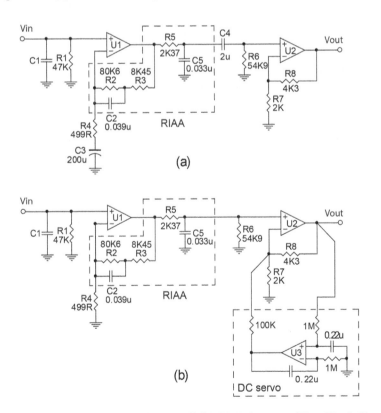

Figure 10.9 Circuit (a) is a semi-passive RIAA equalizing MM phono amplifier. Circuit (b) employs a dc servo so that capacitors C3 and C4 can be removed

If the first amplifier is simplified, as shown in Figure 10.10, the transfer function can be found as follows:

$$H_1(s) = 1 + \frac{Z}{R4}, \text{where } Z = \left(R2 \| \frac{1}{sC2}\right) + R3$$

This, after simplifying, yields

$$H_1(s) = 1 + \frac{R3}{R4} \frac{\left(\dfrac{R2+R3}{R2R3C2} + s\right)}{\left(\dfrac{1}{R2C2} + s\right)} \tag{10.8}$$

The zero of the numerator determines the second breakpoint frequency (f_2). For C2 = 0.039μF, R2 = 80.6kΩ and R3 = 8.45kΩ, we have

$$|s| = |j\omega| = 2\pi f = \frac{R2+R3}{R2R3C2}, \text{ leading to } f = 533\text{Hz, (exact } f_2 = 500\text{Hz).}$$

The pole of the denominator determines the first breaking frequency (f_1),

$$|s| = |j\omega| = 2\pi f = \frac{1}{R2C2}, \text{ leading to } f = 50.6\text{Hz, (exact } f_1 = 50\text{Hz).}$$

The third breakpoint frequency is determined by R5 and C5. It is given as

$$|s| = |j\omega| = 2\pi f = \frac{1}{R5C5}, \text{ leading to } f = 2035\text{Hz, (exact } f_3 = 2122\text{Hz).}$$

The three breakpoint frequencies are off from the exact values by 6.6%. Given the transfer function in Eq. (10.8), the amplitude of the phono amplifier of Figure 10.9(b) can be expressed in the following:

$$A(\omega) = |H(j\omega)| = A_1(\omega) \cdot A_2(\omega) \cdot A_3(\omega) \tag{10.9}$$

where

$$A_1(\omega) = 1 + \frac{R3}{R4} \frac{\sqrt{a^2 + \omega^2}}{\sqrt{b^2 + \omega^2}} \tag{10.10}$$

$$A_2(\omega) = \frac{c}{\sqrt{c^2 + \omega^2}} \tag{10.11}$$

$$A_3(\omega) = 1 + \frac{R8}{R7} \tag{10.12}$$

$$a = \frac{R2+R3}{R2R3C2}, \; b = \frac{1}{R2C2}, \; c = \frac{1}{R5C5}, \; \omega = 2\pi f \tag{10.13}$$

$A_1(\omega)$ = amplitude of the first amplifier, $A_2(\omega)$ = amplitude of the of the low-pass filter formed by R5 and C5, $A_3(\omega)$ = amplitude of the second amplifier. A plot of $A(\omega)$ versus frequency gives the frequency response of the phono amplifier.

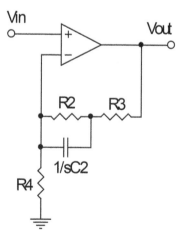

Figure 10.10 A simplified circuit for determining the transfer function of the first amplifier in Figure 10.9(b)

The overall gain of this phono amplifier is 35dB at 1kHz. If a 40dB gain is needed, the gain of the second amplifier is increased from 10dB to 15dB. This can be easily realized by reducing the resistor R7 from 2kΩ to 931Ω. Note that the first and second amplifiers are non-inverted. The gain never goes down to zero at high frequency. Thus, the same output low-pass filter in Figure 10.7 (470R and 5n6) should also be implemented in Figure 10.9.

Passive RIAA MM phono amplifier

Figure 10.11 is a passive RIAA equalizing MM phono amplifier [4]. It contains two non-inverted amplifiers. The first amplifier is implemented by op-amp U1 producing a flat gain of 27.8dB. The second amplifier is implemented by op-amp U2 producing a flat gain of 32dB. Given a –20dB attenuation from the RIAA equalizing network, the overall gain of this phono amplifier is 40dB at 1kHz. The RIAA equalizing network is formed by R4, R5, C2, and C3, which is used to realize the three breakpoint frequencies f_1, f_2, and f_3. A dc servo circuit is not used in Figure 10.11(a). Instead, an output coupling capacitor, C4, is used to block output dc offset voltage. Alternatively, a dc servo circuit is used to eliminate the output dc offset in Figure 10.11(b). Thus, the output coupling capacitor C4 is removed.

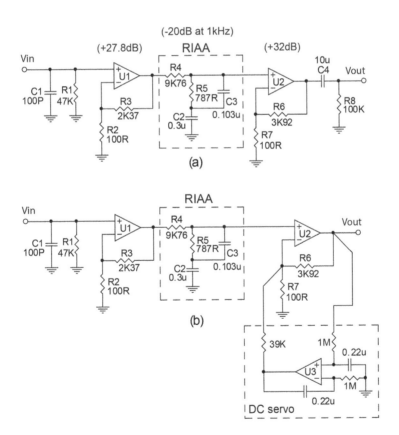

Figure 10.11 Circuit (a) is a passive RIAA equalizing MM phono amplifier. Circuit (b) employs a dc servo so that output coupling capacitor C4 is removed

If the passive RIAA equalizing network is redrawn, as shown in Figure 10.12, the transfer function of the RIAA network can be determined as follows:

$$H_2(s) \equiv \frac{V2(s)}{V1(s)} = \frac{Z}{R4 + Z} \tag{10.14}$$

where $Z = \left(R5 \| \dfrac{1}{sC3}\right) + \dfrac{1}{sC2}$, which, after simplifying, yields

$$Z = \frac{sR5(C2 + C3) + 1}{sC2(1 + sR5C3)} \tag{10.15}$$

Substituting Eq. (10.15) for Eq. (10.14), we have

$$H_2(s) = \frac{sR5(C2 + C3) + 1}{s^2 R4R5C2C3 + s(R4C2 + R5C2 + R5C3) + 1}$$

$$= \frac{R5(C2 + C3)\left[s + \dfrac{1}{R5(C2 + C3)}\right]}{as^2 + bs + c} \tag{10.16}$$

where

$$a = R4R5C2C3, \ b = R4C2 + R5C2 + R5C3, \ c = 1 \tag{10.17}$$

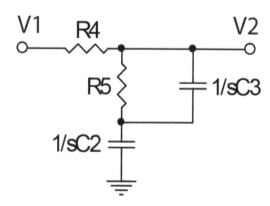

Figure 10.12 A passive RIAA network taken from Figure 10.11 for determining transfer function.

The denominator of Eq. (10.16) is a quadratic equation that has two solutions, which are the poles representing two breakpoint frequencies, 50Hz and 2122Hz. For R4 = 9.76kΩ, R5 = 787Ω, C2 = 0.3μF, C3 = 0.103μF, they are given as follows.

First pole, $s_{p1} = \dfrac{-b + \sqrt{b^2 - 4ac}}{2a}$, so that $|s| = 2\pi f$, leading to f = 50Hz (exact 50Hz).

Second pole, $s_{p2} = \dfrac{-b - \sqrt{b^2 - 4ac}}{2a}$, leading to f = 2126Hz (exact 2122Hz).

The zero of the transfer function gives

$$s = \frac{1}{R5(C2 + C3)}, \text{ so that } |s| = 2\pi f, \text{ leading to } f = 502Hz \text{ (exact 500Hz)}.$$

The three breakpoint frequencies are off from the exact values by only 0.4%. It is clear that passive RIAA equalizing produces breakpoint frequencies closer to the exact values than the active and semi-active RIAA equalization. Given the transfer function of the passive RIAA equalizing network in Eq. (10.16), the amplitude of the phono amplifier of Figure 10.11(b) can be expressed in the following:

$$A(\omega) = A_1(\omega) \cdot A_2(\omega) \cdot A_3(\omega) \tag{10.18}$$

where

$$A_1(\omega) = 1 + \frac{R3}{R2} \tag{10.19}$$

$$A_2(\omega) \equiv |H_2(\omega)| = \frac{(C2 + C3)}{R4C2C3} \frac{\sqrt{d^2 + \omega^2}}{\sqrt{[(s_{p1})^2 + \omega^2)] \cdot [(s_{p2})^2 + \omega^2)]}} \tag{10.20}$$

$$A_3(\omega) = 1 + \frac{R6}{R7} \tag{10.21}$$

$$d = \frac{1}{R5(C2 + C3)}, \omega = 2\pi f \tag{10.22}$$

$A_1(\omega)$ = amplitude of the first amplifier, $A_2(\omega)$ = amplitude of the RIAA equalizing network, $A_3(\omega)$ = amplitude of the second amplifier. Note that the first and second amplifiers are non-inverted. The gain never goes down to zero at high frequency. Thus, the same output low-pass filter in Figure 10.7 (470R and 5n6) should also be implemented in Figure 10.11.

Low noise passive RIAA MM phono amplifier

Figure 10.13 shows a simplified circuit of a low noise MM phono amplifier with passive equalization. It has three selectable gains for 35, 40, and 45dB at 1kHz. The configuration

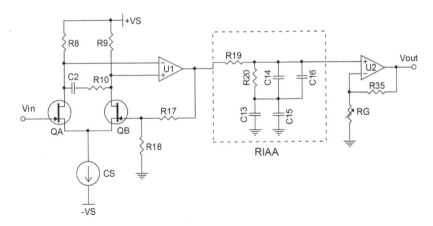

Figure 10.13 A simplified passive RIAA MM phono amplifier with selectable gains

of this phono amplifier is similar to that illustrated in Figure 10.11, in that the first and second amplifiers provide a flat gain. The RIAA equalizing network is placed between the two amplifiers. However, there are several key differences between the two designs. The differences include (i) a low noise composite op-amp is used for the first amplifier, (ii) the gain is selectable for three different levels, and (iii) the output of the dc servo (not shown here yet) is now fed to the first amplifier, while in Figure 10.11, the output of the dc servo is fed to the second amplifier.

Since JFET has much a lower input equivalent noise current density (I_n) than BJT, JFET is a good choice for using in the input stage of an MM phono amplifier. If the input equivalent noise voltage density (E_n) of the JFET is lower than the op-amp, the overall input noise voltage density of the composite op-amp is lower than that of a single op-amp. As a matter of fact, the differential amplifier configuration creates more noise than a single transistor. Before we start reducing the noise, we actually create a little more noise. However, when a sufficient number of transistors are paralleled, the overall SNR is improved. In this MM phono amplifier, six JFETs are paralleled for each side of the differential amplifier, as shown in Figure 10.14. One suggestion is to match the JFETs, if a through-hole package is used, for best performance. See Appendix C for an overview of matching JFETs. The selected JFET in this case is a low cost 2SK209 by Toshiba. It is a surface mounted device (SMD) in an SOT-23 package.

If we look at the composite amplifier in Figure 10.14, transistors Q1–Q6 are paralleled in the input side, while Q7–Q12 in the other side provide the feedback signal. Transistors Q1–Q6

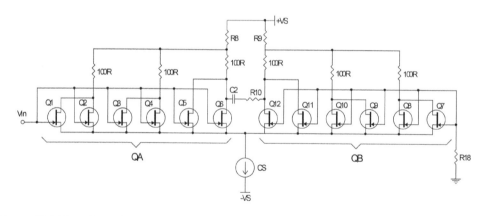

Figure 10.14 Transistors QA and QB are realized by paralleling six pairs of JFETs

are divided into three groups: Q1 and Q2, Q3 and Q4, Q5 and Q6. Each group has two transistors connected in parallel. Then a 100Ω resistor is connected in series to the drain of the two combined transistors before connecting to the load resistor R8. This 100Ω resistor serves two purposes. The first is to help in splitting the dc bias current equally among the three groups. Second, in case there is a group not working properly, by measuring the voltage across the 100Ω resistor, it helps to trouble shoot which group is causing the problem. Here, a total of six 100Ω is used. Ideally, each transistor should connect to an individual 100Ω. By measuring the voltage across each of the twelve 100Ω resistors, we know immediately which transistor is causing a problem, if any. Since the cost of resistors is low, twelve resistors should not be an issue at all.

However, the PCB layout design could be a challenge. Two JFETs sharing with one 100Ω is a good compromise. It has been found that the use of six 100Ω resistors works very well.

The complete schematic of the MM phono amplifier is shown in Figure 10.15. In this phono amplifier circuit, each JFET carries 1mA drain current. Therefore, the total combined tail current from the two cascode current sources (Q13/Q14 and Q15/Q16) is 12mA. If the JFETs in each of the groups are operating correctly, there is 0.2V across each 100Ω resistor. In order to minimize the PCB size as well as noise, a four-layer PCB is used. As shown in Figure 10.16, components are soldered in the top layer. The ground plane and the negative power supply plane (−16.5V) are located in the second layer and third layer, respectively. The bottom layer does not contain components.

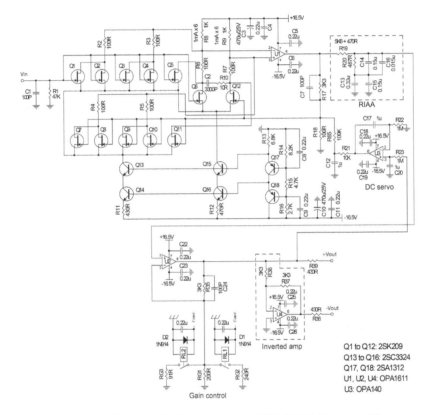

Figure 10.15 The complete schematic of a low noise passive RIAA MM phono amplifier

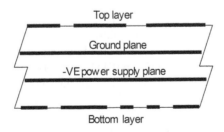

Figure 10.16 A four-layer PCB is used for the phono amplifier shown in Figure 10.15

2SK2145 is a dual version of 2SK209. 2SK2145 is in the SOT-25 package, which is just 0.3mm wider than SOT-23 for 2SK209. If 2SK2145 is used, the differential amplifier will contain only six pieces of 2SK2145. The PCB size for the differential amplifier can be further reduced. Another advantage of using JFET dual is that the two JFETs are more closely matched in electrical properties than two discrete JFETs. Therefore, the dc bias current is more evenly split by the dual JFETs. The 2SK209 and 2SK2145 are low cost JFETs. As can be seen in the next section, for an MC phono amplifier that requires very low noise and high gain, low noise JFETs such as 2SK170 (single) and LSK389 (dual) should be used.

The overall gain of the phono amplifier is determined by the combined gain of the composite amplifier with op-amp U1, the RIAA equalizing network, and the non-inverted amplifier with op-amp U2. The composite amplifier, which comprises Q1–Q12 and op-amp U1, produces a flat gain of 30.6dB, which is determined by resistors R17 and R18. The non-inverted amplifier with op-amp U2 has a switchable gain from 25, 30 to 35dB, determined by resistor R35 and the combined resistances of RG1, RG2, and RG3. For instance, if RL1 and RL2 relays are not switched on, the gain of the second composite op-amp configuration is equal to $(1 + R35/RG1)$ that is 17.5 or 24.8dB. When relay RL1 is switched on, RG1 is paralleled with RG2, giving a combined resistance of 109Ω. Therefore, the gain becomes 31.2 or 29.9dB. Similarly, when RL2 relay is switched on, the gain becomes 53.7 or 34.6dB. When a –20dB attenuation from the RIAA equalization network is taken into account, the three switchable gains at 1kHz for unbalanced output become 35, 40 and 45dB, as shown in Figure 10.17. The distortion is less than 0.01% for all three selectable gains and SNR better than 80dB.

Overload margin is an index that measures the overload headroom of a phono amplifier. It is the maximum input voltage (in rms) at 1kHz before the distortion reaches 1%, i.e., just below where the output starts to clip. The input overload measurements for this phono amplifier vary with gains: 172mV at 35dB gain, 99mV at 40dB gain, and 61mV at 45dB gain. Sometimes, the overload margin is expressed in terms of decibels. For example, if a 5mV MM cartridge is amplified by this phono amplifier at 35dB gain, the overload margin becomes 20log(172/5) = 30.7dB. If cartridges of 3.5mV and 2mV are

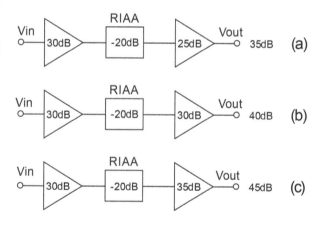

Figure 10.17 Three selectable gains are implemented for the phono amplifier in Figure 10.15

used with the gain of 40dB and 45dB, the overload margins are 29 and 29.6dB, respectively.

The passive RIAA equalization network follows the same arrangement in Figure 10.11 but a different set of component values is used. It should be noted that the series resistor, R19 in Figure 10.15, is equal to 5.6kΩ + 470 Ω = 6.07kΩ, which is smaller than the 9.76kΩ resistor used in Figure 10.11. Even though this series resistor does not appear in the input of the first amplifier, which might have a higher impact, but in the input of the second amplifier, a smaller series resistor always generates less thermal noise.

Several capacitors are used for high frequency compensation that stabilize the amplifier at high frequency. The capacitors are C2, C7, and C24. It turns out that a 10Ω resistor (R10) in series to a 3300pF (C2), which forms a lead-lag compensation, is a very effective way to prevent parasitic oscillation from this phono amplifier.

As usual, a 47kΩ resistor is connected to the input of the phono amplifier so that an MM cartridge is properly damped. In addition, a 100pF capacitor, C1, is connected to the input so as to flatten and to extend the cartridge's frequency response. It should be noted that the total input capacitance of this phono amplifier is a sum of C1 and input capacitance, C_{iss}, of Q1–Q6. For 2SK209, C_{iss} is 13pF. Therefore, the effective input capacitance of the phono amplifier is 100pF + 6×13pF = 178pF. If taking into account the capacitance of an interconnect cable between the cartridge and phono amplifier, the total capacitance seen by the cartridge is about 200pF. If 200pF is what the cartridge manufacturer recommends, keep 100pF for C1. If the recommended capacitance is less than 200pF, C1 capacitance should be reduced accordingly. For example, if the recommended capacitance is just 100pF, remove C1 from the circuit.

Fully discrete MM phono amplifier

Except for vacuum tube phono amplifiers, the MM phono amplifiers discussed above are realized by only op-amps or a semi-discrete approach, i.e., a mix of discrete transistors and op-amps. Figure 10.18 suggests a fully discrete approach to an MM phono amplifier design. A complementary JFET differential amplifier (Q1–Q4) is employed for the input stage. If needed, more JFETs can be paralleled to the differential amplifier for improving the overall SNR. The second stage is a simple complementary common emitter amplifier configuration (Q5 and Q6). Emitter resistors for Q5 and Q6 are partially bypassed so that the desired dc bias current is maintained while a high open-loop gain can be achieved. The output stage is a fixed-bias diamond buffer with unity gain.

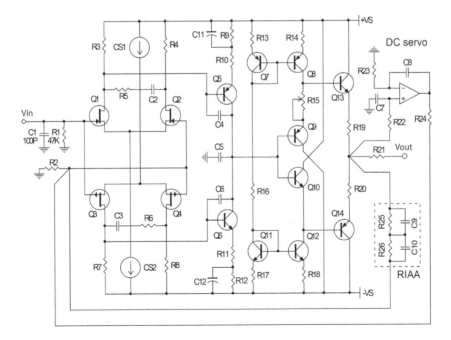

Figure 10.18 This is a potential approach to a fully discrete MM phono amplifier design with active RIAA equalization

The phono amplifier is actively equalized by the RIAA network (R25, R26, C9, C10). The overall gain is determined by resistor R2. Details of determining the three breakpoints frequencies are similar to those for the active RIAA MM phono amplifier discussed earlier. A dc servo circuit is employed to minimize the dc output offset voltage.

If through-hole components are used, the following transistors may be considered: 2SK170 for Q1 and Q2, 2SJ74 for Q3 and Q4, BC550 or 2SC2240 for NPN transistors, and BC560 or 2SA970 for PNP transistors. If surface mounted devices are used: LSK170 for Q1 and Q2, LSJ74 for Q3 and Q4, 2SC3324 for NPN transistors, 2SA1312 for PNP transistors.

If passive RIAA equalization is needed, it is necessary to use two identical amplifiers. These two amplifiers are set to a flat gain of 30dB while a passive RIAA network is placed between them. This will give a fully discrete passive RIAA equalizing MM phono amplifier for 40dB gain at 1kHz. On the other hand, a fully discrete two-stage amplifier can be used to realize a high performance MM/MC amplifier without global feedback [2].

One of the advantages of employing fully discrete components for a phono amplifier is that we can choose the best available components and set the most appropriate dc quiescent conditions for the transistors. This will lead to a level of performance that op-amps or a semi-discrete approach may not be able to achieve. Another advantage is that a high dc power supply voltage can be used. If an op-amp is operated with dual ±15V power supply, the maximum output voltage is around 13V(peak). For fully discrete components, the dc power supply can be easily raised to ±30V or even higher. If the power supply voltage is doubled, the overload margin will be increased by a factor of 6dB.

10.4   MC phono amplifiers

Moving coil (MC) cartridges are often considered to have a better tracking performance than MM cartridges because the moving element is a lightweight coil rather than a magnet, which is made from a heavier alloy. Because the coil must be lightweight and has only a few turns, the output from a MC cartridge is low, often in the range of 0.1mV to 0.5mV. MM cartridges, on the other hand, produce output of around 5mV. Therefore, an MC cartridge's output is at least 20dB lower than that of an MM cartridge. Thus, the gain for an MC phono amplifier has to be at least 20dB greater than MM phono amplifier. The gain of a MC phono amplifier commonly ranges from 60dB to 70dB. At such a high gain, noise may become an issue. But fortunately, the impedance of an MC cartridge is low. It typically ranges from 1Ω to 100Ω. This makes designing an MC phono amplifier with acceptable levels of SNR possible.

Generally speaking, there are two approaches for designing a 60dB gain phono amplifier. The first is to add a step-up transformer, or step-up amplifier, producing 20dB gain in front of a 40dB gain MM phono amplifier. In this approach, we can keep using a good sounding MM phono amplifier, which may be using vacuum tubes, while adding a step-up transformer or amplifier to boost an extra 20dB gain. The other approach is to design a 60dB gain MC phono amplifier right from the beginning. The second approach generally leads to an MC phono amplifier that produces lower noise than that of a 20dB step-up transformer or amplifier and an MM phono amplifier combination. Both approaches are discussed below.

Low noise step-up amplifier

Figure 10.19(a) shows a step-up transformer that is used to boost the required 20 to 30dB gain. Since the signal level is low, the size and weight of the transformer is not a primary concern. But the transformer is susceptible to hum interference, so that good immunity is absolute necessary.

Manufacturers often provide a costly mu-metal shielding can, sometimes two nested cans, in order to achieve this. On the other hand, it is well known that transformers do not have the best frequency response, especially in the low frequency region. Therefore, a step-up amplifier is an attractive alternative, as shown in Figure 10.19(b).

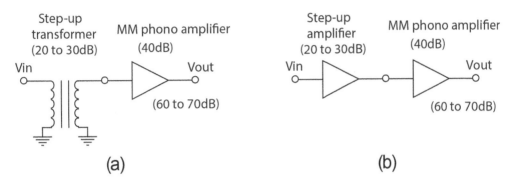

Figure 10.19 Circuit (a) uses a step-up transformer while circuit (b) uses a step-up amplifier

A low noise step-up amplifier is shown in Figure 10.20. This step-up amplifier is also employing a composite op-amp configuration. There are six pairs of differential amplifiers that are paralleled. In order to achieve the lowest possible noise, each differential amplifier is employed with its own current source, which is a low noise cascode current source arrangement. Low noise JFET (LSK389) is chosen for the differential amplifiers. Since LSK389 is in a dual JFET package, six pieces of LSK389 are needed for implementing the differential amplifier input stage. Each JFET (Q1–Q12) is biased at 1mA drain current. Therefore, each cascode current source sets a tail current of 2mA. The low impedance emitters of transistors Q25 and Q26 provide a stable dc potential for each of the cascode current sources.

Note that the differential amplifiers of the composite amplifier in Figure 10.20 are arranged differently from Figure 10.15. Here, in Figure 10.20, each differential amplifier has its own current source, while in the differential amplifier in Figure 10.15 this is not the case. Therefore, every group of two JFETs in the differential amplifier of Figure 10.15 share a 100Ω resistor. As discussed before, this 100Ω resistor helps to split the drain current equally in every group. It also helps to troubleshoot by measuring the dc voltage across the 100Ω resistor. However, in Figure 10.20, 100Ω resistors are no longer needed, as each differential amplifier has its own current source. By measuring the voltage drop across the 1.3kΩ emitter resistor at each of the current sources, R11–R16, we can easily to find out if any differential amplifier is not working properly.

The gain of this step-up amplifier is determined by (1 + R36/RG). In order to achieve low thermal noise, resistance for R36 and RG must be low. However, they cannot be too low, as that may overload the op-amp. As a compromise, 390Ω is chosen for R36. For setting the gain of 20, 25, and 30dB, RG is 39Ω, 22Ω, and 12Ω, respectively.

Components R60 (10Ω) and C60 (3300pF) form a lead-lag frequency compensation. Capacitor C7 is also a frequency compensation capacitor that helps to stabilize the amplifier at high frequency.

A dc servo is added in this step-up amplifier so that the output dc offset voltage can be eliminated and, therefore, an output dc blocking capacitor is not needed. Resistor R65 is a high resistor compared to RG, so as to avoid the loading effect to RG. On rare occasions, the $V_{GS(th)}$ of JFETs Q1–Q12 may be so different that even the dc servo circuit cannot bring

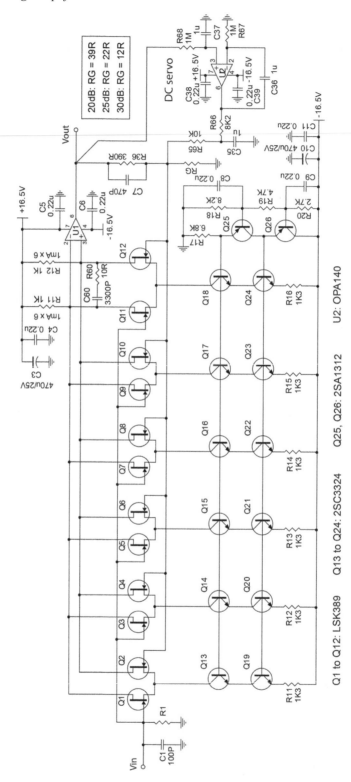

Figure 10.20 A low noise step-up amplifier is suitable for providing a flat gain from 20 to 30dB

the output dc offset down to ground level. If this happens, R66 should be reduced from 8.2kΩ to a lower value.

Passive RIAA MC phono amplifier

Figure 10.21 is a 60dB gain passive RIAA equalizing MC phono amplifier. It is modified from the MM phono amplifier of Figure 10.11(b). The same arrangement is applied to Figure 10.21. The passive RIAA equalizing network is placed between the two amplifiers that have a flat attenuation. Since RIAA equalization has a −20dB attenuation at 1kHz, both amplifiers produce a flat 40dB gain so that the overall 60dB gain is achieved.

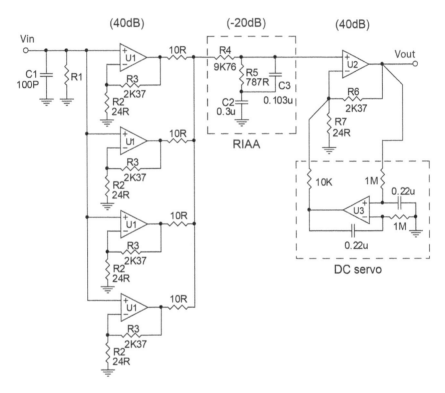

Figure 10.21 A passive RIAA equalizing MC phono amplifier with 60dB gain at 1kHz. The input stage has four identical amplifiers in parallel with a 40dB gain

Gain is increased and so is the noise, if the same amplifier is used. In order to improve the SNR, the number of op-amps used in the first amplifier is increased to four. It is well known that there is an improvement in SNR as a result of doubling the number of amplifying devices. When four amplifiers are now paralleled, there is a greater improvement than is the case with just one amplifier. Each amplifier contains an op-amp U1, resistors R2 and R3. An additional 10Ω resistor is added at the output of each op-amp so that no op-amp can dominate. In this design, ultra-low noise op-amps such as LT1028, LT1038, and LT1115 are recommended. They have a 1nV/$\sqrt{Hz}$ input noise voltage density at 10Hz. (See Table 9.3.)

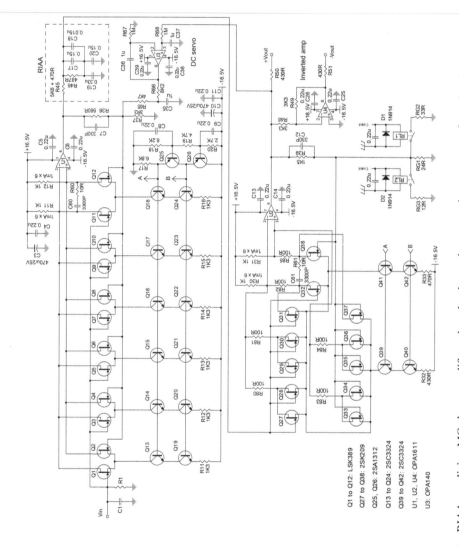

Figure 10.22 A passive RIAA equalizing MC phono amplifier using ultra low noise composite op-amp configuration for the first amplifier. A dc servo is employed so that the output coupling capacitor is eliminated. Unbalanced and balanced outputs are produced

Passive RIAA semi-discrete MC phono amplifier

Figure 10.22 shows a passive RIAA semi-discrete MC phono amplifier with three selectable gains for 60, 65, and 70dB. This phono amplifier employs two composite op-amps. The first composite op-amp comprises transistors Q1–Q26 and op-amp U1. This configuration has a similar arrangement to the step-up amplifier of Figure 10.20. There are six pairs of differential amplifiers formed by Q1–Q12. Low noise JFET dual (LSK389) is used for each of the differential amplifiers. And each differential amplifier has its own current source formed by a cascode current source arrangement with 2mA tail current. Therefore, each JFET is biased to 1mA drain current. The gain of the first composite op-amp is given by (1+ R36/R37). It produces a gain of 170, or 44.6dB.

The second composite op-amp is formed by transistors Q27–Q42 and op-amp U2. This composite op-amp configuration has a similar arrangement to the passive RIAA MM phono amplifier of Figure 10.15. In this composite op-amp configuration, there is only one differential amplifier employing six JFETs in parallel for the non-inverted input, while another six JFETs in parallel operate for the inverted input. In order to avoid running the current source too warm, two cascode current sources (Q13/Q14 and Q15/Q16) are used instead of just one. The two current sources set up a total of tail current 12mA. Unlike the first composite op-amp using ultra-low noise JFET LSK389, the low cost 2SK209 will be sufficient for the second composite op-amp. A total of six pairs of 2SK209 is connected in parallel for the second composite op-amp.

The gain of the phono amplifier is determined by the combined gain of the two composite op-amps and the RIAA equalizing network. The gain of the first composite op-amp is 170, or 44.6dB. The second composite op-amp has a switchable gain from 36, 40.7 to 45.5dB determined by resistor R39 and the combined resistances of RG1, RG2, and RG3. For instance, if RL1 and RL2 relays are not switched on, the gain of the second composite op-amp configuration is equal to (1 + R39/RG1) = 63.5 or 36dB. When relay RL1 is switched on, RG1 is paralleled with RG2, giving a combined resistance of 13.9Ω. Therefore, the gain becomes 108.9 or 40.7dB. Similarly, when RL2 relay is switched on, the gain becomes 188.5 or 45.5dB. When the RIAA equalization network –20dB attenuation is taken into account, the three switchable gains at 1kHz for unbalanced output are approximately equal to 60, 65, and 70dB, as shown in Figure 10.23. The distortion is less than 0.01% for all three selectable gains and the SNR is better than 70dB at 1V output. The overload margins for the gain of 60, 65, and 70dB are 10.5mV, 6.3mV, and 3.6mV, respectively.

A dc servo circuit is employed in Figure 10.22. Since the output from the dc servo is now fed to a small 3.3Ω resistor (R37), resistors R65 and R66 cannot be too high. Otherwise, the potential divider formed by R37, R65, and R66 will not convey sufficient dc voltage from the output of the dc servo to the inverted input of the differential amplifier, base of transistors Q2, Q4, Q6, Q8, Q10, and Q12. Hence, the dc servo circuit may fail to take down the output dc offset voltage of the MC phono

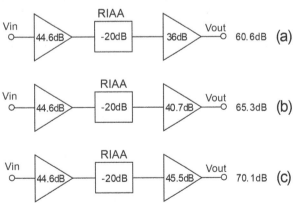

Figure 10.23 Three selectable gains are implemented in the phono amplifier of Figure 10.22

amplifier. R65 (4.7kΩ) and R66 (8.2kΩ) are sufficiently low to work well in most cases. In situations where JFETs have large differences in $V_{GS(th)}$ and I_{DSS}, causing a high dc output offset voltage, then resistor R66 must be reduced to below 8.2kΩ accordingly.

Note that the input capacitance (C_{iss}) of LSK389 is 25pF. When six JFETs are connected in parallel, the input capacitance becomes 6×25pF = 150pF. The capacitance seen from the MC cartridge is the sum of capacitance from the interconnect cable, C1, and 150pF. For example, if the recommended capacitance for the MC cartridge is 200pF, C1 can be a small capacitor around 20 to 30pF. On the other hand, R1 is ranging from 1Ω to a few hundred ohms. Simply follow the recommendation from the cartridge's manufacturer to determine C1 and R1.

As noise is our primary concern, it is recommended to use a four-layer PCB for implementing the MC phono amplifier of Figure 10.22. The arrangement of the four layers in the PCB is the same as in Figure 10.16. Even though the component count is higher than the MM phono amplifier of Figure 10.15, if the PCB layout is carefully designed, a two-channel MC phono amplifier can be fitted into a PCB of the dimension 111×182mm or even smaller. Figure 10.24 shows a photo of a fully assembled PCB for the MC phono amplifier.

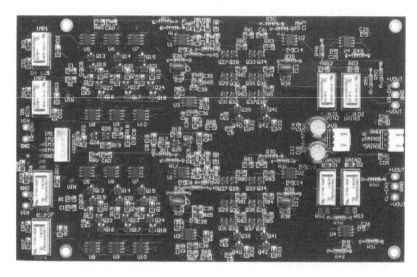

Figure 10.24 Photo of a two-channel MC phono stage amplifier, which contains two MC phono amplifiers for two channels (stereo) of Figure 10.22. (Dimension: 111×182mm)

Low cost MM + MC phono amplifier

The MM phono amplifier shown in Figure 10.15 and the MC phono amplifier in Figure 10.22 are two stand-alone amplifiers for MM and MC cartridges, respectively. The configuration of the two phono amplifiers is similar. Both make use of a complex composite op-amp in the first amplifier and a passive RIAA equalization. The key differences between the MM and the MC phono amplifier are voltage gains and the arrangement of the composite op-amp in the first amplifier. For the MC phono amplifier in Figure 10.22, if we carefully rearrange the composite op-amp in the first amplifier and allow the first and second amplifiers to switch to different gains, it becomes a phono amplifier that is switchable between MM and MC cartridges. Such an MM + MC phono amplifier is shown in Figure 10.25.

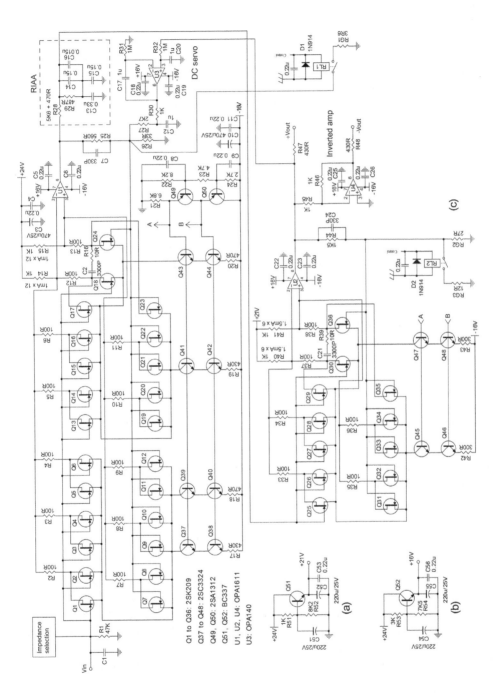

Figure 10.25 A passive RIAA equalized MM + MC phono with three selectable gains: 40, 60, and 70dB

Figure 10.25 follows a configuration similar to the MC phono amplifier shown in Figure 10.22. However, the first composite op-amp now contains a differential amplifier formed by 24 JFETs. The arrangement of this composite op-amp is similar to that in Figure 10.14, except 12 pairs of JFETs are now used. Since the equivalent input noise voltage density of 2SK209 is not as low as LSK389, we have to use many of them in parallel. Fortunately, 2SK209 is an inexpensive device. At 100pcs, the unit price for 2SK209 is $0.23. 24pcs of 2SK209 only cost $5.52. Another choice is 2SK2145, which is a dual version of 2SK209. At 100pcs, the unit price for 2SK2145 is $0.35. A total of 12pcs of 2SK2145 is needed to realize the differential amplifier and the cost is $4.20. The cost is low and, more importantly, the dual JFETs inside a 2SK2145 are more likely to be matched than two discrete 2SK209. As a result, the tail currents produced by the four cascode current sources (total 24mA) are more equally split among the JFETs so that each JFET carries approximately 1mA of drain current.

The MM + MC phono amplifier of Figure 10.25 has three selectable gains. They depend on the combined gains set by the first and second composite op-amps. The gain of the first composite op-amp configuration is determined by (1 + R25/R26). If relay RL1 is not activated, the gain is 25dB. When relay RL1 is activated, the effective resistance for R26 becomes R26||RG1 = 3.25Ω. Thus, the gain of the first composite op-amp configuration becomes 45dB. Similarly, the gain of the second composite op-amp configuration is given by (1 + R44/RG2). If relay RL2 is not activated, the gain is 35dB. When relay RL2 is activated, the effective resistance for RG2 becomes RG2||RG3 = 8.3Ω. Thus, the gain of the second composite op-amp configuration is 45dB. If −20dB attenuation of the RIAA equalization is taken into account, the selectable gains of the MM + MC phono amplifier are 40, 60, and 70dB, as shown in Figure 10.26. The distortion is less than 0.01% for all three selectable gains and SNR is around 70dB for 60dB gain and SNR greater than 80dB for 40dB gain at 1V output. The overload margins for the gain of 40, 60, and 70dB are 112mV, 11.4mV, and 3.6mV, respectively.

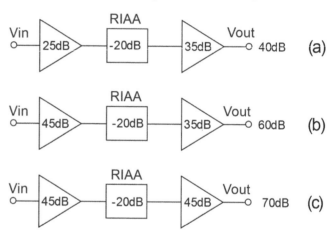

Figure 10.26 There are three selectable gains for the MM + MC phono amplifier in Figure 10.25

Since this phono amplifier can be used for MM and MC cartridges, the input impedance and capacitance are very different as far as equalizing the frequency response of the cartridge is concerned. First of all, the total input capacitances (C_{iss}) of 12pcs of 2SK209 is equal to 12 × 13pF = 156pF. If we take into account the capacitance of the interconnect cable between the cartridge and the phono amplifier, 20 to 30pF will be sufficient for C1 if a total 200pF is recommended by the cartridge manufacturer. On the other hand, 47kΩ is chosen for R1, which is commonly required by MM cartridges. For MC cartridge, the input impedance must be set to a much lower value. Figure 10.27 shows five selectable input impedances for MC cartridges: 10, 33, 100, 250, and 500Ω. These values cover most of the input impedances recommended by MC cartridge manufacturers. Other values can be easily added to the circuit by adding more relays or switches. Figure 10.28 shows the photo of the PCB for a two-channel (stereo) MM + MC phono amplifier, including the input impedance and gain switching relays. Again, a four-layer PCB is used. The dimension of the PCB is 111×188mm. DC power supplies are +24V and −16V. Since +21V and +16V are also needed, Figures 10.26(a) and (b) show how these two voltages are created by the +24V power supply.

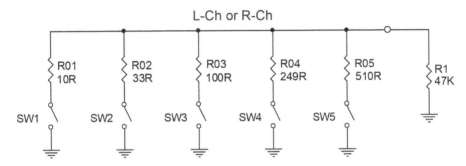

Figure 10.27 Input impedance selection for the MM/MC phono amplifier illustrated in Figure 10.25

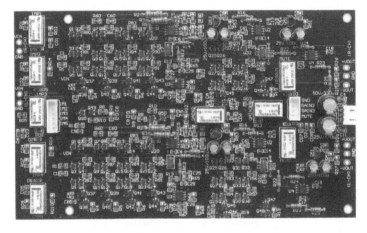

Figure 10.28 A photo of a two–channel MM /MC phono-stage amplifier, which contains the two phono amplifiers shown in Figure 10.25. (Dimension: 111×188mm)

Fully discrete MC phono amplifier

Figure 10.29 shows a potential approach to an active RIAA equalizing fully discrete MC phono amplifier design. It has the same configuration of the discrete MM phono amplifier shown in Figure 10.18. But this MC phono amplifier has a lower noise input and a second stage. The input stage has three complementary differential amplifiers connected in parallel, and each of them has a separate current source. A total of 12 BJTs is used, Q1–Q12. It should be noted that BJTs are used in this MC phono amplifier design. If JFET is preferred, the BJTs can be replaced by JFETs of your choice. The second stage is a complementary common emitter push–pull ampli-fier configuration. Two PNPs are connected in parallel (Q13, Q14) on the top while two NPNs in parallel (Q15, Q16) are on the bottom. This is the basic arrangement for the input and second

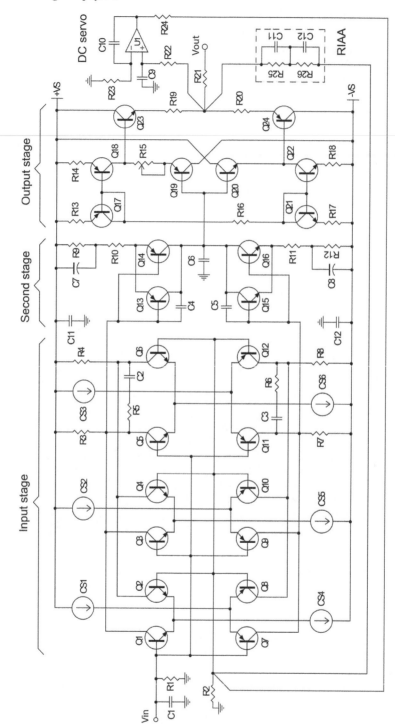

Figure 10.29 A potential approach to a fully discrete MC phono amplifier design with active RIAA equalization

stage. If ultra-low noise is needed, more differential amplifiers and transistors can be connected in parallel to the input and second stage.

This MC phono amplifier configuration is best for active RIAA equalization. The gain can be set to 60dB at 1kHz. A dual ±24V dc power supply is suitable for this amplifier. If the power supply is less than dual ±24V, there may be a compromise in overload margin. Transistor SSM2212 (dual NPNs) is suitable for Q1–Q6 and SSM2220 (dual PNPs) for Q7–Q12. Since the V_{CEO} for both devices is 36V, the dc power supply can be as high as dual ±30V. As there is a dc voltage across the collector resistors R3, R4, R7, and R8, the V_{CE} is much lower than the power supply voltage. For the rest of the transistors, 2SC3324 can be used for the NPNs and 2SA1312 for the PNPs. The V_{CEO} of these two transistors is 120V.

If passive RIAA equalization is needed, we must use a two-amplifier configuration. Figure 10.29 can be used for the first amplifier and Figure 10.18 for the second amplifier. The passive RIAA equalizing network is placed between them. When the gain of the phono amplifier is split properly by the two amplifiers, 70dB gain is achievable. For example, a flat 45dB gain is set for both amplifiers. After a –20dB attenuation from the passive RIAA equalizing network, the overall gain becomes 70dB (at 1kHz). Given a high dc power supply, ±24V or higher, the overload margin of Figure 10.29 should be better than the MC phono amplifier illustrated in Figure 10.22 and Figure 10.25, where the op-amps are operating at dual ±16V power supplies.

We can see that the input stage composite differential amplifier in Figure 10.29 is different from the other differential amplifier configurations that are used in the phono amplifiers discussed earlier. It is worth putting them together for a quick comparison, as shown in Figure 10.30. It should be noted that even though JFETs are shown, BJTs can be used also. Figure 10.30(a) is a composite differential amplifier that has six JFETs connected in parallel for the non-inverted input while another six JFETs are in parallel for the inverted input. The composite differential amplifier has only one (but can be more) current source setting the tail current. This composite differential amplifier configuration has the least component count and it allows the JFETs to be closely packed together so as to save PCB space. As a consequence, the size of the PCB implementing the composite differential amplifier of Figure 10.30(a) should be the smallest among them.

The composite differential amplifier of Figure 10.30(b) has six separate differential amplifiers connected in parallel. Again, a total of 12 transistors is used. However, each differential amplifier has its own current source setting tail current. Thus, the tail current is likely to be equally split between the two transistors of each differential amplifier. In addition, these six current sources are independent and, therefore, the noises that are created by the current sources are uncorrelated. As a consequence, the differential amplifier of Figure 10.30(b) will be likely to produce less noise than the composition shown in Figure 10.30(a).

The composite differential amplifier of Figure 10.30(c) has three complementary differential amplifiers in parallel. Six pieces of N-channel JFETs and six pieces of P-channel JFETs are used. The total number of JFETs is also 12. Each complementary differential amplifier uses two current sources and, therefore, six current sources are also needed, which is similar to Figure 10.30(b). Therefore, the complementary differential amplifier will also achieve a similar noise performance. On the other hand, since complementary design tends to produce low second harmonic distortion, a phono amplifier that makes use of the composite differential amplifier of Figure 10.30(c) may be expected to produce lower distortion.

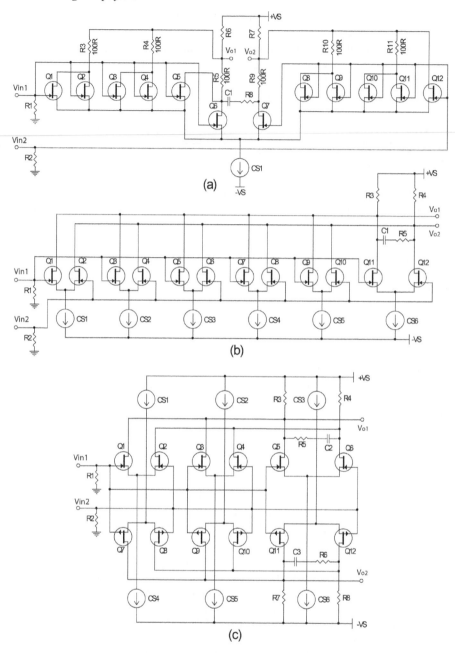

Figure 10.30 Circuit (a) is a composite differential amplifier with six JFETs connected in parallel for the non-inverted input and another six JFETs in parallel for the inverted input. Circuit (b) is a composite differential amplifier formed by six differential amplifiers connected in parallel. Circuit (c) is a composite differential amplifier with three differential amplifiers (N-JFET) connected in parallel on the top while three differential amplifiers (P-JFET) are in parallel on the bottom

10.5 Fully balanced phono amplifier

Figure 10.31 shows a potential approach to a fully balanced MM phono amplifier with passive RIAA equalization. The configuration of this phono amplifier is similar to the unbalanced version shown in Figure 10.11. Instead of using two simple non-inverted amplifiers in the unbalanced version, two balanced amplifiers are now used. Both balanced amplifiers are adopted from the well-known instrumentation amplifier configuration. (See Chapter 4 for details.) The first balanced amplifier is formed by op-amps U1 and U2. It produces a gain equal to $(1 + 2 \times R5/R4)$. The second balanced amplifier is formed by op-amps U3 and U4. It produces a gain equal to $(1 + 2 \times R7/R6)$. Cc is a small frequency compensation capacitor to stabilize the amplifier at high frequency.

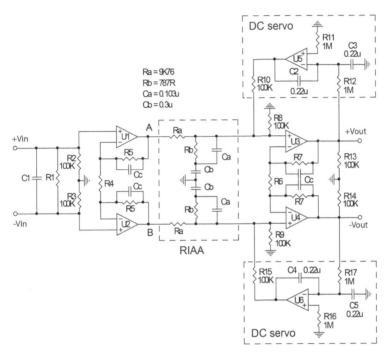

Figure 10.31 A potential approach to a fully balanced MM phono amplifier design with passive RIAA equalization

Since this fully balanced phono amplifier has to deal with balanced signals with two opposing phases, dual RIAA equalizing networks are needed. They are placed between the two balanced amplifiers. Two dc servo circuits are also needed to eliminate dc offset voltages at both outputs. Given the configuration in Figure 10.31, it may be suitable for implementing a fully balanced MM phono amplifier. If a flat 30dB gain is set for both balanced amplifiers, and a −20dB attenuation from the RIAA equalizing network is taken into account, the overall gain of the MM phono amplifier is 40dB (at 1kHz). As required by the MM phono cartridge, R1 is set to 47kΩ. It is also needed for this fully balanced phono amplifier. However, R2 and R3 are connected in the input to provide the necessary dc bias for the op-amps U1 and U2. They produce a loading effect to R1 and, therefore, R1 has to be higher than 47kΩ. For example, if R2 and R3 are 100kΩ each, a total of 200kΩ is loading R1. In order to achieve 47kΩ impedance, as seen from the inputs, R1 must be 62kΩ.

If the configuration of Figure 10.31 is used to implement an MC phono amplifier with 60 to 70dB gain, noise generated from the first balanced amplifier (formed by op-amps U1 and U2)

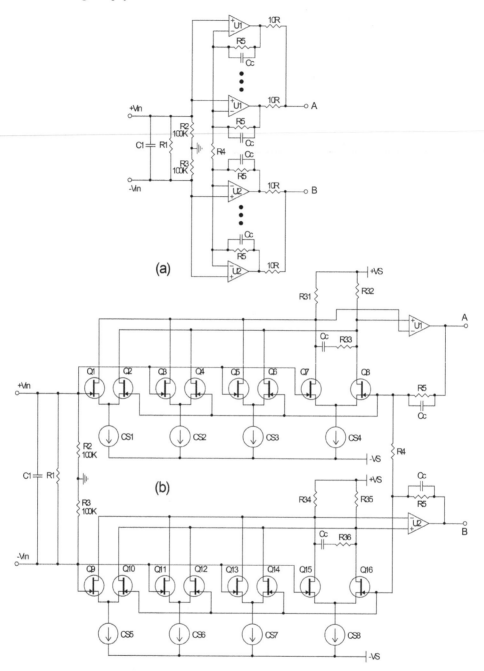

Figure 10.32 Two low noise amplifier configurations that can be used as the input stage for the fully balanced phono amplifier of Figure 10.31. Circuit (a) is a low noise balanced amplifier configuration realized by op-amps. Circuit (b) is a low noise balanced amplifier configuration realized by two composite op-amps

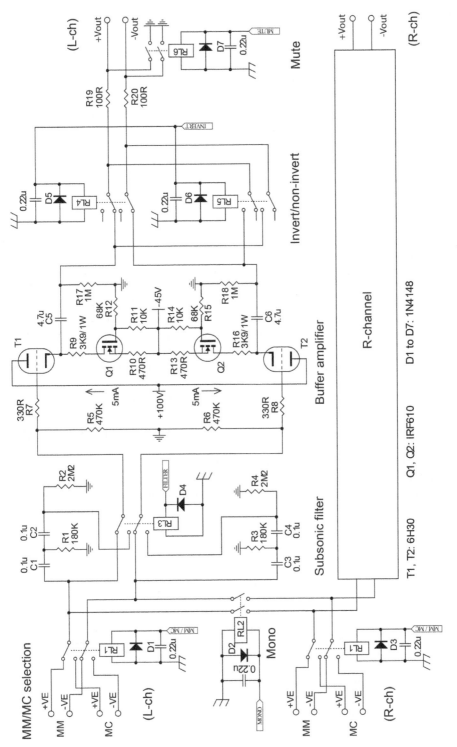

Figure 10.33 A circuit has switches to select MC/MM, mono/stereo, subsonic filter, vacuum tube buffer amplifier, invert/non-invert, and mute

must be taken care of. Figure 10.32 shows two approaches to improve the SNR of the fully balanced phono amplifier. In Figure 10.32(a), several op-amps are connected in parallel. Since the balanced configuration itself already helps to reduce common mode noise, probably we do not need to parallel too many op-amps. Four op-amps for the non-inverted signal and four op-amps for the inverted signal, a total of eight op-amps, may be a good start for the first balanced amplifier. But this should be viewed as the minimum number of op-amps for the first balanced amplifier if low noise is of primary concern.

Figure 10.32(b) is a composite op-amp approach for implementing the input stage balanced amplifier. The composite op-amp comprises eight JFET differential amplifiers (Q1–Q16) and two op-amps, U1 and U2. If LSK389 (JFET dual) is used, Figure 10.32(b) can be used to implement a 70dB gain fully balanced phono amplifier. Instead of JFET, a BJT such as SSM2212 could be also a good choice.

10.6   Subsonic filter and vacuum tube buffer

There are several features that will certainly help to make a phono amplifier more user friendly. These features include a mono, subsonic filter, invert, mute, and vacuum tube buffer stage. All these features are optional and they can be omitted from a phono amplifier altogether. However, if these features are available, it will make life a lot easier for phonograph records playback.

These convenient features are realized by the circuit shown in Figure 10.33. First, assuming that we are using two separate phono amplifiers, an MM phono amplifier as shown in Figure 10.15 and an MC phono amplifier as illustrated by Figure 10.22, the circuit of Figure 10.33 allows us to switch between them via relay RL1. It should be noted that the circuit of Figure 10.33 handles fully balanced signals. Therefore, we are taking the inverted and non-inverted outputs from the MM and MC phono amplifiers.

Mono/stereo function is switchable via relay RL2. When relay RL2 is activated, L-CH and R-CH are merged. However, it should be noted that there is a 430Ω resistor at each of the phono amplifiers' output. Therefore, the 430Ω resistor prevents the outputs of two op-amps from physically shorting together.

A subsonic filter is a second order passive high-pass filter realized by capacitors C1–C4 and resistors R1–R4. The subsonic filter is designed to start rolling off at 20Hz. In order for this subsonic filter to operate without being affected by the output load (i.e., the input impedance of

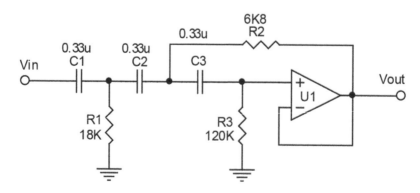

Figure 10.34 A third order Sallen and Key high-pass filter with cut-off frequency at 20Hz

a line-stage amplifier), a buffer amplifier is in place. The buffer amplifier is formed by a 6H30 tube which is biased at a 5mA plane current. If a higher roll-off rate is needed, a third order Sallen and Key high-pass filter, as shown in Figure 10.34, can be used to replace the passive filter. Note that this Sallen and Key filter has a –3dB attenuation at 20Hz.

The invert/non-invert function is implemented by relays RL4 and RL5. When RL4 and RL5 are activated, the outputs swap polarity. Finally, a mute function is also given in the circuit. It is controlled by relay RL6.

10.7 Inverse RIAA equalizing network

Inverse RIAA equalizing network (IREN) is a simple circuit that is useful for phono amplifier designers to verify the RIAA equalization conformance. A reasonably accurate IREN was published by Reg Williamson [6]. It was modified by Lipshitz and Jung [7] for improved accuracy. The improved IREN is shown in Figure 10.35.

If the amplitude of the IREN at 1kHz is normalized to 0dB, the frequency response of the IREN is shown Figure 10.36. The frequency response of the standard RIAA equalization is also shown in Figure 10.36, where the amplitude at 1kHz gain is also normalized to 0dB. It is obvious that the frequency response of the IREN is just the opposite of the standard RIAA equalization. A sum of these two graphs will simply become a flat frequency response.

IREN is a useful tool for verifying the RIAA conformance of a phono amplifier. In addition, it is also a useful gadget to burn in a new phono amplifier. Simply let the output from a CD player first connect to an IREN, then

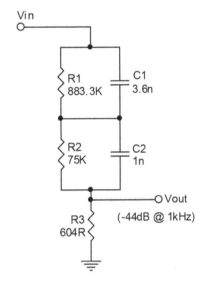

Figure 10.35 An accurate inverse RIAA equalization network by Jung-Lipshitz

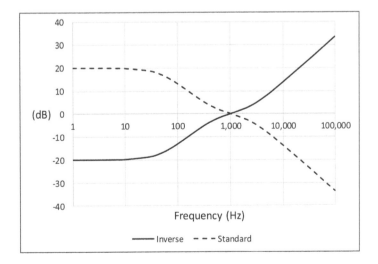

Figure 10.36 Standard and inverse RIAA equalization curves

the output from the IREN feeds to a phono amplifier. IREN is a passive circuit and no dc power supply is required.

10.8　Exercises

Ex 10.1　For the active RIAA equalization in Figure 10.8, determine the values for R3, R4, C3, and C5, such that the total resistance R3 + R4 is less than 100kΩ.

Ex 10.2　The passive RIAA equalizing network used in Figure 10.15 is redrawn in Figure 10.37. Determine the three breakpoint frequencies.

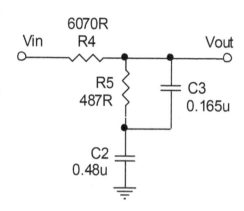

Ex 10.3　An alternative passive RIAA equalizing network is shown in Figure 10.38. Show that the transfer function is given by

$$H(s) \equiv \frac{Vout}{Vin} = \frac{R2C2 \cdot (s + \dfrac{1}{R2C2})}{as^2 + bs + c},$$

where

$a = R1R2C1C2$, $b = R1(C1 + C2) + R2C2$, $c = 1$.

Figure 10.37 A passive RIAA equalizing network for Ex 10.2

Ex 10.4　Show that the transfer function for the inverse RIAA equalizing network in Figure 10.35 is given as

$$H(s) \equiv \frac{Vout}{Vin} = \frac{a \cdot (s + \dfrac{1}{R1C1}) \cdot (s + \dfrac{1}{R2C2})}{as^2 + bs + c},$$

where $a = R1R2R3C1C2$,
$\quad\quad\; b = R1R2(C1 + C2) + R3(R1C1 + R2C2)$,
$\quad\quad\; c = R1 + R2 + R3$.

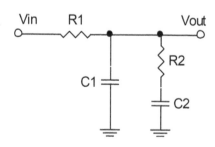

Ex 10.5　In Ex 10.4, the transfer function of the inverse RIAA equalizing network has two zeros in the numerator. These two zeros correspond to the inverse of the two poles from a standard RIAA equalizing network. On the other hand, the denominator of the transfer function has two poles. They are given as

Figure 10.38 An alternative passive RIAA equalizing network for Ex 10.3

$$s_{p1} = \frac{-b + \sqrt{b^2 - 4ac}}{2a}, \; s_{p2} = \frac{-b - \sqrt{b^2 - 4ac}}{2a}.$$

One of the poles corresponds to the inverse of the zero from a standard RIAA equalizing network. The other pole is a high frequency pole that does not affect the frequency response in the audio band. Given the values for R1, R2, C1, and C2, as shown Figure 10.35, show that the two zeros correspond to breakpoint frequencies 50Hz and 2122Hz. The two poles correspond to two breakpoint frequencies 497Hz and 338kHz.

References

[1] Self, D., *Small signal audio design*, 2nd edition, Focal Press, 2015.
[2] Cordell, B., "Vinyltrak – a full featured MM/MC phono preamp," pp. 131–148, Linear Audio, Sept. 2012.
[3] "High-performance audio applications of the LM833," application note AN-346, Texas Instruments.
[4] Jung, W., "Op amp applications," chapter 6, Analog Devices, 2002.
[5] Jung, W., *Audio IC op amp application*, 3rd edition, Howard Sams, 1987.
[6] Williamson, R., Audio Amateur Letters, vol. 2, no. 3, p. 22, 1971.
[7] Lipshitz, S. and Jung, W., "A High Accuracy Inverse RIAA Network", Audio Amateur, 1980.

11 Power amplifiers – vacuum tube

11.1 Introduction

A power amplifier is an audio component that directly drives a loudspeaker to produce sound. Therefore, output power, stability, distortion, and bandwidth are of primary concern. In this chapter, we will study the design of vacuum tube audio power amplifiers, which are regarded as the granddaddy of audio power amplifiers. Even more than a century after the first triode was patented independently by American engineer Lee De Forest and Austrian physicist Robert von Lieben in 1906, vacuum tube power amplifiers are still going strong in the audio market today.

In comparing to solid-state semiconductor power amplifiers, vacuum tube power amplifiers have some obvious shortcomings: they are bulky, have low output power, high distortion, and create high ambient temperature. However, the so-called *tube sound* created by the tubes renders the sound warm and sweet. This makes the sound more appealing to the ears. The sound of vacuum tube power amplifiers notably excels in the mids and highs.

We first discuss a few examples of class A vacuum tube power amplifiers. This is followed by discussion of class AB push–pull type power amplifiers. Types of phase splitters and output power tube dc bias techniques are also explained. Finally, the design of fully balanced vacuum tube power amplifier is discussed in detail.

11.2 Amplifier class (A, B, AB_1, AB_2)

Class A

A class A power amplifier sets the power tubes with the grid dc quiescent point approximately one-half of the cut-off voltage. The input signal never causes the grid-cathode voltage to become positive. In other words, the peak input signal voltage at the grid is limited to a value that does not exceed the dc quiescent voltage. A class A amplifier produces low distortion but the efficiency is as low as 20%. Class A operation is often found in single-ended power amplifiers employing power triodes such as 2A3, 300B, and 211, etc.

Class B

A class B amplifier sets the power tubes with the grid dc quiescent point at or near cut-off. Plate current flows during the positive half of the input grid signal but stops flowing during the negative half-cycle. Class B amplifiers have efficiency as high as 60% but also have high distortion in terms of cross-over distortion. This makes a class B amplifier not the preferred choice for HiFi, but they are designated for public address systems where efficiency is of primary importance.

DOI: 10.4324/9781003369462-11

Class AB₁

The operation of class AB_1 falls midway between class A and class B. The plate current flows for more than 180° but less than 360° of the signal at the grid. Class AB_1 indicates that grid current does not flow during any part of the cycle. With the use of power tetrode and pentode tubes, the efficiency is around 40%, which is twice as much as the class A. Most of the push–pull vacuum tube power amplifiers operate in class AB_1.

Class AB₂

The operation of class AB_2 falls midway between class A and class B. The plate current flows for more than 180° but less than 360° of the signal at the grid. Class AB_2 indicates that grid current is permitted to flow during part of the cycle. The efficiency of class AB_2 is slightly higher than class AB_1. Since grid current is allowed to flow during part of the cycle, the driver stage must be capable of driving a low impedance.

11.3   Single-ended power amplifiers

Figure 11.1 shows a simplified class A power amplifier employing two triodes in a single-ended arrangement. It is a two-stage amplifier configuration. T1 is a small signal triode for the input stage and T2 is a power triode for the output stage. They are both operating in class A. Since the amplifying factor for triode is low, triodes T1 and T2 are working in a common cathode configuration with a capacitor bypassed cathode resistor so as to produce the highest possible voltage gain from the amplifier.

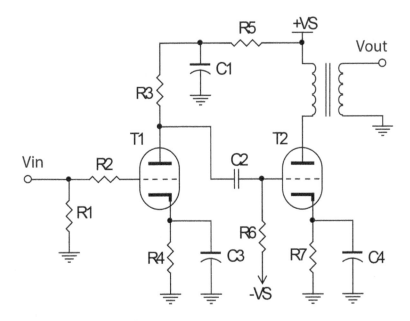

Figure 11.1 A simplified two-stage class A power amplifier using a power triode

The power triode T2 in the second stage is shown with fixed-bias, which has a negative dc supply voltage (–VS) applied to the grid. In this amplifier configuration, R7 is usually a small resistor and, therefore, bypassed capacitor C4 may not be required. However, if self-bias instead of fixed-bias is used, a much higher cathode resistor will be needed. In that situation, a bypass capacitor C4 must be used. If power triode 2A3 is used, the output power is around a few watts. When 300B is used, output power can be around 7W. Since no feedback is used, the distortion and output impedance of the power amplifier is moderately high. A few percentages of distortion near maximum output power is common.

There are two approaches for boosting output power. The first approach is to parallel another power triode on the output stage, as shown in Figure 11.2. The second approach is to use transmitting type power triodes such as 211 and 845. One of these transmitting type triodes can easily deliver 20W output power. Table 11.1 shows a list of some commonly used power triodes for class A audio power amplifiers. Since the power amplifier is operating in class A, a high dc current is flowing through the primary side of the output transformer. In order to prevent the transformer core from magnetic saturation prematurely, C-core transformer is often the preferred choice for single-ended output transformer.

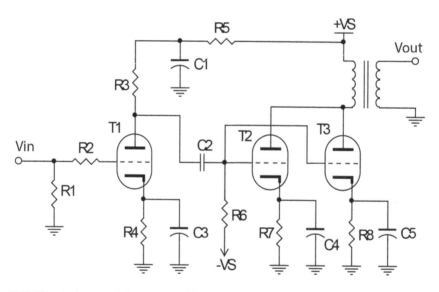

Figure 11.2 Class A single-ended power amplifier with two power triodes in parallel

The reason that C-core (also called cut-core) is often used instead of an E-I transformer is because it is easy to introduce an air gap into a C-core transformer as shown in Figure 11.3. The air gap effectively reduces the magnetic flux flowing in the transformer core generated by the dc current in the primary winding of the output transformer in a single-ended power amplifier. On the other hand, the net dc current flowing in the primary winding of the output transformer in a push–pull power amplifier is nearly zero. Therefore, an air gap is not required. Thus, the E-I transformer, low cost and small size, is often the choice for push–pull power amplifiers.

Table 11.1 A list of representative power triodes. The output power is a typical value for the triode operating in a class A single-ended configuration

Tube	Typical value			Maximum value		Filament				Class-A output power (W)
	Amplifying factor, μ	Transconductance, g_m (mA/V)	Plate impedance, r_p (Ω)	Plate voltage (V)	Plate dissipation (W)	Voltage (V)	Current (A)	Single	Twin	
2A3	4.2	5.25	800	300	15	2.5	2.5	•		3.5
300B	3.9	5	790	400	36	5	1.2	•		7
6AS7G	2	7	280	250	13	6.3	2.5		•	(a)
6C33C	2.7	40	80 to 120	250	> 30W	6.3	6.6	•		(a)
				450	< 30W	12.6	3.2			
211	12	3.15	3800	1250	100	10	3.25	•		20
811	160			1500	65	6.3	4	•		(b)
845	5.3	3.1	1700	1250	100	10	3.25	•		30

Note: (a) 6AS7G and 6C33C were originally designed for series dc regulated power supply application. Due to low plate impedance, they are often found in OTL power amplifiers. (b) 811 was orignially designed for class B, class C and class AB_2 power amplifier application. No class A output power was given in the data sheet.

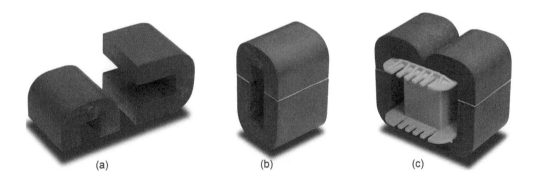

(a)   (b)   (c)

Figure 11.3 C-core silicon oriented transformer core is shown in (a). Two C-core stacks together as shown in (b). Two C-core with a bobbin for winding wires is shown in (c).

In addition to the ease of introducing an air gap, another merit for the C-core transformer is better output efficiency than the E-I transformer. Therefore, a C-core transformer has less winding that produces lower copper loss and lower output dc resistance than an EI transformer. For a high power single-ended audio power amplifier, the output transformer is often provided by a double C-core or even triple C-core.

11.4   Push–pull power amplifiers

The efficiency of a class A single-ended amplifier is around 20%. In other words, if a single-ended amplifier delivers 20W to a load, 80W power is dissipated as loss elsewhere. As mentioned in the preceding section, if additional power triodes are paralleling in a single-ended power amplifier, the output power is increased. Even though output power can be boosted, efficiency remains the same. For example, if the output power is boosted from 20W to 40W by using two power triodes in parallel, the power loss is increased to 160W, giving the same 20% efficiency. At the same time, the size of the output transformer becomes much bigger, as it has to withstand twice the dc plate current.

Figure 11.4 shows the structure of a push–pull amplifier that improves efficiency to around 40% when it is operating in class AB_1 or even higher in class AB_2 mode. In order for a power amplifier to work in a push–pull operation, the power amplifier must contain: (i) a phase splitter, (ii) a pair of output power tubes, and (iii) a push–pull type output transformer. The phase splitter can be placed either in the second stage as shown in Figure 11.4(a) or in the input stage as shown in Figure 11.4(b). The arrangement in Figure 11.4(b) generally requires an extra triode, as the second stage must be a two-triode amplifier. However, the input stage of Figure 11.4(a) can be just a simple single triode amplifier.

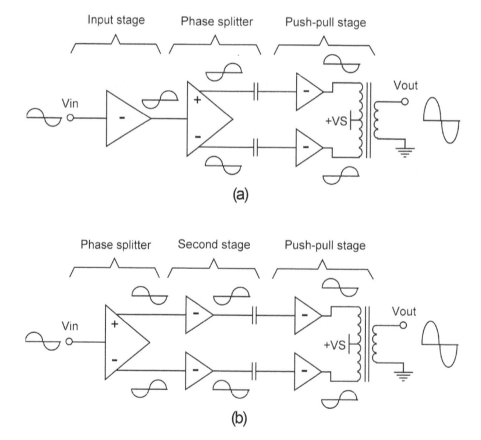

Figure 11.4 A push–pull amplifier contains an amplifying gain stage, a phase splitter and a push–pull output stage. The phase splitter can be placed either in the second stage as shown in circuit (a) or in the input stage as shown in circuit (b)

A phase splitter is an amplifier that amplifies an input signal and produces two 180° out of phase signals. In Figure 11.4, two out of phase signals are passed to the output stage, which contains a pair of power tubes operating in push–pull mode to deliver the output signal. Details of phase splitter and push–pull output stages are discussed in the subsequent sections.

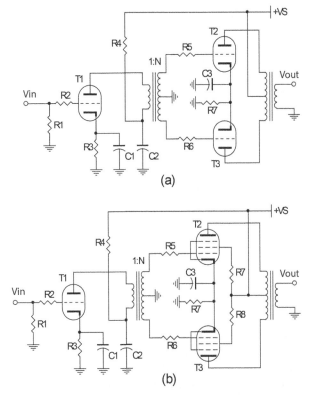

Figure 11.5 shows two examples of a push–pull power amplifier. Both examples use a transformer phase splitter that has a center-tap in the secondary side. The turn-ratio N is much greater than unity so that the transformer phase splitter also produces voltage gain. The input stage is a common cathode amplifier with a bypassed cathode resistor so that a high voltage gain is produced. Therefore, the input stage and phase splitter together produce the necessary voltage gain for the amplifier. Output tubes can be power triode, tetrode, or pentode. In both examples, no feedback is used.

Figure 11.5 Circuit (a) is a push–pull amplifier using a transformer phase splitter and power triodes. Circuit (b) uses power tetrodes or pentode instead of triodes

As shown in Figure 11.5, the output transformer has a center-tap so that the primary winding is equally divided into two halves. The first half is connected to tube T2 while the second half is connected to tube T3. Given the fact that tubes T2 and T3 see two different out of phase input signals, the output transformer is so constructed that it sums up the difference of the two signals. Assume that tube T2 produces an output signal Va, then tube T3 produces an output signal –Va, which is 180° out of phase. The push–pull operation of the output transformer works to produce an output signal proportional to the difference of the two signals, i.e., Va – (–Va) = 2Va. This is twice the output signal of a single triode. Therefore, the efficiency of a push–pull amplifier is an improvement over a single-ended amplifier.

As shown in Figure 11.6(a), there is always a net dc current flowing in the primary winding of an output transformer for a single-ended power amplifier. The net dc current magnetizes the transformer iron core and it may cause magnetic flux saturation prematurely when dealing with high output current. This limits the amplifier's output power and frequency response. In order to counter this problem, a large transformer core is used and an air gap is introduced in the transformer core. As a result, a class A single-ended power amplifier must have a bulky power supply transformer (due to low efficiency) and output transformer (due to dc magnetizing current).

However, in a push–pull power amplifier, since there are two identical, but opposite, dc currents flowing in the primary winding of the output transformer, the net dc current is zero, or

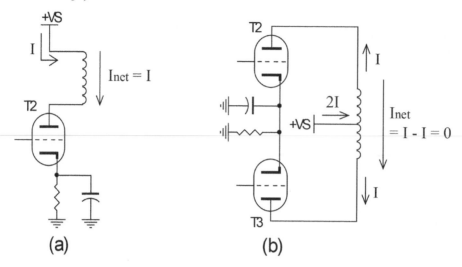

Figure 11.6 Circuit (a) shows that a net dc current flows in the primary winding of an output transformer for a single-ended power amplifier. Circuit (b) shows that no net dc current flows in the primary winding of an output transformer for a push–pull power amplifier

nearly zero when the plate current for the two power tubes is well adjusted. This is shown in Figure 11.6(b). When there is no net dc current in the primary winding, the size of the output transformer can be reduced for the same output power and frequency response. Therefore, it is no surprise to see that a 10W single-ended power amplifier is often heavier and bigger than a 20W push–pull power amplifier.

11.5   Phase splitters

Figure 11.7 shows the input and output relationship of a phase splitter. The primary function of a phase splitter is to amplify an input signal (Vin) with amplitude A1, and to produce two outputs with opposite phase (Vout1 and Vout2) and amplitude A2. An ideal phase splitter should possess the following properties: (i) voltage gain greater than unity (A2 > A1), (ii) identical amplitude for Vout1 and Vout2 ($\Delta = 0$), (iii) exact 180° phase shift between Vout1 and Vout2 (i.e., $\Delta\Phi = 0$),

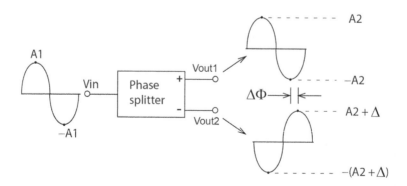

Figure 11.7 A phase splitter amplifies an incoming signal (Vin) and generates two out of phase signals (Vout1 and Vout2)

and (iv) identical output impedance for Vout1 and Vout2. However, in real life, an ideal phase splitter never exists. These ideal properties are always compromised in phase splitter design. The best solution would be to eliminate the phase splitter from the circuit. It turns out that it is indeed possible to eliminate the use of a phase splitter in a fully balanced push–pull power amplifier, which is discussed later. Let us now discuss several commonly used phase splitters.

Split-load phase splitter

Figure 11.8 shows a popular *split-load phase splitter*. When the plate resistor and cathode resistor are equal, the voltage gain for Vout1 with respect to Vin is roughly equal to –1. The negative sign indicates that it is in the opposite phase with respect to the input signal. On the other hand, since Vout2 is taken from the cathode, the voltage gain for Vout2 with respect to Vin is approximately equal to unity. Therefore, both outputs have the same voltage gain but opposite phases.

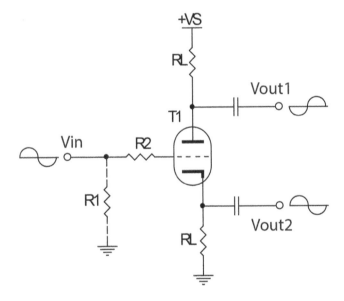

Figure 11.8 This is a split-load phase splitter. Resistor R1 is not needed when the phase splitter is directly coupled to the preceding stage

It is the simplest phase splitter that requires only one triode. However, the split-load phase splitter has only unity gain and produces dissimilar output impedances. This is because a cathode follower produces low output impedance for Vout2 while Vout1 from the plate of T1 has a high output impedance. Another disadvantage for the split-load phase splitter is a low output voltage swing.

A split-load phase splitter is often used in the second stage of a push–pull power amplifier, as shown in Figure 11.9. Because this phase splitter does not produce a voltage gain greater than unity, the total voltage gain of the composite amplifier in Figure 11.9 only comes from the common cathode amplifier of the input stage. In general, the voltage gain for this common cathode amplifier (T1) is not desirable for a push–pull power amplifier that employs global feedback. Therefore, an additional amplifying stage is needed. However, it may defeat the purpose of having a single triode phase splitter. Alternatively, instead of putting in an additional amplifying stage to boost the gain, the triode (T1) can be replaced by a high gain pentode. A good example is the Dynaco Stereo 70 power amplifier to be discussed later.

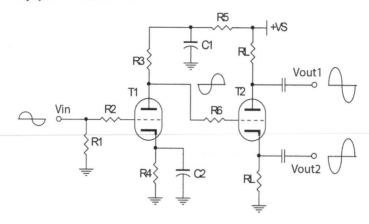

Figure 11.9 A split-load phase splitter is directly coupled to the output of the input stage

Cathode-coupled phase splitter

Figure 11.10 shows the *cathode-coupled phase splitter* and its variants. Figure 11.10(a) is the cathode-coupled phase splitter in its original form. In this cathode-coupled arrangement, the grid of triode T2 is grounded. The signal Vout1 is produced from the plate of triode T1, which is

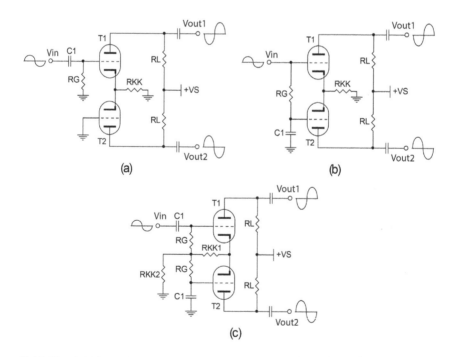

Figure 11.10 Circuit (a) is a cathode-coupled phase splitter in its original form. Circuit (b) is also called a long-tailed pair phase splitter, which is a variant of circuit (a). This phase splitter is designed to be directly coupled to the preceding stage. Circuit (c) is another variant of a cathode-coupled phase splitter. This arrangement allows a larger cathode resistor, RKK2, to be used

in a common cathode amplifier configuration. Therefore, the voltage gain is greater than unity and phase inverted. The signal Vout2 is produced from the plate of T2, which is in a grounded grid amplifier configuration. The gain for Vout2 is greater than unity and non-inverted phase. Note that the grid of T2 is grounded. Thus, the phase splitter of Figure 11.10(a) cannot be directly coupled to the output of the preceding stage. An input capacitor C1 must be used.

Figure 11.10(b) is also called the *long-tailed pair phase splitter*, which is a variant of Figure 11.10 (a). In this arrangement, the phase splitter of Figure 11.10(b) must be dc coupled

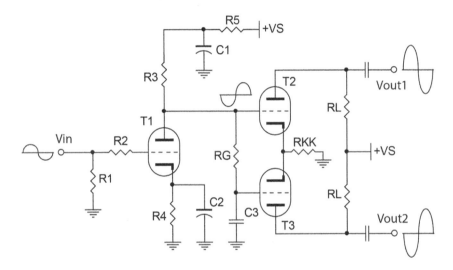

Figure 11.11 A long-tailed pair phase splitter (T1, T2) is directly coupled to the input stage

to the preceding stage. Therefore, no coupling capacitor is needed for the input. Figure 11.11 shows how a long-tailed pair phase splitter is directly coupled to the input stage. RG is usually a large resistor around 1MΩ so that RG and C1 (around 0.1μF) form an RC filter to prevent an ac signal passing to the grid of T2, while the grid of T2 is maintained at nearly the same dc potential of T1's grid.

As shown in Figure 11.11, since Vout1 is produced from a common cathode amplifier (T1) while Vout2 is from a common grid amplifier (T2), the amplitudes of Vout1 and Vout2 are close but not equal. It should be noted that the long-tailed pair phase splitter in Figure 11.11 is directly coupled to the input stage (T3). The cathode resistor RKK carries a high dc voltage. As a result, the long-tailed pair phase splitter produces a lower output voltage swing than Figure 11.10(a) and (c).

Figure 11.10(c) is another popular variant of the cathode-coupled phase splitter. In general, for a cathode-coupled amplifier to perform well, it is desirable to make the cathode resistor as large as possible so that it will minimize the loading effect to the cathode of triode T1 and T2. Since the RKK of Figure 11.10(a) is connected to ground directly, it limits the resistance for RKK. However, the arrangement of Figure 11.10(c) allows a much larger resistor to be used for RKK2. (See Section 6.2 for a discussion of such an arrangement.) Similar to Figure 11.10(a), the phase splitter of Figure 11.10(c) cannot be dc coupled to the preceding stage. An input coupling capacitor C1 must be used.

Floating paraphase phase splitter

Figure 11.12 shows two *floating paraphase splitter* arrangements. It is also called a *floating paraphase inverter*. Figure 11.12(a) is suitable to drive a push–pull output stage where the power tubes are self-biased. When the power tubes are self-biased, the dc voltage at points A and B is at ground level. However, when the power tubes are fixed-bias, there is a negative dc voltage present at points A and B. This will affect the dc quiescent condition of the tubes T1 and T2. In this situation, Figure 11.12(b) is a better choice.

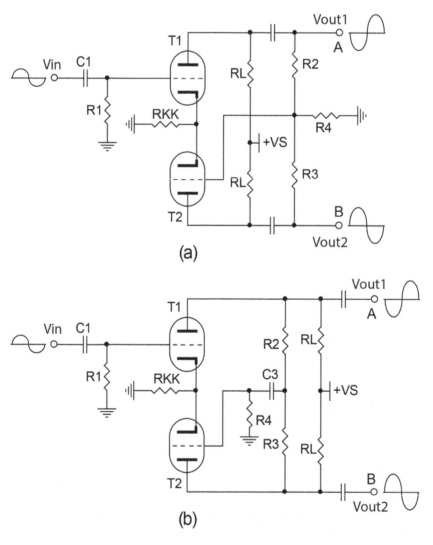

Figure 11.12 Circuit (a) is a floating paraphase phase splitter suitable for the push–pull output stage with self-biased power tubes, while circuit (b) is suitable for fixed-bias and self-bias power tubes

The two floating paraphase splitters of Figure 11.12 have the same operation principle. The incoming signal Vin is amplified by triode T1 and produces output Vout1. Since T1 works in a common cathode amplifier configuration, the output Vout1 has an inverted phase with respect to Vin. When common cathode resistor RKK is un-bypassed, triodes T1 and T2 are working like a cathode-coupled amplifier where the cathode output from T1 becomes the input for T2. Therefore, the output of T2 (Vout2) is phase non-inverted with respect to Vin. At the same time, there is a potential divider network formed by R2 and R3. This feeds back the difference between Vout1 and Vout2 in such a way that the two outputs become stabilized and balanced. In order to avoid the potential divider network creating too much loading effect to the load resistor RL, resistors R2 and R3 are chosen to be much bigger than RL. A floating paraphase splitter produces decent voltage gain, large output voltage swing, and well-balanced outputs. However, the high frequency response may not be the best in class.

Cross-coupled phase splitter

Figure 11.13 shows a *cross-coupled phase splitter*. The cross-couple phase splitter was appeared in 1948 by Van Scoyoc [1]. It should be noted that this amplifier can be used as a differential amplifier by feeding one signal to the grid of T1 and the other signal to the grid of T2. But for a phase splitter application, it has only a single input. The grid of T2 is grounded. The cross-coupled phase splitter was popularized by Joseph Marshal [2–4] in the 1950s and 1960s. He published it in a series of so called "Golden Ears amplifiers."

A cross-coupled phase splitter employs four triodes. Triode T1 works as a cathode follower to buffer the input signal Vin. T1's output is then directly coupled to triodes T3 and T4. Triode T3 works in a common grid configuration and T4 in a common cathode configuration. Therefore, Vout1 is phase non-inverted but Vout2 is phase inverted with respect to the input signal

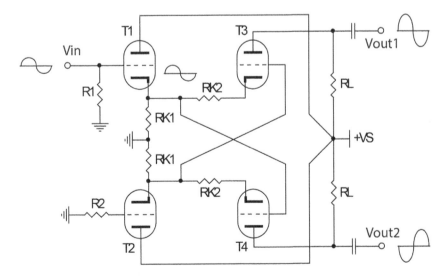

Figure 11.13 A cross-coupled phase splitter that employs four triodes

Vin. Note that T2 does not amplify any signal. However, T2 establishes a dc potential at its cathode that is equal to that of T1. Thus, T2 is helping T3 and T4 to establish identical dc quiescent conditions.

A cross-coupled phase splitter produces decent voltage gain, very close output amplitude and output impedance, as well as good high frequency response. However, it takes four triodes to realize a phase splitter. On the other hand, since cathode resistors RK1 are connected to ground, it is a small resistor. Ideally, a cathode resistor should be as large as possible so as to improve the linearity of the cathode follower. The solution to this problem is to connect the junction between two RK1 resistors to a negative dc power supply instead to the ground. However, this requires an additional negative power supply.

11.6   Push–pull output stages

Figure 11.14 is a simplified circuit showing how power triodes, tetrode or pentode tubes are connected in a push–pull output stage. When power triodes are employed, as shown in Figure 11.14(a), the plates of the triodes are obviously connected to both terminals of the primary side of the output transformer. On the other hand, when power tetrodes or pentodes are employed, as shown in Figure 11.14(b), an extra screen grid becomes available. Therefore, there are several ways of connecting the screen grid to the primary side of the output transformer. Each of these ways will influence the power tube characteristics differently.

The first way is to connect the screen grid to the plate. This becomes a triode connected configuration. The power tetrode or pentode tubes behave like a power triode with a low plate impedance. The amplifier produces low distortion with reduced output power. The second

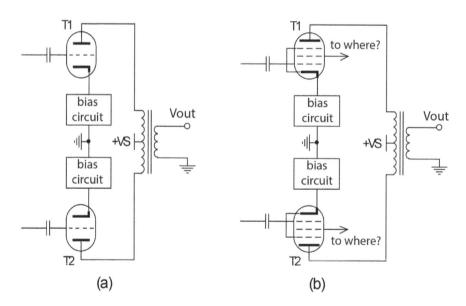

Figure 11.14 Circuit (a) shows power triodes are used in a push–pull output stage. Circuit (b) shows power tetrode or pentode tubes are used in a push–pull output stage

way is to connect the screen grid to the dc power supply. The power tube behaves like a pentode with high plate impedance. The amplifier produces high output power with higher distortion than a triode connected configuration. The third way is to connect the screen grid to a transformer tap which is somewhere between the plate and the power supply. This is called an *ultra-linear* connection with a distributed load to the screen grid. In the ultra-linear connection, the output power is slightly lower than the pentode connection but higher than the triode connected arrangement. However, the distortion is low and it is compatible with the triode connected configuration.

The power tubes in the push–pull output stage need to be biased properly so that the plate currents are nearly the same. There are two common dc bias schemes, self-bias and fixed-bias. In the following, we discuss these two dc bias schemes as well as a self-balanced bias technique. Several transformer configurations for screen grid and cathode loading are also discussed. A list of commonly used power tetrodes and pentodes is shown in Table 11.2.

Table 11.2 A list of representative power tetrodes and pentodes commonly used in push–pull power amplifiers

Tube	Typical value		Maximum rating			Filament		Push-pull output power (W)
	Transconductance, g_m (mA/V)	Plate impedance, r_p (kΩ)	Plate voltage (V)	Plate dissipation (W)	Screen dissipation (W)	Voltage (V)	Current (A)	
6BQ5 / EL84	11.3	40	300	12	2	6.3	0.76	11 to 17
6V6GT	4.1	50	315	12	2	6.3	0.45	10 to 14
5881	5.2	48	400	23	3	6.3	0.9	18 to 26
6L6GC	5.3	35	450	30	5	6.3	0.9	18 to 26
KT66	7	22.5	550	30	4.5	6.3	1.3	35 to 50
6CA7 / EL34	11	15	800	25	8	6.3	1.5	35 to 55
KT77	10.5	23	850	32	6	6.3	1.4	35 to 60
6550	11	15	660	42	6	6.3	1.6	40 to 100
KT88	11.5	12	800	42	8	6.3	1.6	40 to 100
KT120	12.5	10–12.5	850	60	8	6.3	1.95	> 70W
KT150	12.6	10–12.5	850	70	9	6.3	2	> 70W

Self-bias

Figure 11.15 shows several self-bias arrangements for power tubes in a push–pull output stage. To simplify the discussion, the screen grid is connected to the power supply. Note that the self-bias technique works equally well for triode connected and ultra-linear configurations. Figure 11.15(a) is the simplest among the three arrangements. One bypassed cathode resistor (RK) is shared with both power tubes T1 and T2. In the 1950s, when high capacitance and high voltage capacitors were expensive, this was a very common arrangement. It is cost saving to use just one capacitor. However, the downside is that it is difficult to get the balance plate current between the tubes. Unless careful tube matching is affected, T1 and T2 will run at slightly differently plate currents. If the difference is just a few mA, it is generally acceptable. However, it is usually much higher than few mA.

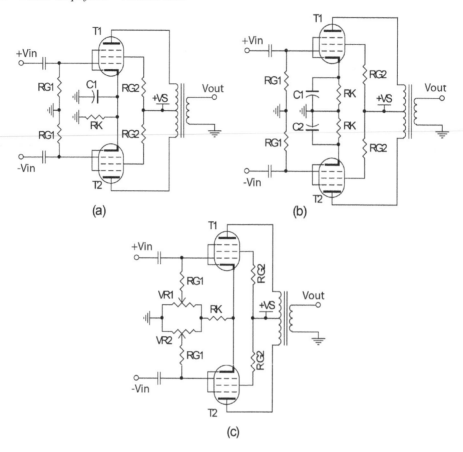

Figure 11.15 Self-bias arrangements for power tubes in a push–pull output stage. Arrangement (a) uses a common single cathode resistor. Arrangement (b) uses two cathode resistors for improved balance. Arrangement (c) uses two variable resistors for finely adjusting the plate current for each power tube

Figure 11.15(b) shows a more balanced self-bias arrangement at the expense of using one more cathode resistor and capacitor. The plate current imbalance is greatly reduced. It is an improved version of Figure 11.15(a). In general, the cathode bypass capacitor (C1 and C2) is at least 220uF and dc 200V rating. At normal operation, the dc potential across the cathode resistor RK is around 40V for EL34 and 50V for KT88. Therefore, it is logical to think that a dc 100V rating is good enough for the bypass capacitors. However, it turns out that when the power tube is near to the end of its life, it runs at much higher plate current and, therefore, it creates a substantial dc voltage across the cathode resistor. Since electrolytic capacitors have become so affordable today, it is strongly recommended to choose a dc rating of no less than 200V.

The arrangement in Figure 11.15(c) employs two variable resistors to finely adjust the plate current for each power tube. This arrangement was popularized by Williamson in the well-known Williamson amplifier [5]. Given two separate variable resistors, the difference between the plate currents can be confined to less than a few mA. When both tubes are well balanced, the cathode resistor RK is not necessarily bypassed by a capacitor. For convenience, a small resistor (10Ω or less) can be inserted in the cathode of each power tube for monitoring the dc bias current.

Fixed-bias

Figure 11.16 shows several fixed-bias arrangements for power tubes in a push–pull output stage. To simplify the discussion, the screen grid is connected to the power supply. However, the fixed-bias technique works equally well for triode connected and ultra-linear configurations. Figure 11.16(a) is the simplest among the three arrangements. One common un-by-passed cathode resistor, RK, is connected to the junction of the cathodes. This is a small

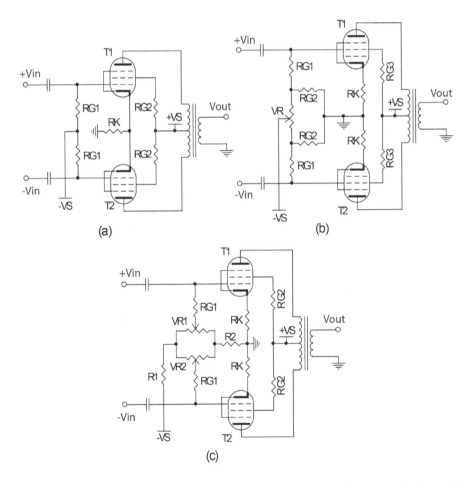

Figure 11.16 A negative dc voltage supply (–VS) is needed in fixed-bias for power tubes in a push–pull output stage. Arrangement (a) uses a single cathode resistor. Arrangement (b) uses two cathode resistors and a variable resistor for improved balance. Arrangement (c) uses two variable resistors for finely adjusting plate current on each power tube

resistor, usually around 10–20Ω, which is used to monitor the total dc bias current. However, it is not possible to monitor the individual plate current for each power tube. A negative dc voltage supply (–VS) must be applied to the junction of two grid resistors, RG1. With a suitable –VS, the total plate current for the power tubes can be set and monitored via the dc potential across RK. In practice, the negative dc voltage supply has to be adjustable for finely adjusting the plate current.

Figure 11.16(b) shows a more flexible arrangement for setting the plate current. Two small cathode resistors, RK, are used so that the plate current for each tube can be monitored. A potential divider network containing four resistors (2×RG1 and 2×RG2) and one variable resistor (VR) is used for adjusting the plate current. Given a suitable negative dc voltage supply (–VS), the plate current of each tube can be set to nearly the same.

Figure 11.16(c) uses two separate variable resistors (VR1 and VR2). Therefore, it is more flexible for setting the plate current to any value. Two small cathode resistors (RK) are also used for monitoring the individual plate current on each tube. In general, if the negative dc voltage is much higher than the required control grid-cathode voltage, resistor R1 is needed. But if the negative dc voltage is low, resistor R1 can be discarded. For example, if –VS is around –60V, resistor R1 is not needed. However, if –VS is –100V or lower, it is recommended to insert a suitable resistor R1 in series with the –VS.

If we compare a push–pull power amp for self-bias and fixed-bias configurations with the same power supply and plate current, the self-bias arrangement in Figure 11.17(a) has a lower plate-cathode dc potential (360V) than the fixed-bias arrangement (399V) in Figure 11.17(b). Therefore, given the same power supply (+VS) and plate current, a fixed-bias power amplifier produces higher output power than a self-bias power amplifier. However, there is a price for producing higher output power. An additional negative dc power supply (–VS) is needed for fixed-bias configuration.

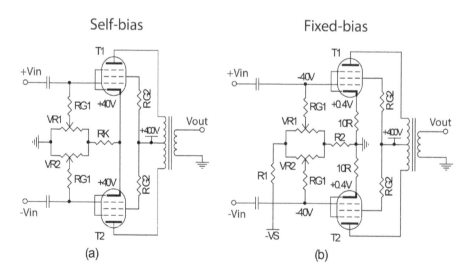

Figure 11.17 Given the same power supply +400V and plate current, say 40mA, the self-bias arrangement in circuit (a) has a lower plate-cathode voltage (360V) than the fixed-bias arrangement in circuit (b), which is 399.6V

Self-balanced bias

For an ideal dc bias arrangement, the plate current on each power tube is equal and stays at the predetermined level at all times when there is no output. As a result, there is no dc magnetization to the transformer core. Thus, the size of the output transformer can be small. However, the self-bias and fixed-bias arrangements discussed earlier can only partially meet the requirements.

The self-bias and fixed-bias arrangements can initially set up equal plate current to a predetermined level for each tube. However, when the tubes are aging and deteriorating over time, the plate current on each power tube may depart from its original predetermined level at a different rate. The result is an imbalanced plate current between the power tubes.

A self-balanced bias arrangement is shown in Figure 11.18. It is considered as an improved arrangement in that the plate currents between the power tubes remain the same, or at least nearly the same, at all times. This arrangement ensures the plate current of T1 and T2 is increasing and decreasing at about the same rate. However, it cannot keep the plate current at the predetermined level when the tubes start deteriorating over time.

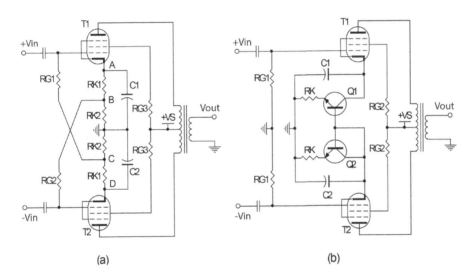

(a)   (b)

Figure 11.18 Circuit (a) shows a self-balanced bias arrangement with cross-coupled grid resistors. Circuit (b) shows a self-balanced bias arrangement using a current mirror

The circuit in Figure 11.18(a) was developed and patented by Alan Dower Blumlein [6,7]. The name "self-balanced bias" was not used by the inventor. However, it is used in this book to point out the nature of this arrangement. It can be seen from Figure 11.18(a) that the circuit resembles the self-bias arrangement of Figure 11.15(b) with two additional cathode resistors. Each tube now has two cathode resistors, RK1 and RK2. A total of four cathode resistors is used. The grid resistors RG1 and RG2 are cross-coupled to the junction of two cathode resistors on each tube. Here is how the self-balanced bias works.

Assume that at the moment the plate current in power tube T1 is higher than the plate current in T2. Therefore, the dc potential at the cathode of power tube T1 (point A) is higher than the cathode dc potential of power tube T2 (point D), $V_A > V_D$. This makes the dc potential at point B higher than the dc potential at point C, $V_B > V_C$. Now, the low dc potential at point C is reflected to the grid of power tube T1 via grid resistor RG1. This reduces the grid dc potential and, therefore, the cathode-grid dc potential difference of T1 increases. As a result, it tends to decrease the plate current in T1. Likewise, the high dc potential at point B is reflected to the grid of power tube T2 via grid resistor RG2. This increases the grid dc potential and, therefore, the cathode-grid dc potential difference of T2 decreases. As a result, it tends to increase the plate current in T2. Hence, the circuit helps to balance the plate current between the two power tubes. The self-balanced bias circuit of Figure 11.18(a) works without a negative dc power supply.

Note that there is always a positive dc voltage present at the grid. For example, if the cathode-grid dc voltage is 50V for a desired plate current, the dc potential at the grid will be at a positive level above ground. Let's say, the circuit rests at a grid dc potential of 20V, $V_C = 20V$. Therefore, given a cathode-grid dc voltage of 50V, the dc potential at point A becomes 70V, $V_A = 70V$. Similarly, the dc potential at point D is around 70V also. With such a high dc potential present at the cathodes, point A and D, the plate-cathode dc potential is reduced. Thus, there is less output power produced for the same power supply compared to the self-biased arrangement of Figure 11.15(b). This may be considered as a trade-off. The plate currents are better balanced but the output power is lowered. Otherwise, this is a clever, simple, and effective self-balanced bias technique.

Figure 11.18(b) is a more effective self-balanced bias circuit. It exploits the advantage of a solid-state current mirror formed by NPN transistors Q1 and Q2. When a dc current flows in the collector of transistor Q2, an equal amount of current also flows in the collector of Q1. Therefore, the plate currents of the two power tubes are kept the same at all times. In addition, the dc potential at the grid is maintained at ground level. Thus, the output power for the arrangement in Figure 11.18(b) is higher than Figure 11.18(a) for the same dc power supply. Transistors Q1 and Q2 should have a high breakdown voltage V_{CEO}.

A negative dc voltage can be introduced to the self-balanced bias arrangement. Let us redraw Figure 11.18(b) in 11.19(a). Assume that given a desired plate current, the cathode dc potential is 50V, so that the cathode-grid dc potential is also 50V. A negative dc voltage, say –30V, is

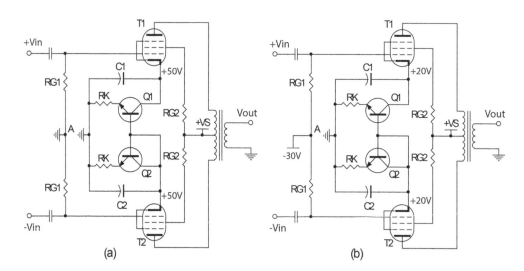

Figure 11.19 Circuit (a) shows that a 50V cathode-grid dc voltage is established when the grid resistors are connected to ground. Circuit (b) shows that the grid resistors are connected to a negative dc voltage (-30V) for the same 50V cathode-grid dc voltage

applied at the grids via resistors RG1, as shown in Figure 11.19(b). If the tube is now biased to the same desired plate current, the cathode-grid dc voltage is also 50V. Since the grid is now biased at –30V, the dc potential at the cathode is reduced to 20V. Therefore, there is an increase in the plate-cathode dc potential. Thus, for the same dc power supply (+VS), output power of Figure 11.19(b) is higher than that shown in Figure 11.19(a).

If now two pairs of power tubes are used for producing high output power, the self-balanced bias arrangement can follow the circuit shown in Figure 11.20. It can be seen that transistor Q1 has its base and collector shorted. The output impedance looking at the collector of Q1 is lower than the collector's output impedance of Q2. The bypass capacitors will generally eliminate the difference in output impedance. However, if now two pairs of tubes are used, we can help to balance out the difference by means of rearranging the current mirrors. When Q1 is placed on top of Q2, it is suggested to place Q4, which has the same configuration of Q1, at the bottom of Q3. Audio driver transistor MJE15034 (V_{CEO} = 350V, $h_{FE(min)}$ = 100) is an excellent choice for Q1–Q4. It is recommended to mount the transistors in a heatsink.

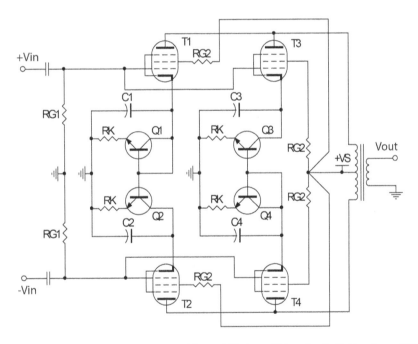

Figure 11.20 Two pairs of power tubes are used in a self-balanced bias circuit. Grid resistors RG1 can be connected to ground as shown above or, alternatively, to a negative dc voltage supply for higher output power

Screen loading and cathode-coupled configuration

When power tetrode or pentode tubes are used, there are several ways to connect the screen grid, leading to different performances. Figure 11.21(a) shows that the screen grid is connected to the power supply. This is called a tetrode or pentode connected configuration (depending on whether a tetrode or a pentode is used), giving the highest possible output power from a pair of power tubes in a push–pull configuration. However, the distortion and the plate impedanceare also high. Alternatively, the screen grid can be connected to the plate in a triode connected configuration, as shown in Figure 11.21(b). In this arrangement, the power tetrode or pentode tube behaves like a triode, producing lower output plate impedance and lower distortion. However, the output power is reduced compared with Figure 11.21(a).

The third configuration is to connect the screen grid somewhere between the plate and the power supply, as shown in Figure 11.21(c). The concept was first presented by Alan Blumlein in his 1937 patent [7] applying to single-ended power amplifiers. Such a configuration was later

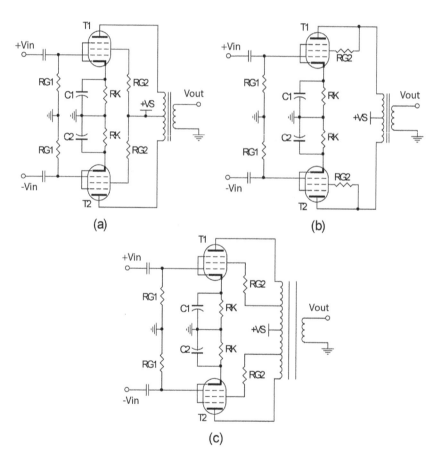

Figure 11.21 Circuit (a) shows the screen grid is connected to the dc power supply. This is often called a tetrode or pentode connected configuration. Circuit (b) shows the screen grid is connected to the plate and this is called a triode connected configuration. Circuit (c) shows the screen grid is connected to the output transformer with partial loading from the transformer. This is called a distributed load, or ultra-linear, connection

called *distributed load*. David Hafler and Herbert Keroes received a patent [8, 9] in 1955 for applying the distributed load in push–pull power amplifiers. They indicated that when the screen grid load impedance was about 18–19% of the plate impedance, this arrangement provides high output power with low internal impedance and low distortion. A transformer with such tapping for distributed load to screen grid was called an ultra-linear transformer – a name popularized by Hafler and Keroes and widely adapted by the audio industry. Given its improved performance over conventional transformers and ease of use, the ultra-linear transformer gained wide popularity in push–pull power amplifier applications.

Figure 11.22 shows an example of an ultra-linear output transformer for 19% screen distributed load on the primary winding. In the secondary winding, outputs are usually designed for 4Ω, 8Ω, and even 16 Ω. Manufacturers seem to produce a somewhat different flavor for the distributed load. It varies from 18% to 40%. There is always a trade-off between power output and distortion for different distributed loading points.

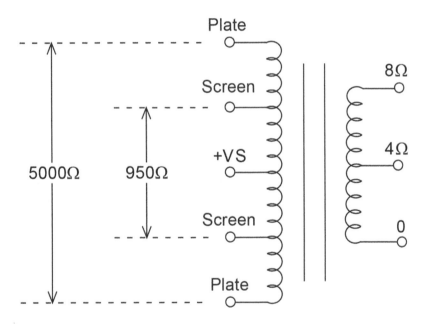

Figure 11.22 A 5kΩ plate-to-plate ultra-linear output transformer uses 19% screen distributed loading

Figure 11.23 shows an extra primary winding is used to provide local feedback to the power tubes. The transformer feedback to the cathode of the power tube effectively reduces the output distortion at the cost of a lowered power output. This is a different approach to the distributed load of the ultra-linear transformer discussed above. Even this is a different approach, but the result is considered similar. The arrangement of Figure 11.23(a) was used in a Quad power amplifier in 1948 and it was the first commercial audio amplifier applying transformer feedback to the power tubes [10].

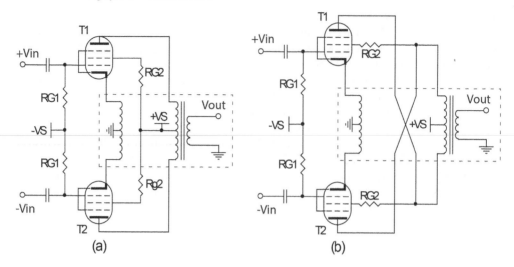

Figure 11.23 The transformer has an extra winding for cathode coupling. Circuit (a) was used by Quad and circuit (b) by McIntosh power amplifiers

Figure 11.23(b) shows a different arrangement for a transformer with cathode-coupled winding. The screen grids are cross-connected to the plate of the other tube. This arrangement was used in the well-known McIntosh power amplifiers. The McIntosh power amplifiers deliver higher power output than conventional push–pull amplifiers with same fixed-bias and power supply. The grid bias of the power tube is so chosen that one or more of the power tubes will be driven into cut-off at some part of the cycle. This usually leads to high cross-over distortion. Frank McIntosh and Gordon Gow cleverly developed a trifilar wound output transformer to achieve a very tight inductive coupling among the windings for the cathodes, plates, and outputs [11]. The leakage inductances are brought to an extremely low level so that cross-over distortion is no longer an issue in this arrangement.

11.7 Classic vacuum tube push–pull power amplifiers

As discussed in the preceding sections, most vacuum tube push–pull power amplifiers follow a three-stage configuration: a voltage amplifying input stage, a phase splitter second stage and a push–pull output stage. We have also discussed various forms of phase splitters, power tube dc bias arrangements and push–pull output transformers. In the following, we are going to discuss several classic push–pull amplifiers and see how the three-stage configuration is realized. The discussion includes Leak 12, Quad II, Williamson, Dynaco Stereo 70, and cross-coupled power amplifiers.

Leak TL12 power amplifier

The schematic of Leak TL12 power amplifier is shown in Figure 11.24. Leak TL12 was launched in the UK in December 1948, together with a matching preamplifier, the pair being called "Point One." The reason for the name is that the power amplifier produces 0.1% distortion at rated output. It was a remarkable figure for the time. This is achieved by using a three-stage amplifier configuration and sufficient negative feedback.

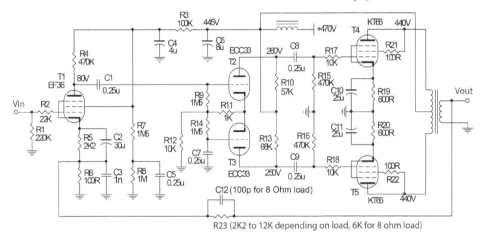

Figure 11.24 Schematic of the Leak TL12 power amplifier with 12W output power

The input stage is a voltage amplifier containing pentode EF36 (T1) with partially bypassed cathode resistors. EF36 is a high gain pentode that is equivalent to 6J7. When EF36 is used, the input stage can easily produce a voltage gain of over 100. Cathode resistor R5 is bypassed by capacitor C2 so that a high voltage gain is retained for the input stage. A small cathode resistor (R6) is used so that global feedback can be applied. A potential divider network is formed by R7 and R8 so as to set up a desired dc bias voltage for the screen grid of tube T1. The output from T1 is capacitor coupled to the phase splitter in the second stage.

The second stage is a cathode-coupled phase splitter consisting of two triodes (ECC33) similar to the configuration discussed in Figure 11.10(c). ECC33 is a medium gain triode that is equivalent to 6SN7. Given the high gain pentode, the input and second stages produce sufficiently high open-loop gain to take down the overall distortion after feedback is applied.

The cathode resistor (R11 + R12) has a total of 12kΩ, which is sufficiently high so that it will not produce too much loading to the cathode of T2 and T3. It should be noted that plate resistor R13 is larger than R10. This boosts the gain so as to compensate for the output from T3, which has a slightly lower gain than T2. The output stage is a push–pull arrangement and the power tubes KT66 are self-biased in a triode connected arrangement.

Quad II power amplifier

The Quad II power amplifier was designed by Peter Walker, founder of Quad Electroacoustics Ltd. Quad II was launched in 1953 in the UK. The amplifier operated in class A and employed a cathode-coupled transformer producing 15W output power. The schematic is shown in Figure 11.25.

The Quad II power amplifier employs a rather simple circuitry that contains only two amplifying stages. The input stage is a phase splitter. The second stage is the output stage, where the power tubes are operated in a class A push–pull configuration.

The phase splitter is a floating paraphase arrangement similar to the one discussed in Figure 11.12(a). Since the phase splitter also needs to provide sufficient open-loop gain for the amplifier, high gain pentodes (EF86) are chosen instead of triodes. The output from the phase splitter is capacitor coupled to the output stage via capacitors C2 and C3. Feedback is taken from a center tap on the secondary side of the transformer and cleverly connected to the junction between resistors R4 and R10.

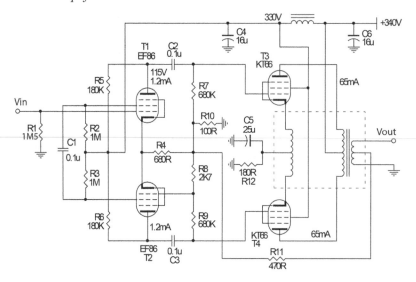

Figure 11.25 Schematic of the Quad II power amplifier with 15W output power

The output stage employs a pair of KT66 working in class A at 65mA plate current. An independent winding in the transformer provides feedback through the cathode of the KT66. Linearity of the power tubes is improved by the cathode feedback and, hence, distortion is reduced.

Williamson power amplifier

The power amplifier shown in Figure 11.26 was developed by D.T.N. Williamson when he was a staff member in the GEC receiving tube development laboratories in London. It was published in *Wireless World* in 1947 [5]. The amplifier offered an excellent performance that was well ahead of the competitors at the time. The article was reprinted on a world-wide scale and the

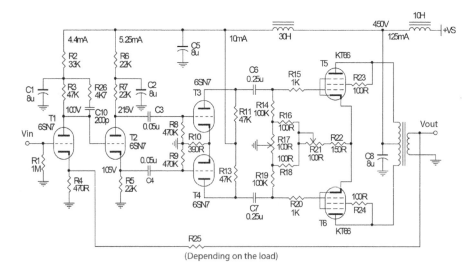

Figure 11.26 Schematic of the Williamson amplifier with 15W output power

amplifier inspired many audio designers. The amplifier was often called the *Williamson amplifier* in recognition of his contribution to the advancement of the design of power amplifiers.

The Williamson amplifier contains four stages. The input stage is a common cathode amplifier configuration formed by triode T1. The output of T1 is directly coupled to the split-load phase splitter formed by triode T2. The split-load phase splitter is similar to the one discussed in Figure 11.8. Here triode T1 is biased to a dc plate potential of 100V. The split-load phase splitter is biased in such a way that each of the resistors R5–R7 has approximately 110V dc across. The plate-cathode dc potential of T2 is also 110V.

Since a split-load phase splitter produces two outputs with unity gain, a third amplifying stage is added so as to boost the open-loop gain for the entire amplifier. A differential amplifier formed by triodes T3 and T4 is used for the third stage to boost the gain.

The output stage contains a pair of triode connected KT66s working in a push–pull configuration. A self-bias arrangement is used for the power tubes. There are two variable resistors, R17 and R21, that allow the matching of plate currents as well as signal drive levels. When the plate currents are closely balanced, the cathode resistors can be left un-bypassed.

Dynaco Stereo 70 power amplifier

The Stereo 70 was introduced in 1959 by Dyna Company (later called Dynaco) as an audio electronic kit that was called Dynakit. The company was formed by audio design engineer David Hafler, who patented the ultra-linear transformer together with Herb Keroes and Ed Lauren. Dynaco was a popular American company producing audio products in the 1960s and 1970s.

The Dynaco Stereo 70 is also called Dynaco ST70, which is a two-channel power amplifier producing 35W of output power per channel. It was intended for assembly by the purchaser and it was also sold as a complete factory-finished unit. It was estimated that 350,000 units of Dynaco ST70 had been sold when the production ceased. This made Dynaco ST70 one of the most popular tube power amplifiers ever made.

Figure 11.27 is the schematic of the Dynaco ST70 amplifier, which contains three stages. The input stage is formed by a pentode (T1), which is taken from the first half of the vacuum tube

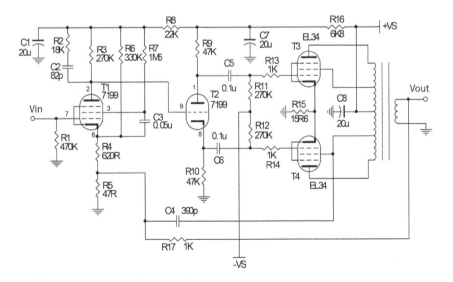

Figure 11.27 Schematic of the Dynaco Stereo 70 with 35W of output power per channel

7199. The second half of 7199 is a triode (T2) with an amplifying factor of 17. The pentode and the triode are housed together in a single vacuum tube (7199).

The output from T1 is directly coupled to the second stage, which is a split-load phase splitter formed by T2. The outputs from the phase splitter are capacitor coupled to the output stage, which has a pair of EL34s operating in fixed-bias and connecting to an ultra-linear output transformer. Feedback is taken from the output via resistor R17 and fed to the cathode resistor R5 in the input stage. A frequency compensation capacitor is usually connected in parallel to R17. However, the frequency compensation capacitor C4 is now connected to the screen grid in this arrangement.

Cross-coupled power amplifier

The power amplifier shown in Figure 11.28 was published by Marshall in 1954 [2]. It was called the "New Golden Ear" amplifier in that article. Marshal published several articles with different variations, making it a popular high quality audio amplifier design among audio enthusiasts in the 1960s. Here, we use the name "cross-coupled" amplifier to highlight the fact that a cross-coupled phase splitter is used. This power amplifier contains four stages.

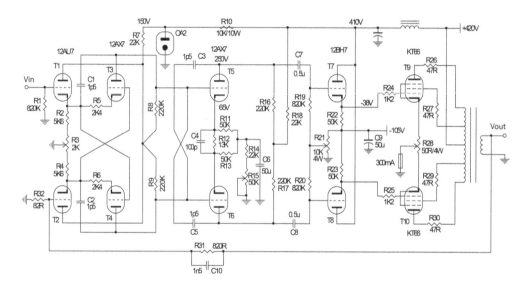

Figure 11.28 Schematic of a cross-coupled amplifier with 15W of output power

The input stage is a cross-coupled phase splitter similar to the one discussed in Figure 11.13. The phase splitter is formed by four triodes, T1–T4. T1 works as a cathode follower providing a low impedance for T3 and T4. Because of this, the amplifier has a rather good high frequency response. Triode T2 is used to handle the feedback and also it sets up the desired dc potential for the cathode of T4. The grids of T3 and T4 are cross-coupled in such a way that T3 is working in a common grid amplifier configuration so that the output from the plate is phase non-inverted. In contrast, T4 is working in a common cathode configuration so that the output from the plate is phase inverted.

The outputs from the phase splitter are directly coupled to the second stage formed by triodes T5 and T6, which are operating in a differential amplifier configuration. Variable resistor R15 can be used for finely adjusting the tail current. The outputs from the second stage are capacitor coupled to the third stage, which has two cathode followers formed by triodes T7 and T8. A variable resistor, R21, can be used to match the incoming signal levels. The cathode followers provide low impedance to drive the push–pull output stage, where fixed-bias is used for a pair of KT66s connecting to an ultra-linear output transformer. C1, C2, C3, and C5 are neutralizing capacitors that minimize the amplifier's phase shift.

11.8   Fully balanced vacuum tube power amplifiers

Figure 11.29(a) shows the configuration of a conventional vacuum tube power amplifier. Since signal source Vin is unbalanced, a phase splitter must be in place to produce two out-of-phase signals for the push–pull output stage. Without two out-of-phase signals, the push–pull output stage simply does not work. Unfortunately, phase splitters always have imperfect properties such that two out-of-phase outputs have different amplitudes, phase shifts, and output impedance. These imperfections affect the push–pull output stage and make it difficult to reproduce the signal faithfully. Since an ideal phase splitter does not exist, the best solution would be the one that eliminates the phase splitter from a push–pull configuration. Today, a vacuum tube power amplifier has advanced to the fully balanced configuration that can take the benefits from the modern audio components – fully balanced CD players and preamplifiers – which produce the necessary two out-of-phase signals.

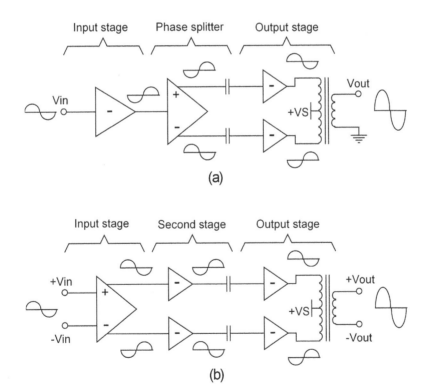

Figure 11.29 Diagram (a): a phase splitter is needed in the conventional tube power amplifier. Diagram (b): no phase splitter is needed in a fully balanced tube power amplifier

Figure 11.29(b) shows a *fully balanced push–pull* (FBPP) power amplifier configuration. The input stage of the FBPP power amplifier reaps the benefit of having two differential signals (out of phase) from a fully balanced audio preamplifier. Given the differential signals, the phase splitter is eliminated from the circuit at last. The major task for the FBPP amplifier is now to amplify the given differential signals so that the push–pull output stage reproduces the signal faithfully. In practice, it is an easier task to amplify differential signals faithfully than to create a phase splitter with two accurate out-of-phase outputs and identical output impedances. In the following, we discuss two examples of the FBPP power amplifier.

Example 1

An FBPP power amplifier is shown in Figure 11.30. The power amplifier contains three stages. The first is an input stage that contains a differential amplifier, formed by triodes T1a and T1b, to amplify the differential input signals. The outputs from the input stage are directly coupled to the second stage, formed by triodes T2a and T2b, which is also a differential amplifier. The first two stages produce the necessary open-loop gain for the entire amplifier. The output stage is formed by a pair of KT88s with fixed-bias and they are connected to an ultra-linear output transformer. Since this is a fully balanced power amplifier, dual feedbacks are employed.

The differential amplifier input stage employs twin triodes, 12AX7, producing high voltage gain. R5 and R6 are cathode degeneration resistors that provide local feedback to the input stage so as to improve the linearity of this stage. At the same time, the cathode degenerating resistors provide a path allowing global feedback to be applied via resistors R40 and R41. In order to ensure the input stage has a good common-mode rejection ratio (CMRR) to handle the fully balanced inputs, a large 150kΩ resistor is used for tail resistor R7. This requires a negative dc power supply (–104V) to be used.

DC self-balanced differential amplifier

The second stage is a differential amplifier formed by twin triodes 6922. As shown in Figure 11.30, the grid resistors R13–R16 are cross-connected to the cathode resistors R17–R22. These grid and cathode resistors appear to be in an unconventional arrangement when compared to a conventional differential amplifier input stage. This is because, if the second stage is also a conventional differential amplifier, the power amplifier will run into a problem producing an asymmetrical output waveform. This is illustrated in Figure 11.31.

Producing an asymmetrical output waveform is due to imbalanced plate currents in the second stage when a signal is present. Since the output from the input stage is directly coupled to the second stage, the dc potentials at the plates of T1a and T1b become the grid dc potentials for T2a and T2b, point A and B. When the ac signals are superimposed on the dc voltage at point A and B, they cause the grid-cathode voltage for one triode (T2a or T2b) to be higher than the other one. Thus, one triode has a higher plate current that drops the plate dc potential, while the plate dc potential of the other triode increases. As a result, the output from one triode will clip much earlier than the other one, creating asymmetrical outputs [12]. In order to avoid imbalanced plate currents, a new dc bias arrangement is needed and it is shown in Figure 11.32.

As shown in Figure 11.32, since the differential amplifier formed by T2a and T2b is directly coupled to the input stage, it is clear that the grids of the triodes T2a and T2b carry both input

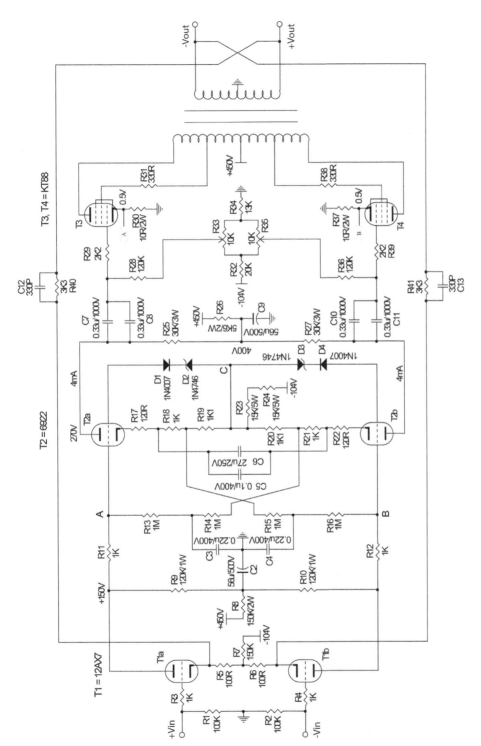

Figure 11.30 A fully balanced push-pull power amplifier for 40W of output power

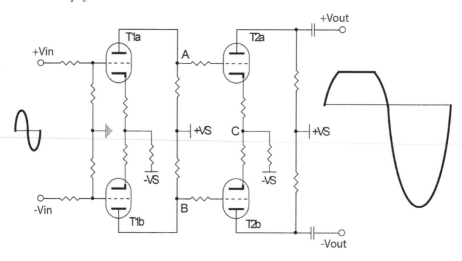

Figure 11.31 An asymmetrical output waveform is produced from a configuration that contains two conventional differential amplifiers

signals and dc bias voltages passing from the input stage. Hence, we should examine the differential amplifier from two different perspectives: (i) an ac signal point of view and (ii) a dc point of view.

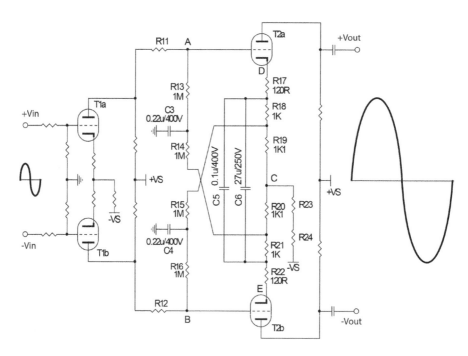

Figure 11.32 The second stage uses a self-balanced differential amplifier so that a symmetrical output waveform is produced

From the ac signal point of view, the operation of the differential amplifier formed by triodes T2a and T2b of Figure 11.32 is as follows. Capacitors C3 and C4 prevent signals from cross-coupling from the grid to the cathode or, from the cathode to the grid between the two triodes. From an ac signal point of view, the junctions between R13 and R14, and between R15 and R16, are shorted to ground by the capacitors C3 and C4. Since R13 to R16 are 1MΩ resistors, they have very little loading effect on the input stage.

Capacitor C6 is employed to bypass resistors R18–R21. Therefore, from the ac signal point of view, resistors R18–R20 are shorted. Only resistors R17 and R22 are left to function as cathode degeneration resistors. Hence, from the ac signal point of view, the differential amplifier formed by T2a and T2b operates identically to the conventional one in Figure 11.31.

On the other hand, from the dc point of view, the operation of the differential amplifier formed by T2a and T2b in Figure 11.32 is as follows. First, it should be noted that when a triode is correctly biased and operates in a steady state condition, the dc potential at the cathode is always higher than the dc potential at the grid. No dc current flows in the grid. DC current only flows from the plate to the cathode.

Assume for the time being that the triodes T2a and T2b are well matched and the dc bias voltages passing from the input stage are also identical ($V_A = V_B$). Let us denote the dc potential difference between cathode and grid as V_{CG}, where $V_{CG} > 0V$. Let us denote the dc plate current as I_p. In order to minimize dissimilar dc plate currents when non-matched triodes T2a and T2b are used, it is best to choose the resistors R17–R22 such that

R17 = R22, R18 = R21, R19 = R20
R17 + R18 = R19
R21 + R22 = R20
$V_{CG} = I_p(R17 + R18) = I_pR19 = I_pR20 = I_p(R21 + R22)$

where R17 and R22 are the desired cathode degeneration resistors that determine the ac signal gain of the differential amplifier.

For instance, if triodes T2a and T2b are operated with the quiescent conditions $V_{CG} = 4.5V$ and $I_p = 4mA$ at a given plate voltage, then the values for R19 and R20 are easily found to be 1.1kΩ. If 120Ω is chosen for the cathode degeneration resistors R17 and R22, then R18 and R21 are 980Ω, or 1kΩ, which is the closest practical resistor value. If the resistors are chosen as above, the differential amplifier will have the dc self-balanced ability that minimizes the mismatch between triodes T2a and T2b, and the unequal dc bias voltages passing from the input stage.

To see how the self-balanced mechanism works, here we assume that the two triodes T2a and T2b are not matched and the dc potentials passing from the input stage are not equal. Let us denote the dc potentials at the grid and the cathode of the triode T2a as V_A and V_D, respectively. Similarly, V_B and V_E denote, respectively, the dc potentials at the grid and cathode of the triode T2b. Assume that at the moment triode T2a operates at a dc potential such that $V_A > V_B$ and $V_D > V_E$, i.e., both grid and cathode dc potential voltages of triode T2a are higher than T2b. The series resistors R13 and R14 will pass along the higher dc potential V_A and lift the dc potential at the junction between resistors R20 and R21. As a result, the cathode dc potential (V_E) of triode T2b is increased and, hence, the grid dc potential (V_B) is also increased. By the same token, the series resistors R15 and R16 will pass along the lower potential V_B and bring down the dc potential at the junction between resistors R18 and R19. As a result, the cathode dc potential (V_D) of triode T2a is lowered and, hence, the grid dc potential (V_A) is also lowered. Since V_E and V_B are raised while V_D and V_A are lowered, V_D and V_E are pulling closer and so are V_A and V_B. Eventually, the differential amplifier formed by triodes T2a and T2b in Figure 11.32 will rest on a dc quiescent point closer than is the case in Figure 11.31. Therefore, the plate currents and dc quiescent conditions between T2a and T2b are brought closer even if a signal is present. Thus, the waveform will not clip prematurely from the outputs of the second stage.

Grid-cathode over-voltage protection

It can be seen that the output from the input stage is directly coupled to the second stage. If solid-state rectifying diodes are used in the dc power supply, when the power amplifier is turned on, a high dc voltage is built up instantly while the triodes' filaments have not even started warming up. The triodes are still in the cold condition. In this instance, the triodes do not draw any plate current. Therefore, there is no dc voltage drop in plate resistors R9 and R10 (see Figure 11.30). At this moment, the plate dc potential of T1a and T1b is just equal to the dc power supply +VS. Here, let us assume +VS = +450V. Since the grids of T2a and T2b are directly coupled to the plate of T1a and T1b, the dc potential at the grid of T2a and T2b is also equal to 450V.

On the other hand, the dc potential at the cathode of T2 and T2b is equal to the negative dc supply voltage, −104V, because there is no cathode current flowing in the cold condition. Therefore, the grid-cathode voltage of T2a and T2b is equal to 450V− (−104V) = 554V. This is an extremely high dc voltage across the grid-cathode. If T2a and T2b warm up faster than T1a and T1b, the high grid-cathode dc voltage will force grid current to flow and may damage the triodes. To a lesser extent, it may also shorten the life span of the triodes. Thus, it is desirable to have an over voltage protection circuit to protect T2a and T2b.

Since there are no plate currents when the triodes T1 and T2 are in a cold condition, we can simplify the first two stages of Figure 11.30 by removing the triodes and redrawing the input and second stage, as shown in Figure 11.33. Given the components as shown, a current *I* flows from the positive dc power supply (+450V) to the negative dc power supply (−104V). The current *I* is approximately 1.1mA, which is sufficient to turn on the diodes (D1 and D4) and zeners (D2 and D3). The zener breakdown voltage for 1N4746 is 18V and the forward voltage for diode 1N4007 is 1V. In other words, the grid-cathode of T2a and T2b is clamped to 19V, which is much lower than the dc potential (554V) without the diode and zener in place. Therefore, the circuit protects the tubes by avoiding a large grid-cathode voltage building up when power is switched on, as the tubes are not yet operational.

After the triodes warm up and start to operate, plate currents begin to flow. If the zener diodes are properly chosen, the zeners will be eventually turned off due to the low grid-cathode dc voltage on a triode, which is much lower than the zener reverse voltage.

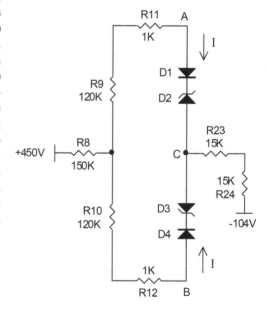

Figure 11.33 The flow of the dc current when the triodes T1 and T2 are in cold condition

After the zeners are turned off, they do not affect the signal that appears on the grid of the triode. In order to ensure that the zeners work properly, they should be chosen such that: diode forward voltage + zener reverse voltage > grid-cathode voltage (V_{GC}) of vacuum tube T2a (or T2b) + voltage drops in resistors R17, R18, and R19 (or R20, R21, and R22).

The above condition will hold true as long as there is no input signal. To prevent the zener from turning on when a signal passes through the grid, the zener reverse should be chosen such that: diode forward voltage + zener reverse voltage > grid-cathode voltage (V_{GC}) of vacuum tube T2a (or T2b) + voltage drops in resistors R17, R18, and R19 (or R20, R21, and R22) + maximum signal's amplitude at the grid of the vacuum tube T2a (or T2b).

If we follow the dc bias condition from Figure 11.30, we have the following:

(a) $V_{GC} = -V_{CG} = -4.5V$;
(b)   Voltage drops in resistors   = cathode current × (R17 + R18 + R19)
                                    = 4mA(120Ω + 1kΩ + 1.1Ω)
                                    = 8.9V

Let us assume that the maximum signal voltage amplitude at the grid of triode T2a = 5V. This can be verified by observing the waveform at the grid via an oscilloscope at maximum output. If we take 1V as the diode forward voltage, the zener reverse voltage (Vz) can be determined by

$$1V + Vz > -4.5V + 8.9V + 5V$$

giving Vz > 8.4V. If we choose an 18V zener (1N4746), the amplifier works in the desired manner in that the zeners are turned on to protect the vacuum tubes when the vacuum tubes are in a cold condition. Then zeners are turned off in the steady state, when the vacuum tubes are in normal operation after being warmed up, so that the zeners do not affect the signals appearing at the grid.

Fully balanced push–pull output transformer

For the FBPP power amplifier of Figure 11.30 to work properly, we need a balance output transformer that has a grounded center tap in the secondary winding. Other than this, the construction of the transformer is more or less similar to a conventional ultra-linear transformer that employs interleaving winding between the primary and secondary coils. However, no such balance output transformers are commercially available in the present market. Therefore, we have to find a way to convert a conventional output transformer into a fully balanced transformer. Fortunately, it is not as difficult as it appears to be.

Figure 11.34(a) is a conventional ultra-linear transformer with 4Ω and 8Ω output taps. Some transformers may also offer a 16Ω output tap. However, the 16Ω output is not needed for this FBPP power amplifier. We simply connect the 4Ω tap to ground, as shown in Figure 11.34(b). This turns a conventional transformer to the one needed for the FBPP power amplifier. However, this limits the transformer to driving only 8Ω loudspeakers. To permit the flexibility of driving both 4Ω and 8Ω loudspeakers, we have to build the transformer. However, the design of audio output transformers is beyond the scope of this book. Interested readers should refer to a design manual written by Robert Wolpert [13]. Given a proper ultra-linear output transformer to match a pair of KT88s, the FBPP power amplifier of Figure 11.30 can deliver over 40W output power with less than 0.2% distortion. The power amplifier has a voltage gain of 30dB and band-width over 100kHz.

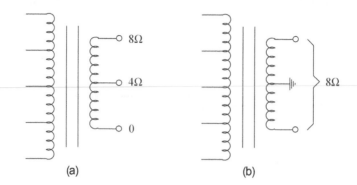

(a)   (b)

Figure 11.34 Circuit (a) is a conventional ultra-linear transformer with 4Ω and 8Ω outputs. Circuit (b) shows how the conventional ultra-linear transformer of (a) is converted to a balance trans-former by grounding the 4Ω output

Bias-level indicator

In the FBPP power amplifier illustrated in Figure 11.30, the output power tubes KT88 are bi-ased to 50mA plate current. Variable resistors R33 and R35 are for adjusting the plate current of T3 and T4, respectively. The cathode resistor R30 and R37 is a 10Ω resistor. When the plate current reaches 50mA, there is a 0.5V dc across the 10Ω cathode resistor. A digital multi-meter can be used to monitor the dc voltage across the cathode resistor when adjusting the plate cur-rent. Alternatively, a bias-level indicator can be handy to indicate whether the plate current has reached 0.5V or not.

Figure 11.35 is a bias-level indicator that can be used for the FBPP power amplifier of Figure 11.30. The indicator circuit uses a comparator and a 1.2V shunt type reference LM285–1.2. The 1.2V reference voltage is first reduced to 0.5V by the potential divider formed by R2 and R3. The comparator is an LM311 with a 15V single supply. A pull-up resistor R4 is needed to ensure the comparator works properly. When the non-inverting input is below 0.5V, the LED is turned off. But when the non-inverting input has reached 0.5V, the LED is turned on and lights up. This circuit can be built into the power amplifier.

Example 2

The second example for the FBPP power amplifier [14] is shown in Figure 11.36. A commercial product, the JE Audio VM60 vacuum tube power amplifier, also uses a similar configuration. This is again a three-stage FBPP power amplifier. The input stage is a cascode differential amplifier in today's terminology. The output from the cascode differential amplifier is directly coupled to the second stage, which is a dc self-balanced differential amplifier similar to the one discussed in Example 1. The output stage uses two pairs of KT88 in a self-balanced bias arrangement. A balanced ultra-linear transformer is used with the power tubes. The self-balanced bias ar-rangement for two pairs of KT88s is similar to the one discussed in Figure 11.20. This amplifier delivers over 75W output power at distortion < 0.2%.

The cascode differential amplifier has a higher voltage gain than a conventional differential amplifier. Therefore, the power amplifier in Figure 11.36 has higher open-loop gain than the one in Figure 11.30. Given the high open-loop gain, we can apply the feedback in two steps,

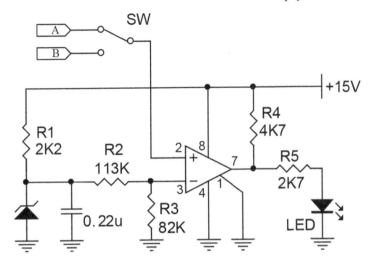

Figure 11.35 This is a bias-level indicator. When a dc potential applied to the non-inverting input of the comparator has reached 0.5V, the LED is turned on

as shown in a simplified diagram in Figure 11.37. A pair of feedback loops is applied to the first two stages. This improves the linearity of the first two stages as well as broadening the bandwidth. This pair of feedback loops contains components R51, R53, C16, C17, C19, and C20. The second pair of feedback loops is applied to the entire amplifier that further improves the linearity of the amplifier as well as taking down the distortion. The second pair of feedback loops contains components R52, R54, C18, and C21.

Distinguished engineer/inventor

Alan Dower Blumlein [15] (29 June 1903–7 June 1942) was born in London. His father was a mining engineer and a naturalized British subject from Germany. He was an electrical engineer, notable for his many inventions in telecommunications, sound recording, stereophonic sound, television, and radar. He received 128 patents and was considered as one of the most significant engineers and inventors of his time.

He died during the Second World War, on 7 June 1942, aged 38, during the secret trial of an H2S airborne radar system then under development. All crew on board the Halifax bomber, including Blumlein, were killed when the plane crashed at Welsh Bicknor in Herefordshire.

The most notable patents that Blumlein received in the audio field include long-tailed pair, stereophonic sound, and distributed load transformer, leading to the so-called ultra-linear transformer. The ideas that were originated from these patents are still playing a significant role in audio applications today.

Alan Dower Blumlein continued to receive distinguished awards and recognitions after his death. On 1 April 2015, an IEEE Milestone Plaque was posthumously awarded to Blumlein for the invention of stereo. In 2017, the Recording Academy posthumously awarded Blumlein with the 2017 Technical Grammy for the invention of stereo and contributions of outstanding technical significance to the recording field.

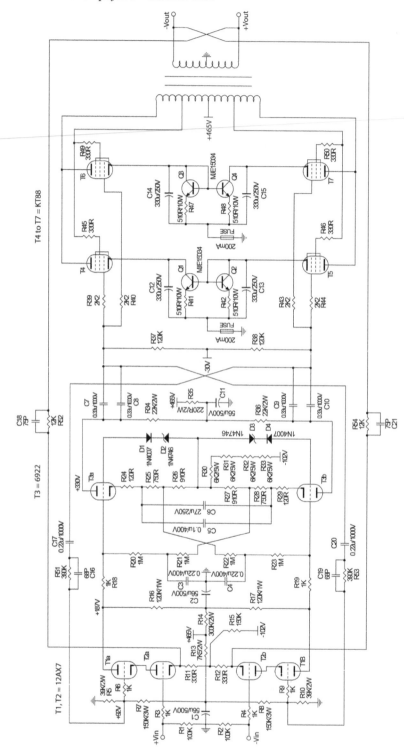

Figure 11.36 A fully balanced push–pull power amplifier employing dual feedback loops

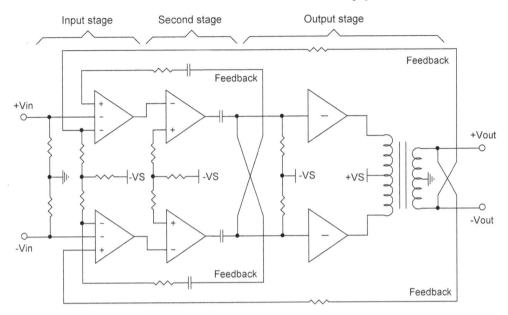

Figure 11.37 A simplified diagram for the power amplifier in Figure 11.38, employing dual feedback loops

11.9 Exercises

Ex 11.1 The circuit in Figure 11.9 contains a common cathode amplifier directly coupled to a split-load phase splitter. Assume T1 = 12A×7, T2 = 12AU7, VS = 350V. If a voltage gain 40 or higher is needed for the common cathode amplifier, determine the resistors, capacitors C1 and C2. Resistor RL must be chosen so as to maximize output swing.

Ex 11.2 The circuit in Figure 11.10(c) is a cathode-coupled phase splitter. Assume T1 = T2 = 6SN7, VS = 400V. If the triodes are biased to a 3mA plate current and a voltage gain of 10 or higher is needed, determine the resistors. Resistor RL must be chosen so as to maximize output swing.

Ex 11.3 The circuit in Figure 11.13 is a cross-coupled phase splitter. Assume T1 = T2 = 6SN7, T3 = T4 = 6SL7, VS = 400V. If a voltage gain of 30 is needed, determine all resistors. What are the plate currents for the triodes?

Exe 11.4 Figure 11.38 is modified from a cross-coupled phase splitter so that it becomes a differential amplifier to handle differential signals. Use this differential amplifier to replace the differential amplifier in the input stage of the FBPP power amplifier in Figure 11.30. Assume T1 = T2 = 12AU7, T3 = T4 = 12A×7. If a voltage gain of 40 or higher is needed, determine all resistors. What are the plate currents for the triodes?

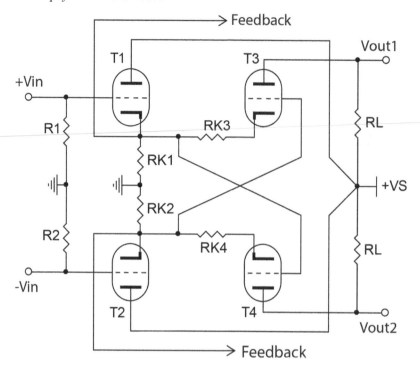

Figure 11.38 This is a differential amplifier for Ex 11.4

References

[1] Van Scoyoc, J. N., "A cross-coupled input and phase inverter circuit," Radio and Television News, Nov. 1948.

[2] Marshall, J., "The new golden-ear amplifier," part 1, pp. 17–18, Audio Engineering, 1954.

[3] Marshall, J., "The laboratory golden ear," part 1, pp. 41–43, Radio Electronics, Aug. 1956.

[4] Marshall, J., "The laboratory golden ear," part 2, pp. 56–58, Radio Electronics, Sept. 1956.

[5] Williamson, D.T.N., "Williamson amplifier," pp. 161–163, Wireless World, May 1947.

[6] Broskie, J., "More auto bias circuits," Tube CAD, May 2005.

[7] Blumlein, A.D., "Thermionc valve amplifying circuits," U.S. patent 2218902.

[8] Hafler, D. and Keroes, H., "Ultra linear amplifiers," U.S. patent 2710312.

[9] Hafler, D. and Keroes, H., "An ultra linear amplifier," Audio Engineering, pp. 15–17, Nov. 1951.

[10] Hood, J.L., *Valve & transistor audio amplifiers*, Newnes, reprinted 2001.

[11] McIntosh, F. and Gow, G., "Description and analysis of a new 50W amplifier circuit," pp. 9–11, Audio Engineering, Dec. 1949.

[12] Lam, C.M.J., "Dc self-biased vacuum tube differential amplifier with grid-to-cathode over-voltage protection," U.S. patent 7733172.

[13] Wolpert, R.G., "Audio transformer design manual," 1989.

[14] Lam, C.M.J., "Balanced amplifier," U.S. patent 7364535.

[15] Wikipedia, "Alan Blumlein," https://en.wikipedia.org/wiki/Alan_Blumlein, n.d.

[16] "Tube circuits for audio amplifiers," reprinted by Audio Amateur Press.

[17] "Audio amateur power amp projects," Audio Amateur Publications, 1996.

[18] Postal, J., "Simplified push–pull theory," Audio Engineering, part 1, pp. 19–20, May, part 2, pp. 21–23, June 1953.

[19] Marshall, J., "High-fidelity drivers and inverters," Radio Electronics, pp. 55–56, March, 1957.

[20] Ravenswood, H., "The fixed-bias story," Radio Electronics, pp. 47–49, Feb., 1958.

[21] Ravenswood, H., "Special amplifier circuits," Audio, pp. 40–42, Aug., 1958.

[22] Hafler, D., "High-power Willamson amplifier for Hi-Fi," Radio Electronics, pp. 42–44, Dec., 1955.

[23] DePalma, B., "A high-power amplifier with minimum distortion," Audio, pp. 15–17, June, 1955.

12 Power amplifiers – solid-state

12.1 Introduction

Vacuum tube power amplifier design is discussed in Chapter 11. The size, weight, and heat dissipation of the vacuum tube power amplifier are the limiting factors that prohibit their use for very high power applications. Even if highly efficient push–pull type vacuum tube power amplifiers are used, the output powers are usually less than 100W. Even so, when a suitable highly sensitive loudspeaker is used, we do not feel that there is a lack of output power from a medium vacuum tube power amplifier, say 50W.

Today there are many floor-standing loudspeakers containing more than one woofer in parallel so that life-like sound-stage and sound pressure can be reproduced. These designs will generally lead to a loudspeaker system with low sensitivity (< 85dB) and low impedance (< 4Ω at low frequency range). When low sensitivity is coupled with low impedance, it will make the loudspeaker a difficult load for a power amplifier to drive. A vacuum tube power amplifier is particularly not the preferred choice for driving these difficult loads. When faced with a difficult load, we turn to solid-state power amplifiers, which are designed for delivering high output current with a high damping factor (low output impedance).

In this section, we present several solid-state power amplifiers of the early days and progressively learn, in the remainder of this chapter, how they evolved into modern designs. Although this chapter is oriented towards audio application, the basic concepts and design techniques also apply to high powered linear amplifiers for industrial and control applications.

Figure 12.1 shows a simple solid-state power amplifier in a single-ended configuration. Transistor Q1 works as an emitter follower to provide a high input impedance to the incoming signal source. Transistor Q2 works in a common emitter configuration. Emitter resistor R4 is bypassed by capacitor C2 so that a moderate voltage gain can be produced. Output is delivered to the load via the output transformer T1. This amplifier closely resembles a vacuum tube single-ended power amplifier, except that vacuum tubes are now replaced by transistors. Output power is around several watts.

The power amplifier in Figure 12.2 has a very similar arrangement to a push–pull vacuum tube

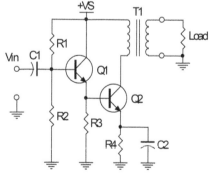

Figure 12.1 A low power solid-state power amplifier in a single-ended configuration

DOI: 10.4324/9781003369462-12

power amplifier. It is a three-stage power amplifier in a push–pull configuration. The input stage contains transistor Q1 working in a common emitter configuration. The output from Q1 is directly coupled to the second stage, formed by transistor Q2. Transistor Q2 is working as a split-load phase splitter. When resistors R5 and R6 are chosen such that R5 = R6, two outputs with equal amplitude but opposite phase are produced at the collector and emitter of Q2. The outputs from Q2 are capacitor coupled to the output stage.

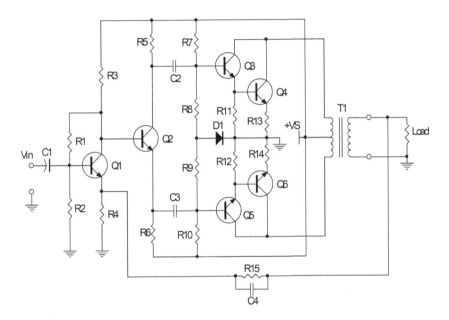

Figure 12.2 A power amplifier in a push-pull configuration

The output stage is formed by transistors Q3–Q6 and output transformer T1. Transistors Q3 and Q4 work in a Darlington transistor configuration producing a high current gain. Similarly, transistors Q5 and Q6 work in a Darlington transistor configuration that handles the signal from the opposite phase. The output signal is produced via the output transformer that operates in a push–pull arrangement the exact same way as in vacuum tube power amplifier.

In an analogy to a vacuum tube push–pull power amplifier, we can imagine that a 12AX7 triode in the input stage is replaced by transistor Q1. A 12AU7 triode in a split-load phase splitter is replaced by transistor Q2. EL34 or KT88 power tubes are replaced by the Darlington transistor Q3/Q4 and Q5/Q6. Given the correct choice of dc power supply, the power amplifier of Figure 12.2 can deliver up to 10W of output power.

The major drawback of the power amplifier in Figure 12.2 is that a bulky output transformer is still needed. However, it did not take too long for designers to recognize that the output stage of the amplifier could be modified to a so-called quasi-complementary configuration so that the bulky output transformer was replaced by a capacitor, as shown in Figure 12.3.

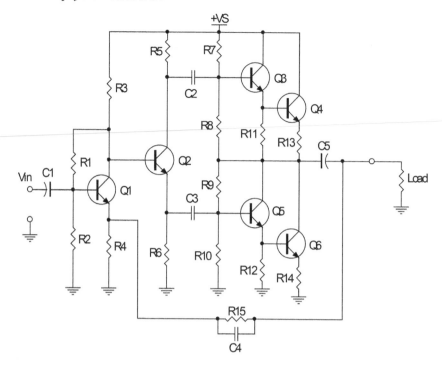

Figure 12.3 A solid-state power amplifier without an output transformer

The power amplifier in Figure 12.3 is a great improvement over that in Figure 12.2 because the bulky output transformer is eliminated. The output transformer is now replaced by a capacitor with much smaller size and weight. However, a capacitor is a frequency dependent device. In order to ensure a good low frequency response, a large capacitor must be used. On the other hand, a large capacitor is not desirable, as it is usually blamed for coloring the sound. Therefore, the ultimate goal for a power amplifier is to get rid of both the output transformer and the output capacitor, as in the modern power amplifiers today.

12.2   Structure of a typical power amplifier

In the 1960s, there was a lack of reliable PNP power transistors. Therefore, many power amplifiers in those years only employed NPN power transistors in the output stage, similar to the amplifiers in Figures 12.1 to 12.3. With the advancement in semiconductor fabrication, complementary power transistors are now produced with almost identical electrical properties and great improvement in current gain linearity, bandwidth, V_{CEO} breakdown voltage, output power, and a safe operating area (SOA).

Figure 12.4 shows the structure of a typical modern power amplifier. It is a three-stage configuration. With the use of dual dc power supply and global feedback, the output transformer and output capacitor are eliminated. Therefore, the low frequency response can be easily extended to dc. In the following, we briefly discuss each functional part of the power amplifier in Figure 12.4. Then detailed discussion is given in subsequent sections in this chapter.

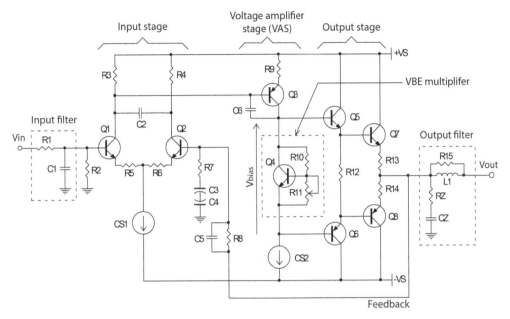

Figure 12.4 Structure of a typical three-stage power amplifier

Input filter

R1 and C1 form a first order low-pass filter to keep out unwanted radio frequency noise. The cut-off frequency of the low-pass filter is determined by $f_H = 1/(2\pi R1C1)$. It is often setting $f_H = 200\text{kHz}$ or higher so that it has little effect on the audio band. Resistor R2 provides a current returning path for the transistor Q1. Since BJT the transistor has a small base current, there is always a small dc input offset voltage present at the base of Q1.

For example, a typical current gain for 2N5551 is 80. If 2N5551 is used for Q1 and biased at 1mA collector current, the base current flowing through R2 is 0.0125mA. If R2 is a 20kΩ resistor, the input dc offset voltage present at the base of Q2 becomes 0.25V. This may create a very small dc offset at the output. As the capacitors C3 and C4 limit the dc voltage gain of the amplifier to unity, the output dc offset voltage is small. It is approximately equal to the input dc offset voltage. Using a dc servo circuit can reduce the output dc offset to nearly ground level.

On the other hand, the input low-pass filter helps to reduce the rapid rising edge of a large input signal. This prevents the power amplifier from producing a high distortion, as the large signal may overload the input stage. However, if the power amplifier has sufficient slew rate, together with the low-pass input filter, the possibility of overloading the input stage will be greatly reduced. Further details are discussed later.

Input stage

The input stage is a differential amplifier formed by transistors Q1 and Q2. Current source CS1 sets the tail current for the differential amplifier. The function of the input stage is to amplify the difference between the input source signal Vin and the feedback signal from the output. The difference is amplified and directly coupled to the second stage.

R5 and R6 are degeneration resistors that serve two purposes. The first is to provide local feedback to the differential amplifier so as to linearize the input stage. As a result, the bandwidth is broadened and distortion is lowered. However, the voltage gain of the differential amplifier, as well as the open-loop gain of the power amplifier, is reduced. The second purpose is to improve the slew rate. More discussion about slew rates is presented later in this chapter. To put it in a simple way, a slew rate is a measure of how fast the power amplifier can respond to a large signal. It is a parameter that indicates how well the power amplifier performs in large signal conditions. The higher the slew rate, the better the performance of a power amplifier handling large signals.

Resistors R7 and R8 form the series-shunt feedback for this power amplifier. Thus, the voltage gain with feedback is given as $A_v = (1 + R8/R7)$. C2 and C5 are frequency compensation capacitors that stabilize the power amplifier at high frequency. C3 and C4 are formed in a non-polarized arrangement. They are used to reduce the voltage gain of the power amplifier to unity at dc. This will help to reduce the output dc offset, especially when a dc servo circuit is not used. In practice, C3 and C4 are often replaced by a single capacitor. Note that components C3, C4, and resistor R7 form a high-pass filter. The cut-off frequency is determined by $f_L = 1/(2\pi R7C)$, where $C = C3\|C4$ in this circuit. R7 and C are chosen so that $f_L \ll 20\text{Hz}$. For example, if $R7 = 1\text{k}\Omega$, $C = 100\text{uF}$, f_L is equal to 1.6Hz.

Voltage amplifier stage (VAS)

The second stage is a voltage amplifier that produces high voltage gain, contributing to the open-loop gain of the entire power amplifier. In Figure 12.4, transistor Q3 works in a common emitter amplifier configuration with a small un-bypassed emitter resistor, R9. In order to achieve high voltage gain, a current source (CS2) having a high impedance is used as the load resistor for Q3. Capacitor C6 is the Miller capacitor that forms a dominant pole to stabilize the power amplifier at high frequency. It should be noted that if C6 is too small, it cannot properly stabilize the amplifier at high frequency. However, if C6 is too large, the slew rate of the amplifier is unnecessarily reduced.

V_{BE} multiplier

It can be seen that transistor Q4 is placed between transistor Q3 and current source CS2. Together with resistors R10 and R11, transistor Q4 forms the so-called V_{BE} *multiplier*. It creates a dc voltage spread between the collector and emitter of Q4. The dc voltage spread is used to compensate the drop of V_{BE} across the driver and power transistors in the output stage. Thus, the output stage is biased in class AB with a low collector dc quiescent current. If the output power transistors are not biased properly without sufficient dc quiescent current, the power amplifier will produce high cross-over distortion.

If Vbias denotes the dc spread across the collector and emitter of Q4, it can be easily shown that Vbias = $V_{BE} (1 + R10/R11)$, where V_{BE} is the base-emitter voltage of transistor Q4. Since there are four V_{BE} across the driver and power transistors, R11 must be adjusted such that the Vbias is greater than $4V_{BE}$. If the V_{BE} of the driver and power transistors is 0.65V each, we must have Vbias > 2.6V.

Since the V_{BE} of bipolar transistor has a negative temperature coefficient, –2.2mV/°C, when the power transistors are heated up, the V_{BE} decreases. If the Vbias remains unchanged when the power transistors are heated up, the power transistors will become over-biased. If a bipolar transistor is over-biased, the junction temperature increases, and the V_{BE} is further reduced. This is a continuous loop that goes in a downward spiral until the transistor is damaged. This is called thermal run-away. In order to prevent the thermal run-away problem from happening, transistor Q4 must be mounted in the same heatsink as the power transistor. It is best to place Q4 near to Q7 and Q8 in this example, so that the temperature of the power transistors is promptly caught up by Q4 and Vbias is properly compensated. Further discussion of a V_{BE} multiplier is presented later.

Output stage

The output stage of the power amplifier comprises complementary Darlington transistors, Q5/Q7 and Q6/Q8. If β_1 is the ac current gain for the driver transistors Q5 and Q6, and β_2 for Q7 and Q8, the ac current gain of the Darlington transistor is $\beta_1 \times \beta_2$. Therefore, this provides a high current gain for the output stage so that a small base current at Q5 will control a huge output current. If a higher current gain is needed, a pre-driver emitter follower can be added. In addition, the high current gain of the Darlington transistor produces a high input impedance. Thus, the output stage will not create too much loading effect on the collector of Q3, which is connected to the current source CS2 with a very high impedance. In contrast, if a very high output current is needed, more complementary power transistors can be paralleled to Q7/Q8. It should be noted that power transistors Q7/Q8 must be mounted in a heatsink, together with transistor Q4, which forms the V_{BE} multiplier. However, no heatsink is needed for pre-driver transistors, if any. Driver transistors (Q5/Q6) do not need to be mounted on the same heatsink with the power transistors, but it is still necessary to mount the driver transistors on a small heatsink.

Zobel network

Some power amplifiers may be unstable under no load conditions. To prevent such an unstable condition from happening, a so called *Zobel network* is placed at the output of a power amplifier. The Zobel network contains a shunt resistor, RZ, and capacitor CZ, as shown in Figure 12.4. The RZ is often a small resistor of several ohms. The CZ is a low capacitor of 0.01μF to 0.1μF. This network stabilizes the power amplifier at high frequency as it sees a low resistor RZ at the output.

Loudspeakers have complex impedance. At some frequency bands, the impedance can be capacitive. In order to stop the capacitive load from destabilizing the power amplifier, an inductor, L1, which is in parallel with a small resistor, R15, is used to isolate the load at high frequency. L1 is a small inductor, usually a few μH, while R15 is a small resistor similar to RZ. Details for constructing the Zobel network and inductor L1 can be found in Thiele [1].

12.3 Input stages

Simple input stage

Figure 12.5(a) is a simple differential amplifier input stage. One of the two inputs (Vin1, Vin2) is used to handle the input signal, and the other input for the feedback signal from the output. R3 and R4 are degeneration resistors that provide local feedback to the input stage, improving the linearity and slew rate of the amplifier. Two outputs (Vout1, Vout2) are generated. If the second stage (VAS) can only take a single input signal, one of the two outputs (Vin1, Vin2) is used while the other output is simply discarded. When the VAS is also a differential amplifier, both outputs will be utilized. The current source CS sets the tail current of the differential amplifier and is usually realized by at least one transistor with the aid of a diode, zener, and transistor. Examples of current sources can be found in Chapter 2. Sometimes a simple resistor is used instead of a current source. However, a current source is preferred as it produces a much higher impedance than a resistor if the dc power supply voltage is low. A high impedance current source improves the CMRR of the differential amplifier.

Figure 12.5(b) is an input stage employing a cascode differential amplifier. There are two advantages of using a cascode differential amplifier. The first is that it has a much wider bandwidth than the simple differential amplifier of Figure 12.5(a). The second is that it allows transistors with a lower V_{CEO} to be used. By choosing the zener D1 to be VS/2, the two transistors can

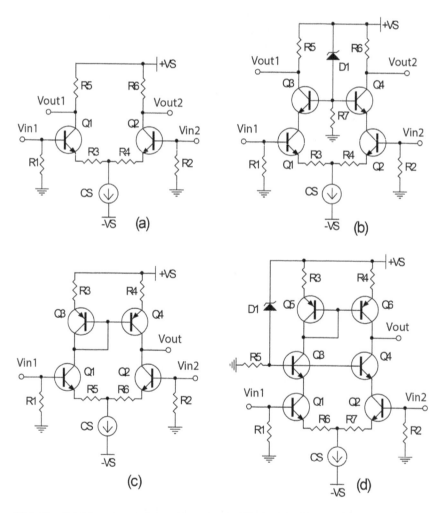

Figure 12.5 Circuit (a) is an input stage using a simple differential amplifier. Circuit (b) is an input stage using a cascode differential amplifier. Circuit (c) is a differential amplifier input stage using an active load. Circuit (d) is an input stage using a cascode differential amplifier with an active load

split the power supply equally. This is very helpful when JFETs are used for the position of Q1 and Q2. It is because JFET's $V_{DS(BR)}$ is usually around 40V, which limits their use in a very high power amplifier unless a cascode configuration is used.

Figure 12.5(c) is a differential amplifier input stage with an active load. The active load is a current mirror formed by transistors Q3 and Q4. It should be noted that the output imped-ance of the current mirror, the collector of Q4, is much higher than that of a resistor (R6 of Figure 12.5(a)); as a result, the differential amplifier produces a much higher voltage gain. The voltage gain of the input stage multiplied by the second stage produces a tremendous open-loop gain for the entire amplifier that can be used to reduce the distortion and extend the bandwidth of the power amplifier by applying feedback. It is the exact same method used in a monolithic op-amp circuit. Note that when an active load is used, there is only one output available instead of two from a simple differential amplifier input stage.

Figure 12.5(d) is a cascode differential amplifier input stage with an active load. It has the benefits of having wide bandwidth from a cascode amplifier configuration, as well as a high voltage gain produced by the active load. The active load is a current mirror formed by transis-tors Q5 and Q6. Again, this input stage produces only one output.

Complementary input stages

Figure 12.6(a) shows an input stage using a pair of complementary differential amplifiers. The first differential amplifier is formed by two NPN transistors, Q1 and Q2. The second differential

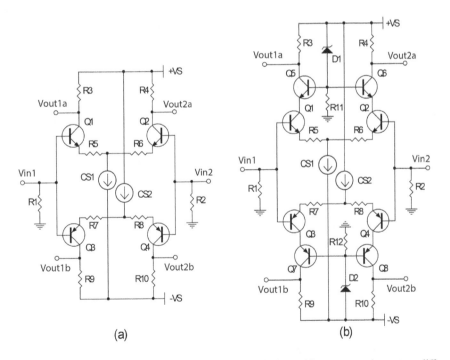

Figure 12.6 Circuit (a) is a complementary input stage that is formed by two complementary differential amplifiers having two sets of inputs connected. Circuit (b) is a complementary input stage that is formed by two complementary cascode differential amplifiers having two sets of inputs connected

amplifier is formed by two PNP transistors, Q3 and Q4. The inputs of the complementary differential amplifiers are connected to form one single composite differential amplifier that has two inputs, Vin1 and Vin2. However, there are up to four outputs, Vout1a, Vout1b, Vout2a, and Vout2b. It can be seen that outputs Vout1a and Vout1b are in phase, but at different dc potentials, as are outputs Vout2a and Vout2b. However, Vout1a,b and Vout2a,b are in opposite phases. When the VAS is a simple complementary push–pull amplifier, only either Vout1a,b or Vout2a,b will be needed. When the VAS is formed by two complementary differential amplifiers, then all four outputs from the complementary differential amplifier input stage will be utilized. Discussion of the various types of VAS is presented in the next section.

When a complementary input stage is paired with a push–pull type VAS, they form a symmetrical structure that tends to reduce secondary harmonic distortion. As a result, the power amplifier produces less distortion. In addition, since there are four transistors used to form the complementary input stage (Figure 12.5[a]), which is twice the number of transistors used in a simple differential amplifier input stage (Figure 12.6[a]), signal-to-noise ratio (SNR) tends to be improved. This is the same principle that applies to the phono stage amplifier for improving SNR. Another interesting finding, which is perhaps merely my opinion, is that the sound of the complementary input stage plus a push–pull VAS configuration is different from that of a simple differential amplifier input stage plus a single input VAS configuration. If the designer can bear the complexity of the complementary input stage with a push–pull VAS, they will be grateful for the sound reward.

Figure 12.6(b) is an input stage formed by a pair of complementary cascode differential amplifiers. The first differential amplifier is formed by four NPN transistors Q1, Q2, Q5, and Q6. The second differential amplifier is formed by four PNP transistors Q3, Q4, Q7, and Q8. The inputs of the complementary differential amplifiers are connected to form one single composite differential amplifier that has two inputs, Vin1 and Vin2, and four outputs, Vout1a, Vout1b, Vout2a, and Vout2b. It should be noted that outputs Vout1a and Vout1b are in phase but at different dc potentials. Similarly, outputs Vout2a and Vout2b are in phase but at different dc bias potentials. However, Vout1a,b and Vout2a,b are in opposite phase. This complementary input stage works similarly to the one in Figure 12.6(a). However, a cascode differential amplifier has wider bandwidth and lower distortion compared with a simple

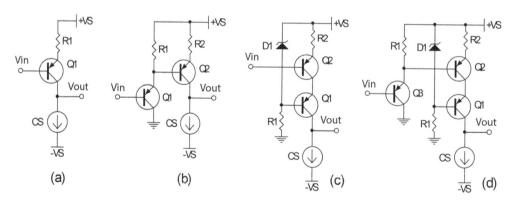

Figure 12.7 Circuit (a) is a VAS formed by a common emitter amplifier. Circuit (b) is a VAS formed by an emitter follower directly coupled to a common emitter amplifier. Circuit (c) is a VAS formed by a cascode amplifier. Circuit (d) is a VAS formed by an emitter follower directly coupled to a cascode amplifier

differential amplifier. These positive characteristics are also carried on to the complementary differential amplifier of Figure 12.6(b).

12.4   Voltage amplifier stages (VAS)

Single input VAS

Figure 12.7(a) is a simple single input VAS formed by transistor Q1 in a common emitter amplifier configuration. Emitter resistor R1 provides local feedback to the VAS. In order not to reduce the voltage gain by too much, resistor R1 is usually a small resistor of less than 1kΩ. The current source CS has a high impedance so that the VAS produces a high voltage gain.

Figure 12.7(b) shows an improved VAS over the one in Figure 12.7(a). Transistor Q2 is an emitter follower that works as a buffer for the common emitter amplifier formed by Q1. In other words, Q2 provides a high input impedance to buffer the input stage and a low output impedance for Q1. Therefore, the VAS will not create too much loading effect at the input stage that may, otherwise, reduce the voltage gain of the input stage. On the other hand, since there are two V_{BE} dropping from Vin to the emitter of Q1, the remaining dc potential across R2 is lower than would be the case without Q2. This allows R2 to be a small resistor and, therefore, the voltage gain of the VAS is very much retained.

Figure 12.7(c) is a VAS in a cascode amplifier configuration. Since it is a cascode amplifier, the VAS has a wider bandwidth and lower distortion compared to the VAS in Figure 12.7(a). This is because the two transistors can split the power supply voltage VS, so BJTs with low V_{CEO} and JFETs with low $V_{DS(BR)}$ can be used for Q2 and Q1.

Figure 12.7(d) is an improved VAS over Figure 12.7(c). Transistor Q3 is in an emitter follower configuration that works as a buffer for the cascode amplifier formed by Q1 and Q2. Therefore, the VAS will not create too much loading effect at the input stage that may, otherwise, reduce the voltage gain of the input stage. Again, since there are two V_{BE} dropping from Vin to the emitter of Q2, the remaining dc voltage across R2 is lower than is the case without Q3. This makes R2 a small resistor and, therefore, the voltage gain of the VAS is very much retained.

Push–pull VAS

Figure 12.8 shows four push–pull VAS configurations. Figure 12.8(a) is a simple push–pull VAS formed by two complementary common emitter amplifiers with their collectors connected, producing the output. There are two inputs Vin that are biased at different dc potentials.

Figure 12.8(b) is a push–pull VAS formed by adding a pair of complementary emitter followers to the VAS of Figure 12.8(a). The complementary emitter followers work as buffers to reduce the loading effect to the input stage.

Figure 12.8(c) is a push–pull VAS formed by a pair of complementary cascode amplifiers with their collectors connected, producing the output. Since a cascode amplifier has a wider bandwidth and lower distortion than a simple common emitter amplifier, the VAS of Figure 12.8(c) also has similar characteristics, with wider bandwidth and lower distortion when compared with Figure 12.8(a).

Figure 12.8(d) is a push–pull VAS formed by adding a pair of complementary emitter followers to the VAS of Figure 12.8(c). The complementary emitter followers work as buffers to reduce the loading effect to the output of the input stage.

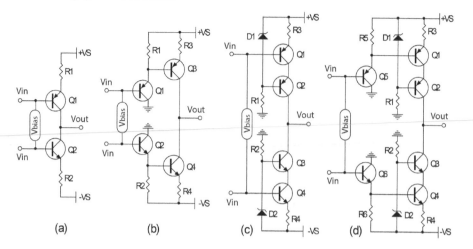

Figure 12.8 Circuit (a) is a simple push–pull VAS with complementary common emitter configuration. Circuit (b) is a push–pull VAS with complementary emitter followers added to circuit (a). Circuit (c) is a push–pull VAS with complementary cascode amplifiers. Circuit (d) is a push–pull VAS with complementary emitter followers added to circuit (c)

Differential VAS

Figure 12.9(a) is a VAS formed by a differential amplifier (Q1 and Q2) with an active load (Q4 and Q5). This VAS is best to pair with a differential amplifier input stage, such as shown in Figure 12.5(a) and (b). Therefore, the two outputs from the differential amplifier input stage are directly coupled to this VAS. The collector of transistor Q5 serves as a load resistor for Q2. Since Q5 works as a current source with high impedance, this VAS produces very high voltage gain. Transistor Q3 is inserted between Q1 and Q4 so that the V_{CE} of Q1 is reduced and shared equally with Q3. Even if Q3 is removed, the VAS still works just fine. However, V_{CE} of Q1 will be exposed to a much higher dc potential.

Figure 12.9(b) is a VAS formed by a pair of common base amplifiers (Q1 and Q2) with an active load (Q4 and Q5). Since common base amplifiers have a good frequency response, this VAS also has wide bandwidth. Again, the collector of Q5 serves as a load resistor for Q2, so that the VAS produces a very high voltage gain. Transistor Q3 works similarly to Figure 12.9(a), which helps to reduce the V_{CE} across Q1. Since the input impedance of a common base amplifier is low, this has a loading effect on the input stage. Therefore, the voltage gain of the input stage is inevitably affected. When the differential amplifier input stage shown in Figure 12.5(a) is used to pair with the VAS of 12.9(b), the composite amplifier is often called a *differential folded cascode amplifier*. The voltage gain of the differential folded cascode amplifier is lower than the voltage gain of the composite amplifier formed by the differential amplifier input stage of Figure 12.5(a) and VAS of Figure 12.9(a). However, a differential folded cascode amplifier has a wide bandwidth and low phase shift.

Figure 12.9(c) is a VAS formed by two pairs of complementary differential amplifiers with their respective collectors connected to produce outputs. This VAS has two pairs of inputs. Therefore, it is obvious that the input stage must be one of the complementary input stages shown in Figure 12.6. If a complementary input stage from Figure 12.6 is used to pair with the VAS of Figure 12.9(c), since the input stage and VAS are in differential configurations, the differential nature of

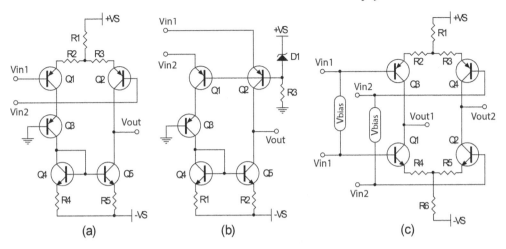

Figure 12.9 Circuit (a) is a VAS formed by a differential amplifier with an active load. Circuit (b) is a VAS formed by a pair of common base amplifiers with an active load. Circuit (c) is a VAS formed by two complementary differential amplifiers with their respective collectors connected

the composite amplifier is very well preserved from inputs to outputs. This differential nature may suggest that the composite amplifier is a potential configuration for developing a fully balanced power amplifier. Discussion of fully balanced power amplifiers is presented later in this chapter.

12.5   Input stage and VAS arrangements

Simple input stage and single input VAS

Figure 12.10(a) is a composite amplifier formed by the input stage shown in Figure 12.5(a) and the VAS in Figure 12.7(b). The input stage is a conventional differential amplifier with

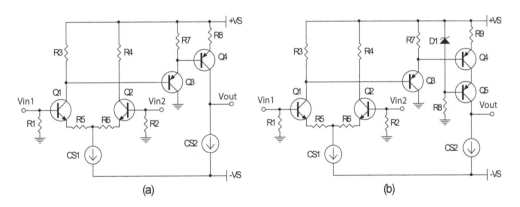

Figure 12.10 Circuit (a) is a composite amplifier formed by the input stage of Figure 12.5(a) and the VAS of Figure 12.7(b). Circuit (b) is a composite amplifier formed by the input stage of Figure 12.5(a) and the VAS of Figure 12.7(d)

degeneration resistors R5 and R6. The degeneration resistors provide local feedback to the input stage, so that distortion is reduced while bandwidth and slew rate are improved. This composite amplifier is a popular input stage plus VAS configuration.

Figure 12.10(b) is a composite amplifier formed by the input stage of Figure 12.5(a) and the VAS of Figure 12.7(d). Since a cascode amplifier is employed by the VAS, the composite amplifier is expected to have a wider bandwidth and lower distortion than the composite amplifier of Figure 12.10(a). This composite amplifier is also a very popular input stage plus VAS configuration.

On the other hand, we can also employ input stages from Figure 12.5(b) to (d) to form various composite amplifier configurations. Examples are shown in Figure 12.11. When composite amplifiers are formed via different input stages and VAS, we must identify the polarity of the output with respect to the two input terminals. For example, the phase of Vout in Figure 12.11(a) is non-inverted with respect to Vin1, but it is inverted with respect to Vin2. However, the phase of Vout in Figure 12.11(b) and (c) is inverted with respect to Vin1 but is non-inverted with respect to Vin2.

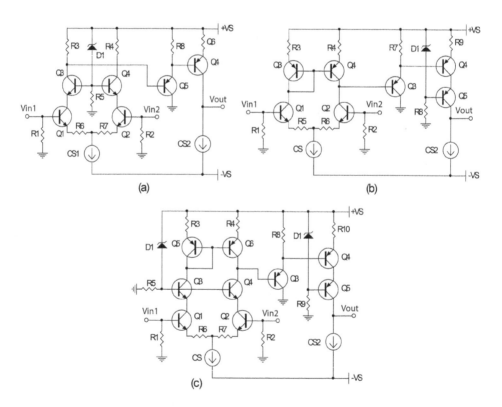

Figure 12.11 Circuit (a) is a composite amplifier formed via the input stage of Figure 12.5(b) and the VAS of Figure 12.7(b). Circuit (b) is a composite amplifier formed by the input stage of Figure 12.5(c) and the VAS of Figure 12.7(d). Circuit (c) is a composite amplifier formed by the input stage of Figure 12.5(d) and the VAS of Figure 12.7(d)

Complementary input stage and push–pull VAS

The composite amplifier of Figure 12.12(a) is formed by the complementary input stage of Figure 12.6(a) and the push–pull VAS of Figure 12.8(b). The composite amplifier has a mirror symmetrical configuration and it tends to reduce secondary harmonic distortion. Thus, when this composite amplifier is employed to form a power amplifier, it may produce lower distortion than that illustrated in Figure 12.10.

Figure 12.12(b) is a composite amplifier formed by the complementary input stage of Figure 12.6(b) and the push–pull VAS of Figure 12.8(d). Again, it has a symmetrical configuration that tends to reduce secondary harmonic distortion. In addition, since the cascode amplifier configuration is used for the input stage and VAS, the composite amplifier has a wider bandwidth than Figure 12.12(a). Even though the composite amplifier of Figure 12.12(b) is a challenge to develop and requires almost twice the number of electronic components, the measured performances and the sound should be rewarding.

Differential input stage and differential VAS

Figure 12.13(a) is a composite amplifier formed by a differential amplifier input stage and a differential VAS with an active load. The input stage is formed by the simple differential amplifier of Figure 12.5(a). The outputs from the input stage are directly coupled to the differential amplifier VAS of Figure 12.9(a). Since an active load (Q6 and Q7) is used, the VAS produces a high voltage gain. Multiplying the voltage gain of the input stage and the VAS, the composite amplifier of Figure 12.13(a) produces a very high open-loop gain. It should be noted that transistor Q5 is added between Q3 and Q6, so that the V_{CE} of Q3 is reduced. If Q5 is not in place, the V_{CE} of Q3 is much higher than the V_{CE} of Q4 and Q6.

Figure 12.13(b) is a so-called *differential folded cascode amplifier* formed by a differential input stage and a pair of common base amplifier VAS. The input stage is given by the simple differential amplifier of Figure 12.5(a). The outputs from the input stage are directly coupled to the common base amplifier VAS of Figure 12.9(b). Again, an active load (Q6 and Q7) is used so that the VAS is expected to produce a high voltage gain. Since a common base amplifier produces a very low input impedance, Q3 and Q4 produce a heavy loading effect to the load resistors (R3 and R4). As a result, the open-loop gain of the composite amplifier in Figure 12.13(b) is lower than that shown in Figure 12.13(a). Technically speaking, the differential folded cascade amplifier is not a two-stage, but a single stage, composite amplifier. The purpose of R3 and R4 is to provide dc quiescent potentials for Q1 and Q2.

Figure 12.13 (c) is a composite amplifier formed by a complementary differential amplifier input stage and a differential VAS. The input stage is formed by the complementary differential input stage of Figure 12.6(a). The outputs from the input stage are directly coupled to the differential VAS of Figure 12.9(c). Since the composite amplifier has a mirror symmetrical configuration, it tends to reduce the secondary harmonic distortion. Therefore, if a power amplifier is realized by Figure 12.13(c), it will be likely to produce lower distortion than a power amplifier of Figure 12.13(a). On the other hand, since the differential mode is preserved throughout the composite amplifier, Figure 12.13(c) may provide a potential approach for developing a fully balanced power amplifier.

Bi-FET input stage and VAS

In addition to BJT, JFET is also a popular amplifying device for low level signal amplification. But there are always pros and cons to using BJT and JFET. Many believe that JFET produces

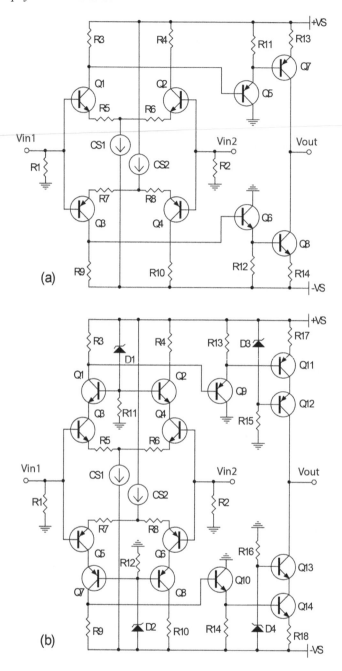

Figure 12.12 Circuit (a) is a composite amplifier formed by the input stage of Figure 12.6(a) and the VAS of Figure 12.8(b). Circuit (b) is a composite amplifier formed by the input stage of Figure 12.6(b) and the VAS of Figure 12.8(d).

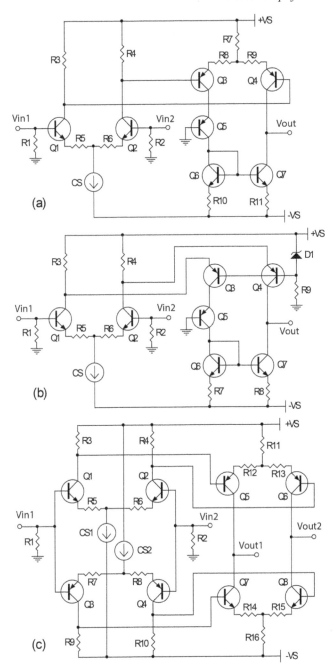

Figure 12.13 Circuit (a) is a composite amplifier formed by a simple differential amplifier input stage and a differential VAS with an active load. Circuit (b) is a differential folded cascode amplifier formed by a differential input stage and a VAS with a pair of common base amplifiers and an active load. Circuit (c) is a composite amplifier formed by a complementary differential input stage and a complementary differential VAS having the respective collectors connected

better sound than BJT. And JFET also has a greater resistance to radio inferences. However, generally speaking, JFET's input equivalent voltage noise is greater than that of BJT. This may make a JFET amplifier produce higher noise than a BJT amplifier. But it is not always true, as there are exceptionally low noise JFETs such as 2SK170 and 2SJ74, or their equivalents, that can be used for this application. These two JFETs are commonly found in low noise phono-stage amplifiers. However, the low V_{DS} break-down voltage for JFETs is a more concerning factor for their use in high power amplifier applications.

Most JFETs have a low V_{DS} break-down voltage, which is around 40V or lower. Therefore, when JFETs are used in a power amplifier, they often work with BJT to form a Bi-FET cascode configuration as shown in Figure 12.14, so that the drain-source voltage is reduced. Figure 12.14(a) shows a simple Bi-FET differential input stage. Figure 12.14(b) shows a composite amplifier formed by a complementary Bi-FET cascode differential input stage and a Bi-FET cascode push–pull VAS.

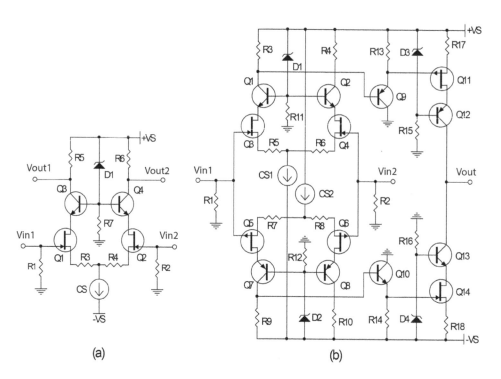

Figure 12.14 Circuit (a) is a Bi-FET cascode differential input stage employing JFET and BJT. Circuit (b) is a composite amplifier formed by a complementary Bi-FET cascode differential input stage and a Bi-FET cascode push–pull VAS

Slew rate

The slew rate is a measure of how fast an amplifier can respond to a large signal. It is a parameter that indicates how well the amplifier (usually a power amplifier) performs in large signal conditions. Therefore, a higher slew rate gives better performance when a power amplifier

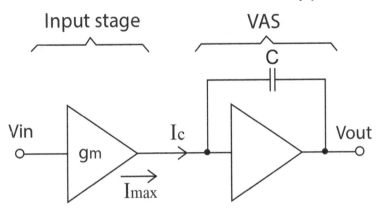

Figure 12.15 A simplified diagram that represents the input stage and VAS of an audio power amplifier

amplifies large signals. The slew rate limitation is due to insufficient bias current from the input stage to quickly charge up the Miller effect capacitor C on the VAS, as shown in Figure 12.15.

The slew rate is measured in V/μs. Since the capacitor C turns the VAS to behaving like an integrator [3–5], the slew rate can be expressed as

$$\text{Slew rate} = \frac{dV_{out}}{dt} = \frac{I_{max}}{C} \tag{12.1}$$

where I_{max} is the maximum current from the input stage. Since we also have

$$\frac{dV_{out}}{dV_{in}} = \frac{dV_{out}}{dI_C} \frac{dI_C}{dV_{in}}, \text{ where } \frac{dI_C}{dV_{in}} = g_m, \frac{dV_{out}}{dI_C} = \frac{1}{sC}$$

therefore, the above expression is simplified to

$$\frac{dV_{out}}{dV_{in}} = \frac{g_m}{sC}, \text{ or } \frac{dV_{out}}{dV_{in}} = \frac{g_m}{j\omega C} \text{ in the frequency domain.}$$

If the gain falls to the unity gain at frequency ω_o, the above expression becomes

$$1 = g_m/\omega_o C, \text{ or } C = g_m/\omega_o$$

Thus, Eq. (12.1) can be written as

$$\text{Slew rate} = \frac{I_{max}}{C} = \frac{I_{max}}{g_m} \omega_o \tag{12.2}$$

The expression for the slew rate suggests that, for a given unity gain frequency of ω_o, in order to increase the slew rate, the ratio (I_{max}/g_m) must be increased. Therefore, one approach to increasing the slew rate is to use JFET in the input stage because, for the same bias current, the transconductance (g_m) of JFET is roughly an order of magnitude lower than that of BJT. Low g_m is often considered a disadvantage of JFET when voltage gain is concerned. However, it becomes an advantage for improving slew rate. Therefore, the JFET input stage of Figure 12.16(b) produces higher slew rate than Figure 12.16(a) when the same bias current I_{max} is given. Alternatively, the degenerating resistor RE can be used to lower the effective transconductance, as shown in Figure 12.16(c), for improving the slew rate.

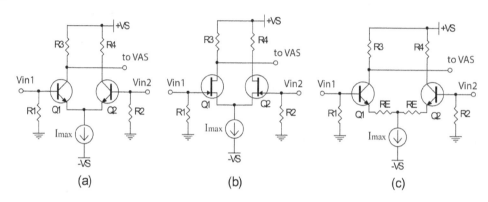

Figure 12.16 Circuit (a) is a conventional differential input stage using BJT. Circuit (b) is a differential input stage using JFET to improve the slew rate. Circuit (c) is a BJT differential input stage using emitter degeneration resistors to improve the slew rate

On the other hand, it is instructive to see how a power amplifier responds to a large sinusoidal signal. Assume that we have a 100W power amplifier with dual dc power supply ±40V. Assume that the power amplifier is rearranged to work with unity gain so that the output $V_{out} = V_{in}$. In other words, the output of the power amplifier has to follow closely to the input sinusoidal signal with 40V peak, V_p. If the input signal is expressed as

$$V_{in} = V_p \sin(\omega t) \tag{12.3}$$

the rate of change of input signal dV_{in}/dt, which is also equal to dV_{out}/dt so that slew rate can be easily determined, is given as

$$\frac{dV_{in}}{dt} = \omega V_p \cos(\omega t)$$

The maximum value of dV_{in}/dt occurs at $|\cos(\omega t)| = 1$ (see Figure 12.17). Thus, we have

$$\left|\frac{dV_{in}}{dt}\right|_{max} = \omega V_p = 2\pi f V_p \tag{12.4}$$

Therefore, we have to make sure that the slew rate of the power amplifier is greater than $2\pi f V_p$. When a 100W power amplifier is to amplify a 20kHz signal with $V_p = 40$V, this corresponds to a

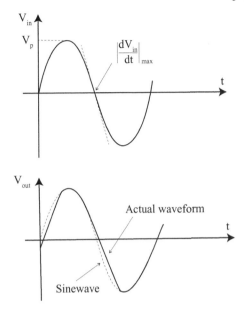

Figure 12.17 (a) A large sinusoidal signal applied to the input. (b) Output waveform from input (a) showing slew limiting

slew rate of $2\pi \times 20 \times 10^3 \times 40$ V/s = 5V/µs. This should be viewed as the absolute minimum slew rate that a power amplifier must have in order to avoid slew rate limited distortion. In reality, a power amplifier may be called upon to amplify a transient signal that has frequency contents up to 100kHz. Thus, the required slew rate for the power amplifier becomes 25V/µs, or greater, for a 100W power amplifier. It should be noted that higher output power requires a higher slew rate as the peak voltage V_p increases.

On the other hand, it is well-known that an R-C low-pass filter placed at the amplifier's input can help to reduce the rapid rising edge of the large input signal (see Figure 12.4). Thus, an R-C filter will soften the requirement for the power amplifier to have a very high slew rate [6–8].

An input stage with sufficiently high slew rate and an R-C input filter will largely prevent the *transient intermodulation distortion* (TIM), which is also called *slewing-induced distortion* (SID), from happening. The TIM and SID are caused by the overloaded input stage that arises before the feedback catches up with the input and corrects the distortion.

12.6   Output stages

Since the output stage of a power amplifier should be capable of delivering high current to a load, it must have a high current capacity as well as a high current gain. For a single transistor, the emitter follower configuration has the highest current gain among the three basic configurations: common emitter, common base, and emitter follower (common collector). The ac current gain of an emitter follower is denoted by β. When two BJTs are cascaded, as shown in Figure 12.18(a) and (b), the compound transistor becomes a Darlington transistor configuration with a current gain of $\beta_1 \times \beta_2$. Therefore, if the current gain of the two transistors are $\beta_1 = 100$

and $\beta_2 = 60$, the current gain of the Darlington transistor becomes 6,000. Given a high current gain, the output stage becomes a very easy load to be driven by the VAS. When the output is an 8Ω load, the VAS sees the Darlington transistor output stage as a load of 48kΩ.

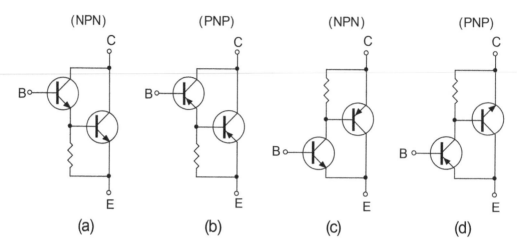

Figure 12.18 Circuit (a) is an NPN type Darlington transistor. Circuit (b) is a PNP type Darlington transistor. Circuit (c) is an NPN type CFP and circuit (d) is a PNP type CFP

The Darlington transistors in Figure 12.18(a) and (b) operate similarly to NPN and PNP type transistors, respectively. The compound transistors in Figure 12.18(c) and (d) are mixed with NPN and PNP transistors [9–11]. It is noted that the compound transistor of Figure 12.18(c) operates as an NPN transistor while that in Figure 12.18(d) as a PNP transistor. They are called a *complementary feedback pair* (CFP) or *Sziklai pair*. The four compound transistors of Figure 12.18 have similar current gain – a product of the current gains of two transistors. Note that driver transistor Q1 in the CFP is arranged such that it compares the output voltage from the emitter with that at the input from the base. This local feedback in CFP produces slightly better linearity than the Darlington transistor. On the other hand, there are $2V_{BE}$ across the base-emitter of the Darlington transistor while there is one V_{BE} in the CFP.

Figure 12.19 shows several output stages using one pair of NPN type and PNP type compound transistors from Figure 12.18. In Figure 12.19(a), it is a complementary emitter follower output stage using a pair of Darlington transistors with typical resistor values. The V_{BE} multiplier, which is needed to set the bias current for the Darlington transistors, is simplified here and denoted as Vbias. For the output stage of Figure 12.19(a), Vbias has to be at least $4V_{BE}$ to properly bias the output transistors. Details of the V_{BE} multiplier design will be discussed shortly.

The output stage of Figure 12.19(b) is an improved version of Figure 12.19(a). It is again an output stage formed by NPN and PNP type Darlington transistors. However, it should be noted that resistor R1 is connected between the emitter of Q1 and Q2 but not to the output. Sometimes, a small capacitor (0.1μF to 1μF) is also used to connect across resistor R1. When the output stage is biased in class AB, at some parts of the signal cycle, output transistor Q3 will be turned off while Q4 is turned on, and vice versa. Resistor R1 helps to speed up discharging the base of the transistor when it is supposed to be turned off. This reduces the so-called *cross-conduction* distortion. Fortunately, modern power transistors are sufficiently fast devices that they rarely exhibit the cross-conduction problem.

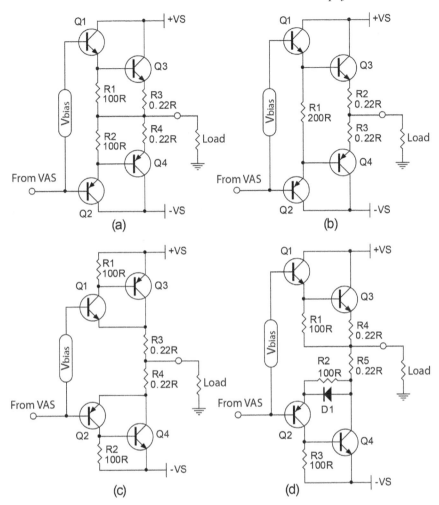

Figure 12.19 Circuit (a) is a complementary push–pull output stage formed by a pair of Darlington transistors from Figure 12.18(a) and (b). Circuit (b) is an improved version of output stage of (a). Circuit (c) is a complementary push–pull output stage formed by a pair of compound transistors from Figure 12.18(c) and (d). Circuit (d) is a quasi-complementary push–pull output stage using NPN output transistors

The output stage of Figure 12.19(c) is formed by a pair of CFPs from Figure 12.18(c) and (d). For this output stage, it requires the Vbias to be at least $2V_{BE}$ for biasing the output transistors. Since feedback is applied from the collector of the output transistor to the emitter of the driver transistor, this may create a stability concern for CFP at high frequency. In general, a CFP output stage may not be as stable as the Darlington transistor output stage.

Figure 12.19(d) is called the *quasi-complementary* (QC) output stage. It is a very popular design going back to the 1960s and early 1970s, when reliable power PNP transistors were not readily available. At that time, designers could only work with NPN transistors. One major problem in the early QC output stage is that the VAS sees a different V_{BE} voltage across transistors Q1/Q3 in the positive cycle and across Q2/Q4 in the negative cycle. By introducing a *Baxandall diode* D1 [12], the VAS now sees an approximate $2V_{BE}$ on both positive and negative cycles. The second problem associated with a QC output stage is the dissimilar output characteristics for the positive and negative cycles. In the positive cycle, the output comes from the emitter follower of Q3, which has a low output impedance. In the negative cycle, the output comes from the collector of Q4, which has a higher output impedance. However, when suitable feedback is applied, power amplifiers with quasi-complementary output stage also produce very low distortion.

The output triples

Figure 12.20(a) is an emitter follower triple, or, more simply, an output triple, which is formed by adding a pre-driver in front of Figure 12.19(b). Note that three emitter followers are cascaded. If the current gain for the pre-driver transistors is $\beta_1 = 120$, driver transistor current gain $\beta_2 = 100$, output transistor current gain $\beta_3 = 60$, the overall current gain of the output triple becomes $\beta_1 \times \beta_2 \times \beta_3 = 720{,}000$. Therefore, when an 8Ω load is connected to the output stage, it is reflected as a 5.76 MΩ load seen by the VAS. It is an easy load for the VAS to drive and does not appear to have any significant loading effect to the VAS. The Vbias has to be greater than $6V_{BE}$ for setting the bias current.

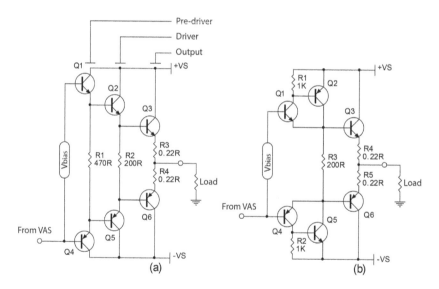

Figure 12.20 Circuit (a) is an output stage formed by a triple complementary emitter follower. Circuit (b) is an output stage formed by CFP then followed by a push-pull emitter follower.

Figure 12.20(b) is another output triple, which is formed by placing an emitter follower after a CFP. In terms of stability, Figure 12.20(b) may not be as good as Figure 12.20(a). However, the required Vbias is reduced from $6V_{BE}$ to $4V_{BE}$.

Another output triple is formed by a fixed-bias diamond buffer, as shown in Figure 12.21. It can be noted that the pre-driver is formed by Q5 and Q6. Unlike the output triple of Figure 12.20(a), in which an NPN transistor is used in the top and a PNP transistor in the bottom, the positions of the NPN and PNP transistors are swapped. In addition, one current source formed by Q1/Q2 is used to set the quiescent current for Q5, while the second current source formed by Q3/Q4 sets the quiescent current for Q6. After the quiescent currents are properly set for Q5 and Q6, the quiescent current for the driver and output transistors is established independent from the VAS. This is different from the output stages discussed earlier, in that they all depend on the VAS to establish the bias current for the output transistors. The diamond buffer output stage does not require the VAS to set the bias current for the output transistors. This offers an added stability to the diamond buffer output stage as well as to the entire power amplifier. More discussion of diamond buffer amplifiers can be found in Chapter 6.

The output triple shown in Figure 12.20(a) has a very high current gain so that it can be easily to be driven by a VAS. However, when high output current is required, additional output transistors should be connected in parallel, as shown in Figure 12.22 to boost output current. If one pair of output transistors (Q5/Q6) can deliver up to 100W to an 8Ω load, three pairs of the same output transistors will be able to deliver up to 300W to the same load. Since three pairs of output transistors are connected in parallel, the power amplifier's output impedance is smaller than would be the case for one pair of output transistors.

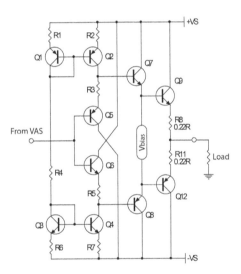

Figure 12.21 This is an output triple formed by a diamond buffer

In the early years, transistors did not come with high breakdown voltages. For example, the V_{CEO} for 2N3055/MJ2955 is only 60V, making them difficult for producing high output power. With the advancement in semiconductor technology, this is no longer an issue for bipolar power transistors. The V_{CEO} is commonly around 250V. For special transistors such as MJL4281A/MJL4302A, the V_{CEO} is 350V.

Figure 12.23 shows how transistors are arranged in a cascode configuration so that transistors with low breakdown voltage can be used with a high power supply. In order to produce high output power, a number of pairs of output transistors (not shown in Figure 12.23) must also be connected in parallel to the output transistors Q9–Q12.

Table 12.1 shows a list of representative bipolar transistors for an audio power amplifier application. They are listed according to the V_{CEO} and output power P_C.

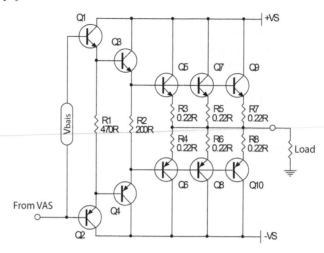

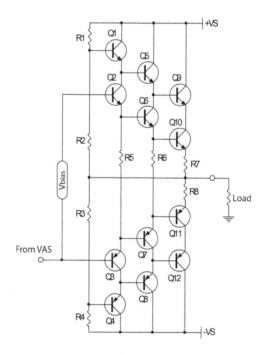

Figure 12.22 An output triple uses three pairs of output transistors (Q5–Q10) in parallel for high output
current demand

Figure 12.23 An output triple can be used for very high voltage application

Table 12.1 A list of representative bipolar transistors for an audio power amplifier application

Parts	V_{ceo} (V)	$I_{c(max)}$ (A)	$I_{b(max)}$ (A)	$h_{FE\,(min)}$	$h_{FE\,(max)}$	I_c (A) for h_{FE}	f_t (MHz)	P_c (W)	C_{ob} (pF)	Built in diode	Package	Manufacturer	Complementary device PNP
NPN													**PNP**
2N3055	60	15	7	20	70	4	2.5	115	—		TO-3	RCA	MJ2955
MJ15015	120	15	7	20	80	8	4	200	500		TO-204	ON Semi	MJ15016
MJ15001	140	15	5	25	150	4	2	200	1000		TO-204	ON Semi	MJ15002
MJ15003	140	20	5	25	150	5	2	250	1000		TO-204	ON Semi	MJ15004
STD01N	150	10	1	5k	20k	6	70	100	—		MT-105	Sanken	STD01P
2SD2561	150	17	1	5k	30k	10	—	200	120	•	MT-200	Sanken	2SB1648
SAP08N	150	10	1	5k	20k	6	—	80	—	•	MT-105	Sanken	SAP08P
SAP15N	160	15	—	5k	20k	—	—	150	—	•	MT-105	Sanken	SAP15P
STD03N	160	15	1	5k	20k	10	—	160	—	•	MT-105	Sanken	STD03P
2SC4388	180	15	4	50	180	3	20	85	300		TO-3P	Sanken	2SA1673
2SC3519A	180	15	4	50	—	5	50	130	250		TO-3P	Sanken	2SA1386A
2SC3856	180	15	4	50	180	3	20	130	300		TO-3P	Sanken	2SA1492
MJ15022	200	16	5	15	60	8	5	250	500		TO-204	ON Semi	MJ15023
2SC3263	230	15	4	40	140	5	60	130	250		TO-3P	Sanken	2SA1294
2SC5200	230	15	1.5	35	60	7	30	150	200		TO-264	Sanken	2SA1943
TTC5200	230	15	1.5	35	—	7	30	150	145		TO-264	Toshiba	TTA1943
2SC6011A	230	15	4	50	180	3	20	160	270		TO-3P	Toshiba	2SA2151A
2SC3264	230	17	5	50	140	5	60	200	250		MT-200	Sanken	2SA1295
MJL3281A	230	15	1.5	50	200	7	30	200	600		TO-264	ON Semi	MJL1302A
MJL21194	250	16	5	20	80	8	4	200	500		TO-264	ON Semi	MJL21193
MJL21196	250	16	5	20	80	8	4	200	500		TO-264	ON Semi	MJL21195
MJL15024	250	16	5	15	60	8	4	250	500		TO-204	ON Semi	MJL15025
MJ21194	250	16	5	25	75	8	4	250	500		TO-204	ON Semi	MJ21193
MJ21196	250	16	5	25	75	8	4	250	500		TO-204	ON Semi	MJ21195
2SC6145A	260	15	4	40	140	5	60	160	250		TO-3P	Sanken	2AS2223A
NJL0281D	260	15	1.5	75	150	3	30	180	400	•	TO-264	ON Semi	NJL0302D
NJL3281D	260	15	1.5	75	150	5	30	200	600	•	TO-264	ON Semi	NJL1302D
MJL4281A	350	15	1.5	80	250	5	35	230	600		TO-264	ON Semi	MJL4302A

Note: Some devices have built in diode for temperature compensation.

V_{BE} *multipliers*

As discussed in the output stages above, a Vbias is needed to set the bias current for the pre-driver, driver, and output transistors. Vbias is a dc voltage created by the transistor Q4 in the typical power amplifier as shown in Figure 12.4. Transistor Q4 and resistors R10 and R11 are arranged in a so-called V_{BE} multiplier configuration. In Figure 12.4, there are $4V_{BE}$ across the base of Q5 and Q6. Therefore, when the output stage is a Darlington transistor, as in Figure 12.4, we need to have Vbias > $4V_{BE}$. However, if the output stage is an output triple, as there is an additional pre-driver, we need to set Vbias > $6V_{BE}$. On the other hand, since the V_{GS} of MOSFET is higher than V_{BE} of BJT, when MOSFETs are used for output transistors, Vbias can be as high as 9V, or $14V_{BE}$. Figure 12.24 illustrates examples of realizing V_{BE} multipliers for different Vbias requirements.

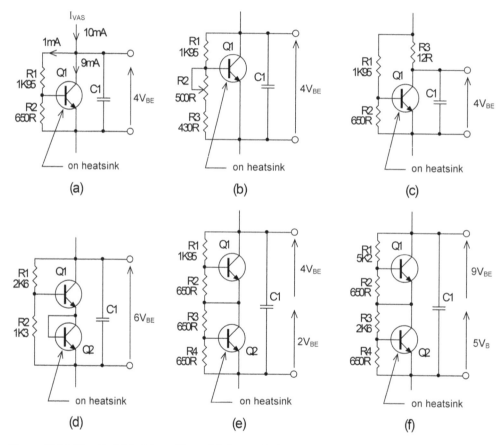

Figure 12.24 Circuits (a) to (c) are V_{BE} multiplier producing $4V_{BE}$. Circuit (c) has an additional resistor, R3, to counterbalance the intrinsic emitter resistor (r_e). V_{BE} multipliers in circuit (d) and (e) offer $6V_{BE}$, but only $2V_{BE}$ are temperature compensated in (e). V_{BE} multiplier in circuit (f) is designed for vertical MOSFET output transistors with $14V_{BE}$ but only $5V_{BE}$ are temperature compensated

Figure 12.24(a) shows a conventional V_{BE} multiplier containing one NPN transistor and two resistors. The resistors are so chosen that most of the current in the VAS, I_{VAS}, flows into the transistor. For typical base-to-emitter voltage $V_{BE} = 0.65V$, this V_{BE} multiplier offers $4V_{BE}$ (2.6V). In order to compensate for the V_{BE} change due to temperature change in the output transistors, transistor Q1 must be mounted in the heatsink near to the output transistors. In practice, an additional variable resistor is used, as shown in Figure 12.24(b) for finely adjusting the bias current. A small capacitor C1 ($0.1\mu F$ to $10\mu F$) can be added to reduce the impedance of the V_{BE} multiplier at high frequency.

It should be noted that BJT has a negative temperature coefficient ($-2.2mV/°C$) for the base-to-emitter voltage, V_{BE}. Thus, it can be seen that V_{BE} decreases as the temperature increases, as shown in Figure 12.25 for a collector current below 8A for a particular power transistor. The collector current around 100mA is the region that we are most concerned with when setting the bias current for the output power transistors.

In Figure 12.24(c), the emitter diode resistance r_e of Q1 is about 2.9Ω for a 9mA collector current. When it is multiplied by the V_{BE} multiplier (with a factor of 4), $r_e \approx 12\Omega$. Thus, there is an added variation in Vbias due to the 12Ω emitter diode resistance. In order to neutralize the effect, a 12Ω resistor (R3) is inserted at the collector of Q1.

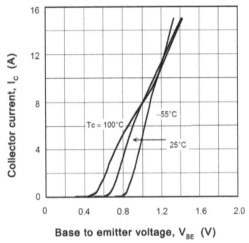

Figure 12.25 A plot of the V_{BE} versus collector current at different temperatures

Figure 12.24(d) shows a diode connected transistor Q2 is used in series with Q1. When Q2 is mounted on the heatsink with the output transistors while Q1 is mounted on PCB, one-half of the total V_{BE} ($3V_{BE}$) is temperature compensated. It may seem that it is over-compensated for a pair of complementary output transistors ($2V_{BE}$). However, it is useful if we consider that there is a thermal attenuation from the transistor's junction to the heatsink, small changes in V_{BE} of the pre-drivers and drivers, and a time delay for Q2 to respond to the temperature change.

Figure 12.24(e) is an alternative way to implement partial temperature compensation. A separate V_{BE} multiplier is formed by transistor Q2 together with R3 and R4. When Q2 is mounted on the heatsink near to the output power transistors, while Q1 is mounted on PCB, only $2V_{BE}$ is temperature compensated. By carefully choosing the resistors for R1–R4, a different level of temperature compensation can be realized.

Figure 12.24(f) shows a V_{BE} multiplier that offers Vbias = $14V_{BE}$ (9.1V). It is a V_{BE} multiplier suitable for vertical MOSFET output transistors. Since the V_{GS} of MOSFET is different from the V_{BE} of BJT, a different level of V_{BE} temperature compensation is used.

Safe operating area

The power transistor (BJT and MOSFET) has a maximum *safe operating area* (SOA) similar to Figure 12.26. For the BJT, the SOA of Figure 12.26(a) shows that the transistor has a maximum allowable collector current of 12A. When the collector current is higher than 12A, the dc current gain will be greatly reduced. And there is a possibility that the transistor will be damaged. The maximum collector-emitter voltage is 120V, which is considered as the minimum value for V_{CEO}. The curve $I_C V_{CE} = 120W$ is the maximum power of the transistor. At any point on this curve, the power dissipation in the transistor is 120W.

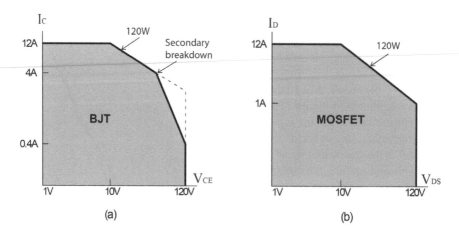

Figure 12.26 Diagram (a) is a typical safe operating area for a bipolar power transistor with a secondary breakdown region. Diagram (b) is the safe operating area for a power MOSFET

Starting from about 50V to 120V, the maximum allowable current is governed by the so-called *secondary breakdown*. Due to the non-uniform thermal properties in the emitter, the junction may create localized heating and hot spots. The temperature dependence of the junction current behaves in such a way that the densities are greater at the hot spots than at other points of the emitter junction. This concentration of current tends to further increase the temperature at the junction. This becomes a downward spiral process leading to thermal run-away and transistor breakdown. Secondary breakdown can be avoided by restricting the operation to the shaded area.

On the other hand, MOSFET does not exhibit secondary breakdown. In general, no secondary breakdown region is given in the SOA of MOSFET, such as in Figure 12.26(b). This benefit is due to the temperature dependence of $R_{DS(ON)}$. The $R_{DS(ON)}$ increases as temperature increases, so that the part of the MOSFET that runs hotter will carry a lower current density. Thus, secondary breakdown is prevented from occurring in MOSFET.

Over current protection

Figure 12.27(a) is a simple current limiting protection circuit that protects the output transistors Q5 and Q6 from exceeding a predetermined current level. Let us first look at the circuit on the top. Resistor R4 is a current sensing resistor. The dc voltage across R4 is proportional to the emitter current of Q5. Resistors R1 and R2 form a potential divider that scales down the dc potential across R4. When Q5's emitter current exceeds a predetermined level, the potential divider has sufficient voltage to turn on transistor Q1. When Q1 is turned on, the collector of Q1 will shunt the base of Q3. Therefore, it limits the signal going into Q3. As a result, the emitter current of Q5 is limited, and the transistor is protected. As soon as the overload is removed, the output stage will return to the normal operation. The complementary part in the bottom half of the circuit works identically to the circuit just described for Q1, Q3 and Q5. Assuming the V_{BE} of Q1 and Q2 is 0.65V, the circuit sets to a current limit of 5A.

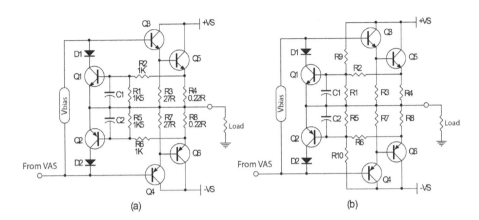

Figure 12.27 Circuit (a) is a simple V-I limiting protection circuit. Circuit (b) is a single slope V-I limiting protection circuit

Blocking diodes D1 and D2 are used to prevent the transistors Q1 and Q2 from conducting during the wrong half-cycle. Small capacitors C1 and C2 (around 10μF) introduce a small time constant that allows higher current to the power transistors for a short interval.

It should be noted that the simple V-I protection circuit of Figure 12.27(a) is set to a fixed current, no matter whether the output level is high or low. At a high output signal level, the V_{CE} of the output transistor is low. Therefore, the output current could be set to a higher level. Contrastingly, for a low output signal level, the V_{CE} of the output transistor is high and the output current should be set to a lower level. Figure 12.26(b) shows a single slope V-I protection circuit [9–11] that allows the current threshold to be higher for a higher output level while the current threshold still remains inside the transistor's SOA.

Power BJT and MOSFET

The pros and cons of using power BJT and MOSFET in audio power amplifier applications are summarized in Table 12.2. Power BJT is noted as a current controlled device. Since there is base current flow in the device, the input impedance is low compared with MOSFET, which is a voltage-controlled device with high input impedance. A power BJT has high transconductance and better linearity compared with MOSFET. Therefore, for the same amount of feedback being used, BJT power amplifiers produce lower distortion. It is well known that a power BJT is associated with secondary breakdown and, therefore, it confines the BJT to a smaller SOA than MOSFET, which is free of secondary breakdown.

The V_{BE} turn on voltage for BJT ranges from 0.6V to 0.7V and, therefore, it is generally not required to match the devices for paralleling. However, the $V_{GS(th)}$ voltage can vary from 4V to 6V for MOSFET. Therefore, it is required to match MOSFET's $V_{GS(th)}$ for paralleling. On the

Table 12.2 A comparison of electrical properties between a power BJT and MOSFET

BJT	MOSFET (Vertical)	Remarks
Current control device	Voltage control device	MOSFET is a voltage control device and no gate current is required.
Low input impedance	High input impedance	Low input impedance requires more complicated driver stage for BJT.
High trans-conductance	Low trans-conductance	Transconductance of MOSFET is around 1/5 to 1/20 of BJT.
Good linearity	Poor linearity	MOSFETs require more feedback to improve the linearity and distortion.
Secondary breakdown	No secondary breakdown	Secondary breakdown confines BJT to be operated in a smaller SOA.
Low V_{BE} (0.6V to 0.7V)	High $V_{GS(th)}$ (4V to 6V)	High variation in $V_{GS(th)}$ makes MOSFET necessary to be matched for paralleling.
Current gain droop	No current gain droop	BJT's current gain decreases at heavy load current.
f_T droop at high current	Increase f_T at high current	BJT's f_T decreases at heavy load current. Thus, bandwidth is reduced.
Lower bandwidth	Higher bandwidth	MOSFET is easier having parasitic oscillation and gate-stopper resistor must be in place.
Input protection not required	Input protection required	Thin gate oxide in MOSFET makes the V_{GS} breakdown voltage around 20V. MOSFET needs to install protection zener between the gate and source.

other hand, current gain droop and f_T droop at high current are also disadvantages for a power BJT. Since the thin gate oxide makes a low V_{GS} breakdown voltage of around 20V, it is required to install zener protection at the gate of a MOSFET.

Vertical and lateral power MOSFETs

Table 12.3 shows a comparison of electrical properties between vertical and lateral power MOS-FETs. It is noted that a vertical MOSFET produces a lower $R_{DS(ON)}$ resistor, higher output current, higher transconductance, and higher speed than a lateral MOSFET. These are important factors for high power applications. However, the $V_{GS(th)}$ voltage for a lateral MOSFET is usually less than 1V. Figure 12.28(a) shows that a lateral MOSFET has a $V_{GS(th)}$ of about 0.4V. For a vertical MOSFET, the $V_{GS(th)}$ voltage can vary from 3V to 4V and, therefore, it is required to select a vertical MOSFET with matched $V_{GS(th)}$ for paralleling.

One interesting feature of the lateral MOSFET is V_{GS}'s temperature coefficient. As shown in Figure 12.28(a), for a particular lateral MOSFET, when the drain current is below 0.08A (i.e., 80mA), V_{GS} decreases when temperature increases from $-25°C$ to $75°C$. This indicates a negative temperature coefficient. However, when the drain current is above 0.08A, the V_{GS} increases for increasing temperature. This indicates a positive temperature coefficient. In other words, at the transition point around 0.08A, the temperature coefficient for V_{GS} is zero. This is a very useful property that makes a lateral MOSFET less crucial for temperature compensation. Since from 0.08A to 0.1A is often the range in which the output transistors are biased, when a lateral MOSFET has a zero temperature coefficient that occurs in this range, it is almost not

Table 12.3 A comparison of electrical properties between vertical and lateral power MOSFETs

Vertical MOSFET	Lateral MOSFET	Remarks
Lower $R_{DS(ON)}$	Higher $R_{DS(ON)}$	The inherent structure makes vertical MOSFET producing lower $R_{DS(ON)}$.
Higher output current	Lower output current	Inherent structure makes it easier for vertical MOSFET to deliver high current.
Higher trans-conductance	Lower transconductance	Vertical MOSFET has higher transconductance than lateral MOSFET.
Higher speed	Lower speed	Vertical MOSFET has higher speed than lateral MOSFET.
High V_{GS} turn-on voltage (4V to 6V)	Low V_{GS} turn-on voltage ($< 1V$)	Lateral MOSFET has a lower V_{GS} turn-on voltage than vertical MOSFET.
Negative temperature coefficient for V_{GS} at low drain current	Positive temperature coefficient for V_{GS} at low drain current	At low drain current (usually less than 0.2A), the V_{GS} of lateral MOSFET has a positive temperature coefficient.
Lower price	Higher price	Lateral MOSFET is higher price and less availability than vertical MOSFET.

necessary for temperature compensating to be installed in the V_{BE} multiplier. As a result, the V_{BE} multiplier can be just a variable resistor without a transistor, as temperature compensation is not required. An example is given in Figure 12.29. It is a lateral power amplifier discussed in a Hitachi application note. This is one of the simplest 100W audio power amplifiers with a very low component count.

On the other hand, vertical MOSFET also has a similar transition point as shown in Figure 12.28(b). However, the transition point is at a much higher drain current. In this example, it is about 3A. Therefore, the V_{BE} multiplier must have temperature compensation.

A list of representative vertical and lateral MOSFET is given in Table 12.4.

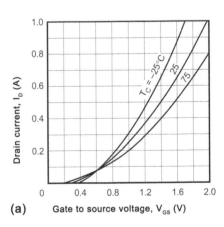

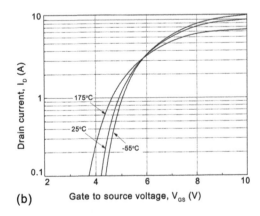

Figure 12.28 Diagram (a) shows the V_{GS} versus drain current of a lateral MOSFET. Diagram (b) shows the V_{GS} versus drain current of a vertical MOSFET

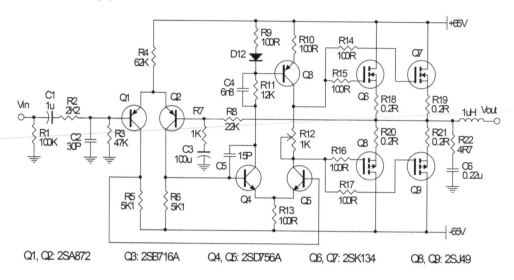

Q1, Q2: 2SA872   Q3: 2SB716A   Q4, Q5: 2SD756A   Q6, Q7: 2SK134   Q8, Q9: 2SJ49

Figure 12.29 This is a 100W audio power amplifier using the lateral MOSFET 2SK134/2SJ49 discussed in a Hitachi application note. The V_{BE} multiplier circuit contains only a simple variable resistor

Table 12.4 This is a list of representative vertical and lateral MOSFETs

	Vertical MOS	Lateral MOS	BV_{DS} (V)	$I_{D(max)}$ (A)	P_D (W)	$V_{G(th)}$ (V)	$|Y_{fs}|$ (S) typ.	C_{iss} (pF)	C_{oss} (pF)	C_{rss} (pF)	Package	Manufacturer	Complementary device
NMOS													**PMOS**
IRF640	•		200	11	125	2 to 4	> 6.7	1300	430	130	TO-220	Vishay	IRF9640
IRFP240	•		200	12	150	2 to 4	> 6.9	1300	400	130	TO-247	Vishay	IRFP9240
FDP12N60NZ	•		600	12	240	3 to 5	13.5	1260	150	12	TO-220	ON Semi	FQP12P20
IXTH15N50L2	•		500	15	300	2.5 to 4.5	6.3	4080	265	68	TO-247	IXYS	IXTH16P20
IXTH24N50L	•		500	24	400	3.5 to 6	7	2500	400	100	TO-247	IXYS	IXTH20P50P
2SK1530	•		200	12	150	0.8 to 2.8	5	900	180	100	TO-247	Toshiba	2SJ201
2SK3497	•		180	10	130	1.1 to 2.1	12	2400	220	30	TO-3P	Toshiba	2SJ618
2SK134		•	140	7	100	0.2 to 1.5	1	600	350	10	TO-3	Hitachi	2SJ49
2SK2221		•	200	8	100	0.2 to 1.5	1	600	800	8	TO-3P	Renesis	2SJ352
2SK1058		•	160	11	100	0.2 to 1.5	1	600	350	10	TO-3P	Renesis	2SJ162
ECW20N20-Z		•	200	16	250	0.1 to 1.5	> 1.4	900	500	16	TO-247	Exicon	ECW20P20-Z

Output stage transfer characteristic

If the output stage is biased in class A operation, ideally the input–output transfer characteristic is similar to that shown in Figure 12.30(a). It is a constant slope, showing a linear relationship between the input and output. When the output swing does not exceed the power supply voltage, there is no distortion. However, when the output swing exceeds the power supply voltage, the output clips, creating severe distortion. The theoretical maximum efficiency for a class A amplifier is 50%. But, in reality, the efficiency can be as low as 25%.

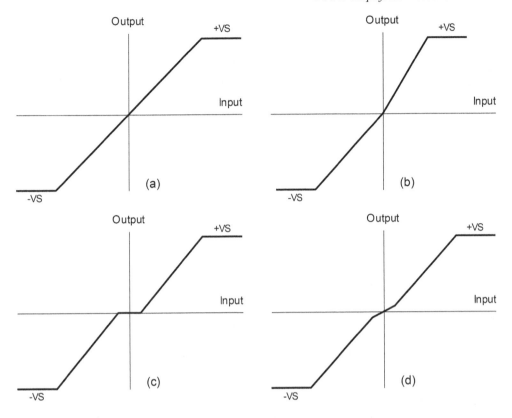

Figure 12.30 Diagram (a) shows the ideal input–output transfer characteristic of a class A output stage. Diagram (b) shows an asymmetrical transfer characteristic of an output stage. Diagram (c) shows the transfer characteristic of a class B output stage producing severe cross-over distortion. Diagram (d) shows the transfer characteristic of a class AB output stage. Some popular power transistors are shown in Figure 12.31

Figure 12.30(b) indicates that there is an asymmetrical transfer characteristic. This can be the result of a quasi-complementary output stage. It may also be due to an unmatched NPN/PNP or NMOS/PMOS output pair. However, by using sufficient global feedback, the asymmetry will be greatly improved and the overall distortion level can be brought to an acceptable level.

Figure 12.30(c) indicates an output stage operating in class B and, therefore, a severe cross-over distortion results. In the cross-over region, the output transistors are turned off. Since feedback cannot turn on transistors from the turn off mode, global feedback is useless in dealing with cross-over distortion. A small amount of dc bias current must be applied to the output transistors so that they operate in class AB.

Figure 12.30(d) indicates an output stage that is slightly biased in class AB. The cross-over region is slightly linearized. When global feedback is applied, the cross-over distortion can be brought to a very low level. Most of the commercial solid-state power amplifiers are operating in class AB. The theoretical maximum efficiency for a class B amplifier is 78.5%. Efficiency for class AB is slightly less than that for class B.

TTC5200   MJL3281A   2SK1530   2SC3519A   2SK2221   2SK3497   IRFP240   FQA10N80C

Figure 12.31 Photos of some power BJTs and MOSFETs

12.7   Classic power amplifiers

Dynaco Stereo 120

Dynaco was founded by David Hafler in Philadelphia in 1955. The company had been a major manufacturer for vacuum tube audio amplifier products. Dynaco's popular vacuum tube power amplifiers included ST70, Mk-II, Mk-III, Mk-IV, and Mk-VI. Mk-II was introduced in 1955 while the final production run of vacuum tube power amplifier Mk-VI occurred in 1976.

Dynaco's first solid-state power amplifier, Stereo 120, was introduced in 1966. Figure 12.32 shows the schematic [15]. It is a quasi-complementary design using NPN output power transistors Q5 and Q6. The input stage and VAS are formed by transistors Q1 and Q2 operating in a common emitter amplifier configuration. Q1 and Q2 produce the necessary open-loop gain for the entire amplifier. C13 creates a Miller capacitor to stabilize the amplifier at high frequency. Components C3, R6, and R7 provide boot-strapping to Q2, so that a large load impedance is seen by the collector of Q2. The boot-strapping is a simple and effective way to increase the voltage gain. In modern days, the input stage is a differential amplifier while the boot-strapping

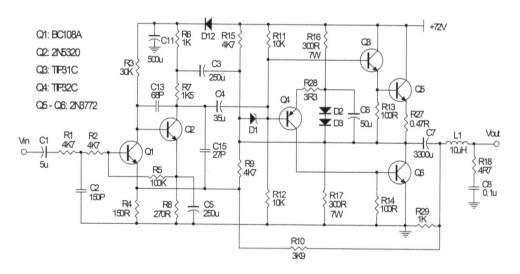

Figure 12.32 Schematic of the Dynaco ST120 amplifier with 60W per channel

circuit is replaced by an active load. It was described in the product manual that the output stage is not biased. In other words, the dc potential across diodes D2 and D3 are below the sum of V_{BE} of Q3, Q4, and Q5. When the power amplifier is warmed up and V_{BE} of Q3 and Q5 is lowered, the output stage begins to bias properly. It is then that the amplifier starts to produce the best sound after warming up.

Phase Linear 400

Phase Linear was an audio manufacturer founded by Bob Carver and Steve Johnston in Seattle in 1970. The first product was the Phase Linear 700, with 350W output power per channel. It was a popular product among professional musicians and audiophiles. The company's second product was Phase Linear 400, with 200W output power per channel, introduced in 1973 and manufactured through 1978.

Figure 12.33 shows the schematic of the Phase Linear 400 [16]. The amplifier took the shape of a modern design using a dual dc power supply so that the output capacitor was eliminated. The input stage is a differential amplifier formed by Q1 and Q2. The VAS comprises Q3–Q5. The output from the input stage is directly coupled to an emitter coupled amplifier formed by Q3 and Q4. The output from Q4 is further amplified by Q5, which it is operating in a common emitter configuration. Components R17, R18, and C11 form the boot-strapping circuit that increases the load impedance for the collector of Q5, so that Q5 produces a very high voltage gain. Components Q6, D6–D8, R19, and R20 form the V_{BE} multiplier that sets the bias current for the quasi-complementary output stage formed by Q7 and Q10–Q18. Transistors Q8 and Q9 form a simple V-I limiting protection circuit to prevent the output transistors from delivering excessive output current.

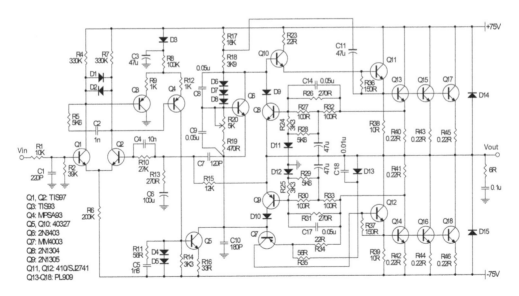

Figure 12.33 Schematic of the Phase Linear 400 amplifier with 200W per channel

Power amplifier with complementary input stage and push–pull VAS

Figure 12.34 shows a power amplifier published by Borbely [2] in 1983. The amplifier contains many features that are still used in modern amplifiers today. The power transistors are lateral MOSFET 2SK134/2SJ49, which were popularized by Hitachi in the 1980s.

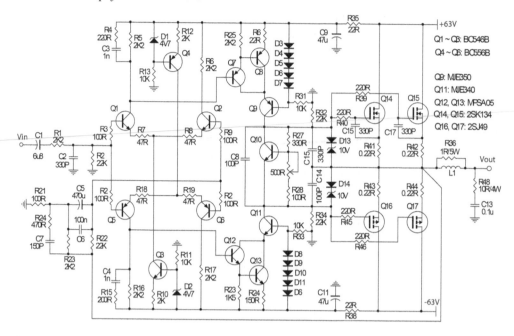

Figure 12.34 A power amplifier developed by Borbely.[2] The input stage is a complementary differential amplifier and the second stage is a push–pull VAS. Output power transistors are lateral MOSFETs. The output power is 120W into 8Ω load

The input stage contains a complementary differential amplifier formed by Q1, Q2, Q5, and Q6 with current sources Q3 and Q4. Capacitor C1 and resistor R2 form a high-pass filter that blocks dc from the source. Resistor R1 and capacitor C2 form a low pass-filter that reduces unwanted radio frequency noise and helps to reduce the rapid rising edge of a large signal. Emitter degeneration resistors (R7, R8, R18, and R19) are used to provide local feedback to the input stage and improve the slew rate of the amplifier. Outputs from the input stage are directly coupled to the push–pull VAS formed by Q7–Q9 and Q11–Q13.

Transistors Q7 and Q12 are emitter followers that buffer the input stage. Transistors Q8/Q9 and Q12/Q13 are push–pull VAS in a cascode amplifier configuration. The cascode VAS has a high voltage gain and wide bandwidth. A string of diodes (D3–D7) is used to set a dc potential at the base of Q9 so that Q8 and Q9 are in a cascode amplifier arrangement while high output swing from the VAS is maintained. Transistor Q10 and several resistors are used to form a V_{BE} multiplier to set the dc bias current for the output transistors. Since MOSFETs are used, the high input impedance of MOSFET eliminates the need for a driver stage. Diodes and zeners (D13, D14) are used to form a protection circuit for power MOSFETs.

The voltage gain of the power amplifier is given by $1 + (R22\|R23)/R21$, which is 21 calculated from the given resistors. The frequency compensation for the input stage is provided by R4, C3, and R15, C4. Components R24 and C7 are frequency compensators used in the global feedback.

It is notable that the David Hafler Company launched a MOSFET power amplifier, DH-200, in 1979, when Erno Borbely was the Director of Engineering. The DH-200 delivered 100W output power into an 8Ω load. The structures of DH-200 and the power amplifier of Figure 12.34 are very similar. The DH-200 has an additional complementary driver in front of the output transistors, which are lateral MOSFET 2SK134/2SJ49. However, the DH-200 has a simple push–pull VAS, while Figure 12.34 uses a cascode push–pull VAS.

12.8 Power amplifier design examples

Figure 12.35 shows the structure of a typical modern class AB power amplifier. The input stage contains a cascode differential amplifier formed by transistors Q1–Q4. A current source (Q5/Q6) is used to set the tail current for the input stage. Typically, the tail current is set to a few mA. The output from the input stage is directly coupled to the VAS formed by Q7 and Q8. Transistor Q7 is working as an emitter follower, which provides a high input impedance to minimize the loading effect to the input stage. It also provides a low output impedance to transistor Q8, which is working as a common emitter amplifier configuration. A small emitter resistor, R14, is used to provide local feedback to the VAS. Transistors Q10 and Q11 form a current source and work as an active load for transistor Q8. Since the current source has high impedance, the collector of Q8 sees a high load impedance and, therefore, the VAS produces a very high voltage gain.

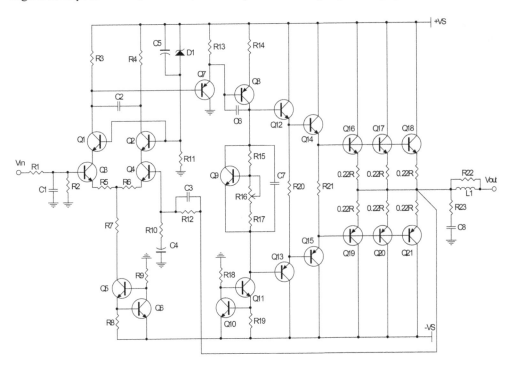

Figure 12.35 This is the structure of a typical power amplifier

Components Q9, R15–R17 form the V_{BE} multiplier that creates sufficient dc voltage spread to bias the output stage. The output stage is formed by an output triple (Q12–Q21) with three pairs of output transistors in parallel (Q16–Q21). Transistors Q12 and Q13 are the pre-drivers while Q14 and Q15 are drivers of the output stage. The emitter triple produces a very high current gain so that the output stage has a very high input impedance. Thus, the output stage's high input impedance minimizes the loading effect to the VAS.

Components C2, C3, and C6 are frequency compensation capacitors that stabilize the amplifier at high frequencies. The input filter (R1 and C1) helps to reduce the radio frequency noise from the input while the Zobel network (R22, R23, C8, and L1) stabilizes the amplifier when driving an inductive load.

It is expected that the power amplifier of Figure 12.35 offers very good performance in terms of sound and measurements. In the following, we will illustrate in several examples how the power amplifier of Figure 12.35 can be further improved.

Example 1: Power amplifier with diamond buffer output stage

Figure 12.35 is a versatile and practical power amplifier that is capable of delivering 150W into an 8Ω load. Note that the output stage dc bias current is set by the V_{BE} multiplier formed by transistor Q9. It is also notable that the dc bias current for Q9 is determined by the current source formed by Q10/Q11 in the VAS. In other words, the dc bias condition of the output stage is dependent on the VAS. On the other hand, transistors Q7 and Q8 are directly coupled to the input stage. The dc voltage across the collector resistor R3 may affect the dc bias current of the VAS. In reality, the input stage and the current source formed by Q10/Q11 are fighting for dominance to control the dc bias current in the VAS. In other words, the input stage can affect the dc quiescent condition of the VAS. In normal operation, the dc voltage spread produced by the V_{BE} multiplier is independent of the dc bias current. However, in an extreme case when the input stage is driven by a large signal such that the input stage reaches saturation, the dc bias condition in the VAS is severely affected. Thus, the dc voltage spread produced by the V_{BE} multiplier and the dc bias condition in the output stage can no longer be maintained. As a consequence, the power amplifier produces high distortion. In the worst case, a high dc offset voltage may be produced at the output that may even damage the loudspeaker. It is true that the above scenario rarely happens. However, if there is a way to make the dc bias condition in the output stage independent from the VAS, the stability of the amplifier will be greatly improved.

Figure 12.36 shows a power amplifier employing a fixed-bias diamond buffer output stage. It uses the same input stage and VAS of Figure 12.35. The diamond buffer output stage is formed by transistors Q11–Q25. MOSFET transistors Q13 and Q14 are working as the pre-driver. Transistors Q11, Q12, Q15, and Q16 form two current sources that set the dc bias current for pre-driver transistors Q13 and Q14. The V_{GS} of Q13 and Q14, together with the dc potential across R20, VR1, and R22 sets the dc bias current of the driver transistors Q17 and Q19. After the dc bias current is established for the driver transistors, Q18 forms a V_{BE} multiplier producing a desired dc voltage spread to bias the output transistors Q20–Q25 in class AB. It is clear that the dc bias condition of the output stage is set up internally and is independent from the VAS. The output from the VAS is directly coupled to the output stage via resistor R17.

It can be seen that MOSFETs are used for the pre-driver and driver. Alternatively, BJTs can also be used. No matter whether MOSFET or BJT is used, the diamond buffer is a triple output stage with very high current gain. Whether to use MOSFET or BJT rather depends on tuning the sound of the power amplifier to your preference. It is a fact that MOSFET and BJT (and JFET, too) produce a different sound. Getting the right mix of components at the right place will definitely be helpful in tuning the sound of a power amplifier.

The dc bias current for the driver transistors Q17 and Q19 is set to around 80mA to 100mA by variable resistor VR1. The dc bias current for the output transistors is set by variable resistor VR2 to around 60mA to 80mA. A dc servo circuit is used to eliminate the output dc offset voltage. Components C2, C3, C5, C6, and C7 are frequency compensation capacitors that stabilize the amplifier at high frequency. Global feedback is set by resistors R10 and R12. Thus, the voltage gain of the amplifier is given by $1 + R12/R10 = 24.5$. When dual ±60V dc power supply is used, the power amplifier delivers 150W output power into 8Ω load.

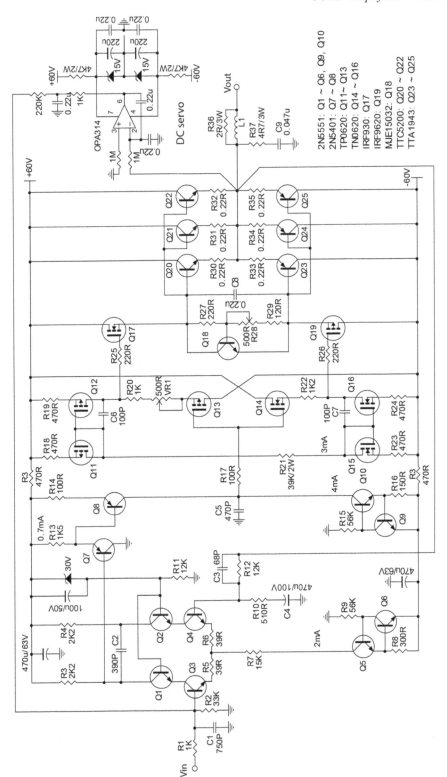

Figure 12.36 A power amplifier has a fixed-bias diamond buffer output stage. It is an improved version of the power amplifier of Figure 12.35. The power amplifier delivers 150W output power to an 8Ω load

Example 2: Power amplifier with improved input stage and VAS

Figure 12.37 shows a variant of the power amplifier described in Example 1. First, this power amplifier now employs a complementary cascode differential amplifier input stage formed by transistors Q1–Q8. Two current sources, Q9/Q10 and Q11/Q12, are used to set the tail current of the complementary cascode differential amplifier. The outputs from the input stage are directly coupled to the complementary emitter follower of the VAS formed by Q13 and Q16. The complementary emitter follower provides a high impedance that reduces the loading effect to the input stage. At the same time, the complementary emitter follower provides low output impedance to the complementary cascode amplifier formed by Q14/Q15 and Q17/Q18. R22 and R24 are small emitter resistors providing local feedback to the VAS. Since the collector of Q15 and Q17 has a high load impedance, the VAS produces a very high voltage gain.

It should be noted that two dual dc power supplies are needed in this power amplifier. A dual ±70V power supply is used for the first two stages plus the pre-driver stage. A dual ±60V power supply is used for the driver and output transistors. The purpose of using a higher power supply for the first two stages is that the output of the power amplifier can swing near to ±60V. Therefore, in principle, if the power transformer has sufficient power and current capacity, the output power produced from the amplifier can reach 225W for an 8Ω load. Another reason for using two power supplies is that a low noise precision regulated dc power supply can be used for ±70V. Since the first two stages produce a very high voltage gain, a good regulated dc power supply will reduce noise as well as improving the sound. If the sound does really matter, it is recommended to use a good regulated dc power supply for the first two stages. Chapter 14 discusses many high performances regulated dc power supplies that are suitable for supplying dual ±70V. However, regulated dc power supply always adds complexity to a power amplifier. Is it worth the trouble to use two dual power supplies and a more complex input stage and VAS? The answer from me is definitely positive. The sound produced from the Example 2 power amplifier simply outperforms that described in Example 1 in every aspect.

The output stage is again a diamond buffer, which is the same design used in Example 1. Variable resistor R31 sets the dc bias current (around 80mA to 100mA) for the driver transistors Q25 and Q27, and R38 sets the dc bias current (around 60mA to 80mA) for the output transistors Q28–Q33 in class AB. The voltage gain of the amplifier is determined by the feedback resistors R3 and R4 and given by $1 + R4/R3 = 24.5$.

Example 3: Power amplifier with dual feedback loops

The power amplifiers in Examples 1 and 2 have distortion levels around 0.01%. It is a respectable figure. However, it is not the lowest distortion that one can achieve. If it is so desired, we can push the distortion level to 0.001% or even lower. However, it should be noted that a low distortion power amplifier may or may not sound better than amplifier with higher distortion. For example, the distortion level of vacuum tube power amplifier is around 0.1% to 1%. But we cannot conclude that solid-state power amplifiers produce better sound than vacuum tube power amplifiers, can we? In the low frequency range, it is true that solid-state power amplifiers produce deeper bass and better control. But at the mids and highs, this is the audio spectrum where the vacuum tube amplifiers excel. Nevertheless, a power amplifier with high distortion (>10%) will definitely have problems producing good sound. We must understand the topology of the amplifier being used and the associated distortion level the power amplifier produces. If it is a solid-state power amplifier producing distortion greater than a few percent, we know there is something wrong. Thus, distortion level is sometimes an indicator for us to gauge whether the power amplifier is correctly designed or working properly. Sometimes it could be a poor PCB layout, a bad

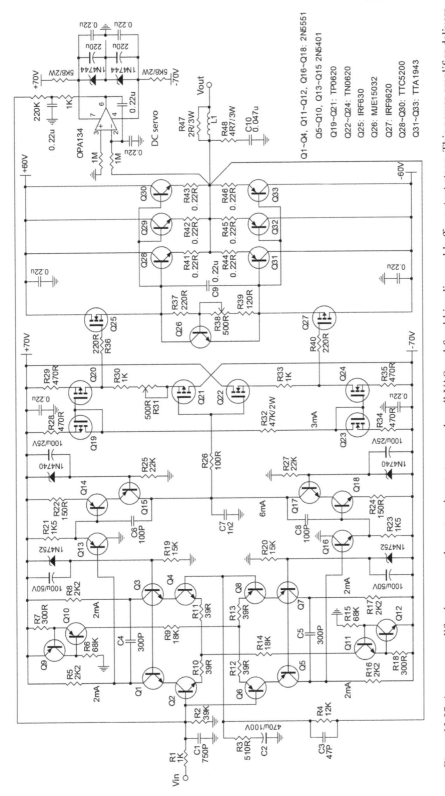

Figure 12.37 A power amplifier has a complementary input stage, push–pull VAS and fixed-bias diamond buffer output stage. This power amplifier delivers 150W output power to an 8Ω load

grounding, or a defective electronic component causing the problem. When the problem is removed, not just the distortion level is reduced, but the sound may be greatly improved.

If we want to reduce distortion in a power amplifier, the most direct way is to employ a greater amount of feedback. However, if the open-loop gain is kept unchanged, employing a greater amount of feedback also reduces the voltage gain of the power amplifier. For example, a power amplifier has a voltage gain of 20 and distortion 0.01% for a certain amount of feedback. If the amount of feedback is doubled, the distortion goes down to 0.005%. But the voltage gain is reduced to 10.

As discussed in Chapter 5, distortion and voltage gain are reduced by feedback to

$$D_f = \frac{D_o}{1 + A\beta} \tag{12.5}$$

where D_o is the original distortion of the amplifier without feedback. D_f is the final distortion after feedback is applied. A is the open-loop gain of the amplifier and β is the feedback factor. The voltage gain with feedback (A_v) is given by the following expression,

$$A_v = \frac{A}{1 + A\beta} \approx \frac{1}{\beta} \text{ for } A\beta \gg 1$$

Therefore, it is clear that for a given open-loop gain A, increasing β will reduce the distortion D_f while the gain A_v is also reduced. However, when the voltage gain is reduced from 20 to 10, for example, the sound quality and stability of the power amplifier will become uncertain. We should always refrain from using an excessive amount of feedback.

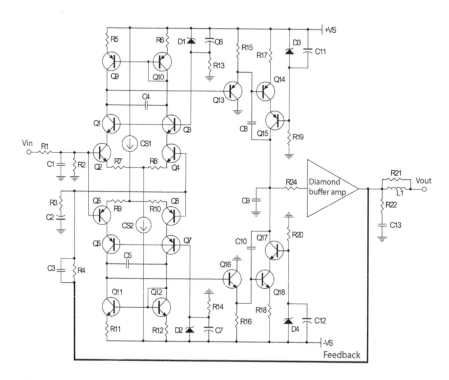

Figure 12.38 The input stage of Example 2 employs an active load to increase its voltage gain as well as the open-loop gain of the entire power amplifier

Alternatively, we can keep the same voltage gain A_v and feedback factor β unchanged while reducing distortion D_r. This can be achieved by increasing the open-loop gain A. To increase the open-loop gain, we make use of active load for the input stage, as shown in Figure 12.38.

As shown in Figure 12.38, transistors Q9/Q10 form an active load for the differential amplifier at the top. Transistors Q11/Q12 form an active load for the differential amplifier at the bottom. Since the collector of Q9 has a high impedance, the input stage produces a very high voltage gain. The voltage gain of the input stage multiplied by the VAS, therefore, produces an immense open-loop gain. When feedback is applied, the distortion can be taken down to a very low level.

There is more than one way of applying feedback. The conventional way is to use one single feedback loop enclosing the entire power amplifier from the output to the input stage, similar to Figure 12.38 and all other power amplifiers discussed earlier. Therefore, the input stage, VAS, and output stage all benefit from the feedback. Another approach is to use two or more feedback loops. Since the output stage is the place that produces the most non-linearity and distortion, if more feedback can be distributed to enclose the output stage, the linearity and distortion of the output stage is greatly improved. As a result, the entire power amplifier benefits. Cherry [13] proposed using nested feedback loops to improve the linearity and distortion of an audio power amplifier. This approach is graphically illustrated in Figure 12.39.

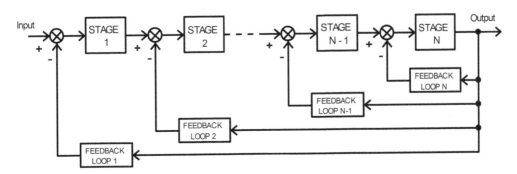

Figure 12.39 Using nested feedback loops to improve linearity and distortion of a power amplifier that contains N stages. Stage 1 is the input stage and stage N is the output stage

Figure 12.40 illustrates a potential approach to applying dual feedback loops. The input stage is a complementary differential amplifier formed by Q1–Q8. Transistors Q9 and Q10 are used as balanced active loads for the differential amplifier at the top while Q11 and Q12 are balanced active loads for the differential amplifier at the bottom. Transistors Q13 and Q14 are emitter followers that work as active load helpers and feed back a common mode voltage to the base of Q9 and Q10. In addition, Q13 and Q14 also provide a low impedance to the VAS, which is a push–pull differential amplifier formed by Q17–Q24. Transistors Q15 and Q16 perform in a similar fashion to Q13 and Q14. Balanced active loads were popularized by Cordell [14].

In addition to the high voltage gain, it is notable that the differential mode is maintained throughout from input stage to VAS. It is suggested in Figure 12.40 that the amplifier works for unbalanced input–unbalanced output. Due to the differential mode the circuit maintains, the circuit may be modified to work for a fully balanced amplifier. However, it requires two diamond buffer amplifiers and modification to the input stage This is discussed in the next section.

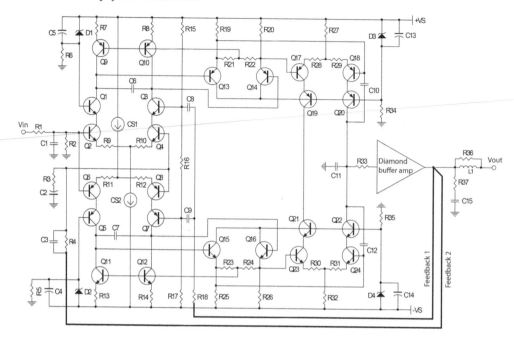

Figure 12.40 A complementary differential amplifier input stage employs balanced active loads. A push–pull differential amplifier is used for the VAS. The differential mode is very well preserved from input stage to VAS. Dual feedback loops can be applied

12.9 Fully balanced power amplifiers

The simplest approach for developing a fully balanced power amplifier is by adding a balanced input–unbalanced output conversion device or circuit in front of a conventional power amplifier, as shown in Figure 12.41. A balanced input–unbalanced output signal transformer is added in Figure 12.41(a), and a simple difference amplifier in Figure 12.41(b). The composite amplifier is capable of amplifying balanced input signals. Since the output is still unbalanced, it is not a true fully balanced amplifier.

There are shortcomings for using a transformer and an op-amp. A transformer does not have a very good low frequency response. It is susceptible to hum noise and electromagnetic interference. Both the transformer and the op-amp may also bring coloration to the sound. However, these conversion devices or circuits are easy to implement.

In Section 7.4, we saw that by applying a Tee-feedback network, a conventional unbalanced input amplifier can be turned into a balanced input amplifier. However, it is found that this technique only works for line level applications where the signal amplitude is just a few volts. Unfortunately, the Tee-feedback network does not work for power amplifiers where the output voltage swing may reach ±50V and even higher.

Figure 12.42 suggests several ways to realize a fully balanced power amplifier. Figure 12.42(a) shows the simplest way. One power amplifier is used to amplify the non-inverting signal +Vin while a second power amplifier is amplifying the inverting signal –Vin. Thus, it takes twice the number of electronic components than an unbalanced input–unbalanced output amplifier to realize a fully balanced power amplifier. It should be noted that there is no interaction or

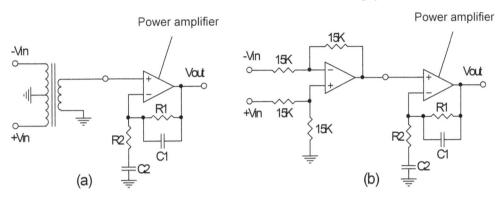

Figure 12.41 Circuit (a) shows a balanced input–unbalanced output small signal transformer added to a power amplifier. Circuit (b) shows a balanced input–unbalanced output small signal amplifier added to a power amplifier

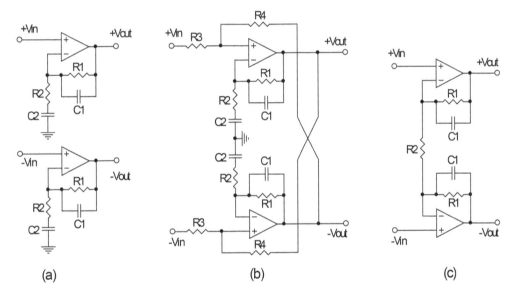

Figure 12.42 Several approaches for realizing a fully balanced power amplifier. Circuit (a) uses two independent power amplifiers. Circuit (b) uses two amplifiers with cross-coupled feedback. Circuit (c) uses a simplified instrumentation amplifier configuration

cross-coupling between the two amplifiers. The non-inverted signal does not interact with the inverted signal through the amplifier, and vice versa. This arrangement is not a true balanced amplifier. It is simply using two independent power amplifiers to handle separate non-inverted and inverted signals. Note that CMRR is poor.

Figure 12.42(b) allows cross-coupled feedback between two power amplifiers. There is interaction between the inverted and non-inverted signals in the amplifier. The CMRR is better than that of Figure 12.42(a). However, care should be taken in applying cross-coupled feedback from two outputs. Applying too much of cross-coupled feedback may create instability in the amplifier. It is recommended to make sure that the condition R1/R2 ≪ R4/R3 is observed. In other words, a small amount of cross-coupled feedback is applied via R3 and R4. The main feedback is applied via resistors R1 and R2, which also determine the voltage gain of the amplifier.

Figure 12.42(c) is a simplified version taken from an instrumentation amplifier configuration. It produces a fully balanced power amplifier. It is also a mature and stable design. To illustrate an example, we make use of the power amplifier described in Example 1 (Figure 12.36) with some changes in component values. When two power amplifiers of Figure 12.36 are used to implement the configuration of Figure 12.42(c), the result is a fully balanced power amplifier, as shown in Figure 12.43(a).

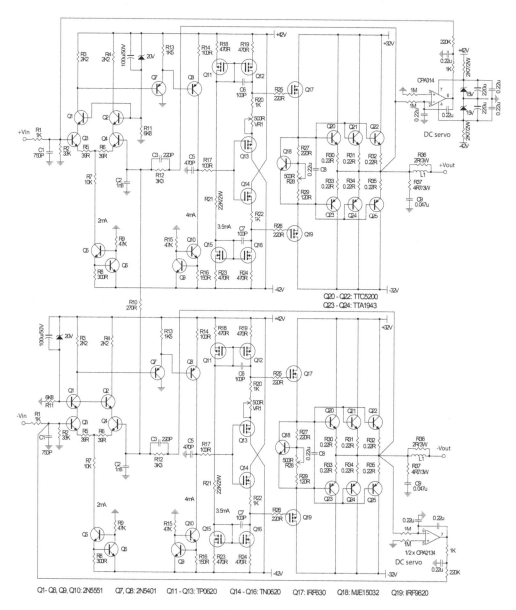

Q1- Q6, Q9, Q10: 2N5551   Q7, Q8: 2N5401   Q11 - Q13: TP0620   Q14 - Q16: TN0620   Q17: IRF630   Q18: MJE15032   Q19: IRF9620

Figure 12.43 (a) A fully balanced power amplifier is constructed by using two of the power amplifiers in Figure 12.36

It is notable that two dual dc power supplies are used. A dual ±42V power supply is used for the input, VAS, and pre-driver stage. In contrast, a dual ±32V power supply is used for the driver and output transistors. Since this is a fully balanced amplifier, the differential output swing can reach a peak voltage of around 60V. In other words, if a power transformer with sufficient power is used, the output power can reach 225W into an 8Ω load. On the other hand, a dc servo circuit is employed to eliminate output dc offset voltage.

There is one improvement that we can apply to the fully balanced power amplifier of Figure 12.43(a). I have noticed that the power amplifier in Figure 12.37 (Example 2) sounds better than the one in Figure 12.36 (Example 1). If the two amplifiers that are used to implement the fully balanced power amplifier shown in Figure 12.43(a) are replaced by the two power amplifiers in Figure 12.37, it is very likely to bring significant improvement to the sound. However, it adds complexity to the construction of the power amplifier. But it is worth trying. Anyway, the photo of the fully assembled PCB for the power amplifier of Figure 12.43(a) is shown in (b).

Figure 12.43 (b) Photo of the fully balanced amplifier illustrated in Figure 12.43(a)

Potential approaches to fully balanced power amplifiers

Figure 12.44 shows two potential approaches to fully balanced power amplifiers. From the ac signal point of view, Figure 12.44(a) and (b) are identical. In both cases, two output buffer amplifiers are needed. One buffer amplifier is used for the non-inverting output +Vout and the second one for inverting output –Vout. Figure 12.44(a) has feedback cross-coupled to the input stage, which is a complementary differential amplifier. The outputs from transistors Q2 and Q4 are directly coupled to a push–pull VAS formed by Q6 and Q8. Then, output from the VAS is directly coupled to a buffer amplifier for output –Vout. Likewise, the outputs from transistors Q1 and Q3 are directly coupled to a second push–pull VAS formed by Q5 and Q7. Then, output from the second VAS is directly coupled to a buffer amplifier for output +Vout.

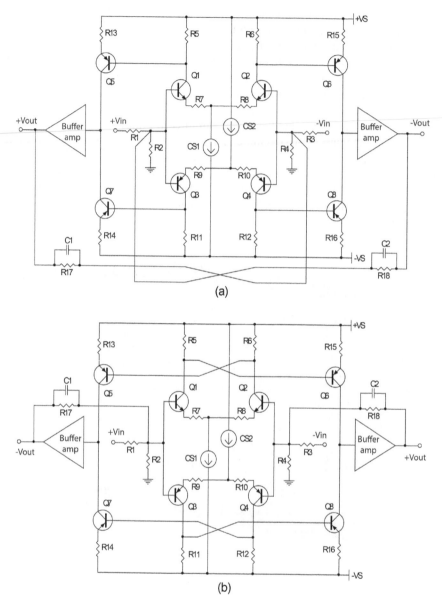

Figure 12.44 Two potential approaches for realizing a fully balanced power amplifier. Circuit (a) has the outputs cross-coupled to the input stage. Circuit (b) has the outputs from the input stage cross-coupled to the VAS

C1 and C2 are frequency compensation capacitors. In reality, more frequency compensation capacitors are required. For example, a small capacitor may be needed to connect between the collectors of Q1 and Q2. A similar capacitor is used between the collectors of Q3 and Q4. On the other hand, a small capacitor may be needed to connect between the base and collector of each of the transistors Q5 to Q8.

Figure 12.44(b) shows a different arrangement for realizing a fully balanced power amplifier. The outputs from the input stage are cross-coupled to two VAS. However, the feedback from the outputs is not cross-coupled to the input. From the ac electrical point of view, Figure 12.44(a) and (b) are identical. However, whether the arrangement (a) or (b) may produce the better sound and stability is difficult to judge by just looking at the two circuit arrangements.

Even though Figure 12.44 looks promising, I did not have the chance to verify them. There is one decisive factor that determines whether Figure 12.44 will work or not. It is the dc output offset voltage. A dc servo circuit has to be applied to each output and the dc output offset must be taken down to ground level. This has yet to be proved.

Figure 12.45 shows another approach to a fully balanced power amplifier. The input stage is a complementary differential amplifier formed by transistors Q1–Q4. The VAS comprises a complementary push–pull differential amplifier formed by transistors Q5–Q8. It is noted that the differential mode is very well preserved from the input stage to the VAS. As a matter of fact, Figure 12.45 is a simplified version of Figure 12.40, which has much higher open-loop gain and, therefore, a fully balanced power amplifier based upon Figure 12.40 will produce much lower distortion. However, Figure 12.45 is a simpler configuration to begin with. If it works and the output dc offset voltage can be taken down to ground level by a dc servo circuit, then it is worth exploring Figure 12.40, whether it can be turned into a fully balanced power amplifier or not.

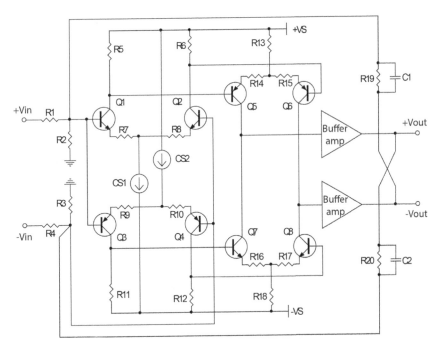

Figure 12.45 This is a potential approach to realizing a fully balanced power amplifier

Mix-and-match of output power transistors

Assume that we are given two different pairs of complementary power BJTs, NPN1/PNP1 and NPN2/PNP2. The sound produced by NPN1/PNP1 excels in the highs, while NPN2/PNP2 excels in the lows. Unfortunately, complementary power BJTs that produce great sound across the entire audio spectrum from lows to highs simply are hard to find. Therefore, for an unbalanced input–unbalanced output power amplifier, a compromise has to be made. It is very often a personal decision, depending on the sonic preferences of the designer. In this scenario, the designer has a choice of using either NPN1/PNP1 or NPN2/PNP2, but not a mix of the two. However, when we move to a fully balanced power amplifier, there are two output stages that we can manipulate. We can mix and match the output transistors as shown in Figure 12.46. In Figure 12.46(a), this is the conventional way of employing the same complementary power BJTs for both outputs. Figure 12.46(b) and (c) show different complementary power transistors are used. The mixed arrangements do produce a blend of the sound from the two different pairs of complementary transistors. The sound produced from a mix of power transistors can be remarkable if we find the right combination.

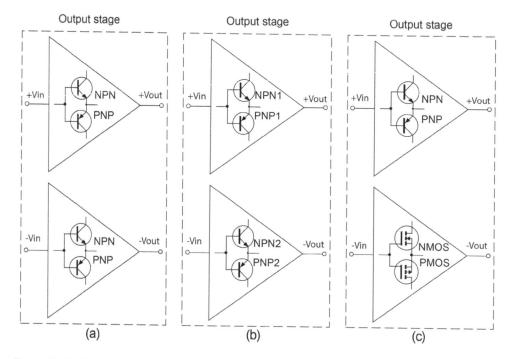

Figure 12.46 Two identical pairs of complementary power BJTs (NPN/PNP) are used in two output stages for a fully balanced power amplifier in circuit (a). Two different pairs of complementary power BJTs (NPN1/PNP1) and (NPN2/PNP2) are used for two output stages in circuit (b). One pair of complementary power BJTs and one pair of complementary power MOSFETs are used for two output stages in circuit (c)

12.10   Loudspeaker protection

UPC1237 is a versatile loudspeaker protection integrated circuit in a SIP-8 package. It was developed by NEC and has been a popular choice for loudspeaker protection application. Since

the production of UPC1237 was discontinued by NEC, several manufacturers have begun producing the integrated circuit as a second source. These second source products are working
fine. The UPC1237 is an easy-to-use, low cost and reliable solution for loudspeaker protection.
It provides the following functions:

- output dc offset voltage protection;
- over current protection;
- time delay function;
- fast amplifier disconnecting from power off.

Figure 12.47 shows the functional block diagram of the UPC1237. Pin-8 is to connect to
the dc power supply VCC ranging from 25V to 60V. Pin-5 is connected to ground. Pin-6 is to
drive a dc relay with up to a maximum 80mA current. Pin-3 provides an option to select latch or
automatic reset after the protection is triggered. The functions of the other pins would be better
be explained by an example.

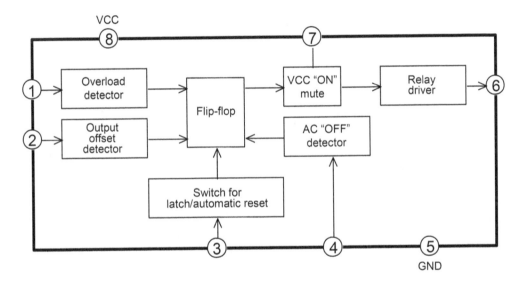

Figure 12.47 Functional block diagram for speaker protection integrated circuit UPC1237

A typical application is shown in Figure 12.48. It is used to protect a 50–60W power amplifier into an 8Ω load. The main power transformer employs bridge rectifiers producing a dual dc
power supply ±VCC, which is approximately ±30V to ±32V. The power amplifier's output is fed
to a relay RL1 and to the pin-2 via resistor R6. Components R6 and C4 produce a time delay so
that a small dc is permitted by the circuit before the protection kicks in. To alter the delay time,
R6 can be changed. But it is best to work from 56kΩ to 75kΩ.

When an emitter current flows through emitter resistor RE, a potential is developed across it.
This potential is monitored by pin-1 via transistors Q3 and Q4. When emitter currents exceed a
predetermined level for over-current protection, transistor Q3 will conduct so that its collector
current produces a dc potential across R3 to turn on Q4. As long as there is 0.11mA flowing into
pin-1, protection is triggered by UPC1237 to disconnect the loudspeaker. It is noted that resistors R3 and R7 both affect the over-current threshold. Diode D3 prevents the signal from the
negative cycle from turning on Q3 accidentally.

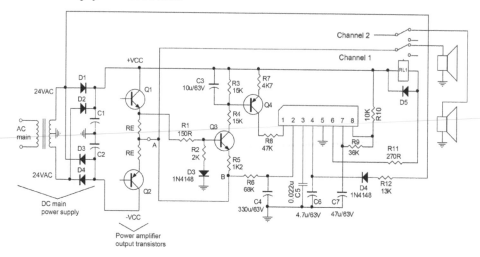

Figure 12.48 A typical loudspeaker protection circuit by UPC1237 for a 50–60W power amplifier

When the protection is triggered, pin-3 determines whether the protection is latched or automatically reset. If pin-3 is connected to a small capacitor, 0.022μF, to ground, the protection is latched until the main power is switched off and the over-current problem is resolved. When pin-3 is connected directly to the ground, the protection is automatically reset after the over-current problem is resolved.

Components R12, D4, and C6 produce a small dc voltage at pin-4. When the main power transistor is switched off, the small dc voltage will drop quickly and pin-4 will trigger to disconnect the loudspeaker immediately. This avoids a loud, unpleasant pop noise passing to the loudspeaker when the power is switched off. It may require trial and error to select the right resistor value for R12.

Components R9, R10, and C7 determine a time delay to connect the loudspeaker when the power transformer is switched on. Note that a dc potential is gradually developed at pin-7 until the UPC1237 is triggered to connect the loudspeaker. Resistor R10 limits the dc potential to pin-8, which has a maximum limit of 8V. Therefore, the value for R10 depends on the dc supply VCC.

If the ac/dc power supply is different from that given in Figure 12.48, resistor values for R10, R11, and R12 have to be changed. The resistor values are shown in Table 12.5.

Table 12.5 Suggested components for different ac/dc supply voltages

VCC (V)	24	30	36	42	48	54	60
R10 (kΩ)	6.5	10	12	13	16	18	20
R11 (Ω)	0	160/1W	330/1W	490/2W	680/2W	820/3W	910/3W
AC (V)	17	21	26	30	34	38	43
R12 (kΩ)	10	12	15	18	22	25	27

We can make use of a signal generator to determine resistors R3 and R7 for setting the over-current threshold as shown in Figure 12.49. The 1kHz sine wave is set to a peak voltage, say V_p. It is equivalent to the voltage across emitter resistor RE of Figure 12.49(b), i.e., $V_p = I_p RE$. For 100W output power to an 8Ω load, we have $(I_p/\sqrt{2})^2(8\Omega) = 100W$, i.e. $I_p = 5A$. If emitter resistor RE = 0.22Ω, $V_p = (5A)(0.22\Omega) = 1.1V$. Thus, we feed a 1kHz sine wave with

1.1V$_p$ amplitude to the circuit of Figure 12.49(a). First, keep R7 = 4.7kΩ. Then choose a small resistor (say 5kΩ) for R3 to begin with. When the circuit is switched on and the 1kHz signal is connected to the circuit, wait for a few seconds to see if the relay RL1 is triggered. If it is not, increase the resistor for R3. And repeat the process until the relay is triggered. This process gives us a method to determine the resistor R3 for setting the over-current threshold.

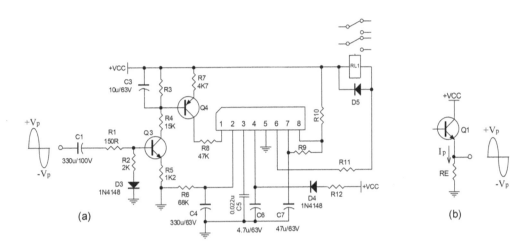

Figure 12.49 Circuit (a) uses a 1kHz signal from a sine wave generator to determine R3 for over-current protection. In circuit (b), the 1kHz signal is equivalent to the signal across RE

To verify if the dc offset protection works properly, the power amplifier must be disconnected from the loudspeaker protection circuit. In Figure 12.48, point A (power amplifier's output) and point B should be open circuit. After the dc power supply for the loudspeaker protection circuit is turned on, feed a +3V dc to point B. The relay RL1 should be triggered in a few seconds. Then repeat the test with −3V.

12.11   Exercises

Ex 12.1 Figure 12.24(a) to (f) shows different configurations for V_{BE} multiplier. Verify the resistors and the dc spread voltages for each figure.

Ex 12.2 Figure 12.27(a) is a simple V-I limiting protection circuit. Assume that the V_{BE} turn on voltage for Q1 and Q2 is 0.65V. Verify that, for the given resistors, when the emitter current of Q5 and Q6 exceeds 5A, transistors Q1 and Q2 will be turned on to limit the output current by shunting the base of Q3 and Q4.

Ex 12.3 Figure 12.50 shows a fixed-bias diamond buffer amplifier with all bipolar transistors. Assume that this diamond buffer amplifier is used to replace the one in Figure 12.36, which contains MOSFETs and BJTs. Assuming the transistors are given as: Q11 to Q13 = 2N5401, Q14 to Q16 = 2N5551, Q17 = MJE15034, Q19 = MJE15035, if the dc bias current for Q13 and Q14 is set to 4mA, determine the resistors R18 to R24.

Ex 12.4 Figure 12.38 is a potential power amplifier that utilizes dual feedback loops. Verify whether or not this circuit arrangement works by completing the design.

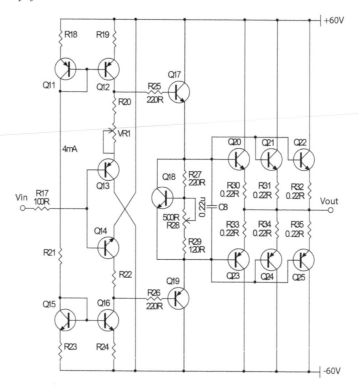

Figure 12.50 This is a diamond buffer amplifier for Ex 12.3

Ex 12.5 Figure 12.44 and Figure 12.45 are potential fully balanced power amplifiers. Verify if these circuit arrangements work by completing the design.

Ex 12.6 Figure 12.43 shows a fully balanced power amplifier which is realized by the two power amplifiers in Figure 12.36. Now replace the two amplifiers with the one shown in Figure 12.37.

References

[1] Thiele, A.N., "Load circuit stabilizing networks for audio amplifiers," Proc. IREE pp. 297–300, Sept. 1975.

[2] Borbely, E., "High power high quality amplifier using MOSFET," Wireless World, March 1983.

[3] Solomon, J.E., "The monolithic op amp: a tutorial study," *IEEE Journal of Solid-state Circuits,* vol. SC-9, pp. 314–332, Dec. 1974.

[4] Solomon, J.E., et al., "A self-compensated monolithic operational amplifier with low input current and high slew rate," *Int. Solid-state Circuits Conf. Digest Tech. Papers,* vol. 12, pp. 14–15, Feb. 1969.

[5] Gray, P.R. and Meyer, R.G., *Analysis and design of analog integrated circuits,* Wiley, 3rd edition, 1992.

[6] Garde, P., "Amplifier first-stage design for avoiding slew-rate limiting," *Journal of Audio Engineering Society,* vol. 34, no. 5, pp. 349–358, May 1986.

[7] Garde, P., "Slope distortion and amplifier design," *Journal of Audio Engineering Society,* vol. 26, no. 9, Sep. 1978.

[8] Jung, W.G., et al., "An overview of SID and TIM, part I," Audio, June 1999.

[9] Slone, G.R., *High-power audio amplifier construction manual*, McGraw Hill, 1999.

[10] Self, D., *Audio power amplifier design*, Focal Press, 6th edition, 2013.

[11] Cordell, B., *Designing audio power amplifiers*, McGraw-Hill, 2011.

[12] Baxandall, P.J., "Symmetry in class B," Letter, pp. 416–417, Wireless World, Sept. 1969.

[13] Cherry, E.M., "Nest differentiating feedback loops in simple audio power amplifiers," *Journal of Audio Engineering Society*, vol. 30, no. 5, pp. 295–305, May 1982.

[14] Cordell, R.R., "A MOSFET power amplifier with error correction," *Journal of the Audio Engineering Society*, vol. 32, no.1, pp. 2–19, Jan. 1984.

[15] Dynaco Stereo 120 power amplifier user manual.

[16] Phase Linear 400 power amplifier user manual.

[17] Otala, M., "Transient distortion in transistorized audio power amplifiers," *IEEE Trans. Audio Electroacoust*, vol. AU-18, pp. 234–239, Sept 1970.

[18] Otala, M. and Leinonen, E., "The theory of transient intermodulation distortion," *IEEE Trans. Acoustic, Speech and Signal Processing*, vol. ASSP-25, pp. 2–8, Feb. 1977.

13 Power amplifiers – hybrid

13.1 Introduction

A simplified vacuum tube push–pull power amplifier is shown in Figure 13.1. It is a three-stage configuration. The input stage contains triode T1 working as a common cathode amplifier that produces a moderate voltage gain. The output from the input stage is directly coupled to the second stage, which contains triode T2 working as a split-load phase splitter. Two out-of-phase signals are produced by T2 and they are coupled to the output stage via capacitors C2 and C3. The output stage comprises power tubes T3, T4, and an ultra-linear output transformer. The power tubes are working in a push–pull configuration that may produce 10–40W output power from this circuit.

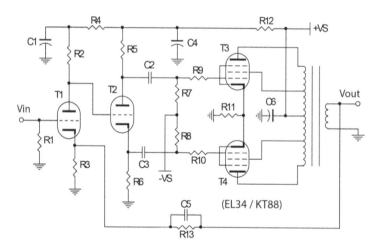

Figure 13.1 A simplified vacuum tube push–pull power amplifier

The vacuum tube push–pull power amplifier of Figure 13.1 can be entirely replaced by solid-state devices. Such a solid-state push–pull power amplifier can be found in Figure 12.2 of Chapter 12. Even though the power amplifiers of Figure 12.2 and Figure 13.1 have a similar configuration, they produce a different sound. It has to be because one is using solid-state devices and the other is using vacuum tubes. It is commonly believed that a vacuum tube amplifier produces good mid-range sound while solid-state amplifier produces good bass. If a hybrid

DOI: 10.4324/9781003369462-13

power amplifier is made with a combination of vacuum tubes and solid-state devices, the amplifier may benefit from both worlds. In this chapter, we discuss several examples of hybrid power amplifier design.

Since power vacuum tubes produce a considerable amount of heat, it is natural to replace them with solid-state power transistors such as the depletion type power MOSFETs in Figure 13.2(a) and the power JFET in Figure 13.2(b). Since MOSFET and JFET are three-terminal devices, the output transformer does not need to have an ultra-linear configuration. On the other hand, power MOSFET and JFET have different output impedance compared with power tubes. For instance, the primary impedance for a push–pull output transformer using KT88 is typically 5kΩ. It is very unlikely that a 5kΩ output transformer provides good matching impedance for power MOSFETs and JFETs. When an output transformer does not have matching impedance, the output power is significantly reduced. Thus, it is necessary to find the optimized impedance for the output transformer to match with the power MOSFET or JFET to produce optimal output power and distortion. It may take a long process to find the optimal impedance through numerous tests. Table 13.1 shows a list of representative depletion type power MOSFETs and power JFETs.

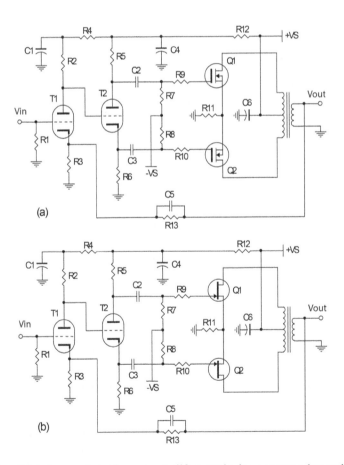

Figure 13.2 Circuit (a) shows a hybrid power amplifier employing vacuum tubes and depletion type power MOSFETs, while circuit (b) is using vacuum tubes and power JFETs

Table 13.1 A list of representative depletion type power MOSFETs and power JFETs

Part number	Depletion MOSFET	Power JFET	V_{DSS} (V)	I_D at $T_C =$ 25°C (A)	$V_{GS\text{-}off}$ max (V)	$R_{DS(ON)}$ (Ω)	P_D (W)	C_{iss} (pF)	C_{rss} (pF)	Package type	Manufacturer
IXTH16N10D2	•		100	16	-4.5	0.064	830	5700	940	TO-247	IXYS
IXTH16N20D2	•		200	16	-4.5	0.08	695	5500	607	TO-247	IXYS
IXTH6N50D2	•		500	6	-4	0.5	300	2800	64	TO-247	IXYS
IXTH16N50D2	•		500	16	-4	0.3	695	5250	130	TO-247	IXYS
IXTH6N100D2	•		1000	6	-4.5	2.2	300	2650	41	TO-247	IXYS
UJ3N065080K3S		•	650	32	-14	0.095	190	630	88	TO-247	United Silicon Carbide
UJ3N120070K3S		•	1200	33.5	-14	0.09	254	985	95	TO-247	United Silicon Carbide
SJEP120R100A		•	1200	17	1.25	0.075	114	670	97	TO-247	SemiSouth[a]
IJW120R100T1		•	1200	18	-15	0.08	190	1200	30	TO-247	Infineon[b]

Notes: (a) This device was aimed for audio application with a low $V_{GS(th)}$. It is suitable for single-ended class A power amplifiers and class AB push-pull power amplifiers in quasi-complementary configuration. However, SemiSouth Laboratories was closed in 2013. (b) This device was introduced by Infineon on 2012. However, it is now announced end-of-life by the manufacturer when this table is being compiled by the author.

13.2   Unbalanced hybrid power amplifiers

The hybrid power amplifier of Figure 13.2 is an improvement over the vacuum tube power amplifier of Figure 13.1, in the sense that power tubes are replaced by robust and reliable solid-state power transistors. However, a bulky output transformer is still needed. In the following examples, we are going to discuss several hybrid power amplifiers that do not require output transformers.

Example 1

Figure 13.3 shows a hybrid power amplifier employing vacuum tubes for signal amplification and solid-state devices for output current amplification. It is in a three-stage power amplifier configuration. The input stage is a differential amplifier that contains triodes T1a and T1b. The output from the input stage is directly coupled to the second stage that contains triode T2a working as a common cathode amplifier. The cathode resistor is partially bypassed so as to maximize the voltage gain while a small 10Ω resistor (R15) is used as degeneration resistor that provides local feedback to help linearize the second stage. The output from the second stage is coupled to the output buffer amplifier via capacitor C4.

The output stage is a solid-state semiconductor buffer amplifier [1] as shown in Figure 13.4. It is referred to as a floating-bias buffer amplifier, discussed in Chapter 6. Here, a dual ±45V dc power supply is used. Two pairs of power complementary BJT transistors (Q8–Q11) are used for the output. A dc servo circuit is used to eliminate output dc offset voltage. If a sufficient power supply is given, the buffer amplifier and, therefore, the hybrid power amplifier, is capable of delivering 100W output power to an 8Ω load.

As shown in Figure 13.3, a relay is placed at the input of the buffer amplifier. The relay is arranged in such a way that the input of the buffer amplifier is momentarily bypassing to ground via a 220Ω resistor (R19) when the hybrid power amplifier is powered up. After the hybrid power amplifier is switched on, it takes about 20–30 seconds for the vacuum tubes to warm up before they can operate properly. However, the solid-state buffer amplifiers operate almost instantly when the power supply is switched on. Therefore, during the 20–30 seconds

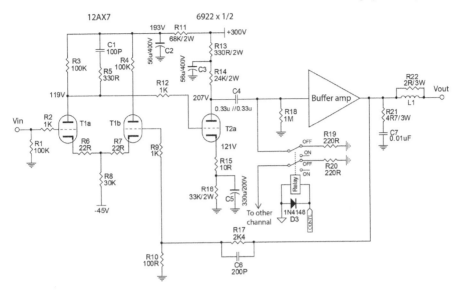

Figure 13.3 A hybrid power amplifier employing vacuum tubes for signal amplification and a solid-state output buffer amplifier for current amplification

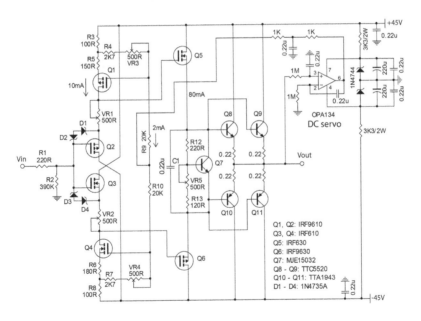

Figure 13.4 A floating-bias diamond buffer amplifier for the hybrid amplifier of Figure 13.3

warm-up process, the buffer amplifier is already starting to amplify the signal coming from the plate of triode T2a.

Suppose that the steady state dc potential at the plate of T2a is 207V, as shown in Figure 13.3. At the moment the power amplifier is switched on, the triode is in a cold condition and it draws no plate current. Therefore, the dc potential at the plate of T2a is 300V. When the triode starts to warm up and is drawing plate current, the plate dc potential starts to drop from 300V to 207V.

This is a drop of 93V. However, the drop of 93V plate dc voltage is not steady and gradual. The plate dc voltage may bounce up and down a few times before resting at 207V. Therefore, during the 20–30 seconds warm-up process, the buffer amplifier sees a transient signal with 93V amplitude and it tries to amplify it. Since the dc power supply for the buffer amplifier is 45V, when the buffer amplifier responds to a 93V transient signal, the output of the buffer amplifier clips heavily. This should be avoided.

A time-controlled relay is therefore required and it is operated in such a way that during the first 30 seconds after the power amplifier is switched on, the input of the buffer amplifier is bypassed to ground via a 220Ω resistor. This limits the signal feeding into the buffer amplifier. After 30 seconds, when the vacuum tubes are properly warmed up, the relay disconnects the 220Ω resistor. Then the buffer amplifier is ready to take on the input.

In Figure 13.3, the feedback network is formed by resistors R10 and R17. Thus, the voltage gain is approximately equal to $(1 + R17/R10) = 25$. C6 is a frequency compensation capacitor that stabilizes the power amplifier at high frequency. The network formed by C1 and R5 also helps to stabilize the amplifier at high frequency.

Example 2

A differential amplifier is used for the input stage of the hybrid power amplifier of Figure 13.3. It should be noted that the use of differential amplifier is popular in a solid-state amplifier but not in a vacuum tube amplifier. Figure 13.3 serves as an example that illustrates how an unbalanced input–unbalanced output hybrid power amplifier can be realized. In practice, however, the input stage is often a simple common cathode amplifier employing just one triode, as shown in Figure 13.5.

The hybrid power amplifier of Figure 13.5 is again a three-stage configuration. The input stage is a common cathode amplifier formed by triode T1. The second stage and output stage are identical to Figure 13.3. If comparing Figure 13.3 and Figure 13.5, the hybrid power amplifier of Figure 13.5 uses one less triode and the negative dc power supply is no longer needed. Obviously, Figure 13.5 is a better approach.

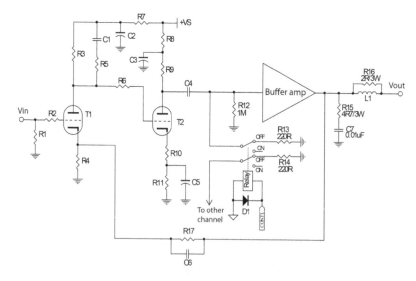

Figure 13.5 A hybrid power amplifier using fewer components than the one in Figure 13.3. Negative power supply is not required

13.3 Fully balanced hybrid power amplifiers

Figure 13.6 is a potential design for a fully balanced hybrid power amplifier. However, there is a problem associated with this design. When there is no input signal, the dc quiescent potentials at the plate of T2a and T2b are biased properly. But if an input signal is present, the amplified signal will cause the two plate voltages to shift in opposite directions. As a consequence, the output signal is clipped at one cycle earlier than the other cycle, as shown in Figure 13.6. Discussion of this asymmetrical behavior is presented in Section 11.8. This problem can be resolved by replacing the simple differential amplifier in the second stage with a dc self-balanced differential amplifier in the following examples.

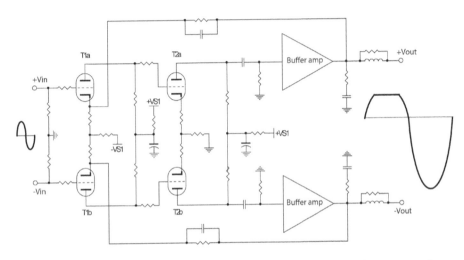

Figure 13.6 A fully balanced hybrid power amplifier producing an asymmetrical output waveform

Example 1

Figure 13.7 shows a fully balanced hybrid power amplifier. The second stage employs a dc self-balanced differential amplifier [2]. Given this new second stage arrangement, the problem of an asymmetrical output is resolved. Diodes and zeners (D1–D4) are employed to protect the triodes T2a and T2b by limiting the grid-cathode dc voltage during the warm-up process after the amplifier is switched on. Details can be found in Section 11.8.

Referring to Figure 11.30, it is a vacuum tube fully balanced power amplifier. The output stage employs a pair of KT88s and an ultra-linear output transformer. When comparing this with the fully balanced hybrid power amplifier of Figure 13.7, the power tubes and output transformer are replaced by two solid-state buffer amplifiers. This has several advantages.

The first is improved reliability. A power tube such as KT88 has a lifespan of several thousand hours. Depending on how much output power is often driven, it is a common practice to replenish the KT88 power tubes after around one thousand hours in service. Even though the power tubes can be operated much longer than that, the sound quality usually starts deteriorating after a thousand hours or so. But when power tubes are replaced by solid-state buffer amplifiers, the power transistors do not need to be replaced for many years unless the output is deliberately shorted while the amplifier does not have over-current protection.

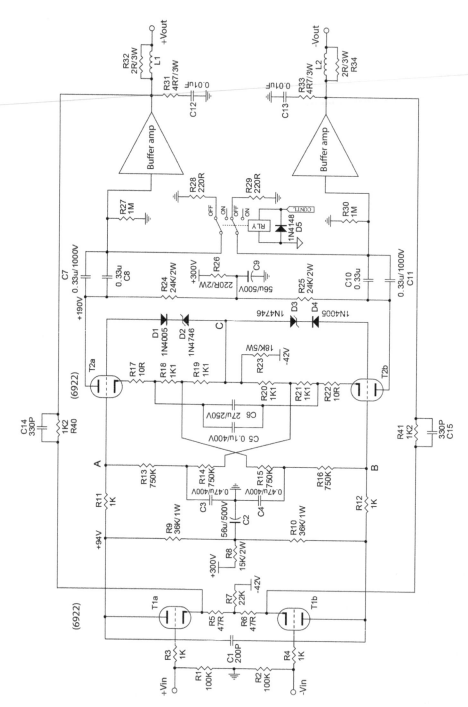

Figure 13.7 A hybrid power amplifier using a simple differential amplifier input stage and a dc self-balanced differential amplifier second stage

The second advantage of using a solid-state buffer amplifier is that a hybrid power amplifier produces a much deeper bass than a pure vacuum tube power amplifier. Since vacuum tubes are still being used for signal amplification, a hybrid power amplifier can retain the so-called warm tube sound for the mid-range while a solid-state buffer amplifier excels in the bass with tremendous output current and low output impedance. A well-designed hybrid power amplifier may benefit from both worlds.

The output stage contains two buffer amplifiers, as shown in Figure 13.8. One buffer amplifier handles the non-inverted signal (+Vin) while the other one does so for the inverted signal

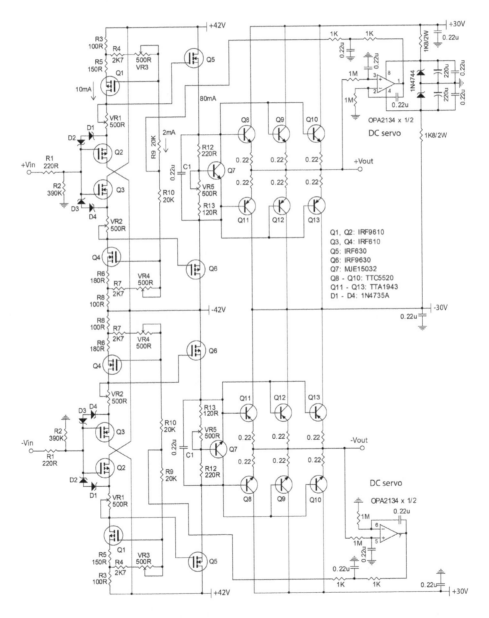

Figure 13.8 A pair of buffer amplifiers for Figure 13.7

(-Vin). Note that two dual dc power supplies, ±42V and ±30V, are used. Given this arrangement, each buffer amplifier delivers nearly 30V peak output. Since the hybrid power amplifier is operating fully balanced, it is capable of delivering nearly 60V peak output, which is 225W output power to an 8Ω load.

Example 2

Figure 13.9 shows a different approach for realizing a fully balanced hybrid power amplifier. It employs dual balanced feedback loops [3]. The first feedback loop, marked as "Feedback 1," encloses the first and the second stage. This helps to broaden the bandwidth and improve the linearity of the first two stages. The second feedback loop, marked as "Feedback 2," encloses the entire power amplifier. This sets the voltage gain of the amplifier and improves overall linearity, distortion, and bandwidth of the entire power amplifier.

The input stage contains a cascode differential amplifier. It amplifies the input signal as well as the feedback signal. The difference between the input and feedback signals is amplified and directly coupled to the second stage for further amplification. The second stage is similar to the one used in Example 1. It is a dc self-balanced differential amplifier so that it will not produce an asymmetrical output.

A resistor network formed by R33 and R34 is used to scale down the plate dc potential of triode T3a. For instance, if triode T3a's best operating plate dc potential ranges from 270V to 390V, this corresponds to a dc potential at the junction between R33 and R34, ranging from 2.38V to 3.44V. The same working principle also applies to the resistor network formed by R35 and R36 that scales down the dc potential from the plate of triode T3b. If an external detecting circuit is employed to monitor the dc potential at the resistor network, a warning light or a power shut-down will take place when the dc potential exceeds the range of 2.38V to 3.44V. When one of the vacuum tubes, T1–T3, is near to the end of life, it will cause the dc plate potential of T3a or T3b to exceed the safe operating dc potential. Thus, an external detecting circuit will jump in and take action.

Figure 13.10 shows a pair of buffer amplifiers for the fully balanced hybrid power amplifier in Figure 13.9. A dual ±45V dc power supply is used. The buffer amplifier is similar to the one in Figure 13.4, except it has greater output power. Note that in the positive phase buffer amplifier on the top, there is a dc 3V dropped across source resistors R3 and R5 at Q1, and also across R6 and R8 at Q4. Therefore, the maximum output voltage from this buffer amplifier becomes 42V. Similarly, this also applies to the negative phase buffer amplifier. Since the differential output is given by +Vout – (–Vout), the maximum output voltage from this fully balanced power amplifier is 42V × 2 = 84V. In other words, the fully balanced hybrid power amplifier of Figure 13.9 is capable of delivering over 400W into an 8Ω load.

As discussed in section 12.9 of Chapter 12, when two buffer amplifiers are used for a fully balanced power amplifier, we can mix and match the output power transistors for the two buffer amplifiers. This may help to get a blend of the sound from two types of output power transistors.

Example 3

It should be noted that the fully balanced hybrid power amplifier in Figure 13.7 originates from the vacuum tube power amplifier in Figure 11.32, where power tubes KT88 and the output transformer are now replaced by two solid-state buffer amplifiers. A different approach is to keep the output transformer, but replace the power tubes with solid-state power transistors, as shown in Figure 13.11.

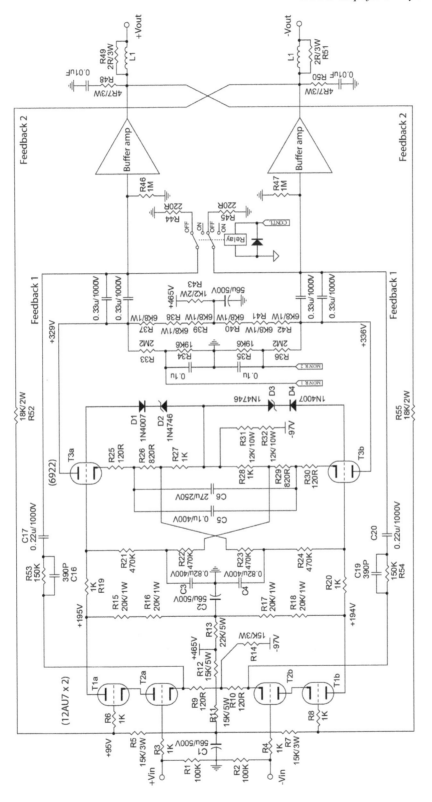

Figure 13.9 The second example for a fully balanced hybrid power amplifier

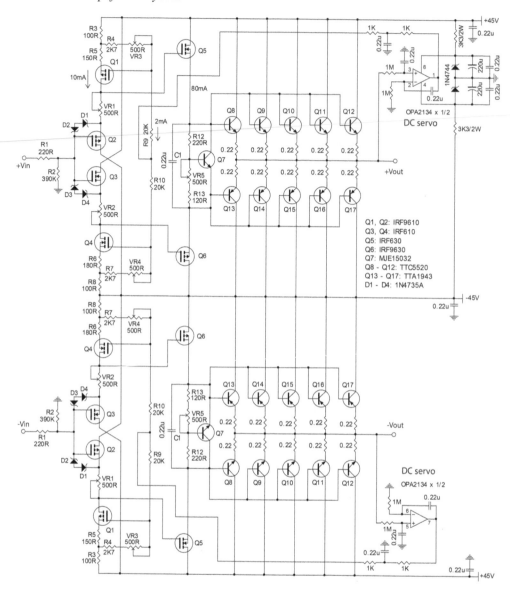

Figure 13.10 A pair of floating-bias buffer amplifiers for the fully balanced hybrid power amplifier of Figure 13.9

Note that depletion type power MOSFETs are used in Figure 13.11. Alternatively, power JFETs can also be used instead of depletion type power MOSFETs. As shown in Figure 13.11, a total of six power MOSFETs is used. Depending on how much output power the amplifier delivers, paralleling more power transistors can always boost the output power. It is recommended

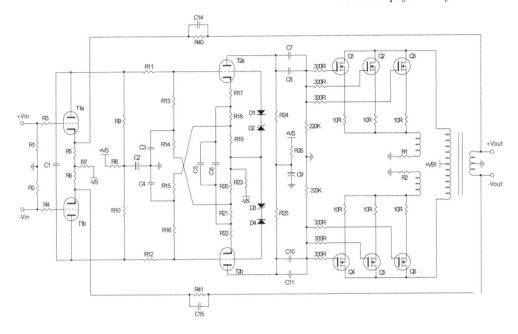

Figure 13.11 A potential approach to a fully balanced hybrid power amplifier employing depletion type power MOSFETs and an output transformer with an extra winding for source feedback

to use a small source resistor, around 10Ω, for improving current sharing among the power transistors. It should also be noted that the output transformer provides an extra winding for source feedback for improving the linearity of the output stage.

The major drawback for this approach is that no such output transformer is commercially available. We can imagine that it is not an easy task to design and develop an output transformer. A great deal of trial and error must be involved in optimizing the matching impedance, output power, and distortion.

13.4 Exercises

Ex 13.1 The hybrid power amplifier in Figure 13.5 is a simplified version of that in Figure 13.3. The input stage and second stage are common cathode amplifier configurations using triode T1 and T2, respectively. Assume that T1 = 12AX7, T2 = 6922, +VS = +300V. The plate-cathode dc potential for T2 must be lower than 95V. Determine the unknown resistors R1–R11.

Ex 13.2 The buffer amplifier of Figure 13.4 contains both MOSFETs and BJTs. The same buffer amplifier is now redesigned as shown in Figure 13.12, employing only BJTs. Assume Q1 and Q2 = MJE15031, Q3 and Q4 = MJE15030, Q5 = MJE15033, Q6 = MJE15032. The dc bias currents are given in the figure. Determine the resistors for R1 to R5 and VR1 to VR4.

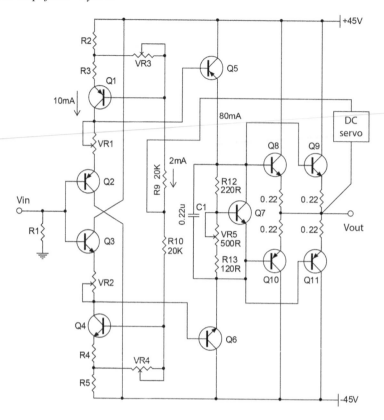

Figure 13.12 A buffer amplifier employing all BJTs

13.5   References

[1] Lam, C.M.J., "Buffer amplifier," US patent 8854138 B2.
[2] Lam, C.M.J., "Dc self-biased vacuum tube differential amplifier with grid-to-cathode over-voltage protection," U.S. patent 7733172.
[3] Lam, C.M.J., "Balanced amplifier," U.S. patent 7364535.

14   DC regulated power supplies

14.1   Introduction

DC regulated power supplies play an important role in audio amplifiers. They have a direct impact on an amplifier's sonic performance. No matter how good an amplifier design is, a poor dc regulated power supply will degrade the audio quality, and it may even produce unwanted hum and noise. A poor dc regulated power supply will simply make an otherwise good amplifier produce bad sound. In contrast, a good dc regulated power supply can easily elevate the sonic performance of an amplifier to a higher level.

Speaking from my own experiences, I come across many cases in which a dc regulated power supply plays a crucial role in affecting an audio amplifier's overall sonic performance. It is especially true for low level signal amplification, such as line-stage amplifiers and phono-stage amplifiers. Therefore, it is recommended to invest time to develop a good quality dc regulated power supply for a low level amplifier. It is definitely worth the effort for a circuit design engineer and audio enthusiast to learn the skills and techniques. Generally speaking, the attributes for a good dc regulated power supply include low noise, stabilized line regulation against ac mains variation, loading variation, and temperature variation. Most of the dc regulated power supplies discussed in this chapter meet these requirements.

Various types of linear dc regulated power supplies are discussed in this chapter. First, we discuss the basic ac-to-dc unregulated power supplies by means of half-wave, full-wave rectifications, L-C and R-C filtering. Second, we discuss some commonly used three-terminal dc regulators, including fixed voltage and adjustable voltage types. Third, we discuss shunt type and series type regulated power supplies employing discrete solid state devices, vacuum tubes, and a hybrid approach. Finally, we discuss how to prevent power transformers from producing unwanted noise.

Half-wave and full-wave rectifying

Let us consider a simple 10VA power transformer with a secondary output of 5V, as shown in Figure 14.1. Given a 10VA power rating, the transformer is capable of delivering 2A to a load. Since transformer voltages are expressed in root mean square (rms) values, the peak of the secondary voltage is $5V \times \sqrt{2} = 7.071V$ at no load condition. The waveform of the secondary side voltage is a simple sinusoidal wave with frequency of 50Hz or 60Hz (depending on in which country the transformer is used) alternating from $-7.071V$ to $+7.071V$.

Figure 14.2 shows a simple half-wave rectifying circuit using one single solid-state semiconductor rectifier, D1. For full-wave rectification, there are two approaches as shown in Figure 14.3. When four rectifiers are used, as shown in Figure 14.3(a), it is called bridge rectifying. When there is a center tap in the secondary side of the transformer, as shown in

DOI: 10.4324/9781003369462-14

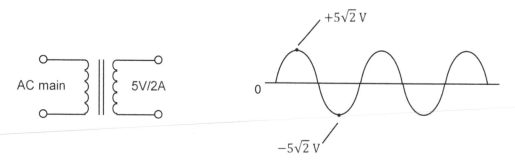

Figure 14.1 A 5V power transformer producing alternating ac voltage from −7.071V to +7.071V

Figure 14.3(b), the number of rectifiers can be reduced to two. The outputs from the secondary side of the transformer, half-wave and full-wave rectifying, are shown in Figure 14.4(a), (b), and (c), respectively. In real life, a rectifier will not conduct until the voltage is greater than the forward voltage V_F. Therefore, the rectifier is turned on at slightly less than half of the cycle. And the peak voltage is reduced from $5V \times \sqrt{2}$ to $(5V - V_F) \times \sqrt{2}$, i.e., a drop of $V_F \times \sqrt{2}$. Forward voltage for solid-state rectifiers range from 0.8V to 2V. See Table 14.1 for a list of some representative rectifiers.

When the voltage of the power transformer's secondary side is high, for example 100V or higher, a drop of $V_F \times \sqrt{2}$, i.e., around 1.1V to 2.8V, is relatively small. However, for a low voltage secondary side such as in this example, 5V, a drop of 1.1V to 2.8V should be a concern. In view of this, we need to take into account a rectifier's forward voltage V_F in low voltage power supply design.

Given the same voltage and current rating, the Schottky rectifier has lower forward voltage V_F compared with a silicon type rectifier. From Table 14.1, Schottky rectifier's V_F is around 0.8V. In applications such as dc regulated power supplies for a vacuum tube filament, i.e., 6.3V or 12.6V, the Schottky rectifier is an excellent choice. An example of a 6.3V filament dc regulated power supply is given in Figure 14.40.

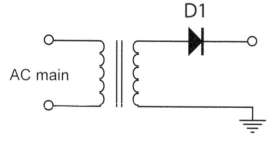

Figure 14.2 Half wave rectifying circuit

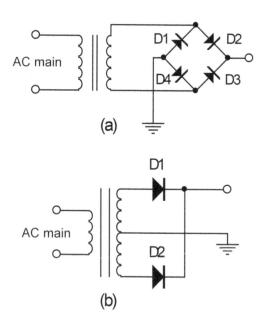

Figure 14.3 Two full-wave rectifying arrangements. Circuit (a) is also called bridge rectifying

The output waveform of full-wave rectifying is shown in Figure 14.4(c). Again, since a rectifier will not conduct until the voltage is greater than the forward voltage V_F, the peak voltage is also $(5V - V_F) \times \sqrt{2}$. And the notches appearing on the horizontal axis indicate the time for one rectifier to start conducting (on half of the cycle) and the other conducting rectifier to lose conduction. This process switches on the two rectifiers alternately on every half cycle. It is obvious that there are double the rectified waveforms in full-wave rectifying than in half-wave rectifying. This gives full-wave rectifying twice the energy to power the load than half-wave rectifying. As a result, full-wave rectifying produces lower ripple voltages and, therefore, better line regulation than half-wave rectifying. Today, now that the cost of a solid-state semiconductor rectifier is no longer a concern, it is a common practice to employ full-wave rectifying in dc regulated power supplies for audio applications.

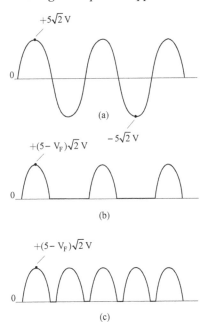

Figure 14.4 (a) Waveform of a 5V power transformer, (b) half-wave rectifying and (c) full-wave rectifying

Unregulated dc power supplies

Figure 14.5 shows that a capacitor is used for smoothing out the unregulated dc power supplies. C1 is often called a *reservoir capacitor*. Here, Rload is a resistive load for both half-wave and full-wave rectifying circuits. When the voltage of the secondary side of the transformer rises to a dc voltage V_F above the voltage holding at the capacitor, the rectifier conducts and starts charging the capacitor.

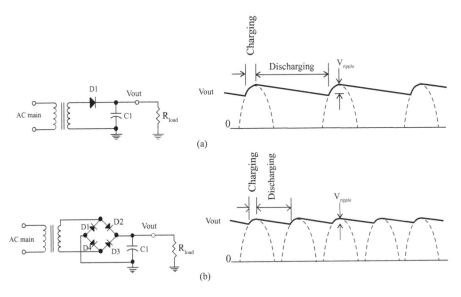

Figure 14.5 Filtering capacitor in (a) half-wave and (b) full-wave rectifying

Table 14.1 A list of representative solid-state rectifying diodes

Parts	(a) I_F (A)	(b) V_R (V)	(c) I_R (mA)	(d) V_F (V)	(e) trr (ns)	(f) No. of diodes 1	2	4	Package	Comments
Silicon										
1N4002	1	100	0.01	1.1	high	•			Axial lead	standard recovery
1N4007	1	1000	0.01	1.1	high	•			Axial lead	standard recovery
UF4002	1	100	0.01	1	50	•			Axial lead	fast recovery
UF4007	1	1000	0.01	1.7	75	•			Axial lead	fast recovery
EGP20B	2	100	5	0.95	50	•			Axial lead	fast recovery
EGP20K	2	800	5	1.7	75	•			Axial lead	fast recovery
BYV27–100	2	100	0.15	0.88	25	•			Axial lead	fast recovery
BYV27–600	2	600	0.15	1.15	40	•			Axial lead	fast recovery
1N5401	3	100	0.01	1	high	•			Axial lead	standard recovery
1N5408	3	1000	0.01	1	high	•			Axial lead	standard recovery
EGP30B	3	100	5	0.95	50	•			Axial lead	fast recovery
EGP30K	3	800	5	1.7	75	•			Axial lead	fast recovery
MUR410	4	100	0.005	0.89	35	•			Axial lead	fast recovery
MUR4100E	4	1000	0.9	1.85	75	•			Axial lead	fast recovery
BYW80	8	200	0.01	0.85	35	•			TO-220–2	fast recovery
MUR880E	8	800	0.01	1.8	100	•			TO-220–2	fast recovery
RHRP8120	8	1200	0.1	3.2	70	•			TO-220–2	fast recovery
MUR2020R	20	200	0.05	1.1	95	•			TO-220–2	fast recovery
MUR3020WT	30	200	0.01	1.05	35	•			TO-247–3	fast recovery
FFPF30UP20S	30	200	0.1	1.15	50	•			TO-220 FP	fast recovery
FFPF30UA60S	30	600	0.1	2.2	90	•			TO-220 FP	fast recovery
FFPF20UP20DN	10	200	0.1	1.15	45		•		TO-220 FP	fast recovery
FFPF20UP60DN	10	600	0.1	2.2	70		•		TO-220 FP	fast recovery
FFA60UP20DN	30	200	10	1.15	40		•		TO-3P-3L	fast recovery
FEP30DP	30	200	0.5	0.95	35		•		TO-247-AD	fast recovery
RURG3060CC	30	600	0.25	1.5	80		•		TO-247–3	fast recovery
FEP30JP	30	600	0.5	1.5	50		•		TO-247-AD	fast recovery
GBSB20	1.5	200	0.3	1	high			•	GBL	standard recovery
GBSB80	1.5	800	0.3	1	high			•	GBL	standard recovery
GBL01	4	100	0.5	1	high			•	GBL	standard recovery
GBL10	4	1000	0.5	1	high			•	GBL	standard recovery
GBPC1502	15	200	0.5	1.1	high			•	GBPC	standard recovery
GBPC1505	15	500	0.5	1.1	high			•	GBPC	standard recovery
GBPC1510	15	1000	0.5	1.1	high			•	GBPC	standard recovery
GBPC2502	25	200	0.5	1.1	high			•	GBPC	standard recovery
GBPC2505	25	500	0.5	1.1	high			•	GBPC	standard recovery

Parts	(a) I_F	(b) V_R	(c) I_R	(d) V_F	(e) trr	(f) No. of diodes			Package	Comments
	(A)	(V)	(mA)	(V)	(ns)	1	2	4		
GBPC2510	25	1000	0.5	1.1	high			•	GBPC	standard recovery
GBPC3502	35	200	0.5	1.1	high			•	GBPC	standard recovery
GBPC3505	35	500	0.5	1.1	high			•	GBPC	standard recovery
GBPC3510	35	1000	0.5	1.1	high			•	GBPC	standard recovery
Schottky (g)										
MBR1100	1	100	5	0.79	N/A	•			Axial lead	low VF
SB1100	1	100	10	0.85	N/A	•			Axial lead	low VF
MBR3100	3	100	20	0.79	N/A	•			Axial lead	low VF
SB3100	3	100	10	0.85	N/A	•			Axial lead	low VF
SB5100	5	100	10	0.85	N/A	•			Axial lead	low VF
VT5202	5	200	5	0.88	N/A	•			TO-220AC	low VF
MBR10100CT	5	100	6	0.85	N/A		•		TO-220AB	low VF
MBR20100CT	10	100	6	0.8	N/A		•		TO-220AB	low VF
V20202G	10	200	8	0.92	N/A		•		TO-220AB	low VF
V30202C	15	200	14	0.88	N/A		•		TO-220AB	low VF

Notes: (a) I_F is average forward current (max). (b) V_R is reverse voltage (max). (c) I_R is reverse current (max). (d) V_F is forward voltage (max). (e) trr is reverse recovery time (max). (f) Device can be a single, dual or bridge rectifying diode. (g) Schottky diodes have zero reverse recovery time and low forward voltage.

When the voltage of the secondary side of the transformer start to decline from the peak to a dc voltage V_F below the voltage holding at the capacitor, the rectifier stops conducting. This forces the capacitor to start to discharge through Rload until the next cycle begins. The charging and discharging process of the capacitor repeats.

The ripple voltage can be determined by the following expression,

$$C \frac{\Delta V}{\Delta t} = I \tag{14.1}$$

where C = capacitor, I = load current, ΔV = ripple voltage, $\Delta t = 1/f$, f = frequency of charging up the capacitor. For half-wave rectifying, the ripple voltage is given as,

$$V_{ripple} = \frac{I}{C \times f} \tag{14.2}$$

For full-wave rectifying, since the charging rate of capacitor is twice the rate of half-wave rectifying, the ripple voltage is reduced by half such that,

$$V_{ripple} = \frac{I}{C \times 2f} \tag{14.3}$$

Ripple voltages are always present at the output of unregulated dc power supplies regardless of the types of rectifying devices used. When a vacuum tube rectifier is used, as shown in Figure 14.6, the ripple voltage has the same form and can be determined in the same way. Table 14.2 gives a list of some vacuum tube rectifiers. From their inherent properties, vacuum tube rectifiers deliver much lower output current than solid-state semiconductor rectifiers. The average output current for vacuum tube rectifiers is about 100mA to 300mA. As a matter of fact, output current is not the vacuum tube's strength, but voltage is. The reverse voltage is commonly over 1200V for vacuum tube rectifiers.

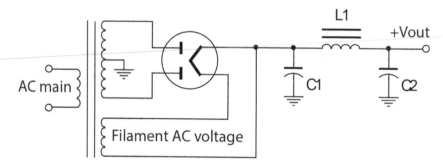

Figure 14.6 Full-wave rectifying circuit employing a directly heated vacuum tube rectifier

Table 14.2 A list of representative vacuum tube rectifying diodes

		5AR4/GZ34	5U4G	5V4G	5Y3G	6CA4/EZ81	6X4
Rectifier type		Full wave	Full wave	Full wave	Full wave	Full wave	Full wave
Reverse voltage (max)	(V)	1700	1550	1400	1400	1200	1250
Average output current (max)	(mA)	250	350	175	150	150	90
Filament voltage	(V)	5	5	5	5	6.3	6.3
Filament current	(A)	1.9	3	2	2	1	0.6
Tube base		Octal	Octal	Octal	Octal	Mininature noval	Mininature 7 pin

Figure 14.7 shows L-C and R-C filtering circuits that are commonly found in ac-to-dc power supplies for both vacuum tube and solid-state rectifiers. However, L-C filters are more commonly used in vacuum tube rectifying circuits. The main reason is that a filtering capacitor with high capacitance is not recommended for use with a vacuum tube rectifier. Therefore, inductors are used together with capacitors to form a better filtering circuit.

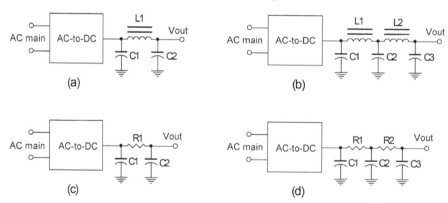

Figure 14.7 L-C and R-C filtering networks

What is the reason that vacuum tube rectifiers are not recommended to work with high filtering capacitors? It is because vacuum tubes have limited current handling capability. It should be noted that a large capacitor is capable of holding a voltage longer before it needs to be charged up again. Therefore, the charging time for the capacitor is very short while the discharging time is long. However, it takes a high current to charge up a large capacitor in such a short charging time. Delivering high current in a very short time is not something that vacuum tube devices are capable of handling. Therefore, many vacuum tube manufacturers recommend the first capacitor, C1, which is connected right next to the rectifier, should be around 40μF or smaller. Given an inductor that limits rapid current change, then the second capacitor, C2, and third capacitor, C3, can be increased to 100μf. Inductors with inductance of 20H or higher are commonly used so as to achieve small ripples and stable dc output voltage.

Choosing the rectifier's reverse voltage

If a load draws a certain amount of current from a dc power supply, it is clear that a rectifier with current rating higher than the load's demand should be selected. For example, if the maximum load current is 1A, it would be wise to use a 2A or higher current rating rectifier. On the other hand, if the unregulated dc voltage is 100V, for example, what reverse voltage should we select for the rectifier? Will a rectifier with V_R of 150V to 200V be sufficient? To answer this question, let us examine the simple half-wave rectifying circuit shown in Figure 14.8.

If the output from the secondary side of the transformer is ac 100V, the peak voltages $\pm V_p$ will be $\pm 100V \times \sqrt{2} = \pm 141V$. In no load condition, the dc voltage right after capacitor filtering is $V_p - V_F \approx 140V$ assuming $V_F = 1V$. The rectifier will face the greatest challenge when the anode sees the negative peak voltage of $-141V$. In this instance, the rectifier must withstand a dc potential difference of $140V - (-141V) = 281V$, which is 2.81 times of 100V, in no load condition. If there is a load, the transformer output will drop slightly and, therefore, the dc potential across the rectifier will be less than 281V.

In order to ensure a rectifier is working safely from no load to full load condition, it is recommended to select a rectifier with reverse voltage V_R of at least three times the ac voltage to be rectified. For instance, if the output of a transformer is ac 100V, select a rectifier of $V_R = 300V$ or higher. By the same token, select a rectifier with $V_R = 1500V$ for rectifying ac 500V.

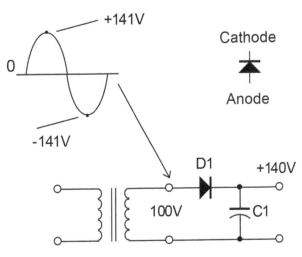

Figure 14.8 Dc potential across a rectifier

14.2 Three-terminal fixed voltage regulators

Unregulated dc power supplies, as discussed in section 14.1, will be served as the dc voltage source (+Vin or –Vin) in a complete dc regulated supply in the remainder of this chapter. In this section, we begin discussing three-terminal fixed voltage regulators.

The most popular three-terminal fixed voltage regulators are the positive voltage 78XX series (also called LM78XX, L78XX, and MC78XX) and negative voltage 79XX series (also called LM79XX, L79XX, and MC79XX). The suffix XX denotes the voltage output, i.e., 7805 for +5V and 7912 for −12V. These three-terminal fixed voltage regulators are based on the series type regulated power supply configuration, as shown in Figure 14.9(a), with all components integrated into one single monolithic integrated circuit. TO-220 is a popular package for the three-terminal regulator, including fixed and variable voltage regulators. Only two external capacitors are required for a three-terminal regulator, as shown in Figure 14.9(b).

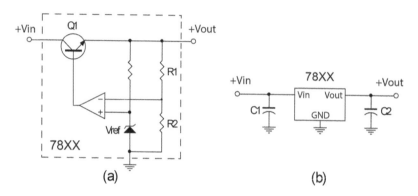

Figure 14.9 Internal structure of a 78XX three-terminal positive voltage regulator is shown in circuit (a) and its application in circuit (b). Resistors R1 and R2 are built internally for some predetermined discrete output voltages

The 78XX series offers discrete output voltages from +5V to +24V and −5V to −24V for the 79XX series. The output current is up to 1A. If higher current is needed, an external transistor can be added to boost the current. Owing to their stability and ease of use, they are extensively used everywhere from industrial, control, to consumer electronic applications.

78XX and 79XX

Figure 14.10 shows typical applications for a 78XX regulator [1]. Even though the 78XX series features a so-called fixed voltage regulator, Figure 14.10(a) shows that it is possible to get a voltage higher than the designated fixed output V_{XX}. The output voltage can be easily set by two external resistors, R1 and R2. Note that Vxx is always across resistor R1. Thus, the regulator can be used as a current source, as shown in Figure 14.10(b). On the other hand, when higher output current is needed, an external PNP power transistor can be used to boost the current, as shown in Figure 14.10(c). When a 78XX regulator delivers current to a load, there is a dc potential voltage generated across resistor R1. This dc voltage starts increasing until it reaches a $V_{BE(Q1)}$, i.e., around 0.65V. Then Q1 is turned on to supply additional current to the load. If R1 is 3Ω, transistor Q1 will be turned on when 78XX delivers current beyond 0.65V/3Ω, i.e., 0.217A.

It should be noted that the current boosting transistor Q1 of Figure 14.10(c) does not have any current limit protection. If transistor Q2 and current sensing resistor R_{SC} are added, as shown in Figure 14.10(d), transistor Q1 is protected against delivering excessive output current. The current limitation is determined by $V_{BE(Q2)}/R_{SC}$.

Negative voltage three-terminal regulator 79XX works in a similar fashion. Figure 14.11 shows the internal structure of 79XX and its application. If dual regulated power supplies are

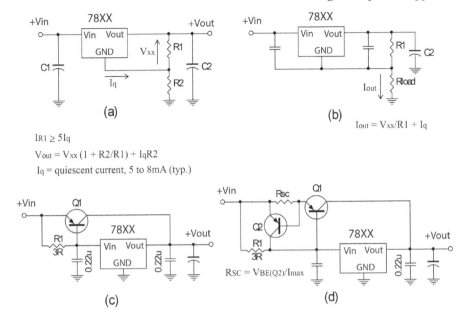

$I_{R1} \geq 5I_q$

$V_{out} = V_{xx}(1 + R2/R1) + I_qR2$

I_q = quiescent current, 5 to 8mA (typ.)

Figure 14.10 Applications for three-terminal regulator 78XX. Variable output voltage in circuit (a). Deliver constant current to a load in circuit (b). Boost output current in circuit (c). Boost output current and set current limitation in circuit (d)

needed, 78XX and 79XX work together to offer a practical and easy to use solution. Since there are only a few external components needed to implement a dc regulation power supply, three-terminal regulators can be found almost everywhere for low to medium current applications where a simple and reliable power solution is needed.

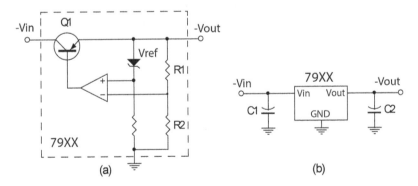

Figure 14.11 Internal structure of a 79XX 3-terminal negative voltage regulator in circuit (a) and its application in circuit (b). Resistors R1 and R2 are built internally for some predetermined output voltages V_{xx}

14.3 Three-terminal variable voltage regulators

If we cannot find a fixed voltage regulator for a desired output voltage, or if a better line regulation and lower noise alternative is needed, three-terminal variable voltage regulators will be the solution. Figure 14.12(a) shows the popular LM317 positive voltage variable regulator and a high voltage version, TL783. Both devices work in the same way and set the output voltage by the same formula. Standard LM317 is rated at 1.5A (max) output current and (Vin – Vout) = 37V (max). TL783 is rated at 0.7A (max) output current and (Vin – Vout) = 125V (max). For applications with a load current of less than 0.7A and (Vin – Vout) lower than 37V, both LM317 and TL783 can be used. In fact, we will discuss in section 14.6 a technique to float a regulator above the ground. Then, both LM317 and TL783 can be used in a power supply that delivers above 37V. In other words, both LM317 and TL783 may be used for most output voltages as long as they do not exceed the device's maximum output current and voltage ratings.

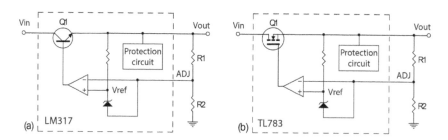

Figure 14.12 Circuit (a) is a standard three-terminal variable voltage regulator, LM317, and circuit (b) is a high voltage version, TL783

The reference voltage Vref, temperature variation, ripple rejection, and line regulation are very similar between the two devices. Nevertheless, one key difference is the type of transistor Q1 being used. Q1 is often called a *series pass transistor*, which is connected in series between the input and output. An NPN transistor is used in LM317, while an N-MOS transistor is used in TL783. Speaking from my experiences, I have found that BJT and MOSFET produce slightly different sonic characteristics when they are used as series pass transistors in a regulated power supply for audio amplifiers. Therefore, it is recommended that you try both the LM317 and TL783 regulators and judge for yourself which one produces a better synergy with your audio amplifier.

LM317, TL783 and LM337

Typical applications for LM317 are shown in Figure 14.13. Similarly, these applications also are applied to TL783. The output voltage is determined by the expression,

$$V_{out} = V_{ref}\left(1+\frac{R2}{R1}\right)+I_{adj}R2 \tag{14.4}$$

where V_{ref} is the reference voltage and I_{adj} is the bias current flowing to the ADJ pin. For a quick comparison, Vref for LM317 is 1.25V (typ.) and TL783 is 1.27V (typ.). I_{adj} for LM317 is 0.05 to 0.1mA and 0.08 to 0.11mA for TL783. Since both devices have the same pin assignment, these two devices are interchangeable except for a small difference in output voltage. But there is a need to bear in mind that TL783 has a lower current rating, while LM317 has a lower (Vin – Vout) rating.

Figure 14.13(a) shows that by using two resistors, R1 and R2, we can get a variable output voltage governed by Eq. (14.4). Figure 14.13(b) has included several components for improvement. Capacitor C3 helps to reduce noise and ripple coming from V_{out} before passing to the ADJ pin. A 47μF to 100μF capacitor will be sufficient. A small capacitor, C1, is placed in the input and a small capacitor, C5, is placed in the output so as to lower the impedance at high frequency.

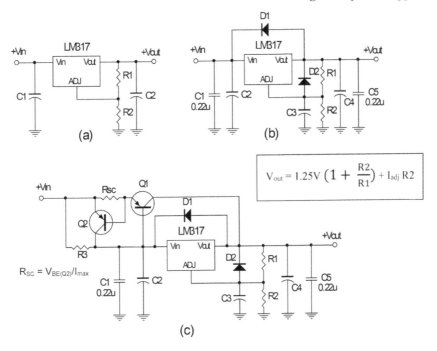

$$V_{out} = 1.25V \left(1 + \frac{R2}{R1}\right) + I_{adj} R2$$

$$R_{SC} = V_{BE(Q2)}/I_{max}$$

Figure 14.13 A simple dc regulated power supply using LM317 in circuit (a). Circuit (b) shows that protective diodes are used to protect the LM317 and additional capacitors to reduce noise. Current boosting transistor Q1 and current limiting transistor Q2 are used in circuit (c)

Diodes D1 and D2 are used to protect the three-terminal regulator in case of a short circuit at the output [2].

In an application where high output current is needed, we can follow Figure 14.13(c), boosting the output current and setting the current limit. Similar to Figure 14.10(d), we have $R_{SC} = V_{BE(Q2)}/I_{max}$.

LM317 and TL783 are three-terminal variable positive voltage regulators. The complementary negative voltage three-terminal variable regulator is LM337. The operation and the output voltage setting for LM337 is exactly the same as LM317. Figure 14.14(a) shows a typical application

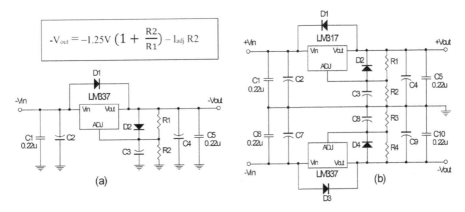

$$-V_{out} = -1.25V \left(1 + \frac{R2}{R1}\right) - I_{adj} R2$$

Figure 14.14 (a) A typical application for negative variable regulator LM337. (b) A dual output dc regulated power supply using LM317 and LM337

for LM337 in a negative regulated power supply. One important application for LM337 is to pair with LM317, forming a dual regulated power supply, as shown in Figure 14.14(b).

Example 14.1

Op-amps are often used in audio applications for signal conditioning and amplification. They often require dual ±15V regulated power supplies. The dc supply current for an op-amp is usually around 5–20mA. Therefore, LM317 and LM337 are very suitable for this application. In this example we follow the circuit in Figure 14.14(b) to develop a dual ±15V regulated power supply.

Let us start with LM317 for realizing the +15V power supply. The output voltage is given by Eq. (14.4). Since the second term ($I_{adj} \times R2$) is small compared with the first term, it is neglected for the time being. Therefore, we have

$$V_{out} = V_{ref}\left(1 + \frac{R2}{R1}\right)$$

where $V_{ref} = 1.25$V and $V_{out} = 15$V. Simplifying the above equation yields

$$R2 = 11 \times R1 \tag{14.5}$$

This is one equation with two unknowns. Therefore, we need to have one more condition to set up the second equation. That condition arises when we examine the no load condition. When the power supply is in no load condition, it is required to have a small current flowing through the resistor network formed by R1 and R2. This is not just a convenience for setting the second equation, it is also needed for the internal circuit of LM317 to set up a proper dc bias current. In common practice, we choose 5mA. Therefore, we have,

$$\frac{V_{out}}{R1 + R2} = 5\text{mA}$$

where $V_{out} = 15$V. Simplifying the above equation yields

$$R1 + R2 = 3\text{k}\Omega \tag{14.6}$$

Solving Eqs. (14.5) and (14.6) for R1 and R2, we have R1 = 250Ω and R2 = 2750Ω. We choose a practical value 240Ω for R1 and, therefore, R2 becomes 2640Ω. We now come back to examine the second term, I_{adj}R2, where I_{adj} is 0.1mA given by the data sheet, and R2 is 2640Ω. Therefore, I_{adj}R2 gives 0.264V, which is about 1.7% out of 15V. This is not a big difference, even if we dismiss it from the calculation. R2 has an odd resistance of 2640Ω. Since we have discarded the 0.264V from the calculation, at this point we can pick a slightly larger resistance for R2 to compensate for it, so we take 2700Ω, a practical value and greater than 2640Ω. Another approach is to use a variable resistor for R2. We take a 5kΩ variable resistor for R2. Since the 5kΩ variable resistor can be adjusted to below and above 2640Ω, we can be sure that an exact 15V can be achieved. And this is what we select for the final circuit, as shown in Figure 14.15.

Except for a change in polarity for the output of LM337, we can follow the exact same calculation to determine R3 and R4 for the −15V power supply. And it is no surprise that we get the same values for R3 and R4. Finally, it should be clear that V_{in} has to be at least a few volts higher than V_{out}, for example $|V_{in}| \geq 20$V.

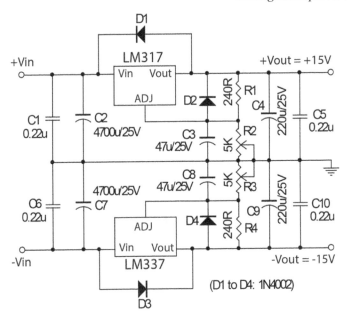

Figure 14.15 A dual ±15Vregulated power supply using LM317 and LM337

14.4 Shunt type regulated power supplies

The three-terminal dc regulators discussed in the preceding sections are series type regulators. In a series type regulator, there is always a series pass transistor across V_{in} and V_{out}. (See Figure 14.9(a) and Figure 14.11(a).) Later, we will discuss more series type dc regulated power supplies producing high voltages. However, a different type of dc regulated power supply is now discussed in this section: shunt type regulated power supply.

The simplest form of shunt type power supply is a zener diode in series with a current limiting resistor, as shown in Figure 14.16. Output is simply equal to the zener voltage, Vout = Vz. However, when a load is connected to the output, the Vout starts to drop when the load current increases. Therefore, a zener diode cannot provide good line regulation for a high load current demand. They are more commonly used as a reference voltage rather than a voltage supply regulator.

TL431 (or LM431) is a three-terminal shunt type regulator, as shown in Figure 14.17. This device can produce up to 36V output and sink up to100mA. It of-

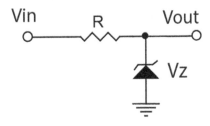

Figure 14.16 A simple zener shunt type power supply

fers far better line regulation than a zener diode. The basic working principle of a shunt type regulator is to sink as little current as possible by the regulator itself while the load draws current. One important specification for a shunt regulator is the maximum sinking current for the

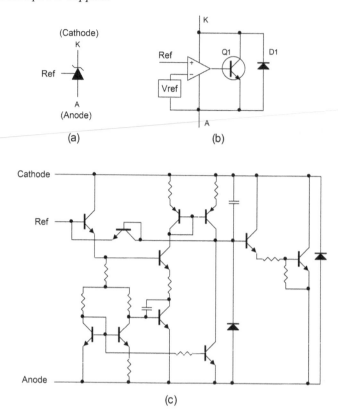

Figure 14.17 The electrical symbol for TL431 in (a), functional block diagram in (b), and equivalent circuit in (c)

device. For a series type regulator, we do not pay too much attention to how much it can sink because the current sinking in a series type regulator, which is actually the dc bias current of the regulator itself, is almost independent of the load current. So, as long as the load current is below the maximum allowable current from the series type regulator (1.5A for LM317), and within the allowable power dissipation of the device, we generally do not need to worry about how much current the device will sink. At no load or full load condition, current sinking in a series type regulator remains nearly unchanged. But for a shunt type regulator, for example, designed for 100mA output current, in a loaded condition the external load consumes nearly all 100mA. However, in no load condition, the regulator has to consume all 100mA by sinking the current and dissipating the power by itself. Therefore, the working principle of a shunt type regulator is inherently different from a series type regulator.

There are two inherent properties of a shunt regulator [3]:

1) The operating efficient is very low at no load to light loads because the shunt regulator then dissipates maximum to near maximum power.
2) The shunt regulator is protected from overload or short circuit of the output. (However, if the shunt regulator is involved in a complex circuit with electrolytic capacitors, it will not always be free from damage.)

TL431/LM431 shunt regulator

TL431 is a three-terminal shunt regulator, as shown in Figure 14.17, with the pins named "cathode," "anode" and "Ref." The output voltage is controlled by external resistors R1 and R2 while the output current is controlled by resistor R3, as shown in Figure 14.18 (a). It is given as follows:

$$V_{out} = V_{ref}\left(1 + \frac{R1}{R2}\right) + I_{ref}R1 \tag{14.7}$$

$$R3 = \frac{Vin - Vout}{I_{in}} \tag{14.8}$$

where $I_{in} = I_k + I_r + I_{out}$, $V_{ref} = 2.5V$, $I_{ref} =$ a few µA (can be neglected in the calculation for V_{out}).

Figure 14.18(a) is a basic configuration for a shunt regulated power supply employing TL431 (or LM431) [4]. If an output current higher than 100mA is needed, a PNP transistor Q1 is used to boost the current, as shown in Figure 14.18(b). TL431 is limited to an output voltage of 36V. By stacking up two TL431s, as shown in Figure 14.18(c), a higher output voltage can be achieved [5].

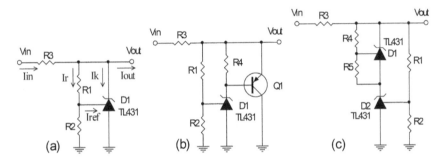

Figure 14.18 (a) A simple shunt type dc power supply using TL431. (b) Using transistor Q1 to boost output current. (c) Stacking up two TL431 for higher output voltage

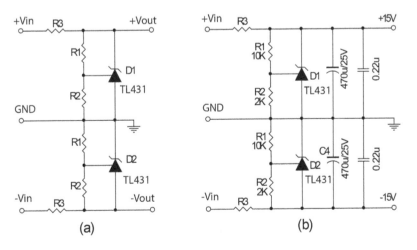

Figure 14.19 (a) A typical shunt type dual regulated power supply using two TL421. (b) A practical shunt-type dual ±15V power supply

TL431 is intended for use as a positive regulator. However, with a different circuit arrangement TL431 can work as a negative regulator also. See Figure 14.19. It is noted that when using TL431 for negative voltage supply, the cathode is connected to the ground while the anode is connected to the negative output. The same expressions can be used for determining Vout and resistors (R1, R2, and R3) that set the output voltage and current.

Example 14.2

In analog mixed signal applications, we often come across situations where integrated circuits require dc regulated power supplies ranging from ±2.5V to ±5V. For example, the digital potentiometer AD5260/AD5262 requires a dual ±5V power supply and AD5241/AD5242 requires a dual ±2.5V power supply. Three-terminal variable voltage regulators LM317 and LM337 can deliver a dc voltage as low as ±1.2V. They are excellent choices for a load current up to 1.5A. We illustrate in this example an alternative by using TL431 to implement two low voltage dual power supplies for a low output current requirement. In this example, we assume the desired output current requirement for ±5V is 30mA while it is 20mA for ±2.5V (see Figure 14.20).

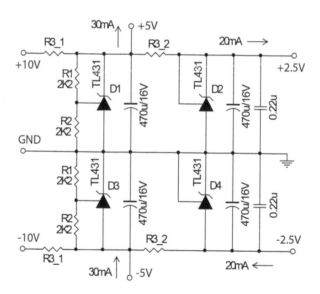

Figure 14.20 Shunt type power supply employing TL431 for dual ±5V/30mA and ±2.5V/20mA outputs

First, in order to simplify the calculation, we separate the circuit into two parts, as shown in Figure 14.21. Since current I_{in1} in the 5V power supply also depends on the current I_{in2} of the 2.5V power supply, it makes sense to start working out circuit (b) first. The current setting resistor R3b is given as follows:

$$I_{in2} = \frac{5V - 2.5V}{R3b}, \text{ and } I_{in2} = I_{ref} + I_k + I_{out2}$$

where $I_{out2} = 20mA$, $I_{ref} = $ few μA (can be neglected), I_k is the current flowing into the cathode of TL431. In practice, assuming 1mA to 2mA for I_k will suffice. Let us say $I_k = 2mA$. Then we have

$I_{in2} \approx 2mA + 20mA = 22mA$. Therefore, R3b is given by $2.5V/22mA = 114\Omega$. We can pick the nearest practical value, say R3b = 110R. Now we use R3b to recalculate the current I_{in2}, leading to $I_{in2} = 2.5V/110\Omega = 22.7mA$.

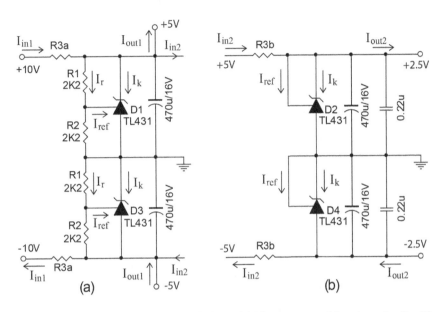

Figure 14.21 The regulated power supplies of Figure 14.19 are separated into two circuits. Circuit (a) shows dual ±5V power supplies and circuit (b) dual ±2.5V power supplies. Calculation shows R3a = 86.6Ω and R3b = 110Ω

On the other hand, since the built in reference voltage Vref in TL431 is 2.5V, the lowest Vout2 that we can get out from this design is also 2.5V. Simply connecting the Ref pin of the TL431 to the cathode gives the desired output of 2.5V.

To determine the resistors R1, R2, and R3a of Figure 14.21(a), we have

$$I_{in1} = \frac{10V - 5V}{R3a}, \text{ and } I_{in1} = I_r + I_k + I_{out1} + I_{in2}$$

where I_r = current flowing into the resistor network formed by R1 and R2, plus the current into the Ref pin of TL431. Again, we can ignore the small current flowing into the Ref pin, and set $I_r = 1mA$. Taking $I_k = 2mA$, $I_{out1} = 30mA$, $I_{in2} = 22.7mA$, we have $I_{in1} = 55.7mA$. Hence, R3a is given by $5V/55.7mA = 89.7\Omega$. Let us take the practical value 86.6Ω (E48 series). Now we use R3a = 86.6Ω to recalculate the incoming current I_{in1}, and we have $I_{in1} = 5V/86.6\Omega = 57.7mA$.

Resistors R1 and R2 can be determined by

$$V_{out1} = 2.5V\left(1 + \frac{R1}{R2}\right), \text{where } V_{out1} = 5V$$

The above equation is simplified to R1 = R2. Since we have set a dc bias current $I_r = 1mA$ for the resistor network formed by R1 and R2, we must have $5V/(R1 + R2) = 1mA$. This leads to R1 + R2 = 5kΩ. Given these two conditions, we find R1 = R2 = 2.5 kΩ. Pick the nearest practical value, 2.4kΩ, for R1 and R2. In fact, even if we pick 2.2kΩ or 2.7kΩ, the circuit will work just fine.

The final step is to check if the TL431s can operate within their current limitation when there is no load. When no load is connected to the 2.5V output, the TL431 (D2) has to sink all current for I_{in2} (22.7mA). This is within the maximum current (100mA) for TL431. The power dissipation of D2 is 2.5V × 22.7mA = 57mW. Then let us examine the situation when no load is connected to both 2.5V and 5V outputs. The TL431 (D1) has to sink all current for I_{in1} except I_r, giving I_{in1} = 57.7mA – (5V/4.8kΩ)mA = 56.6mA. This is still within the 100mA maximum current limit. The power dissipation by D1 is 5V × 56.6mA = 0.28W.

As seen from the above example, the input current (Iin) of a shunt regulated power supply is constant, regardless of whether no load or a load is connected. Therefore, the input series resistor can be replaced by a current source. Figure 14.22(c) shows a current source comprising two PNP transistors, Q1 and Q2. This current source is identical to the one discussed in Chapter 2, where NPN transistors are used in Figure 2.47(e). The constant current I_{in} is determined by $V_{BE(Q1)}/R2$.

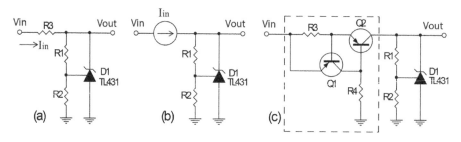

Figure 14.22 The input setting resistor R3 in circuit (a) is replaced by a current source in circuit (b). Current source is formed by two PNP transistors in circuit (c)

Before we move on to the discrete version for shunt regulated power supplies in the next section, let us examine a practical dual ±24V shunt regulated power supply, as shown in Figure 14.23. Let us first examine the +24V power supply. Transistors Q1 and Q2 form a current source for input current I_{in}. Instead of using PNP transistors for both Q1 and Q2, a P-channel MOSFET is used for Q2. The input current I_{in} is determined by $V_{BE(Q1)}/R1 = 0.65V/4.7Ω = 138mA$. Given a few mA flowing through the resistor network formed by 15kΩ, 500Ω, and 1.5kΩ resistors and the cathode current through TL431, the rest of the input current is delivered to the output load and transistor Q3. If the load takes more current, less current will be taken by Q3, and vice versa. This power supply can deliver 100mA to an output load.

Note that a 68Ω resistor is connected in series to the cathode of TL431. This resistor monitors the current going into D2. When the voltage across the 68Ω resistor goes beyond 0.7V, transistor Q3 is turned on to sink the excess current that was not taken by the load. This happens when the current through the resistor is greater than 0.7V/68Ω = 10.3mA. In a low or no load situation, transistor Q3 will sink most of, or all, available currents. When sinking 100mA at 24V, Q3 will dissipate 2.4W power and temperature rise can be substantial without a suitable heatsink. A 1Ω is placed at the emitter of Q3 so as to prevent V_{BE} thermal run-away from happening. Nevertheless, transistor Q3 must be mounted on a heatsink.

The 330Ω resistor placed at the gate of Q2 is a gate stopper resistor to prevent the MOSFET from parasitic oscillation. Usually, a few hundred ohms, up to 1kΩ, will be sufficient for a gate

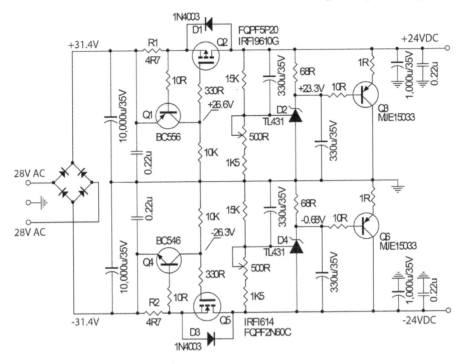

Figure 14.23 Dual ±24V shunt regulated power supply using TL431 delivering 100mA output current. The dc quiescent voltages are measured when ±40mA current is driven by the load

stopper resistor. A 10Ω base resistor is used in Q1 and Q2 to limit base current. The working principle of the −24V power supply follows the exact same way of the +24V power supply. However, it should be noted that the V_{BE} of Q4 can be slightly different to the V_{BE} of Q1, similarly for the V_{BE} between Q6 and Q3. Therefore, if this small variation of V_{BE} is taken into account, there should be a small change for R2 and the 68Ω cathode resistor on the TL431, D4. Anyway, we have kept the same values 4.7Ω and 68Ω. The −24V power supply works just fine with similar output current capability.

Discrete shunt type regulated power supplies

Here, we are going to discuss two examples of shunt type regulated power supplies. Instead of using TL431, we use op-amp and low noise reference zener. For convenience, they are called discrete shunt type regulated power supplies.

In terms of cost and components count, these are not advantages for discrete shunt type regulated power supplies. The first example is shown in Figure 14.24. It is easy to see that the number of components is greater than that of Figure 14.23 for the same dual ±24V output. However, the discrete shunt type regulated power supply has several advantages over Figure 14.23. Since LM329 is a low noise and good thermal stability reference voltage, it improves the noise and thermal stability of the entire power supply. An op-amp has a higher open-loop gain compared with the embedded error amplifier in a TL431. Higher open-loop gain gives the discrete power supply better line regulation and lower noise.

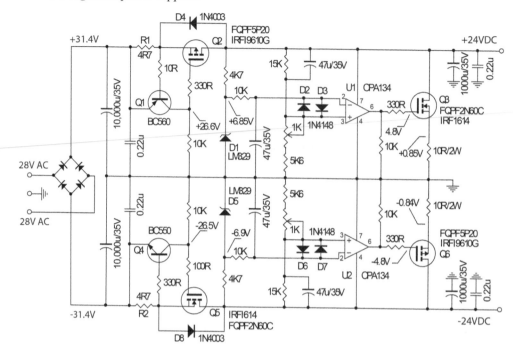

Figure 14.24 A discrete dual ±24V shunt type regulated power supply is capable of delivering 100mA output current. The dc quiescent voltages are measured when ±40mA is driven by the load

The operation of the discrete shunt type regulated power supply in Figure 14.24[6] is similar to Figure 14.23. Again, the input current is determined by resistor R1, which is given by $V_{BE(Q1)}$/R1 = 0.65V/4.7Ω = 138.3mA, neglecting the base current through the 10Ω resistor. The dc bias current for D1 is (24V – 6.9V)/4.7kΩ = 3.6mA, while the bias current for the resistor network formed by 15kΩ, 1kΩ and 5.6kΩ is 24V/(15kΩ + 1kΩ + 5.6kΩ) = 1.1mA. The actual position of the center tap of the 1kΩ variable resistor will affect this current slightly. The bias current for U1, OPA134, is 5mA (max.). Therefore, the current available for the output load and transistor Q3 is 138.3mA – 3.6mA – 1.1mA – 5mA = 128.6mA. If a load takes up 100mA, transistor Q3 will consume the remaining 28.6mA. When the load takes less than 100mA, transistor Q3 has to consume more than 28.6mA. Therefore, a suitable heatsink is needed for Q3. The components for the −24V power supply follow the exact same way of the +24V power supply.

What is the maximum output voltage that we can obtain from discrete shunt type power supply in the configuration of Figure 14.24? To answer this, let us take a look of the op-amp U1. The power supply rails of U1 (pin-4 and pin-7) are connected to +24V output and the ground. Similarly, the power supply rails of U2 are connected to −24V and the ground. Since most general purpose op-amps are limited to a maximum supply voltage of 36V, the shunt type regulated power supplies in the arrangement as in Figure 14.24 will be limited to less than ±36V output. In practice, in order to prolong the lifespan of the op-amp, we always operate the device below its maximum allowable supply voltage. For a 36V (max.) op-amp, it is often operated at 30V or below. Thus, it is not recommended to develop the power supply from Figure 14.24 to exceed ±30V by using op-amps with 36V maximum supply voltage.

Therefore, when OPA134 is used, do not set the output voltages over ±30V. If higher output voltage is needed, high voltage op-amps must be used. For example, OPA552 has a maximum supply voltage of 60V and 180V for OPA462. See Table 14.3 for a list of high voltage op-amps. For safety reasons, it is recommended to operate the op-amp at 15% below its maximum allowable supply voltage. Thus, OPA462 can be operated in a discrete shunt type regulated power supply for an output voltage up to dual ±150V. This is perhaps the highest output voltage we can achieve in a discrete shunt type regulated power supply using monolithic op-amps from the market today. Hybrid type op-amps can go to an even higher voltage at the expense of cost. For example, PA97 (from Apex Microtechnology) has a maximum supply voltage of 900V.

Table 14.3 A list of representative high voltage op-amps

Parts	Mfg	Total supply min (V)	max (V)	I_Q typ (mA)	GBW typ (MHz)	Slew rate typ (V/μs)	I_{out} max (A)	Package
low power								
LTC2057HV	LTC	4.8	65	0.8	1.5	0.45	0.02	SO-8
ADA4700	Analog	10	100	1.7	3.5	20	0.03	SO-8
OPA445	TI	20	100	4.2	2	10	0.015	DIP, SO-8
OPA454	TI	10	100	3.2	2.5	13	0.12	SO-8
LTC6090	LTC	9.5	140	2.8	12	21	0.05	SO-8, TSSOP
OPA462	TI	12	180	4	6.5	25	0.03	SO-8
medium power								
OPA552	TI	8	60	7	12	24	0.2	DIP, DDPak
OPA547	TI	8	60	10	1	6	0.5	TO-220 (7)
OPA548	TI	8	60	17	1	10	3	TO-220 (11)
LM675	TI	20	60	18	5.5	8	3	TO-220 (5)
OPA549	TI	8	60	26	0.9	9	8	TO-220 (11)
OPA452	TI	20	80	4.5	7.5	23	0.05	TO-220 (7)
OPA453	TI	20	80	4.5	7.5	23	0.05	TO-220 (7)
OPA541	TI	20	80	20	2	10	10	TO-3 (8), TO-220 (11)

Notes: (a) I_Q, quiescent current; (b) GBW, gain bandwidth product; (c) I_{out}, maximum output current.

Can we float the op-amp, similar to the way we handle series type regulated power supplies, to be discussed in Section 14.6, so that an even higher output voltage can be achieved? Unfortunately, the floating technique does not work for shunt type power supplies. However, why do we need high voltage shunt type regulated power supplies? The operation efficiency is very poor at low load to no load condition. The op-amp must be able to withstand high output voltage and the output transistor has to dissipate all power at no load condition. A shunt type regulated power supply does not seem to be an attractive solution for high voltage application. In contrast, a series type power supply is a natural choice for high voltage application, where the "floating" technique helps the op-amp to stay within its safety region while producing a high output voltage.

Nevertheless, Figure 14.25 illustrates a discrete shunt type regulated power supply with dual ±45V outputs that is capable of producing 100mA to a load. With a maximum supply voltage of 60V, OPA554 should be safe and comfortable for taking a 45V supply voltage.

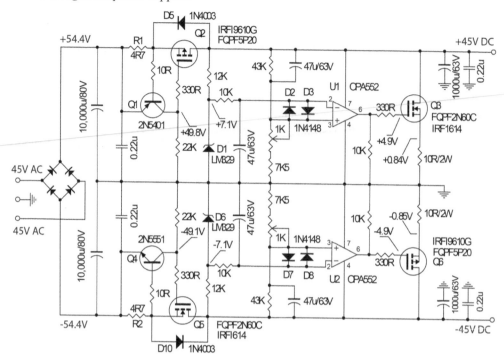

Figure 14.25 A discrete dual ±45V shunt type regulated power supply. The dc quiescent voltages are measured when ±40mA current is driven by the load

14.5   Series type regulated power supplies (fixed-mode)

Three-terminal fixed voltage regulators, 78XX, 79XX, and variable voltage regulators LM317 and LM337, have been discussed early in this chapter. They all belong to the series type regulated power supplies. Figure 14.26(a) shows the structure of series type regulated power supply with an NPN series pass transistor, Q1, across the unregulated input and regulated output. Since the base of Q1 is controlled by the error amplifier, which compares a portion of the output voltage Vout to the reference voltage Vref, the output can be easily determined by assuming that $V_- = V_+$ at the inputs of the error amplifier. Therefore, we have,

$$Vref = \frac{R2}{R1 + R2} Vout, \text{ or}$$

$$Vout = \left(1 + \frac{R1}{R2}\right) Vref \tag{14.9}$$

The output voltage is determined by resistors R1 and R2. Resistor R3 is to limit the bias current, which is usually a few mA, for the reference zener. Resistor R is connected between Vin and the base of the transistor Q1, where the output of the error amplifier is also connected. The function of resistor R is to jump-start the circuit during power up. Without resistor R, this series regulated power supply will not work properly.

When we look at the unregulated voltage Vin closely, high ripples are always superimposed on the dc voltage, as shown in Figure 14.26(b). Because of the low output impedance of the

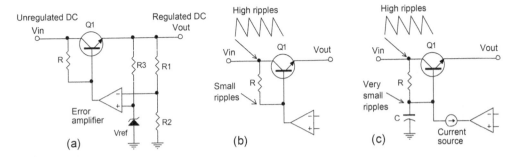

Figure 14.26 A series type regulated power supply is shown in circuit (a), with high ripple voltages present at Vin and small ripples at the base of Q1 in the simplified circuit (b). A capacitor C is added in circuit (c) to reduce the ripple voltage at the base of transistor Q1. A current source is added between the base of Q1 and the error amplifier

error amplifier, ripples at the base of Q1 are relatively small. By adding a capacitor, C, at the base of Q1, the ripples are further reduced. In addition, a current source is added to help jump-start the circuit during power up, as shown in Figure 14.26(c). By putting all these elements together, the structure of a practical series type regulated power supply is shown in Figure 14.27, which is adopted from Jung [7] with some modifications.

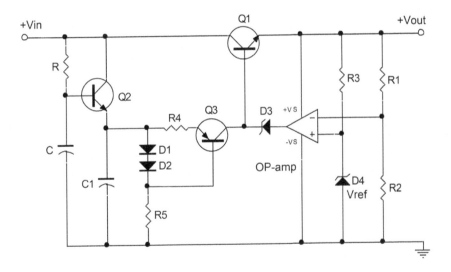

Figure 14.27 Structure of a discrete series type dc regulated power supply. Components R, C, Q2, and C1 form a filter to reduce the ripples at the emitter of Q2. Components D1, D2, Q3, R4, and R5 form a current source to jump-start the circuit during power up. Zener D3 is added to level-shift the dc potential at the output of the op-amp to Vout/2

The simple R-C filter is now further assisted by an additional transistor, Q2 and capacitor, C1. In principle, a large capacitor for C1 should be better for reducing ripples. But, in a number of designs, it is found that it is best to keep C1 no larger than 100μF. Capacitance 10μF to 47μF will be a good compromise. Components D1, D2, R4, R5, and Q3 form a current source. Resistor R5 sets up a few mA bias currents for diodes D1 and D2. Once diodes D1 and D2 are forward biased, the diode's forward voltage will force transistor Q3 to deliver a constant current to bias the zener diode D3 and jump-start the op-amp during power up. The constant current generated by Q3 is given as $(2 \times V_F - V_{BE})/R4$, where V_F is the forward voltage of diode D1 and D2, and V_{BE} is the voltage across base and emitter of Q3.

The series pass transistor Q1 is a BJT in this circuit. But MOSFET is also a popular choice because MOSFET does not require gate current so that it will relax the current output requirement for the op-amp. Resistor R3 sets up a few mA dc bias currents for the reference zener D4. The resistor network formed by R1 and R2 samples a portion of the output voltage Vout and feeds back to the op-amp. The op-amp compares the feedback voltage with the reference voltage Vref, then it produces an output voltage according to Eq. (14.9). The resistor network is often biased to around 1–2mA.

We notice that there is a zener D3 inserted between the op-amp and transistor Q1. The purpose of D3 is to level-shift the dc potential from the output of the op-amp. Without this zener D3 in place, the dc potential at the output of the op-amp may be leaning towards its positive power supply rail +VS, which is equal to Vout. When an op-amp is biased with its output's dc quiescent potential near to its positive supply rail, or negative supply rail, some of the internal circuits of the op-amp are working near to the cut-off or saturation region. When this happens, the output voltage (Vout) will drift away from its designated value. To avoid this problem, zener D3 is inserted into the output of the op-amp. As a result, the op-amp's output dc potential is level-shifted by zener D3. If a suitable zener D3 is used, the output (Vout) will be rested at the designated value governed by Eq. (14.9). Note that the ideal choice for setting the dc potential for the op-amp's output is in the middle, between Vout and ground (i.e., Vout/2). There is no universal zener D3 that works for all power supplies of different Vout. For example, D3 is a 11V zener (1N4741) for a 15V power supply in Figure 14.28. But for a 45V power supply in Figure 14.29, D3 is a 27V zener (1N4750). It takes trial and error in the actual circuit to determine the right zener for D3.

Dual ±15V regulated power supply

Figure 14.28 shows a discrete series type dual ±15V regulated power supply. This power supply follows the configuration of Figure 14.27. Some capacitors are added. Series pass transistors Q2 and Q6 are MOSFET instead of BJT. Since MOSFET does not have gate current, this reduces the current requirement for the op-amp. Let us look at the +15V power supply first. The current source is formed by the components D2, D3, R1, 180Ω, and Q3. The constant current is determined by $(2 \times V_F - V_{BE(Q3)})/180\Omega = (2 \times 0.65V - 0.65V)/180\Omega = 3.6mA$, where 0.65V is assumed for V_F, forward voltage of D2 and D3, and for V_{BE} of Q3. Resistor R1 is 6.8kΩ so that diodes D2 and D3 are biased to 3.1mA.

There are several capacitors used to help reduce noise and ripples. A 47μF is connected across the zener D4 to suppress zener noise. Another 47μF capacitor is connected across resistor R2 so that the capacitor is also across the V_ and output of the OPA134 via transistor Q2. Given this 47μF connection, the op-amp works like a low-pass filter that helps to filter out high frequency noises. A third 47μF is in series with a 10kΩ resistor, forming a simple R-C filter that reduces noise coming from the reference zener D7. Even though D7 is a low noise reference Zener, LM329, this simple R-C filter is essential for reducing noise in the entire power supply. Imagine if there is only a very tiny noise created by the reference zener D7, the op-amp with huge open-loop gain (around 120dB) will inevitably amplify the tiny noise to a much greater level if the R-C filter is not used.

1N4741 is chosen for zener D4 so that the dc bias potential at the output of the OPA134 is 7.8V, which is around the desired dc potential, i.e., Vout/2 = 7.5V. A 0.3V difference is not

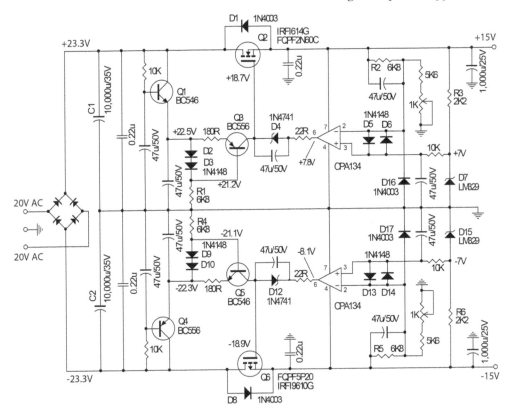

Figure 14.28 A discrete dual ±15V dc regulated power supply. The dc quiescent voltages are measured
        when ±50mA current is driven by the load

crucial. Because of the discrete values offered by commercial zener diodes, sometimes the difference can be over 1V. Again, this is acceptable. Just choose the nearest zener diode so that the
dc quiescent potential of the op-amp's output is sitting near to Vout/2.

There are many enhancement type MOSFETs available on the market that can be used for
Q2 and Q6. Look for power dissipation P_D around 50W. Most enhancement type MOSFETs
have gate threshold voltage $V_{GS(th)}$ around 3V–4V. In Figure 14.28, the $V_{GS(th)}$ for Q2 is 3.7V
and for Q6 is 3.9V. This is much higher than the V_{BE} of BJT, which is around 0.7V. Therefore,
in order to ensure Q2 working in the linear region for a MOSFET, the voltage between drain-
source (V_{DS}) of the MOSFET must be higher than a certain level. It is found that for most cases,
drain-source $V_{DS} > 7V$ must be maintained. If V_{DS} is below 7V, we will lose the control of the
dc quiescent condition at the output of the op-amp. When this happens, no matter what zener
voltage we choose for D4, the dc quiescent potential at the output of the op-amp always stays
near to Vout.

On the other hand, it is not advised to use a power transformer with very high secondary
voltage so that the unregulated input voltage is much higher than output voltage (Vin ≫ Vout)
because a high unregulated input voltage produces a high drain-source voltage, V_{DS}, leading
to high power dissipation in Q2 and Q6. Therefore, choose a power transformer with the right
secondary voltage for your needs. In this example, a 40V ac secondary voltage with center tap is

used. When the output load draws 40mA, the input unregulated dc voltage appears to be 23.3V. Therefore, $|V_{DS}|$ is equal to $23.3V - 15V = 8.3V$ for both Q2 and Q6. Even though there is not much power dissipation in the transistors, it is recommended to mount Q2 and Q6 on a suitable heatsink.

This dual $\pm 15V$ power supply can deliver well over several hundred mA to a load. However, you must have the right power transformer to cope with the current demand. The actual limitation to output current is the power transformer and the series pass transistors Q2 and Q6. For instance, the maximum continuous drain currents $I_{D(max)}$ for IRFI614G and FQPF2N60C are 2.1A and 2A, respectively, at $T_C = 25°C$. Therefore, in reality, output current can never go up to its maximum 2A when the transistor is heated up.

In practice, if we need 2A output current, it is recommended that you look for MOSFETs with $I_{D(max)}$ of 4A or higher, and mount them in a suitable heatsink. Then choose a power transformer so that the condition $V_{DS} > 7V$ for both Q2 and Q6 is maintained when 2A current is driven by the load. Running at 2A output current, it is obvious that you do not want too high a V_{DS} across the transistor because transistors Q2 and Q6 will have to dissipate substantial power (i.e., $2A \times V_{DS}$). To reduce the power dissipation in the transistor at high current output, BJT may be a better choice than MOSFET for Q2 and Q6 as this circuit still works well for $V_{CE} < 7V$. But then we have to make sure the op-amp can deliver sufficient current to drive the BJT. A Darlington pair, which requires a much lower base current, may be a suitable choice for very high output current.

Dual ±45V regulated power supply

By following the dual $\pm 15V$ version of Figure 14.28, it is very straightforward to develop a dual $\pm 45V$ power supply. Obviously, a higher output power transformer is required. Other than this, the $\pm 45V$ power supply circuit is almost identical to the $\pm 15V$ version. See Figure 14.29.

At 45V output, OPA134 cannot be used because the maximum supply voltage for the device is only 36V. Therefore, we have to use a high voltage op-amp. OPA552, which has a maximum supply voltage of 60V, is chosen for this application. Resistors R1 to R8 are changed to accommodate a $\pm 45V$ output. The same MOSFETs are kept for Q2 and Q6. However, we select higher voltage transistors, 2N5401 and 2N5551, for Q1, Q3, Q4, and Q5. Zener diodes D4 and D12 are changed from 1N4741 to 1N4750 so that the dc potential at the output of the OPA552 is around 22V–25V, which is approximately equal to one-half of 45V. Capacitors are also needed to tolerate a higher voltage rating. Given all these changes, the dual $\pm 45V$ power supply is shown in Figure 14.29.

When comparing with shunt type regulated power supply, series type consumes much less power under low load to no load condition. We can compare what the power consumes internally at no load condition between Figure 14.25 (shunt type) and Figure 14.29 (series type) for the same output voltage $\pm 45V$. Let us first examine the series type power supply in Figure 14.29. At no load condition, the internal circuit consumes dc bias currents I1 to I5. Let I_Q denote the total dc bias currents given as

$$I_Q = I1 + I2 + I3 + I4 + I5$$

where I1 is the dc bias current for the current source containing two current components. The first component, I1_1, is the dc bias current for D1, D2. The second component, I1_2, is the emitter current of Q3. They are determined as follows. $I1\_1 = (55.9V - 2 \times V_F)/18k\Omega = 3mA$,

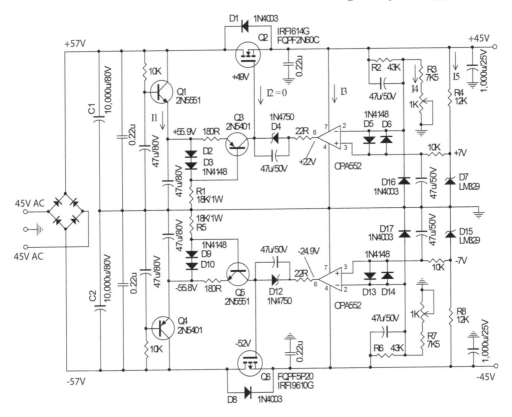

Figure 14.29 A discrete series type dual 45V dc regulated power supply. The dc quiescent dc voltages are measured when ±40mA current is driven by the load

where V_F is forward voltage of the diode (0.65V). And $I1\_2 = (2 \times V_F - V_{BE(Q3)})/180\Omega = 3.6mA$, where $V_{BE(Q3)} = 0.65V$. Therefore, we have $I1 = 3mA + 3.6mA = 6.6mA$. Since MOSFET does not have any gate current, we can assume current $I2 = 0$. For op-amp OPA552, $I3$ is the supply current for the device. It has a maximum value of 8.5mA. Let us take $I3 = 8.5mA$. Assuming the op-amp's input dc bias current flowing into V_- and V_+ terminals can be neglected, the dc bias current $I4$ is determined by $45V/(43k\Omega + 7.5k\Omega + 1k\Omega) = 0.87mA$. Finally, dc bias current $I5$ is equal to $(45V - 7V)/12k\Omega = 3.2mA$. Hence, the internal dc bias current for the +45V series type regulated power supply of Figure 14.29 is given as $I_Q = 6.6mA + 0 + 8.5mA + 0.87mA + 3.2mA = 19.2mA$. Similarly, the -45V power supply will also have approximately the same dc bias currents. Therefore, the total dc bias current for the dual ±45V power supply of Figure 14.29 is $19.2mA \times 2 = 38.4mA$. For the shunt type power supply of Figure 14.25, the total internal current consumption at no load has been found to be $138.3mA \times 2 = 276.6mA$. It is obvious that for the same dual ±45V output, at no load condition the shunt type power supply consumes more than six-fold. Because of this high internal current consumption in shunt type power supply at low to no load condition, it is not a popular choice for an application where the load may vary frequently. Again, series pass transistors Q2 and Q6 of Figure 14.29 must be mounted on a heatsink.

14.6   Series type regulated power supplies (floating-mode)

We have discussed several three-terminal adjustable regulators in section 14.3. They include LM317, LM337, and TL783. These devices do not have a ground pin. The dc bias current of the device, instead of flowing to the ground directly, flows to the ground through the output resistor network. On the other hand, what matters most to the regulator is the difference between input and output voltages (Vin − Vout). The difference is 37V maximum for a standard LM317 and 125V maximum for a TL783. Since it is not required to connect the regulator to the ground directly, this makes them, ideally, float above the ground. As a consequence, they are an excellent choice for realizing high voltage regulated power supplies. In the following, we discuss a +250V power supply using LM317/TL783, and a −150V power supply using LM337 in floating mode.

+250V regulated power supply using LM317/TL783

In a National Semiconductor application note [8], a 1.2V to 160V power supply is developed by using an LM317 together with two NPN transistors, which are in a Darlington transistor configuration. The circuit is modified here so that the output is now +250V, as shown in Figure 14.30. One major change is replacing the Darlington pair in the original design with a single transistor Q1, MJE15034. It has a maximum V_{CEO} and V_{CBO} of 350V; dc current gain of 100. These make MJE15034 very suitable for a high voltage power supply.

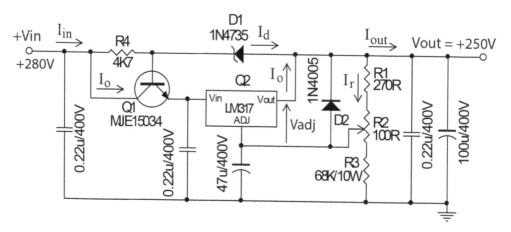

Figure 14.30 A +250V dc regulated power supply using LM317 operating in floating mode. Alternatively, TL783 can also be used instead of LM317

A zener diode, D1, is connected across the base of Q1 and LM317's Vout. Therefore, LM317 sees a (Vin − Vout) difference equal to $V_Z - V_{BE}$, where V_Z of 1N4735 is 6.2V and V_{BE} for Q1 is around 0.7V. Therefore, we have Vin − Vout = 6.2V − 0.7V = 5.5V. This is well below the 37V (max) for a standard LM317. Diode D2 provides a path to discharge the 47μF capacitor so as to protect the LM337 in case of an output short. Resistors R1, R2, and R3 are so selected that V_{adj} is equal to the internal reference voltage of LM317, 1.25V. Hence, the output voltage is given by

$$Vout = \left(1 + \frac{r2 + R3}{R1 + (R2 - r2)}\right)1.25V \qquad (14.10)$$

where r2 is the center tap resistance of R2 such that $0 \leq r2 \leq R2$. If we take R1 = 270Ω, R3 = 68kΩ, then adjust R2 for Vout = 250V.

Note that input current, I_{in}, is equal to $I_o + I_d$, neglecting the small base current to Q1. I_d is the reverse bias current for zener diode D1. I_o is equal to the emitter current of Q1 when the base current is neglected. At the same time, I_{in} is also equal to $I_{out} + I_r$, where I_{out} is the output load current and I_r is the dc bias current for the resistor network. Therefore, we must have $I_o + I_d = I_{out} + I_r$. In practice, it is designed such that I_r and I_d are just several mA. Thus, the load current I_{out} is almost entirely supplied by the output current I_o from LM317.

As an example, let us assume the output current I_{out} is 30mA. First we determine I_r = 250V/(270Ω + 100Ω + 68kΩ) = 3.66mA. We also find I_d = (280V – 6.2V – 250V)/4.7kΩ = 5.06mA, by neglecting the base current for Q1. Since $I_{out} + I_r = I_o + I_d$, we have 30mA + 3.66mA = I_o + 5.06mA. This leads to I_o = 28.6mA, which is the current delivered by LM317. Note that transistor Q1 and LM317 must be mounted on heatsinks.

–150V regulated power supply using LM337

Figure 14.31 shows a –150V dc regulated power supply by operating a LM337 regulator in floating mode. Since this is a negative power supply, the diodes and capacitors are reversed in polarity compared to the positive power supply of Figure 14.30. However, the working principle is the same. Q1 is a PNP transistor, MJE15035. Similar to its NPN complementary part, MJE15034, V_{CEO} and V_{CBO} are equal to 350V and dc current gain 100. By choosing R1 = 330Ω, R2 = 100Ω and R3 = 47kΩ, –Vout is equal to –150V. Diode D2 provides a path to discharge the 47μF capacitor so as to protect the LM337 in case of an output short. Transistor Q1 and LM337 must be mounted on a heatsink.

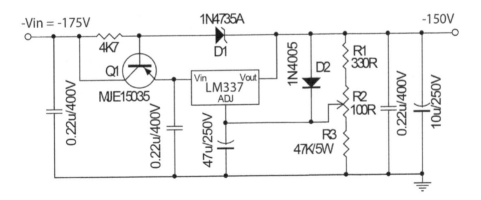

Figure 14.31 A –150V regulated power supply using LM337 in floating mode

Discrete floating-mode regulated power supply

An op-amp is one of the core components for a discrete regulated power supply. It helps to set up the desired output voltage and maintains good line regulation. However, today's monolithic op-amps can only take a maximum supply voltage of up to 180V. (See Table 14.2.) If we follow the fixed-mode power supply from section 14.5, the output voltage can never exceed 180V. We now simplify the fixed-mode power supply and redraw it in Figure 14.32(a) for comparison.

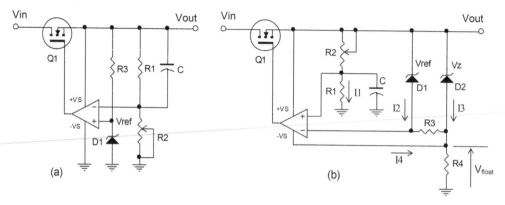

Figure 14.32 (a) A simplified series type fixed-mode dc regulated power supply. (b) A simplified floating-mode dc regulated power supply

Since the power supply rails of the op-amp (+VS and −VS) are connected to Vout and ground, respectively, it is clear that the output of a fixed-mode regulated power supply is limited to the maximum supply voltage an op-amp can take.

However, if we allow the op-amp's power supply rails to float above the ground, in principle there is no limit for setting a high output voltage as long as the op-amp's power supply rails are kept within the maximum limit. It is called a floating-mode regulated power supply, as shown in Figure 14.32(b). The power supply rails of the op-amp are clamped to V_z by the zener diode D2. It is common to choose V_z of less than 36V so that most commercial op-amps can be used.

The dc bias currents are determined in the following. I1 is the dc bias current of the resistor network R1 and R2. Setting I1 to around 1mA will be sufficient. I2 is the dc bias current of reference zener D1 and I3 is the dc bias current for zener D2. Vref must be lower than Vz so that dc bias current I2 can be determined by (Vz − Vref)/R3. Setting I2 and I3 to a few mA is sufficient. I4 is the dc supply current for the op-amp. Depending on what device we use, I4 is usually around 10mA or less for most general purpose op-amps. The floating voltage, V_{float}, that the op-amp is sitting on, is the dc potential produced by currents flowing through R4. Assuming that the tiny input current to the op-amp inverted input can be neglected, V_{float} is given by (I2 + I3 + I4) × R4. Therefore, we have

$$V_{float} = Vout - Vz = (I2 + I3 + I4) \times R4.$$

Thus, resistor R4 can be determined by

$$R4 = \frac{Vout - Vz}{I2 + I3 + I4} \tag{14.11}$$

And Vout is given by

$$Vout = \left(1 + \frac{R1}{R2}\right) Vref \tag{14.12}$$

In the following, we discuss two examples of discrete series type floating-mode high voltage regulated power supplies.

+250V discrete regulated power supply

We have discussed a 250V power supply in the preceding section by operating an LM317 in floating mode. The components count is low for such a high voltage power supply. It is a simple and effective solution. However, if we want to achieve better performance in terms of low noise, low temperature variation, and good line regulation, the discrete floating-mode approach is a much better solution.

Figure 14.33 is a 250V discrete floating-mode regulated power supply. Let us analyze the dc bias conditions of the circuit. First, the 22kΩ/15W resistor creates a dc voltage at OPA134's pin-4, which is the –VS power supply pin for the op-amp, so that the op-amp floats above the ground. The dc potential at OPA134's pin-4 is measured at 226V, which is 24V (zener voltage of 1N5359, D9) below the output voltage 250V. Therefore, the current flowing through the 22kΩ resistor is 226V/22kΩ = 10.3mA. This is the sum of the dc supply current of OPA134 (I4), dc bias current for reference zeners D7 and D8 (I2), and zener D9 (I3). DC bias current I2 is given by $(24V – 2 \times Vref)/3.3k\Omega = (24V – 2 \times 7V)/3.3k\Omega = 3mA$, assuming the dc bias current to the OPA134's inverted input can be neglected. And this is true, because OPA134 has a JFET input stage. As the op-amp's input dc bias current is in the order of pA, it can be ignored in the calculation for I2. The supply current for OPA134 is I4 = 4mA typically. Therefore, the bias current for zener diode D9 is given by $I3 = 10.3mA – I2 – I4 = (10.3–3–4)mA = 3.3mA$.

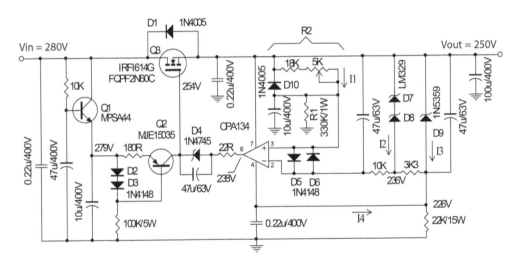

Figure 14.33 (a) A simplified series type fixed-mode dc regulated power supply. (b) A simplified floating-mode dc regulated power supply

On the other hand, the dc bias current for the resistor network, I1, is determined by 250V/(18kΩ + 5kΩ + 330kΩ) = 0.7mA. Depending on the center tap position of the 5kΩ variable resistor, I1 can be slightly higher than 0.7mA. The dc bias current I1 is designed to run low so that the 330kΩ resistor does not dissipate too much power. At current 0.7mA, the 330kΩ resistor dissipates 0.16W. Generally speaking, a resistor with a 0.5W power rating can be used. However, in order to minimize a resistor's temperature rise, it is recommended to choose a resistor with a power rating three to five times its actual power dissipation. For example, if a resistor dissipates 0.16W power in a circuit, choose a 1W resistor. Likewise, if a resistor dissipates 1W power, choose a 5W resistor. This is the reason why 15W power rating is chosen for the 22kΩ resistor, even though it only dissipates 2.3W of power. Since the resistance of resistor has a positive temperature coefficient, a resistor running at a cooler temperature will help to keep the dc quiescent conditions and, thus, the output voltage unchanged.

It should be noted that two reference zeners, LM329, are used in this 250V power supply. This will help to avoid using a high value resistor for R1. Imagine that if we have just one LM329, in order to get the same output voltage, we must double the value for resistor R1 (it becomes 660kΩ), if the same R2 resistor is used. The change in resistance for a 660kΩ resistor due to temperature rise is always higher than that of a 330kΩ resistor if they have the same temperature coefficient. Thus, using two reference zeners will get us an output voltage that is more temperature stabilized. For 100V or lower power supply, it is found that one LM329 is sufficient. When output voltage is over 100V, two LM329s are recommended. If the output voltage is over 300V, three LM329s may be necessary. But then D9 has to be changed to a higher voltage zener because three LM329 will have a total of 21V, causing a 24V zener (D9) marginally to bias them. This can be overcome if we choose a high voltage op-amp and, therefore, a higher voltage (>24V) zener D9 can be used.

Several 47μF capacitors are used to suppress noises from zener diodes. A capacitance ranging from 47μF to 100μF will do the job. However, we should avoid attempting to increase the 10μF capacitor, which is connected to R1. In this power supply, any attempt to increase the capacitor beyond 10μF will result in an unstable output. Again, transistor Q3 must be mounted on a heatsink of suitable size.

Dual ±45V discrete regulated power supply

The floating-mode technique works equally well for a negative power supply. We illustrate this in an example for a dual ±45V power supply in Figure 14.34. Compare it to the fixed-mode

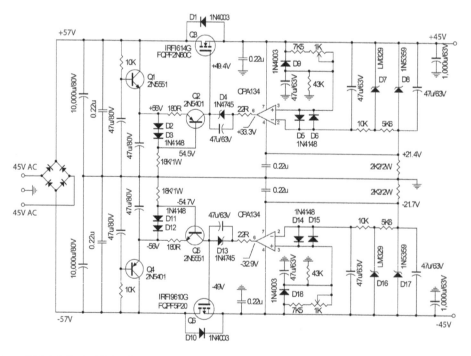

Figure 14.34 A dual ±45V series type floating-mode dc regulated power supply. The dc quiescent voltages are measured when 40mA output current is driven by the load

power supply of Figure 14.29, where a high voltage op-amp must be used for the same output. However, by floating the op-amp above the ground, we can employ any standard op-amp with 36V maximum supply voltage. This opens up a great variety of choices for op-amps. For illustration purposes, we chose OPA134 again for this example.

As a matter of fact, I have found that OPA134 is a very stable op-amp for dc regulated power supplies ranging from 3.3V to 300V. OPA134 is low cost, reliable, and has a JFET input stage. The supply current for the op-amp is only around 4–5mA. Another good thing is that there is no parasitic oscillation problem when it is used for the power supplies discussed in this chapter. However, if the PCB layout design work is poor, the power supply may still run into a parasitic oscillation problem. When this happens, the oscillation problem can be solved by inserting a small capacitor, 1nF to 10nF, connecting it between the inverted input V− or non-inverted input V+ and the op-amp's output, depending on whether it is a fixed-mode or floating-mode design. This will usually solve the parasitic oscillation problem. If the problem persists, the PCB layout design has to be revisited.

Since the output voltage is less than 100V, only one reference Zener, LM329, is needed, as shown in Figure 14.34. Here, let us examine the +45V power supply first. Resistor 2.2kΩ/2W establishes a +21.4V to float the OPA134 above the ground. The current flowing through this resistor is given by 21.4V/2.2kΩ = 9.7mA, which is the total sum of currents from the supply current of OPA134 and the dc bias current for zener D7 and D8. In other words, each of these devices contributes about a few mA.

Zener diode D4 is chosen in such a way that OPA134's output dc bias potential (33.3V) is around the mid-point between the op-amp's power supply pins, pin-7 (45V) and pin-4 (21.4V). The −45V power supply is developed in a similar fashion. It should be noted that transistors Q3 and Q6 must be mounted on a heatsink.

14.7 Temperature variation

A regulated power supply implemented by a three-terminal variable voltage regulator has the advantage of low component count and ease of use. Three-terminal variable voltage regulators LM317 and LM337 have all key components embedded in the integrated circuit. It only requires two external resistors for setting the output voltage. The key components include a reference voltage, error amplifier, and a series pass transistor. Since these components are embedded in the device, when the series pass transistor is heated up, the rise in temperature will inevitably affect the reference voltage and error amplifier. As a result, a power supply using a three-terminal variable voltage regulator has a lower temperature stability than the discrete power supplies.

Figure 14.35 shows the graph of LM317's reference voltage versus temperature. At a room temperature of 25°C, the reference voltage is 1.25V. When the temperature rises to 50°C, the reference voltage drops to 12.48V. There is a drop of 0.02V for a rise of 25°C. This may appear to be a small change, but it will bring a more significant impact to the output when the three-terminal variable regulator is used in a high voltage power supply. Let us take the 250V power supply of Figure 14.30 as an example. We know that the output voltage is given as

$$\text{Vout} = \left(1 + \frac{\text{r2} + \text{R3}}{\text{R1} + (\text{R2} - \text{r2})}\right)\text{Vref} \tag{14.13}$$

At room temperature, Vref = 1.25 and the power supply of Figure 14.30 gives Vout = 250V. In other words, the first term on the right-hand side shown in brackets in the above expression is a multiplier of 200. Assuming the resistors R1, R2, and R3 have low temperature coefficient

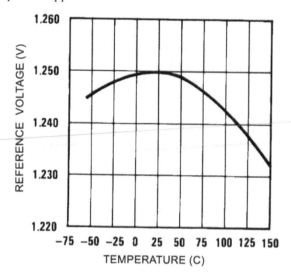

Figure 14.35 Temperature variation of LM317's reference voltage. Courtesy of Texas Instruments

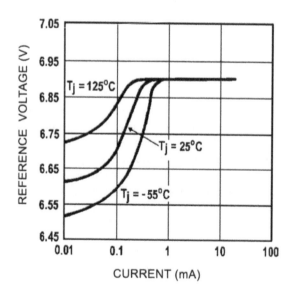

Figure 14.36 Temperature variation of LM329's reference voltage. Courtesy of Texas Instruments

and the resistance does not change much for a rise of 25°C, the change of output voltage is, therefore, caused by the change in Vref alone. Thus, we have ΔVout = ΔVref × 200 = −0.02V × 200 = −4V for a rise of 25°C. In other words, there is a drop of 4V when the temperature of the

LM317 in Figure 14.30 is increased by 25°C from room temperature. Thus, a supposedly 250V power supply becomes 246V at 50°C, a 1.6% change in voltage.

Figure 14.36 shows the graph of LM329's reference voltage versus temperature. This reveals a much superior temperature stability to Figure 14.35. When an LM329 is biased at 1mA or higher, the reference voltage (6.9V) remains flat for a junction temperature ranging from −55°C to 125°C. Additionally, LM329 is a discrete through-hole device that is soldered into a PCB separately from the series pass transistor. Therefore, in a discrete regulated power supply, the temperature rise of the series pass transistor, which is mounted on a separate heatsink, does not affect LM329. Thus, there is little change to the reference voltage. The reference voltage of LM329 remains at almost a constant 6.9V for a bias current over 1mA.

If you are looking for a simple regulated power supply that can tolerate a few percent change in output voltage due to temperature rise, a three-terminal regulator will definitely suit your needs. However, if you are looking for very low noise and low output voltage variation due to temperature change, a discrete regulated power supply is a superior solution.

14.8   Vacuum tube regulated power supplies

Vacuum tubes had been the exclusive choice for regulated power supplies long before solid-state semiconductor devices became commercially available. However, a vacuum tube regulated power supply is inevitably bulky and it generates a considerable amount of heat dissipation. In addition, a vacuum tube has a much shorter lifespan and is less reliable when compared with its solid-state counterpart. As a result, after solid-state devices became commercially available, vacuum tube regulated power supplies were removed from most audio and commercial electronic applications. However, it is worth studying some examples of a vacuum tube regulated power supply.

Two-tube regulated power supply (200V to 300V)

Figure 14.37 shows a two-tube regulated power supply [9]. Since twin triodes 12AX7 are housed in the same tube, it is considered as one tube. Together with power tube EL34, it is a two-tube design. Power tube EL34 is used to provide the necessary current drive to the output load. A differential amplifier (long-tailed pair) is formed by triodes T2a and T2b. The differential amplifier works as an error amplifier comparing a reference voltage Vref to a portion of the output voltage fed from the resistor network formed by R8, VR, and R9. The output from the error amplifier is directly coupled to the power tube T1, which works as a series pass tube, similar to a series pass transistor in a solid state series type regulated power supply.

It is clear that the higher voltage gain of the differential amplifier produces better line regulation. Triode 12AX7 is chosen because of its high amplification factor, $\mu = 100$. However, the open-loop gain is very limited in this design. Since 12AX7 is already a high gain triode, if we want to further increase the open-loop gain for the error amplifier, we have to cascade a second voltage gain stage.

Three-tube regulated power supply (250V to 350V)

By cascading a second stage to the error amplifier of Figure 14.37, this forms a three-tube regulated power supply, as shown in Figure 14.38. Two differential amplifiers are used to boost the

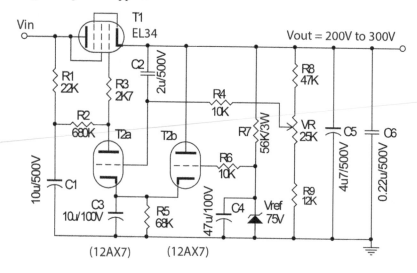

Figure 14.37 A two-tube dc regulated power supply capable of producing 50mA output current

open-loop gain. The first stage is formed by tube T3, which works as an error amplifier. It compares the reference voltage Vref to a portion of the output fed from the resistor network formed by R12, VR, and R13. The outputs from the error amplifier are directly coupled to the second stage formed by tube T2. The output from T2 is directly coupled to the series pass tube T1 that provides the necessary output current to a load.

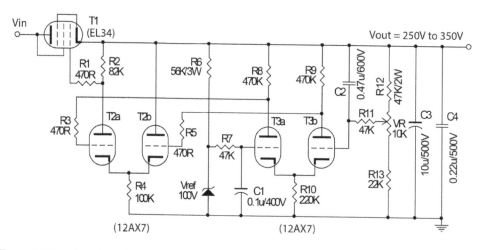

Figure 14.38 A three-tube dc regulated power supply capable of producing 50mA output current

Since this three-tube design employs two amplifying stages, it produces a much higher open-loop gain than the two-tube design of Figure 14.37, which has only one amplifying stage. As a result, this three-tube power supply has better line regulation and produces lower noise.

14.9 Hybrid regulated power supply (190V)

In recent years, vacuum tube regulated power supplies can be found deployed in some high-end commercial vacuum tube line-stage amplifiers. They are not an all-tube power supply design. Instead, it is a hybrid approach that employs both vacuum tubes and solid-state devices. If we examine the hybrid approach closely, it can be seen that it follows the same topology as the discrete series type floating-mode regulated power supply discussed in the preceding sections. In this section, we discuss a 190V hybrid regulated power supply.

There are two approaches for using solid-state devices to improve the performance of the all-tube power supply of Figure 14.38. The first approach is to replace power tube T1 by a power transistor but keep tubes T1 and T2. This approach helps to reduce the size of the power supply. The second approach is to replace tubes T2 and T3 by an op-amp but keep the power tube T1. This approach helps to improve line regulation and lower noise. If we want to achieve both reducing size and improving regulation and noise, we may perhaps give up the idea of using vacuum tubes altogether. We could simply choose the series type regulated power supplies that contain all solid-state devices discussed earlier in this chapter. Well, not exactly. A hybrid approach may give us what we want. And it may retain the so-called tube sound produced by the hybrid power supply.

The first approach is not a very attractive solution as it only concerns the size reduction without addressing the line regulation and noise. When a high voltage regulated power supply is used in an audio application, it is very likely to power a vacuum tube line-stage amplifier. Therefore, good line regulation and low noise are of primary concern. Hence, the second approach, replacing the vacuum tube error amplifier by an op-amp, is a more desirable solution. In the following, we demonstrate the second approach through a 190V power supply design.

Figure 14.39 shows a hybrid regulated power supply employing vacuum tube and solid-state semiconductor devices. T1 is a popular power tube, KT88, which works in a triode connected arrangement with the screen grid (pin 4) connected to the plate (pin 3) via a 100Ω resistor. A 1kΩ grid-stopper resistor is in series with the control grid of KT88 to prevent high frequency oscillation. Components Q1, 10KΩ resistor, 47μF and 10μF capacitors form a filter to reduce the ripples of the input voltage before passing to a current source. The current source is formed by Q2, 180Ω, D1, D2, and an 180kΩ resistor. The emitter current of Q2 is given

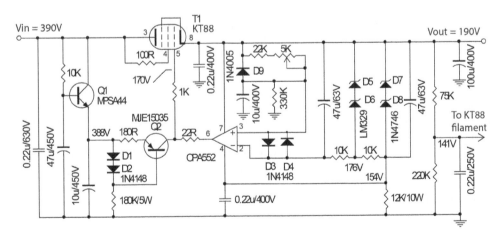

Figure 14.39 A hybrid dc regulated power supply for 190V output using high voltage op-amp OPA552. The dc quiescent voltages are measured when 20mA of current is driven by the load

by $(2 \times V_F - V_{BE(Q2)})/180\Omega = 3.6\text{mA}$, assuming $V_F = 0.65\text{V}$ and $V_{BE(Q2)} = 0.65\text{V}$. As discussed in series type regulated power supplies, Q2 is a current source that helps to jump-start the circuit during power up.

Since KT88's grid-cathode dc potential difference is 20V, compared to around 4V for the V_{GS} of a MOSFET, we need a high voltage op-amp to drive the tube. The power supply for the op-amp is set by the zener diodes D7 and D8, a total of 36V. Given this supply voltage, many standard op-amps that have a maximum of supply voltage 36V cannot be used because operating an electronic device at its maximum allowable voltage will shorten the lifespan drastically. OPA552, which has a maximum supply voltage of 60V, appears to be a suitable choice. OPA552 operates in Figure 14.39 with its pin-4 floating at 154V above the ground.

Power tube KT88 needs a 6.3V filament voltage. In general, an ac power supply is used for the power tube's filament. However, if we want to reduce the dc power supply noise to the lowest possible level, a dc regulated filament voltage should be used. Since the dc potential at KT88's cathode is 190V, in order to avoid any unwanted current flowing from the filament to the cathode, the filament has to be level-shifted to near the cathode's dc potential. A simple potential divider formed by the 75kΩ and 220kΩ resistors create 141V, which is to be connected to the 6.3V filament circuit as shown in Figure 14.40. When the 6.3V filament is lifted up by 141V, the dc potential difference between the filament and cathode of the KT88 now becomes 190V – 141V = 49V, which is well below the design maximum 200V allowed for the KT88.

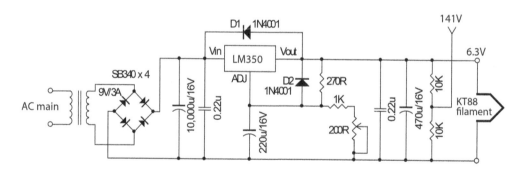

Figure 14.40 A 6.3V regulated power supply for filament of KT88. The dc 141V is taken from a potential divider in Figure 14.39 or Figure 14.41

KT88's filament requires 6.3V at 1.6A. Three-terminal variable voltage regulator LM317 cannot be used for this application because its maximum output current is limited to 1.5A. We choose LM350, which can deliver up to 3A current. Since the 6.3V output is lifted by 141V from the 190V high voltage power supply, it should be noted that no part of this 6.3V power supply should be allowed to connect to the ground of the high voltage power supply.

The power supply of Figure 14.39 is strictly limited to working with a high voltage op-amp. If we want to allow a standard op-amp with maximum 36V supply voltage to be used, the circuit must be modified. By changing the high voltage op-amp to a composite op-amp configuration, as shown in Figure 14.41, we can use any standard op-amp for this hybrid power supply. For illustrative purpose, OPA134 is used again.

The composite op-amp consists of components R1–R4, D3, D4, Q3, Q4, and OPA134. Resistors R1–R4 and diodes D3 and D4 work as a potential divider that reduces the supply voltages for the OPA134. The dc bias current for the potential divider is about 1.1mA. The

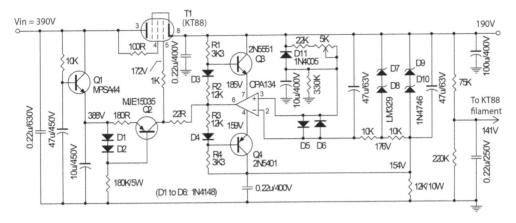

Figure 14.41 A hybrid dc regulated power supply for 190V output using a composite op-amp. The dc
quiescent voltages are measured when a 20mA current is driven by the load

potential divider sets up a supply voltage, 185V – 159V = 26V, for OPA134. This is a safe
supply voltage for OPA134 as well as for many other standard op-amps. The dc potential at
the OPA134's output is 172V, which is at the mid-point between 185V and 159V. Other than
the extra components that form the composite op-amp, the rest of the circuit is identical to
Figure 14.39.

In order for a power vacuum tube to work in the linear region, the plate of the tube should be
maintained at a voltage substantially higher than the cathode. For example, when KT88 works
in a push–pull power amplifier, the plate-cathode voltage is usually higher than 350V. When
KT88 is used in a regulated power supply, even though 350V is not necessary, it is recommended to set the plate-cathode voltage to around 200V. As shown in Figure 14.39 and 14.41,
the input voltage Vin (390V) is set to be 200V higher than Vout (190V).

In addition to KT88, there are many other power tubes that can be considered. They include
6550, KT77, KT66, EL34, 6CA7, and 6L6G. They are single tubes suitable for a single power
supply. I found that these tubes work equally well in Figure 14.39 and 14.41. If you want to
develop dual mono power supplies, using two of these tubes is a good choice. Alternatively, you
can also consider the twin triodes power tube 6AS7G. This is a power tube specifically designed
for regulated power supply applications, but nowadays it is also found to be very popular in
OTL tube power amplifiers.

14.10   Regulated power supplies ≤ 5V

Most audio line-stage amplifiers employ op-amps and discrete transistors that require power
supplies ranging from dual ±15V to ±35V. Vacuum tube line-stage amplifiers require an even
higher power supply, starting from 100V to a few hundred volts. Therefore, we will find the
power supplies discussed in the preceding sections very suitable, as they produce output voltages ranging from +15V to +250V and from −15V to −150V.

We have discussed dual ±2.5V and ±5V power supplies by using shunt regulator TL431
in Example 14.2. It provides a simple and easy to use solution for low voltage applications.
However, a discrete series type power supply often outperforms a TL431 power supply in terms
of lower noise and better line regulation. Therefore, it is worth developing a series type power

supply for low voltage applications. Let us consider a regulated power supply voltage for 5V or lower. However, if noise and line regulation are not crucial issues in the application, many low drop-out regulators as well as LM317 and LM337 can easily satisfy the need. Some low drop-out regulators can even offer output voltage down to about +1.2V and −1.2V. They are surface-mounted devices that save tremendous printed circuit board space compared to a discrete regulated power supply.

A dual ±3.3V discrete series type regulated power supply is illustrated in Figure 14.42. The dual ±3.3V power supply follows closely the dual ±15V of Figure 14.28. However, there are three major differences. First, if the output is 3.3V, it cannot directly power the op-amp OPA134, which has a minimum supply voltage requirement of 5V. Therefore, pin-7 of OPA134 must be connected to the emitter of Q1. In the −3.3V power supply, pin-4 must be connected to the emitter of Q4. Given this arrangement, OPA134 operates at around 11V, which is a sufficient supply voltage for the op-amp.

Second, the reference zener LM329 produces a reference voltage of 6.9V–7V. Therefore, the 3.3V output cannot be used directly to power the zener. Again, LM329 is connected to

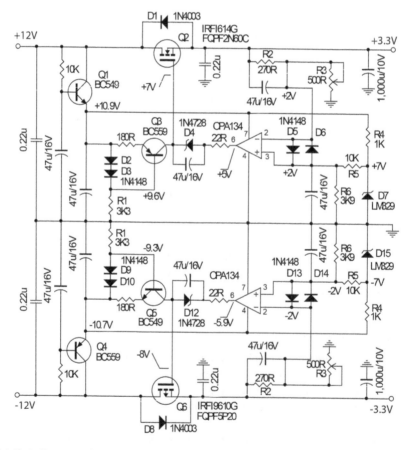

Figure 14.42 A discrete series type dual ±3.3V dc regulated power supply. The dc quiescent voltages are measured when ±50mA current is driven by the load

the emitter of Q1 for the +3.3V power supply and to the emitter of Q4 for the −3.3V power supply via the separate resistor R4. The potential divider formed by R2 and R3 is biased at a few mA that is also the dc bias current for Q2 in the +3.3V power supply and for Q6 in the −3.3V power supply at no load condition. We select the values for R2 and R3 in such a way that it takes about 5mA. This will be sufficient for biasing the series pass transistor Q2 and Q6 at no load.

Note that LM329 produces a reference voltage of 6.9V. Since the output of the power supply is only 3.3V, the reference voltage must be scaled down to below 3.3V when it reaches the non-inverted input (pin-3) of OPA134. Thus, a simple potential divider formed by R5 and R6 is used to scale down the reference voltage to around 2V. Alternatively, a low noise reference zener offering a voltage lower than 3V can be used instead of LM329. This can eliminate two resistors, R5 and R6.

At this low dc level, the components are often SMDs so as to shrink the size of the PCB. Here are some SMDs that can be considered: Q1 and Q5 = MMBT5088; Q3 and Q4 = MMBT5087; Q2 = FQT4N20L or IRLL110; Q6 = FQT3P20 or IRFL9110.

14.11 Reducing transformer vibration noise

A power transformer may create a buzzing vibration noise when the primary winding sees a dc voltage. The resistance of the primary winding of a power transformer can be as low as a few ohms. Even if just a few volts of dc are present in the primary winding, they magnetize the transformer core, making the transformer susceptible to vibration. A toroid power transformer is more easily prone to buzzing noise than E-I and C-core power transformer.

The cause of this dc issue is usually uneven harmonics on the ac mains when the loads are highly non-linear, and if half-wave rectification is used. In cities with heavy industries, the ac main is more likely to be corrupted by a dc. In this section, we examine several approaches to reduce the transformer vibration noise caused by an ac main corrupted by a dc. Even though these approaches may not cure the problem completely, they may reduce the buzzing noise to an acceptable level.

Isolation transformer

It happens very often that we cannot alter or remove the power transformer, which generates the buzzing vibration noise, from an audio amplifier. Therefore, we can only rely on using external devices to reduce this noise. One of those external devices is an isolation transformer, which has the primary to secondary windings turns ratio of 1:1. Therefore, it will not alter the output ac voltage. Figure 14.43 shows an isolation transformer placed between the ac mains and the power transformer of the audio component (i.e., power amplifier), which is producing a buzzing noise. As shown in the figure, there is a dc 1V and dc 5V unevenly corrupted to the line "N" (neutral) and "L" (live), respectively. Therefore, the primary winding of the isolation transformer sees a net dc $\Delta 1 = 4$V. Without the isolation transformer, this dc 4V will appear in the primary winding of the power transformer in a power amplifier. However, when an isolation transformer is used, the power transformer of the audio amplifier now sees only pure ac without a dc component superimposed on it.

In order for the isolation transformer to reduce the buzzing noise effectively, the choice of transformer is important. If it is a toroid transformer, it is likely that the isolation transformer itself will also run into the same problem, producing a buzzing noise. Therefore, it is

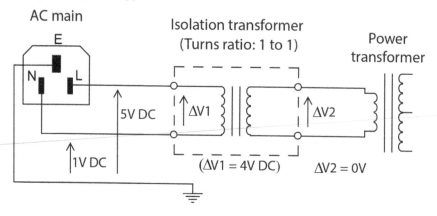

Figure 14.43 Using E-I or C-core type isolation transformer to isolate the dc from the ac mains

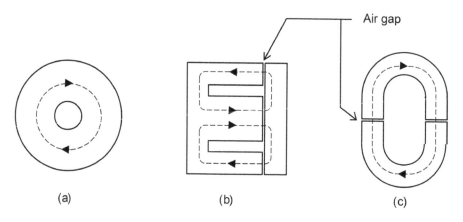

Figure 14.44 Magnetic flux flows in the cross-section of iron core in (a) a toroid transformer, (b) an E-I transformer and (c) a C-core transformer. There is a small air gap in the E-I and C-core transformer cores

recommended to choose either an E-I or C-core type isolation transformer. Figure 14.44 shows the cross-section of the toroid, E-I, and C-core transformers. In E-I and C-core transformers, there exists a small air gap even though the transformer core is very closely packed. This small air gap will create a resistance to the magnetic flux generated by the dc on the primary winding of the transformer. Therefore, this makes the E-I and C-core transformers less susceptible to magnetic saturation to the transformer iron core and, therefore, they produce lower buzzing noise. When the corrupted dc level on the ac mains is small, an E-I or C-core isolation transformer can produce promising results.

DC neutralizer

In recent years, there are products on the market that work as dc neutralizers for ac mains by injecting a dc voltage to one of the ac lines in such a way that it neutralizes the dc voltage difference between the lines "N" and "L". If this works as promised, it will be a very simple solution to get rid of the transformer buzzing noise. Figure 14.45 shows how a dc neutralizer works.

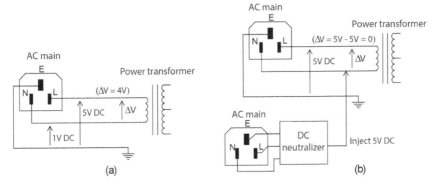

Figure 14.45 (a) A net 4V dc presenting in the primary winding of power transformer. (b) A net 0V dc after the neutralizer injects a counterbalance dc to the ac line

In Figure 14.45(a), assuming there is a 1V dc and a 5V dc present in the neutral (N) and live (L), respectively. The primary winding of the transformer sees a net 4V dc on the ac lines. Even this small level dc, toroid power transformer, usually having a low primary winding resistance, will create a buzzing vibration noise. When a dc neutralizer is placed on an ac power outlet near to where the transformer is powered, the dc neutralizer will first determine the dc voltages in the lines. Then it injects a counterbalance dc voltage into one of the lines. In Figure 14.45(b), a 5V dc is injected into the neutral (N). As a result, the primary of the transformer now sees a net dc $\Delta V = 0$ on the ac lines. In principle, a dc neutralizer looks very promising. Since I have never had a chance to use one, I cannot say how effective it is in reducing the transformer buzzing noise.

Blocking dc with a bridge rectifier

A direct approach to remove dc voltage is to use a dc blocking capacitor. However, it may be an unreliable approach if the capacitor is directly applied to an ac line. Another approach [10,11] is to use a bridge rectifier to block a dc voltage up to two rectifying diodes of forward voltage ($2 \times V_F$), which is around 1.4V to 2V (Figure 14.46[a]).

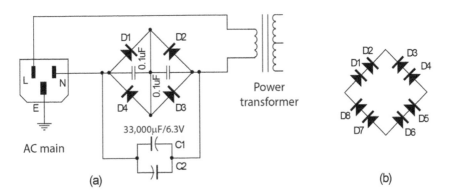

Figure 14.46 (a) A circuit to block dc from the ac mains up to two diodes forward voltage, $2 \times V_F$. (b) Blocking up to four diodes forward voltage, $4 \times V_F$

A bridge rectifier is connected in such a way that the positive and negative outputs are shorted together. Therefore, the ac line always drops two diode forward voltages to the primary winding of the transformer. This approach can also combine using large capacitors C1 and C2 connected to the bridge rectifiers. The capacitors should have very low ESR and high ripple current rating. Since the voltage across the capacitors is limited to two diode forward voltages, the voltage rating for the capacitors is low. When the rectifying diodes are doubled up, as shown in Figure 14.46(b), the circuit may block dc up to four diodes forward voltages ($4{\times}V_F$), around 2.8V to 4V.

Power transformer dual secondary windings

Here, we assume the ac main is free from dc voltage. Therefore, there is no need to isolate or neutralize the dc from the ac mains. In other words, the power transformer will generally not produce an unpleasant buzzing vibration noise.

Let us take a look at a common dual voltages power supply as shown in Figure 14.47(a). A power transformer with a center tap on the secondary side winding is used for producing a dual voltages power supply. It employs a bridge rectifying arrangement for full-wave rectification. If the load currents in the positive output (Iout1) and negative output (Iout2) are identical, or nearly identical, the power transformer should work without vibration noise. However, if one of the output currents is much higher than the other, and especially if it is a toroid power transformer, the power transformer may have a greater chance of producing a buzzing vibration noise even if the ac mains is free of corrupted dc voltage.

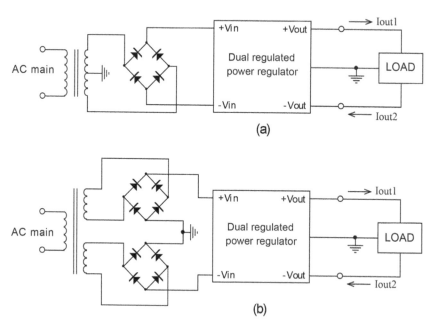

Figure 14.47 In circuit (a), a power transformer with a center tap on the secondary winding may produce buzzing noise in a dual dc power supply when output load currents are very different, Iout1 ≫ Iout2 or Iout1 ≪ Iout2. In circuit (b), a power transformer with two separate secondary windings can eliminate the buzzing noise problem even if output load currents are very different

When two output load currents are very different, there is a net dc current flowing into the winding of the secondary side of a transformer and magnetizing the iron core. As a result, this will be likely to cause a buzzing vibration noise if it is a toroid power transformer. However, this problem can be eliminated by using two separate secondary windings, as shown in Figure 14.47(b). It should be noted that no winding of the secondary side of the transformer is now connected to the power supply ground. In an application that requires dual power supply, and where the load currents are very different, it is strongly suggested to use a power transformer with two separate secondary windings. However, there is price for this solution, as we have to use twice the number of rectifiers and a transformer with two secondary windings. However, it is absolutely worth doing.

Distinguished products

2N3055 is a solid-state silicon NPN power transistor intended for general purpose applications. It was introduced in the early 1960s by RCA. The hometaxial base process was first employed and later changed to the epitaxial base process in the mid-1970s. A modern 2N3055 is rated at P_D = 115W, h_{FE} = 20 to 70, V_{CEO} = 60V, I_C = 15A and f_T = 2.5MHz. TO-3 package is used and the device has become a popular "work-horse" since the 1960s. 2N3055 was the first power transistor device sold for less than one dollar. It remains a very popular low-cost device today as a series pass transistor for linear regulated power supply applications.

MJ2955 is a PNP transistor designed as a complementary device for 2N3055. They were used as complementary power transistors for audio power amplifier applications. However, the low V_{CEO} breakdown voltage limits the output power to around 40W. In addition, current gain (beta) droop and f_T droop limit the current gain and increasing phase shift at high frequency. Modern power BJT transistors have much higher P_D, V_{CEO}, current gain, f_T, and less beta and f_T droop, making them more suitable for audio power amplifier applications.

14.12 Exercises

Ex 14.1 By modifying the circuit in Figure 14.15, design a dual ±24V power supply.

Ex 14.2 A variable resistor is commonly used for setting a more precise output voltage. Figure 14.48 shows a three-terminal variable regulator LM317 in two different arrangements using a variable resistor. Is the arrangement in Figure 14.48(a) or (b) a better solution?

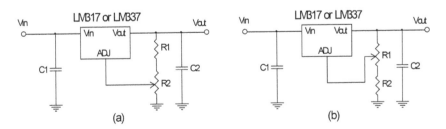

Figure 14.48 A variable resistor is used in a three-terminal variable regulator circuit at two different positions for Ex 14.2

Ex 14.3   Determine the value for R3 in Figure 14.19(b) for 30mA output currents.

Ex 14.4   By modifying the series type dual ±15V power supply in Figure 14.28, design a dual ±24V power supply.

Ex 14.5   Figure 14.49 shows a simplified series type fixed-mode power supply in circuit (a) and a simplified shunt type power supply in circuit (b). In circuit (a), the reference zener is placed at the non-inverted input of the error amplifier. Explain why, in circuit (b), the reference zener has to be placed at the inverted input of the error amplifier.

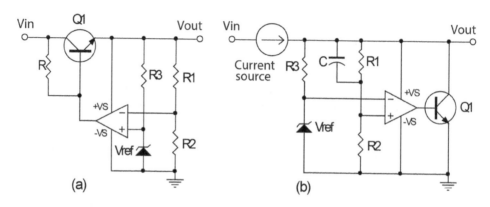

Figure 14.49 Circuit (a) is a simplified series type fixed-mode power supply. Circuit (b) is a simplified shunt type power supply

Ex 14.6   By modifying the +250V power supply in Figure 14.30, design a +100V power supply.

Ex 14.7   By modifying the dual −150V power supply in Figure 14.31, design a −50V power supply.

Ex 14.8   By modifying the +250V power supply in Figure 14.33, design a +150V power supply.

Ex 14.9   By modifying the dual ±3.3V power supply of Figure 14.42, design a dual ±2V power supply.

References

[1] "3-terminal regulator is adjustable," Application Note 181, National Semiconductor.
[2] "Improving power supply reliability with IC power regulators," Application Note 182, National Semiconductor.
[3] Gottlieb, I., *Regulated power supplies*, Howard W. Sams & Co., 3rd edition, 1990.
[4] Bode, P.A., "Designing with references – shunt regulation," Application Note 58, Zetex Semiconductors.
[5] Bode, P.A., "Designing with shunt regulators – extending the operating voltage range," Application Note 61, Zetex Semiconductors.

[6] Waagbo, A., "Shunt or not," pp. 30–35, 2/2008, AudioXpress.

[7] Jung, W., "Improved positive/negative regulators," pp. 8–19, 4/2000, Audio Electronics.

[8] "High voltages adjustable power supplies," Linear Brief 47, National Semiconductor.

[9] Bicknell, T., "Valve (tube) regulated power supplies," AudioXpress, 2008.

[10] Bryston 9B-SST audio power amplifier schematic.

[11] Cordell, B., *Designing audio power amplifiers*, McGraw-Hill, 2010.

Appendix A

Basic network theory

A.1 Ohm's law

Ohm's law states that a voltage across a resistor is proportional to the current flowing through the resistor. This current–voltage relationship was published in 1827 by a German physicist, Georg Simon Ohm [1] (see Ohm's law in Wikipedia). As a result, the unit for the resistor is called the *ohm* (abbreviated Ω). When the voltage in volts and current is in amperes, the current–voltage relation is expressed by

$$V = I \cdot R \tag{A.1}$$

For example, if the voltage is 1V and current is 1A, the resistor becomes 1Ω. Figure A.1 shows a circuit that represents the expression in Eq. (A.1).

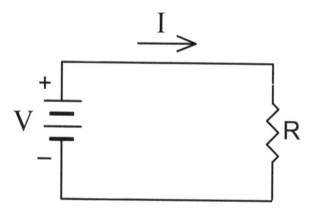

Figure A.1 A circuit represents current–voltage relationship $V = I R$

A.2 Kirchhoff's current law

Kirchhoff's current law (KCL) and *Kirchhoff's voltage law* (KVL) were stated by the German physicist Gustav Kirchhoff in 1845 (see Kirchhoff's circuit laws in Wikipedia) [2]. Kirchhoff's current law states that

The algebraic sum of the currents entering a node must be zero.

This can be illustrated by Figure A.2. The algebraic sum of the currents entering the node A is expressed by

$$I_1 + I_2 + I_3 + \dots + I_n = 0 \tag{A.2}$$

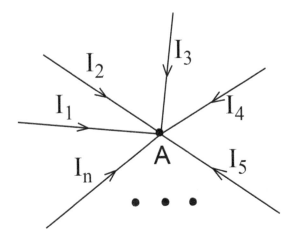

Figure A.2 The algebraic sum of currents entering node A is zero

For example, let us apply the KCL in node A and B of Figure A.3. By employing the KCL in node B, we have

$$I_4 + (-I_2) + (-I_3) = 0 \tag{A.3}$$

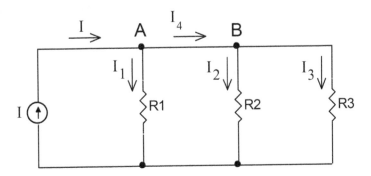

Figure A.3 Apply KCL in node A and B

It should be noted that current I_4 carries a positive sign. It denotes that I_4 is flowing into the node B. However, currents I_2 and I_3 carry a negative sign because they are flowing out of the node B.

Similarly, by employing KCL in node A, we have

$$I + (-I_1) + (-I_4) = 0 \tag{A.4}$$

Substituting I_4 from Eq. (A.3) with (A.4), we have

$$I_1 - I_2 - I_3 - I_4 = 0$$

A.3   Kirchhoff's voltage law

Kirchhoff's voltage law states that

The algebraic sum of voltages around a closed path must be zero.

The closed path is often referred to as a loop or mesh. For example, let us consider the loop A-B-C-D in Figure A.4(a). First, let V_{AB} denote the voltage difference between point A and B such that

$$V_{AB} = V_A - V_B$$

where V_A = dc voltage at point A, V_B = dc voltage at point B.
Therefore, applying KVL around the loop A-B-C-D, we have

$$V_{AD} + V_{BA} + V_{CB} + V_{DC} = 0$$

And the above voltages can be expressed by

$$V_0 + V_1 + V_2 + V_3 = 0 \tag{A.5}$$

Let us take a look at the polarities for V_1 to V_3. Here we have assumed that the dc voltage at point B is higher than at point A, so that V_1 carries a positive sign at point B and a negative sign at point A. The same applies to V_2 and V_3. On the other hand, we can assume the polarities to be just the opposite. This becomes the situation in Figure A.4(b). By applying the KVL, we have

$$V_0 + (-V_1') + (-V_2') + (-V_3') = 0 \tag{A.6}$$

In a simple circuit with just a single voltage source, we know Figure A.4(b) is showing the correct polarities. But in a circuit containing more than one source, getting the correct polarities by inspection is more difficult. However, in both cases of Figure A.4, they will give us the same results. Therefore, if we find $V_1' = 2V$ in Figure A.4(b), then it must be $V_1 = -2V$ in Figure A.4(a).

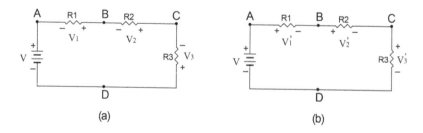

Figure A.4 Voltages across R1 to R3 are denoted by V1 to V3 in circuit (a). Voltages across R1 to R3 are denoted by V_1' to V_3' in circuit (b)

A.4 Series and parallel combinations of resistors

Series combination

If we apply a voltage V to the two terminals of Figure A.5(a), a current I flows through three resistors R1 to R3. From KVL, we have

$$V - IR1 - IR2 - IR3 = 0$$

And if the three series resistors are now replaced by an equivalent resistor, R_{eq}, and the same voltage, V, is applied, by using KVL we obtain the equivalent resistor as follows

$$R_{eq} \equiv \frac{V}{I} = R1 + R2 + R3 \tag{A.7}$$

It should be noted that Eq. (A.7) can be generalized to any number of resistors.

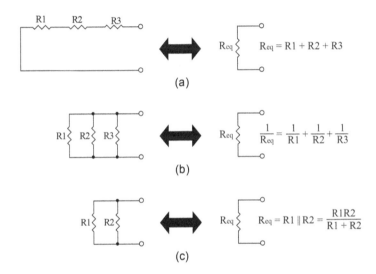

(a)

(b)

(c)

Figure A.5 Three series resistors in (a). Three parallel resistors in (b). Two parallel resistors in (c)

Parallel combination

If we apply a voltage, V, to the two terminals of Figure A.5(b), a total current, I, flows through the three parallel resistors R1–R3. From KCL, the sum of the currents in each resistor must be equal to the total current I. Therefore, we have

$$I = \frac{V}{R1} + \frac{V}{R2} + \frac{V}{R3} \tag{A.8}$$

If the three parallel resistors are now replaced by an equivalent resistor, R_{eq}, and the same voltage, V, is applied, the total current remains the same. Therefore, we have

$$I \equiv \frac{V}{R_{eq}} \tag{A.9}$$

By equating Eqs. (A.8) and (A.9), the equivalent resistor can be expressed by

$$\frac{1}{R_{eq}} = \frac{1}{R1} + \frac{1}{R2} + \frac{1}{R3} \qquad\qquad (A.10)$$

It should be noted that the number of parallel resistors should not be limited to three. Eq. (A.10) can be generalized to any number, two or more. For a special case of two resistors in parallel, as shown in Figure A.5(c), the equivalent resistor is written as

$$R_{eq} \equiv R1 \| R2 = \frac{R1R2}{R1 + R2} \qquad\qquad (A.11)$$

where the symbol $\|$ is read as "in parallel with."

A.5   Superposition

In any linear resistive circuit containing two or more independent sources (e.g., voltage source or current source), any circuit voltage or current can be calculated as the algebraic sum of the individual voltages or currents caused by each independent source acting alone, with all other independent sources dead. This is called the method of superposition, or, simply, superposition. By using superposition, we can analyze linear circuits with more than one independent source by analyzing separately one single source at a time. Since there is only one single source active, the circuit can be simplified and it is easier to analyze.

In a linear circuit $V_{out} = (V_1 + V_2 + V_3)$, we can say

$$2V_{out} = 2(V_1 + V_2 + V_3) = 2V_1 + 2V_2 + 2V_3$$

However, in another circuit we cannot say

$$2\left(V_1 + V_2 + V_3\right)^2 = 2V_1^2 + 2V_2^2 + 2V_3^2$$

because it is quadratic and not a linear expression. Thus, we have to remind ourselves that superposition only works for linear circuits. It is best to illustrate how the superposition works in examples, as shown in the following.

Example A.1

Note that there are two independent sources, V_1 and I_1, as shown in Figure A.6(a). We want to determine I_2, the current in resistor R2. By applying superposition, first we assume the current source is dead, and determine current I_b in R2. This is shown in Figure A.6(b). Then we assume the voltage source V_1 is dead, and determine the current I_c in R2 as shown in Figure A.6(c). Thus, the required current is given as

$$I_2 = I_b + I_c \qquad\qquad (A.12)$$

It should be noted that, when a voltage source is dead, there is no potential difference between the voltage source terminals. Therefore, a dead voltage source is replaced by shorting the source terminals, i.e., a short-circuit. On the other hand, when a current source is dead, there is no current flowing into or out from it. Therefore, a dead current source is replaced by an open circuit.

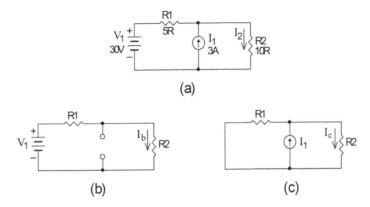

(a)

(b)   (c)

Figure A.6 Voltage source V_1 and current source I_1 are given in circuit (a). V_1 is alive but I_1 is dead in circuit (b). V_1 is dead but I_1 is alive in circuit (c)

From Figure A.6(b), we have

$$I_b = \frac{V_1}{R1+R2} = \frac{30V}{5\Omega+10\Omega} = 2A$$

From Figure A.6(c), we form an expression for the voltage across R1 and R2

$$(I_1 - I_c)\,R1 = I_c\,R2$$

After simplifying yields,

$$I_c = \frac{R1}{R1+R2}I_1 = \frac{5\Omega}{5\Omega+10\Omega} \times 3A = 1A$$

From Eq. (A12) we finally obtain

$$I_2 = 2A + 1A = 3A$$

Example A.2

The circuit of Figure A.7 contains two voltage sources and one current source. It is required to determine the voltage across resistor R3, V_{BC}. Since we have three independent sources, by applying the superposition we have three cases as shown in Figure A.8. In Figure A.8(a), V_1 is alive but the other two sources are dead. In Figure A.8(b), V_2 is alive but the other two sources are dead. In Figure A.8(c), I_1 is alive but the other two sources are dead.

We determine the voltage across R3 in three cases so that the required voltage is given as

$$V_{BC} = V_a + V_b + V_c \tag{A.13}$$

where

$$V_a = \frac{R3}{(R1\,\|\,R2)+R3}V_1 = \frac{15\Omega}{(20\,\|\,5)\Omega+15\Omega} \times 10V = 7.89V$$

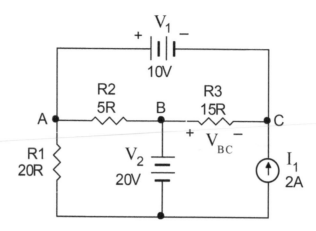

Figure A.7 A circuit contains two voltage sources and one current source.

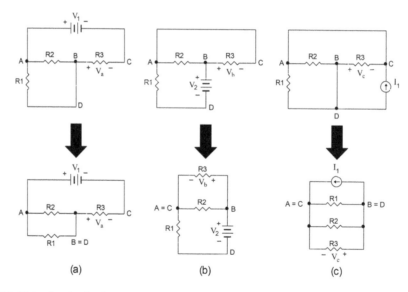

Figure A.8 V_1 is alive in circuit (a). V_2 is alive in circuit (b). I_1 is alive in circuit (c)

$$V_b = \frac{(R2 \| R3)}{R1 + (R2 \| R3)} V_2 = \frac{(5 \| 15)\Omega}{20\Omega + (5 \| 15)\Omega} \times 20V = 3.16V$$

$$V_c = -(R1 \| R2 \| R3)I_1 = -(20\Omega \| 5\Omega \| 15\Omega) \times 2A = -6.31V$$

From Eq. (A.13) we finally obtain

$$V_{BC} = 7.89V + 3.16V - 6.31V = 4.74V$$

A.6 Thévenin and Norton's theorems

Figure A.9(a) shows a linear circuit, A, connected to circuit B. The circuit A is a linear circuit that contains resistors, independent and dependent voltage sources, and current sources. It is restricted so that no part of circuit A is controlled by circuit B. In other words, a dependent source in circuit A is not controlled by circuit B, and vice versa. The goal is to find an equivalent circuit to replace circuit A such that the voltage–current relations at terminals A–B remain the same. The equivalent circuit contains a voltage source, V_{oc}, and a resistor, R_{th}, as shown in Figure A.9(b). It is called the Thévenin equivalent circuit.

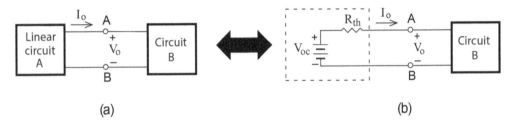

(a)   (b)

Figure A.9 A linear circuit A is connected to circuit B in (a). The linear circuit A is replaced by a Thévenin equivalent circuit which contains a voltage source V_{oc} and resistor R_{th} in (b)

To determine the Thévenin equivalent circuit, we first replace circuit B in Figure A.9(a) by a voltage source, V_0, such that the voltage–current relations are not affected. This is shown in Figure A.10(a). By applying the superposition, the linear circuit A can now be analyzed in two cases: (i) circuit A is dead and (ii) the output is shorted, as shown in Figure A.10(b) and (c). Thus we have

$$I_0 = I_1 + I_{sc} \tag{A.14}$$

where I_1 is the current produced by the voltage source V_0 when the independent sources in circuit A are dead and I_{sc} is the short-circuit current produced by any sources inside circuit A when voltage source V_0 is dead.

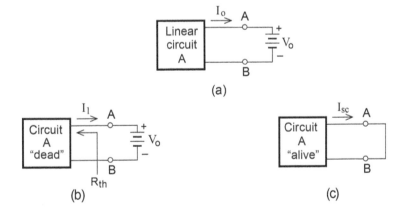

(a)

(b)   (c)

Figure A.10 Circuit B is replaced by a voltage source V_0 in (a). The output resistance R_{th} is determined when the sources are dead in (b). The short circuit current I_{sc} is determined when the terminals A–B are shorted

In Figure A.10(b), since the independent sources are dead, the impedance looking into the circuit A is a resistive circuit, which is the equivalent resistance called R_{th}. Thus, by Ohm's law, we have

$$I_1 = -\frac{V_o}{R_{th}} \tag{A.15}$$

Substituting Eq. (A.15) for (A.14), we have

$$I_o = -\frac{V_o}{R_{th}} + I_{sc} \tag{A.16}$$

Since Eq. (A.16) describes the current generally, it must hold for any condition at the terminals. Now, supposing the terminal is open so that $I_o = 0$, we denote the voltage by $V_o = V_{oc}$ (the open circuit voltage). Substituting these for Eq. (A.16), we have

$$0 = -\frac{V_{oc}}{R_{th}} + I_{sc}$$

and rewritten as

$$I_{sc} = \frac{V_{oc}}{R_{th}} \tag{A.17}$$

Substituting Eq. (A.17) for (A.16), we obtain

$$V_o = V_{oc} - R_{th}I_o \tag{A.18}$$

The relations of Eq. (A.18) can be used to express an equivalent circuit for the linear circuit A. It is called the Thévenin equivalent circuit in honor of the French engineer, Charles Leon Thévenin, who published the result in 1883. It is noted that the German scientist Hermann von Helmholtz also derived it independently in 1853. Figure A.11(a) is the Thévenin equivalent circuit that represents the expression of Eq. (A.18). The statement that Figure A.11(a) is equivalent to the terminals A–B of Figure A.9(a) is known as the *Thévenin theorem*.

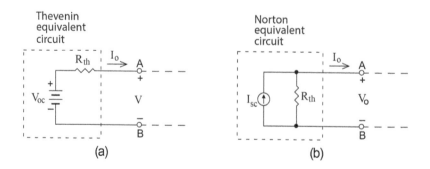

Figure A.11 Thévenin equivalent circuit (a) and Norton equivalent circuit (b)

Since voltage and current are dual attributes that are correlated in a linear circuit, Eq. (A.16) can be used to express an equivalent circuit, as shown in Figure A.11(b). It is called the Norton equivalent circuit after the American engineer E.L. Norton. The statement that Figure A.11(b) is equivalent to the terminals A–B of Figure A.9(a) is known as the *Norton theorem*.

Several examples are given below to illustrate how the Thévenin and Norton theorems help us to solve problems in linear circuits.

Example A.3

Example A.1 has illustrated the use of superposition. Now we re-examine the same circuit by using the Thévenin and Norton theorems. The circuit is now redrawn in Figure A.12(a). It is required to determine the current I_2.

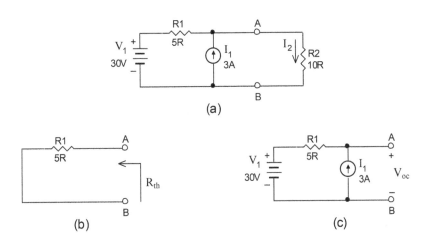

Figure A.12 Voltage source V_1 and current source I_1 are given in circuit (a). V_1 and I_1 are dead in circuit (b). V_{oc} is the open circuit voltage between terminals A–B in circuit (c)

First, we simplify the circuit by placing resistor R2 to the right of terminals A–B. Then we proceed to find the Thévenin equivalent circuit by determining the equivalent resistor R_{th} and open circuit voltage V_{oc} from Figure A.12(b) and (c). When the voltage and current sources are dead, the equivalent resistor is easily found as

$$R_{th} = R1 = 5\Omega \qquad\qquad (A.19)$$

In Figure A.12(c), since the circuit to the right of terminals A–B is open, the current I_1 must flow into resistor R1, the open circuit voltage is determined by

$$V_{oc} = I_1 R_1 + V_1 = 3A\cdot5\Omega + 30V = 45V$$

The Thévenin equivalent circuit is expressed in Figure A.13(a). Thus, the required current is determined by

$$I_2 = \frac{V_{oc}}{R_{th} + R2} = \frac{45V}{5\Omega + 10\Omega} = 3A$$

It is, of course, the same result as found in Example A.1. Alternatively, we determine the short circuit current I_{sc} by using Eq. (A.17). Thus, the Norton equivalent circuit can be obtained as shown in Figure A.13(b).

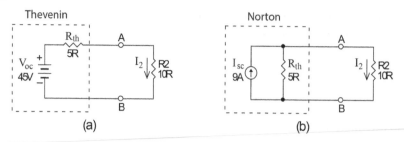

(a) (b)

Figure A.13 Thévenin equivalent circuit (a) and Norton equivalent circuit (b)

Example A.4

The circuit of Figure A.14(a) contains an independent current source, I_1, and a dependent voltage source, kI_2, where k is a proportional constant equal to 5V/A. The dependent voltage source generates a voltage that is dependent upon the current I_2. For example, if $I_2 = 1A$, the generated voltage is 5V, and 11V for $I_2 = 2.2A$. In order to determine the equivalent resistor, R_{th}, the independent current source I_1 must be dead, as shown in Figure A.14(b). It can be noted that the dependent voltage source is kept in the circuit as it is not allowed for dependent sources to be dead. Otherwise, the calculation will be incorrect.

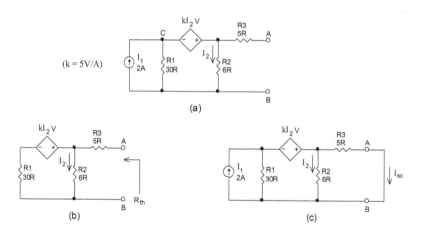

Figure A.14 Independent current source I_1 and dependent voltage source are given in circuit (a). The current source I_1 is dead in (b). Output is shorted in circuit (c) for determining I_{sc}

Since the dependent voltage source is presented in the circuit, we cannot determine the equivalent resistor R_{th} directly from Figure A.14(b). Thus we have to use a different approach. If Eq. (A.17) is rewritten as,

$$R_{th} = \frac{V_{oc}}{I_{sc}}$$

(A.20)

we can determine R_{th} by first finding V_{oc} and I_{sc}, where I_{sc} is determined from Figure A.14(c).

To determine the open voltage V_{oc}, we examine circuit (a) of Figure A.14. If V_{oc} is the open voltage between terminals A–B, then the voltage across the resistor R1 is kI_2 below V_{oc}, i.e., $(V_{oc} - kI_2)$. Thus, by examining the currents at node C we have,

$$I_1 = \frac{V_{oc} - kI_2}{R1} + I_2 \tag{A.21}$$

where

$$I_2 = \frac{V_{oc}}{R2} \tag{A.22}$$

Substituting Eq. (A.22) for (A.21) and solving for V_{oc}, we have

$$V_{oc} = \left[\frac{R1R2}{R1+R2-k}\right]I_1 = \left[\frac{30 \times 6}{30+6-5}\right] \times 2V = 11.6V \tag{A.23}$$

To determine the short-circuit current I_{sc}, we examine circuit (c) of Figure A.14. We know that the sum of the currents in R1, R2, and I_{sc} must be equal to current source I_1. Thus, we have

$$I_1 = \left[\frac{I_2 R2 - kI_2}{R1}\right] + I_2 + I_{sc} \tag{A.24}$$

and the voltage across resistor R2 is equal to the voltage across R3 such that

$$I_2 R2 = I_{sc} R3$$

rearranging to

$$I_2 = \frac{I_{sc} R3}{R2} \tag{A.25}$$

Substituting Eq. (A.25) for (A.24) and solving for I_{sc}, we have

$$I_{sc} = \frac{I_1}{1 + \dfrac{(R1 + R2 - k)R3}{R1R2}}$$

$$= \frac{2}{1 + \dfrac{(30 + 6 - 5) \times 5}{30 \times 6}} = 1.07A \tag{A.26}$$

By substituting Eq. (A.23) and (A.26) for (A.20), we determine the Thevenin equivalent resistor

$$R_{th} = \frac{V_{oc}}{I_{sc}} = \frac{11.6V}{1.07A} = 10.84\Omega \tag{A.27}$$

Having found R_{th}, V_{oc} and I_{sc}, therefore, we obtain the Thévenin and Norton equivalent circuit.

From Example A.4, we have seen that if there is a dependent source (voltage or current) in the circuit, we cannot determine the equivalent resistor R_{th} directly by making the independent sources dead because the dependent source is still in the circuit. However, we can get around this by first finding out V_{oc} and I_{sc} so that R_{th} can be eventually determined. We also note that both the method of superposition and the Thévenin theorem (or Norton theorem) can be used to determine the voltage or current in linear circuits. However, if a dependent source appears in the circuit, the method of superposition will be difficult to apply, as the dependent source cannot be dead. In such a situation, the Thévenin theorem (or Norton theorem) is the preferred choice for tackling the problem.

References

[1] Ohm, G.S., *Die Galvanische Kette, Mathematisch Bearbeitet*, Kessinger, 1827.
[2] Kirchoff, G., Kirchhoff's Laws, 1845, Wikipedia.

Appendix B
Transfer function

B.1 Complex impedances

When a resistive circuit, which only contains resistors, is driven by a sinusoidal signal, the voltage–current relationship always remains the same because a resistor is a frequency independent device. In reality, we know that at very high frequencies of 10MHz and beyond, the intrinsic inductance and capacitance will make a resistor become frequency dependable. However, in the audio band it is safe to assume a resistor is frequency independent so as to make circuit analysis much easier to deal with.

However, inductors and capacitors are frequency dependent devices. When inductors and capacitors are driven by a sinusoidal signal, the voltage–current relationship varies at different phase. To account for this phase shift, we need to turn to *complex impedance* (or *reactance*):

$$Z_L = sL, \text{ for inductor L} \tag{B.1}$$

$$Z_C = \frac{1}{sC}, \text{ for capacitor C} \tag{B.2}$$

where

$$s = j\omega \text{ (a complex variable)}, \omega = 2\pi f \text{ (f = frequency)} \tag{B.3}$$

such that

$$j^2 = -1 \tag{B.4}$$

and

$$e^{j\phi} = \cos\phi + j\sin\phi \tag{B.5}$$

It should be noted that symbol i is usually used instead of j in mathematics textbooks. However, in electrical and electronic engineering, i is often used to denote current. To avoid confusion, the symbol j is preferred and it will be used throughout in the discussion below.

At low frequency, an inductor becomes bulky and not cost effective. Except in two areas, inductors are not commonly used in most audio applications. One area is an inductor used in the *Zobel* filter on the output of a power amplifier. The other area is in a loudspeaker cross-over network. However, at very high frequencies, the size of the inductor is substantially reduced and it will become a desired passive component for shaping the frequency response of amplifiers and filters. On the other hand, capacitors and resistors are the preferred passive components for audio applications. Let us see how to derive the transfer function of an R–C network. Once the transfer function is found, we can determine the characteristic of the circuit by means of amplitude and phase response.

B.2 Transfer function

The output–input relation expressed in the s complex variable is called the transfer function of the circuit. It is commonly used in active filter analysis. Let us determine the transfer function for the simple high-pass and low-pass filter in Figure B.1. To determine the transfer function, we will employ the standard tools such as Ohm's law, Kirchhoff's current law, and Kirchhoff's voltage law and superposition.

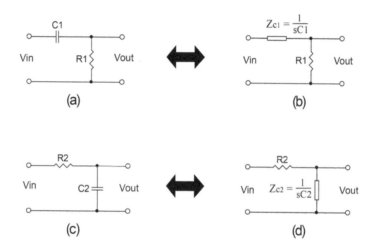

Figure B.1 A high-pass filter in (a) and (b) is the same circuit showing the impedance of C1. A low-pass filter in (c) and (d) is the same circuit showing the impedance of C2

When the capacitor C1 of Figure B.1(a) is expressed in terms of the capacitor impedance Z_{C1} in (b), the transfer function, denoted by $H_1(s)$, is given by

$$H_1(s) \equiv \frac{Vout(s)}{Vin(s)} = \frac{R1}{R1 + Z_{C1}} = \frac{R1}{R1 + \dfrac{1}{sC1}} = \frac{sR1C1}{1 + sR1C1} \tag{B.6}$$

Similarly, the transfer function for Figure B.1 (d) is given by

$$H_2(s) \equiv \frac{Vout(s)}{Vin(s)} = \frac{Z_{C2}}{R2 + Z_{C2}} = \frac{\dfrac{1}{sC2}}{R2 + \dfrac{1}{sC2}} = \frac{1}{1 + sR2C2} \tag{B.7}$$

Note that the transfer functions $H_1(s)$ and $H_2(s)$ can be expressed in the form of

$$H(s) = \frac{N(s)}{D(s)} \tag{B.8}$$

where N(s) and D(s) are polynomials of s with real coefficients, and the order of N(s) never exceeds that of D(s). The order of D(s) is called the *order of the filter* (first-order, second-order, etc.). The order of the filter is equal to the number of independent capacitors used in the circuit. The transfer functions $H_1(s)$ and $H_2(s)$ reveal that both of them are first-order filter, as there is only one independent capacitor in each circuit.

For a third-order active filter, the transfer function of Eq. (B.8) can be expressed in the form

$$H(s) = \frac{N(s)}{D(s)} = \frac{K(s + s_{z1})(s + s_{z2})}{(s + s_{p1})(s + s_{p2})(s + s_{p3})} \tag{B.9}$$

where K is a constant. The values of s, for which H(s) = 0, are called the *zeros* of the transfer function. In the above transfer function, H(s) has two zeros, s_{z1} and s_{z2}. The values of s, for which H(s) = ∞, are called the *poles* of the transfer function. The above transfer function has three poles, s_{p1}, s_{p2} and s_{p3}.

The transfer function of the simple first-order high pass filter of Eq. (B.6) has one zero at s = 0 and one pole at s = −1/(R1C1). The pole gives the breakpoint frequency at

$$|s| = |j\omega| = |j2\pi f| = \left| \frac{-1}{R1C1} \right|$$

Therefore, we have

$$f = \frac{1}{2\pi R1C1} \tag{B.10}$$

For example, if R1 = 1kΩ, C1 = 0.159μF, we get f = 1kHz. Similarly, the pole of the low-pass filter of Eq. (B.7) is given by s = −1/(R2C2). Thus, the breakpoint frequency is given by f = 1/(2πR2C2). If R2 = 200Ω, C2 = 7.95nF, we get f = 100kHz.

It is noted that at the breakpoint frequency caused by a zero, the amplitude starts to increase at a rate of +20dB/decade as frequency increases. On the other hand, at the breakpoint frequency caused by a pole, the amplitude starts to decrease at a rate of −20dB/decade as frequency increases. In Figure B.2(a), the zero starts at 0Hz and the amplitude continues to increase at +20dB/decade until reaching the breakpoint frequency f = 1/2πR1C1. Then amplitude at the breakpoint starts to increase at a combined rate of 20dB/decade − 20dB/decade = 0dB/decade. In other words, the amplitude becomes flat right after the breakpoint frequency. Then the amplitude continues to remain flat at a gain of unity. Similarly, in Figure B.2(b) the amplitude is flat

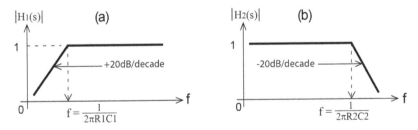

Figure B.2 (a) Amplitude response of a first order high-pass filter. (b) Amplitude response of a first-order low-pass filter

at a gain of unity until reaching the breakpoint frequency caused by a pole, $f = 1/2\pi R2C2$. Then the amplitude starts to roll off at -20dB/decade.

For example, the amplitude response of an active filter that contains two zeros and three poles is shown in Figure B.3. At low frequency, the amplitude is flat until reaching the breakpoint frequency caused by the first pole, s_{p1}. Then the amplitude continues to roll off at -20dB/decade until reaching the breakpoint frequency caused by the first zero, s_{z1}. The amplitude continues to remain flat until reaching the breakpoint frequency caused by the second pole, s_{p2}. Then the amplitude continues to roll off at -20dB/decade until reaching the breakpoint frequency caused by the second zero, s_{z2}. The amplitude continues to remain flat until reaching the breakpoint frequency caused by the third pole, s_{p3}. Then the amplitude continues to roll off at -20dB/decade.

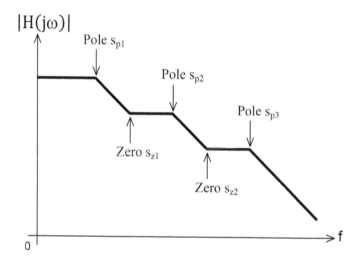

Figure B.3 Amplitude response for an active filter with two zeros and three poles

B.3   Amplitude and phase response

Since transfer function is a complex number, it can be written in the following expression.

$$H(s) = H(j\omega) = R(\omega) + jX(\omega) \tag{B.11}$$

where $R(\omega)$ is the real part of the complex number $H(s)$ and $X(\omega)$ is the imaginary part. Then the amplitude response is given by

$$A(\omega) \equiv |H(j\omega)| \tag{B.12}$$

To determine the amplitude response, we employ the complex conjugate $H^*(j\omega)$, which is defined as

$$H^*(j\omega) \equiv R(\omega) - jX(\omega) \tag{B.13}$$

Then we have

$$\left|H(j\omega)\right|^2 = H(j\omega)H^*(j\omega) = \{R(\omega) + jX(\omega)\}\{R(\omega) - jX(\omega)\}$$

$$= R^2(\omega) + X^2(\omega) \tag{B.14}$$

Thus, the amplitude response in equation (B.12) is expressed by

$$A(\omega) = \sqrt{\left|H(j\omega)\right|^2} = \sqrt{R^2(\omega) + X^2(\omega)} \tag{B.15}$$

Once we obtain the amplitude response, the transfer function $H(j\omega)$ can be expressed in the form of

$$H(j\omega) = R(\omega) + jX(\omega) = A(\omega)e^{j\phi(\omega)} \tag{B.16}$$

where $\phi(\omega)$ is the phase response of the transfer function. Substituting Eq. (B.5) with (B.16) yields

$$R(\omega) + jX(\omega) = A(\omega)\{\cos\phi(\omega) + j\sin\phi(\omega)\}$$

By equating the real and imaginary parts we have

$$R(\omega) = A(\omega)\cos\phi(\omega) \tag{B.17}$$

$$X(\omega) = A(\omega)\sin\phi(\omega) \tag{B.18}$$

Dividing Eq. (B.18) by (B.17) yields

$$\frac{X(\omega)}{R(\omega)} = \frac{\sin\phi(\omega)}{\cos\phi(\omega)} = \tan\phi(\omega)$$

By taking the inverse tangent on both sides of the above equation, we finally obtain the phase response

$$\phi(\omega) = \tan^{-1}\left(\frac{X(\omega)}{R(\omega)}\right) \tag{B.19}$$

Therefore, once the real part $R(\omega)$ and imaginary part $X(\omega)$ of the transfer function are found, the amplitude response $A(\omega)$ and phase response $\phi(\omega)$ can be determined by Eq. (B.15) and (B.19), respectively. For a first-order filter, the real and imaginary part and imaginary part of the transfer function is relatively simple so that the phase response can be easily found. However, for the second- and third-order filters, it may require a great deal of efforts for determining the phase response though. On the other hand, there are situations that can simplify our work finding the amplitude response.

For example, if the transfer function is expressed in terms two or more individual transfer functions, the amplitude response can be expressed in terms of the individual transfer function. When a transfer function $H(j\omega)$ is expressed by cascading three transfer functions such that

$$H(j\omega) = H_1(j\omega)H_2(j\omega)H_3(j\omega) \tag{B.20}$$

then the amplitude response for H(jω) becomes

$$A(\omega) = \left|H(j\omega)\right| = |H_1(j\omega)| \times |H_2(j)| \times |H_3(j\omega)| \tag{B.21}$$

If the transfer function is expressed in a fractional form such that

$$H(j\omega) = \frac{H_1(j\omega)}{H_2(j\omega)H_3(j\omega)} \tag{B.22}$$

then the amplitude response for H(jω) becomes

$$A(\omega) = \left|H(j\omega)\right| = \frac{|H_1(j\omega)|}{|H_2(j\omega)|\,|H_3(j\omega)|} \tag{B.23}$$

Example B.1

(Eg. 1)

$$H_1(j\omega) = a + j\omega$$

$$A_1(\omega) = \left|H_1(j\omega)\right| = \sqrt{a^2 + \omega^2} \tag{B.24}$$

(Eg. 2)

$$H_2(j\omega) = \frac{1}{a + j\omega} = \frac{1}{H_1(j\omega)}$$

$$A_2(\omega) = \left|H_2(j\omega)\right| = \frac{1}{\left|H_1(j\omega)\right|} = \frac{1}{\sqrt{a^2 + \omega^2}} \tag{B.25}$$

(Eg. 3)

$$H_3(j\omega) = (a + j\omega)(b + j\omega)$$

$$\left|H_3(j\omega)\right|^2 = \left|a + j\omega\right|^2 \left|b + j\omega\right|^2 = (a^2 + \omega^2)(b^2 + \omega^2)$$

$$A_3(\omega) = \sqrt{\left|H_3(j\omega)\right|^2} = \sqrt{(a^2 + \omega^2)(b^2 + \omega^2)} \tag{B.26}$$

(Eg. 4)

$$H_4(j\omega) = \frac{c + j\omega}{(a + j\omega)(b + j\omega)} = \frac{c + j\omega}{H_3(j\omega)}$$

$$A_4(\omega) = \left|H_4(j\omega)\right| = \frac{|c + j\omega|}{|H_3(j\omega)|} = \frac{\sqrt{c^2 + \omega^2}}{\sqrt{(a^2 + \omega^2)(b^2 + \omega^2)}} \tag{B.27}$$

Example B.2

To determine the transfer function for the low-pass filter in Figure B.4, we replace the combined resistor R2 and capacitor C by impedance Z.

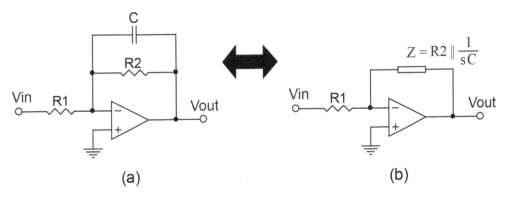

Figure B.4 Circuit (a) is a low-pass filter while circuit and (b) is the same circuit except capacitor C and resistor R2 are replaced by an impedance Z

$$Z = R2 \,\|\, (1/sC) = \frac{R2}{1 + sR2C} \qquad (B.28)$$

The transfer function of the circuit is expressed in the form similar to an inverted amplifier.

$$H(s) = \frac{Vout(s)}{Vin(s)} = -\frac{Z}{R1} = -\frac{\dfrac{R2}{1 + sR2C}}{R1} = -\left(\frac{R2}{R1}\right)\left(\frac{1}{1 + sR2C}\right)$$

$$= -\left(\frac{1}{R1C}\right)\left(\frac{1}{\dfrac{1}{R2C} + s}\right) \qquad (B.29)$$

Thus the amplitude response is given by

$$A(\omega) = |H(j\omega)| = \left(\frac{1}{R1C}\right)\frac{1}{\sqrt{(\dfrac{1}{R2C})^2 + \omega^2}} \qquad (B.30)$$

To see the amplitude response at the frequency extremes, we put $\omega = 0$ and $\omega = \infty$, getting $A(0) = R2/R1$ and $A(\infty) = 0$. It is clear that the amplitude response behaves like a low-pass filter. The breakpoint frequency occurs at $s = 1/(R2C)$, i.e., $f = 1/(2\pi R2C)$.

To determine the phase response, we must express the transfer in terms of real part and imaginary part so that it is determined by Eq. (B.19). Let us rewrite the transfer function of Eq. (B.29) in the following.

$$H(j\omega) = -\left(\frac{R2}{R1}\right)\frac{1}{1 + j\omega R2C} = -\left(\frac{R2}{R1}\right)\frac{1}{1 + j\omega R2C}\left[\frac{1 - j\omega R2C}{1 - j\omega R2C}\right]$$

$$= -\left(\frac{R2}{R1}\right)\frac{1 - j\omega R2C}{1 + (\omega R2C)^2} = -\left(\frac{R2}{R1}\right)\frac{1}{1 + (\omega R2C)^2} + j\omega\left(\frac{R2}{R1}\right)\frac{\omega R2C}{1 + (\omega R2C)^2} \qquad (B.31)$$

If we express the transfer in terms of real $R(\omega)$ and imaginary part $X(\omega)$, we have

$$H(j\omega) = R(\omega) + jX(\omega) \qquad (B.32)$$

Equating Eq. (B.31) and (B.32) yields

$$R(\omega) = -\left(\frac{R2}{R1}\right)\frac{1}{1 + (\omega R2C)^2} \tag{B.33}$$

$$X(\omega) = \left(\frac{R2}{R1}\right)\frac{\omega R2C}{1 + (\omega R2C)^2} \tag{B.34}$$

Thus, the phase response is determined by Eq. (B.19) such that

$$\phi(\omega) = \tan^{-1}\left(\frac{X(\omega)}{R(\omega)}\right) = -\tan^{-1}(\omega R2C) \tag{B.35}$$

Appendix C

Matching junction field effect transistors

C.1 Background

In audio applications, it is very popular to use a JFET differential amplifier in the first stage of a line-stage amplifier, phono-stage amplifier, and even a power amplifier. Figure C.1(a) is a typical example of a JFET differential amplifier. The ac characteristic of JFET is determined by its gate transfer characteristic, similar to Figure C.1(c). Given the graph of gate transfer characteristic, we can graphically plot the output and, thus, determine the gain of the amplifier. If transistors Q1 and Q2 are matched, the two outputs from the differential amplifier should have identical gain. This improves the CMRR as well as lowering noise and distortion. Therefore, it is desired to have two closely matched JFETs for the differential amplifier configuration. In order for two JFETs to be matched, their I_{DSS} should be identical or, in real life, at least very close.

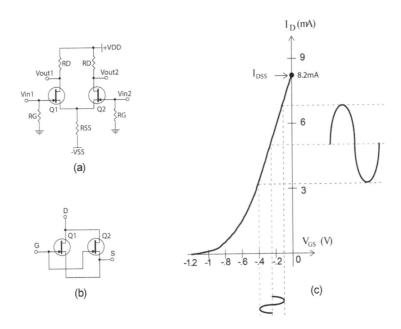

(a)

(b)

(c)

Figure C.1 Circuit (a) is a JFET differential amplifier. Circuit (b) is a compound transistor containing two JFETs in parallel. The gate transfer characteristic of a JFET is shown in (c).

In phono-stage applications, in order to improve the SNR, we often use a number of JFETs or BJTs in parallel to form a compound transistor. (See Chapter 10 for some examples.) Figure C.1(b) shows two JFETs in parallel. Again, it is important to select matched JFETs if low noise and low distortion is of primary concern.

Modern fabrication processes can produce BJTs with very close electrical properties in terms of current gain β and $V_{BE(on)}$. As a result, we do not need to match BJTs for most applications. However, the structure of JFET is different. It has a relatively long channel between the source and drain. This makes it difficult to precisely fabricate JFETs for very close electrical properties. As a result, the I_{DSS} can vary immensely even if the JEFTs are produced from the same wafer. For example, let's take a look at the popular 2SK170, an N-channel JFET. This is a single device in the TO-92 package. It is offered in three I_{DSS} grades, (i) GR: 2.6 ~ 6.5mA, (ii) BL: 6.0 ~ 12mA, and (iii) V: 10 ~20mA. Even if we order 2SK170 from the same grade, the I_{DSS} can differ by 200% or higher. This makes matching I_{DSS} a necessity.

C.2   Matching JFETs

As we can see from the gate transfer characteristic in Figure C.1(c), I_{DSS} is the drain current when $V_{GS} = 0$. Therefore, the gate and source leads are shorted together, as shown in Figure C.2. However, we do not physically solder the gate and source leads together. Since this is a simple circuit, the use of a breadboard is very suitable for this setup. A regulated dc voltage supply around 9V–12V can be used for VDD. If the supply voltage, say 10V, is used, all the rest of the JFETs must be measured under the same supply voltage.

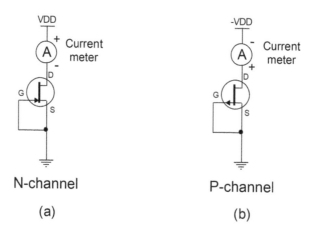

(a) N-channel   (b) P-channel

Figure C.2 Circuit (a) determines the I_{DSS} of an N-channel JFET. Circuit (b) determines the I_{DSS} of a P-channel JFET

When current is conducted, the junction of a JFET behaves like a resistor and the device will be heated up slightly. Therefore, the I_{DSS} measured in the first few seconds is slightly different from the measurement taken a few minutes later. It is suggested to count for five minutes before taking the measurement from each JFET. Shorter time can be used. But the same period of time should be consistently used for all devices. After all devices are tested, match the devices for I_{DSS} within 1mA range. This should be good enough for most applications.

Index

Locators in **bold** refer to tables and those in *italics* to figures, though please note that where interspersed with text discussion, these are not distinguished from regular locators

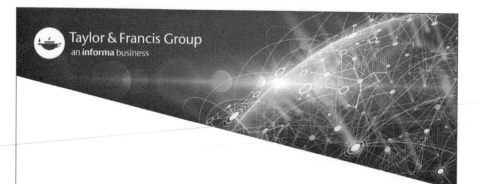